THE RISE OF THE
NEW PHYSICS

ITS MATHEMATICAL AND PHYSICAL THEORIES

formerly titled 'DECLINE OF MECHANISM'

A. d'Abro

in two volumes

VOL

2

DOVER PUBLICATIONS, INC.
NEW YORK

Published in Canada by General Publishing Company, Ltd., 30 Lesmill Road, Don Mills, Toronto, Ontario.
Published in the United Kingdom by Constable and Company, Ltd., 10 Orange Street, London WC 2.

This Dover edition, first published in 1952, is an unabridged republication of the work originally published in 1939 by D. Van Nostrand Co., Inc. under the title *Decline of Mechanism*. This Dover edition contains corrections by the author and a new section of portraits.

International Standard Book Number: 0-486-20004-3
Library of Congress Catalog Card Number: 52-8895

Manufactured in the United States of America
Dover Publications, Inc.
180 Varick Street
New York, N. Y. 10014

CONTENTS

PORTRAITS

QUANTUM PHYSICISTS

following page 962

MAX WIEN, MAX PLANCK, ALBERT EINSTEIN, HENRY MOSELEY, WALTHER RITZ, NIELS BOHR, ARNOLD SOMMERFELD, WOLFGANG PAULI, LOUIS VICTOR DE BROGLIE, ERWIN SCHRÖDINGER, WERNER HEISENBERG, PAUL ADRIEN MAURICE DIRAC

PART III
THE QUANTUM THEORY

CHAPTER XXIII

THE THEORY OF RELATIVITY AND THE QUANTUM THEORY

THE Theory of Relativity and the Quantum Theory are the two latest products of natural philosophy in physics. They are not opposing doctrines between which we must choose; they are complementary, covering different fields.

The Special Theory of Relativity—Inasmuch as some understanding of the special theory of relativity is required in the quantum theory, we shall recall the more elementary features of Einstein's theory. A few preliminary definitions will be necessary. A frame of reference in space is usually defined by three mutually perpendicular axes along which spatial measurements are made. Consider two such frames moving relatively to each other with a uniform translational motion, and suppose that a body is moving with respect to both frames. If we refer the successive positions of the body to the first frame (*i.e.*, if our measurements are made in the first frame), the body is found to be describing a certain path with a certain motion. As referred to the second frame, the path and motion will be different. A problem of frequent occurrence is to deduce the path and the motion relative to one frame when the path and motion in the other frame are known. More generally, we are required to establish a correlation between the space and time measurements performed in the two frames. The mathematical relations which determine the correlation are called a "space and time transformation." To obtain them, we must perform space and time measurements in the two frames and then compare results. Commonsense, however, seems to obviate the necessity of direct measurement, for it appears to impose a definite transformation, called the Newtonian, or classical, space and time transformation. The commonsense feature of this classical transformation is revealed in any of its consequences, *e.g.*, in the rule of composition of velocities. A simple illustration of this rule follows.

Let us suppose that two frames are attached respectively to the earth and to a train moving along a straight track with a velocity v of 3 miles an hour. Suppose also a ball is rolling along the track with a constant velocity u of 5 miles an hour in the same direction as the train. Commonsense tells us that, with respect to the train, the ball should be moving with the velocity $u - v$ (*i.e.*, $5 - 3 = 2$ miles an hour). This conclusion illustrates the rule of composition of velocities that are codirectional: they add and subtract like ordinary numbers. The intuitive appeal of the classical transformation is not restricted to this special rule. In all cases the transformation betrays the dictates of commonsense. But the development of natural philosophy, whether it pertain to mathematics or to physics, has taught that the dictates of commonsense can be accepted only tentatively and that they often lead to incorrect conclusions. Elementary caution would therefore demand that, before accepting unreservedly the classical transformation, we should verify its correctness for all possible relative velocities between the two frames. Whether the founders of mechanics realized this necessity or whether they were guided solely by commonsense is a matter of minor importance, for it so happens that, in view of the smallness of the velocities at their disposal, their commonsense would have been vindicated by their measurements; nothing in the field of their experience could have suggested the inappropriateness of the classical transformation. Small wonder then that the transformation was assumed to be valid, however great the relative velocity between the two frames. Only today, thanks to an experimental technique and to an accumulation of facts undreamed of by the earlier scientists, do we know that the classical transformation becomes increasingly incorrect when the relative velocity of the two frames is increased. It was the recognition of this fact which constituted the starting point of the special theory of relativity. The original aim of the relativity theory was to discover the amended transformation which would supersede the classical one. To understand how the new transformation, that of Lorentz, was derived, we must revert to the Classical Principle of the Relativity of Motion, propounded by Galileo and Newton.

We mentioned in Chapter XVII that the classical laws of dynamics of Galileo and Newton appeared to be in agreement with experimental results provided measurements were referred to so-called Galilean, or inertial, frames. If one of these frames is known, any other Galilean frame is found to be moving with respect to the first with uniform speed. As a result, there is never any relative acceleration or rotation as between Galilean frames. From a phenomenological standpoint, Galilean frames are recognized by the fact that, for an observer situated in one of them, very distant objects, such as the stars, appear to be at rest. In Newton's language, Galilean frames are at rest or are moving with uniform translational speed through absolute space, and hence without acceleration or rotation. As for deciding which particular one of the Galilean frames is at rest in absolute space, Newtonian science gives no answer. The reason for this may easily be understood from the following considerations.

The laws of mechanics being exactly the same in all Galilean frames, a mechanical experiment performed in one Galilean frame or in another will be governed by the same laws. We cannot therefore, by means of mechanical experiments, differentiate between the various Galilean frames; and so we cannot determine which one of the Galilean frames is at rest in absolute space. In this conclusion consists the physical expression of the "principle of relativity of Galileo and Newton."

We must now give mathematical expression to this principle. To do so we first note that if we apply the classical space and time transformation to the differential equations which express the laws of dynamics in one Galilean frame, we obtain thereby the laws such as they will appear in another Galilean frame. Now, when the classical transformation is applied to the equations of dynamics, these equations retain exactly the same mathematical form. In the language of the mathematician the differential equations of dynamics are "covariant" under the classical transformation. We here have the mathematical transcription of the fact that the classical laws of dynamics are the same regardless of the Galilean frame in which they are expressed. Thus, the covariance of the dynamical equations constitutes the mathematical statement of the principle of relativity of Galileo and Newton. The foregoing considerations show that the classical laws of mechanics, the classical space and time transformation, and the Newtonian principle of relativity are consistent.

We now pass to Maxwell's laws of electromagnetism *in vacuo*. These laws are expressed by partial differential equations which establish relations connecting space, time, and the electromagnetic field-magnitudes. A number of experiments, culminating in the ultra-refined ones undertaken at the close of the nineteenth century, showed that electromagnetic processes were controlled by exactly the same laws regardless of the particular Galilean frame in which they were taking place. The situation thus appeared the same as in mechanics, and as a result the Newtonian principle of relativity seemed to be valid for electromagnetic processes just as it was already known to be for mechanical ones. But this belief led to a contradiction. Quite generally, in order to obtain the laws holding in a Galilean frame B when the laws holding in a Galilean frame A are known, we must apply to the latter laws the classical space and time transformation. If this is done for the laws, or equations, of electromagnetism, we find that these laws are modified by the adjunction of terms involving the relative velocities of the two frames. In other words, the equations of electromagnetism are not covariant under the classical space and time transformation. But then electromagnetic experiments would be expected to betray a difference when the Galilean frame in which they were performed was changed, and the principle of relativity would appear to be untenable. We are thus led to the important conclusion that the laws of electromagnetism, the classical space and time transformation, and the Newtonian principle of relativity are incompatible.

If, then, we wish to retain the Newtonian principle of relativity in electromagnetism, as seems to be required by the most accurate experiments, either we must amend the laws of electromagnetism or else modify the space and time transformation. The first alternative was soon found to be impossible, for the electromagnetic laws, amended so as to be compatible with the principle of relativity and with the classical transformation, were unable to express even the simplest electromagnetic phenomena. The second alternative was therefore followed by Lorentz: Maxwell's equations were retained, and a revised space and time transformation was sought. Lorentz's investigations resulted in the discovery of the celebrated space-time transformation which bears his name. Maxwell's equations, the new transformation, and the Newtonian principle of relativity were compatible.

The new transformation was subsequently rediscovered by Einstein, who followed a different method. To understand Einstein's procedure, we first recall one of the direct consequences of Maxwell's equations: Electromagnetic waves (*e.g.*, waves of light), propagated through a frame in which Maxwell's equations are valid, will proceed with the same velocity c in all directions (*in vacuo*). This consequence of Maxwell's equations is sometimes referred to as the "law of the constant velocity of light." Now, if the laws of electromagnetism are to retain an invariant form in all Galilean frames, the same must be true of their consequence, the law of the constant velocity c of light. In other words waves of light, and more generally any existent moving with velocity c through a Galilean frame, should also be moving with this same velocity with respect to any other Galilean frame in motion relatively to the first. But then it follows that the space and time transformation which ensures the invariance (or rather covariance) of Maxwell's equations should also be consistent with the invariance of the velocity c, and conversely. Einstein therefore sought to obtain the transformation which would leave the velocity c unaltered, and he thus rediscovered the Lorentz transformation.

There are several important differences between the treatments of Lorentz and of Einstein. In the first place, Einstein's derivation of the transformation is the simpler. Furthermore, when we follow Einstein's procedure we do not have to consider the electromagnetic magnitudes present in Maxwell's equations. All we need do is to express the invariance of the velocity c (186,000 miles per second). The net result is that, whereas in Lorentz's electromagnetic treatment the validity of the new space-time transformation might be thought to hold only for electromagnetic processes, in Einstein's treatment we are forced to the conclusion that the new transformation is applicable to velocities in general, regardless of the particular existents to which the velocities refer. The new transformation thus acquires a universal instead of a restricted significance. From Einstein's standpoint the new transformation betrays fundamental characteristics of space and time, characteristics erstwhile unsuspected. In particular, the invariant velocity c, which is at the basis of the Lorentz transformation, becomes interwoven, as it were, into the very fabric of the world, and for this reason it appears frequently in the equations of the theory of relativity.

The difference in the viewpoints of Lorentz and Einstein has important consequences when we consider mechanical phenomena. Thus, if we follow Lorentz, we can see no particular reason for supposing that the new transformation should necessarily apply to mechanical processes or indeed to any processes which are not purely electromagnetic; in the case of mechanical processes the classical transformation would still appear to be valid. Einstein's attitude, on the other hand, compels us to discard the classical transformation entirely and to state that, for physical processes in general and for mechanical ones in particular, the new transformation must be valid. Quite aside from the reasons mentioned, there are also others which prompt us to assume the validity of the new transformation for mechanical processes. Let us examine one of these reasons. We recall that when we are dealing with an electromagnetic field, and we pass from one Galilean frame to another, we must submit the field-magnitudes to the new transformation. Suppose, then, an electron is situated in the field. The electron is acted upon by the field and is thereby subjected to a force. This force, being built up from the field-magnitudes, is necessarily transformed along with these magnitudes when we change Galilean frames, and hence is transformed according to the new transformation. Now the present force is generated by the field, and it might therefore be viewed as non-mechanical. However, it is difficult to see how a force of the electromagnetic variety can differ essentially from a mechanical force, such as is developed by a compressed spring. We are thus led to suspect that any force, whatever its origin, must be transformed in accordance with the new transformation. But we cannot stop here; for force in mechanics is defined by the time rate of change of momentum, and hence momentum and more generally all the mechanical magnitudes should, like force, be transformed according to the new transformation. The foregoing discussion shows that if we accept the new transformation for electromagnetic processes, we can scarcely avoid extending it to mechanical processes.

Suppose, then, we apply the new transformation to the classical equations of mechanics. We find that these equations do not remain covariant: they change in form. This fact implies that the classical laws of mechanics, the new transformation, and the Newtonian principle of relativity are incompatible. As a result, mechanical experiments should enable us to detect absolute motion. Thus, the principle of Newtonian relativity, which, thanks to the Lorentz transformation, has just been justified on theoretical grounds in the case of electromagnetic processes, now appears to be untenable for mechanical processes. The classical situation is thereby reversed.

At this point Einstein postulates that the principle of Newtonian relativity must hold for all physical processes without exception, whether these processes be electromagnetic, mechanical, or of any other kind. His assumption appears justified by other considerations. First, we cannot well suppose that an electro-mechanical phenomenon involving electromagnetic fields and material electrons should, insofar as it is electromagnetic, satisfy the Newtonian principle of the relativity of motion and, insofar as it is mechanical, violate this same principle. The problem of differentiating between what is electromagnetic and what is mechanical would in itself create a difficulty. Besides, since the mechanical part of the electro-mechanical phenomenon would violate the Newtonian principle of relativity, the phenomenon as a whole would violate it likewise; and this would conflict with some of the negative experiments which are precisely of the complex kind just discussed.

Let us, then, accept Einstein's views and see where they lead. If we assume the Newtonian principle of relativity for mechanical processes, and if we retain the new transformation, we must reject the classical mechanical laws. Einstein was thus led to modify these laws so as to render them compatible with the principle of relativity and with the new transformation. He was guided by the clue that at low velocities the new mechanical laws should pass over into the classical ones (for we know that at low velocities the classical laws appear to be correct). As a result of these considerations the revised mechanical laws were obtained. The new mechanical laws require that the concept of mass, formerly regarded as absolute, be viewed as relative, the mass of a body in a Galilean frame increasing with the body's velocity. In addition, mass and energy are identified; the total energy associated with a particle at rest is m_0c^2, where m_0 is the mass at rest and c the velocity of light *in vacuo*. Even before the advent of the theory of relativity, the behavior of particles moving at extremely high speeds (β-particles) indicated that the laws of classical mechanics could not be correct. The new mechanical laws of the special theory of relativity have proved themselves able to account for the discrepancies. To this extent, therefore, the new laws have been verified experimentally.

It is of interest to contrast the sequence of steps that has been followed up to this point. The impossibility of reconciling the Newtonian principle of relativity with Maxwell's electromagnetic laws and the classical transformation was overcome by sacrificing the classical transformation while retaining the electromagnetic laws. But the impossibility of reconciling the Newtonian principle of relativity with the classical mechanical laws and the new transformation was overcome by rejecting the laws and retaining the new transformation. Thus, whereas the theory of relativity has entailed a revision of the classical mechanical laws, it has imposed no change on Maxwell's electromagnetic laws.

Important consequences are derived from the Lorentz transformation. The transformation shows that the resultant velocity we obtain when we compound parallel velocities is not the sum of the component velocities, as classical kinematics supposed. The transformation also shows that, if two events take place at different points of a Galilean frame and occur at the same instant of time as estimated in this frame, these events will not usually occur simultaneously when referred to another Galilean frame moving with respect to the first. The transformation further yields the exact dislocation in simultaneity that will be generated by the relative velocity of the two frames. The relativity of simultaneity thus reveals itself as an immediate consequence of the Lorentz transformation. The relativity of simultaneity was never stressed by Lorentz, because he did not attribute any universal significance to his transformation. For him the new transformation had but a local electromagnetic validity. As such, it did not refer to real physical space and time; the classical transformation was still regarded as the correct one in this respect.

Einstein's definition of the simultaneity of events measured in a given Galilean frame is an immediate consequence of the prémises of the theory and, in particular, of the principle of the invariance of the velocity c. To verify this point, we first recall the classical view. According to classical science the velocity of light (*in vacuo*) is the same in all directions only if it is measured in the privileged Galilean frame in which the laws of electromagnetism are valid. For argument's sake let us first accept the correctness of the classical view. We are then justified in saying that, if two events occurring at points A and B emit light signals simultaneously in the privileged frame, these signals will be received simultaneously at the midpoint C on the line AB. But obviously this statement would be untenable if our frame of reference were not the privileged one, for in this event the velocity of light through the frame would not be the same in all directions. To obtain correct results, we should have to take into account the difference in the velocity of light in the two opposite directions, and this could be done only if we knew the velocity of our frame through the ether. Since the latter velocity had never been ascertained, a rigorous physical definition of the simultaneity of distant events could not be given on the basis of classical ideas. The entire situation changes when we accept Einstein's principle of the invariance of the velocity c, for now the arguments that were valid in the privileged frame become valid automatically in all Galilean frames. In short, Einstein's definition of simultaneity does not introduce any new postulate; it is a direct consequence of the premises on which the theory is based.

Minkowski's contribution consists in having shown that the Lorentz transformation, viewed as dealing with real physical space and time, illustrates the existence of a 4-dimensional world in which space and time are no longer independent, but intimately fused. To this 4-dimensional world the name *space-time* is given. This world is just as real (if the theory of relativity be accepted) as was the world of a separate space and time on the basis of classical conceptions. It is therefore utter nonsense to accept the theory of relativity, on the one hand, and to deny a fourth dimension to the world-continuum, on the other. The concept of 4-dimensional space-time entails a remarkable simplification in our understanding of physical processes. One of the most striking examples was given in Chapter X. We mentioned that in space-time there is no such thing as an electric field existing by itself, any more so than there is a privileged space. The electric and the magnetic fields are always fused, just as space and time are fused, and only by adopting a local standpoint restricted to one particular Galilean frame can a separation be made to appear between space and time or between an electric and a magnetic field. If we wish to adopt the more general impersonal standpoint instead of restricting our attention to one particular frame, we must not speak of space and time, but only of 4-dimensional space-time. Similarly, we must not consider an electric or a magnetic field but must confine ourselves to the 4-dimensional electromagnetic tensor situated in 4-dimensional space-time. However, the impersonal standpoint is not always the one which is the most convenient. For instance, if we are performing an electromagnetic experiment in a particular Galilean frame, our immediate interest lies in the local standpoint, *i.e.*, in the space and the time of our frame and in the particular field existing in this frame. Only when we wish to discuss the underlying reality which may manifest itself in one way or the other according to the conditions of observation, must the impersonal 4-dimensional conception be adopted.

The Transition from Relativistic to Classical Science—The invariance of the velocity c in all Galilean frames may be taken as the cardinal assumption of the special theory of relativity. From this initial assumption the Lorentz transformation may be deduced, and therefrom the entire theory follows.* Classical science likewise recognized the invariance of a certain velocity, but this velocity was infinite; no finite velocity was invariant. Inasmuch as the invariant infinite velocity plays in classical science much the same rôle as does the invariant finite velocity in the special theory of relativity, we may regard the two invariant velocities as characteristic of the respective theories. From this standpoint it is correct to say that the fundamental difference between the two theories is that in classical science the invariant velocity is infinite whereas in the special theory of relativity it is finite (though very great).

Now, an extremely high velocity will, to all practical purposes, appear infinite if we contrast it with a relatively small velocity; the smaller the latter velocity, the smaller the relative error we shall make when we confuse the very great velocity with the infinite one. Thus, we may infer that when we are dealing with low velocities, the anticipations and the laws of relativity will not differ perceptibly from the classical results. This conclusion may also be proved rigorously by noting that, for small velocities, the Lorentz-Einstein transformation passes over into the classical one. We here have a characteristic feature of the theory of relativity, *i.e.*, for small velocities the relativistic laws pass over into the classical ones. We mentioned that Einstein availed himself of this clue when he sought to determine the revised mechanical laws. The same situation holds for the composition of velocities. The classical rule for the addition of velocities is refuted by the theory of relativity; nevertheless for low velocities the rule becomes almost perfectly correct. Similarly, for low relative velocities, the relativity of simultaneity gives way to the classical concept of absolute simultaneity. Finally, when low velocities are considered, the discrepancies between the 4-dimensional space-time of Minkowski and

---

* The relativity of motion must also be assumed.

the 3-dimensional space and separate time of classical science cease to become apparent.

All these considerations show that unless we be dealing with very high velocities, the anticipations of the theory of relativity will differ so slightly from those of classical science that no differences will be detected. This fact has an important bearing on the nature of commonsense. Many of the conclusions derived from the theory of relativity conflict strongly with the dictates of commonsense. For instance, the relativistic rule for compounding parallel velocities seems to imply that $2 + 3$ is not equal to 5, a conclusion which contradicts commonsense. But the theory of relativity shows that the dictates of commonsense are contradicted only when the velocities compounded are extremely high, and thus when these velocities differ considerably from those with which we are familiar on the commonplace level of ordinary experience. For low velocities, the relativistic rule of composition tends to differ less and less from the classical rule, which is supported by commonsense. Thus, the theory violates commonsense only when we consider unfamiliar situations. What else, then, is this commonsense but an expression of the belief that things cannot be otherwise than we have observed them to be in the limited field of our immediate experience?

The theory we have sketched in the previous pages is called the "Special Theory of Relativity." In it we restrict our attention to Galilean frames; and the principle of relativity that is relevant is the Newtonian principle of the relativity of motion. Thus, the special theory is merely an application of the Newtonian principle to all processes, whether electromagnetic or mechanical. In the general theory of relativity, the restriction to Galilean frames is removed. For our present purpose an understanding of the general theory is unnecessary.* Today, the special theory is accepted unanimously by all the outstanding theoretical physicists. Its great merit is that it is the only theory which has been able to account for the phenomena associated with moving bodies and moving fields in electromagnetism. The general theory is also accepted by most physicists, but it is more speculative.

The Quantum Theory—The quantum theory has been extended and modified repeatedly since its original formulation by Planck, and even today it is still in a state of rapid change. Consequently, we cannot follow the historical sequence of discovery without mentioning assumptions which in the light of further study have been found untenable. However,

---

* The general theory is discussed in our book, "The Evolution of Scientific Thought." Boni and Liveright. 1927.

it would be misleading to omit all reference to the earlier form of the theory and start with a presentation of the modern quantum mechanics and of Heisenberg's Principle of Uncertainty, for, by so doing, the theory would appear gratuitous. We shall therefore follow a middle course, adhering to the historical sequence while correcting, as far as possible, the erroneous assumptions which were made from time to time and which were not essential. The entire theory revolves around the discovery of a new constant of nature, called h by Planck. This constant has the dimensions of "action" (energy × time), *i.e.*, of that abstract dynamical concept which enters into the principle of Least Action. The original formulation of the theory by Planck as subsequently extended by Bohr may be understood from the following special illustration.

The hydrogen atom is viewed as a mechanical system formed of a positively charged proton around which a negative electron circulates under the electrostatic attraction of the proton. The electron describes an ellipse, just as the earth does round the sun; the motion is thus periodic, repeating after each cycle. The action developed when the electron moves over an element of length ds on its orbit is given by the product of its momentum mv and the length ds of the tiny stretch covered. By summing, or integrating, such amounts over the entire orbit, we obtain the total action developed by the electron over each cycle. We may refer to this amount of action as the action of the dynamical system in the state of motion considered. Corresponding to each value of the action, there is a definite value for the total energy of the dynamical system, so that when the action is stated, the energy is known, and vice versa. Prior to the discovery of the quantum theory the laws of classical mechanics and of electromagnetism were thought to control this model of the hydrogen atom. According to classical ideas the electron, owing to its accelerated motion round the nucleus, would radiate energy, and the attendant loss of energy would entail a corresponding lowering of the action. The system might also absorb radiation, in which event the energy and the action would increase. But the classical treatment was unable to account for the precise radiations which observation detects in the hydrogen spectrum. Bohr was thus led to modify the classical treatment by introducing restrictions known as "quantum restrictions."

For instance, classical theory requires that the emission and absorption of radiation should take place continuously. The action of the system must then vary likewise in continuous fashion, and hence at any given instant any positive value should be possible for the action. The quantum theory modifies these conclusions. In the original form of the theory, the action of any periodic system is assumed to be susceptible of taking only such values as

(1) $0, h, 2h, 3h, \ldots nh \ldots,$

i.e., some integral multiple of the unit of action h. These values for the action of the system determine its so-called "stable," or "stationary" states; the corresponding values for the energy of the system are called "energy levels." The reason why the stationary states are defined in terms of the action rather than in terms of the energy is that the successive values of the action form a regular step-ladder progression, whereas the energy levels usually differ from one another by unequal amounts. According to the quantum theory, when the energy decreases through radiation (or increases through absorption of energy), the action can pass only from one to another of the privileged values (1). In any case the transition between states is in the nature of a jump of finite magnitude nh ($n = 1$, $2, 3 \ldots$) and is not continuous, as it was assumed to be in classical science. Similar considerations may be applied to all dynamical systems in which the motions are periodic or quasi-periodic. Planck summarizes the situation by saying that *action is atomic* instead of continuous as was formerly believed. He still adheres to this notion of the atomicity of action in a recent popular book. We shall adopt his attitude for the present, because it is convenient in certain cases; but, as we shall see later, it is misleading and must not be taken too seriously.

The quantum theory and the theory of relativity are totally different theories dealing with different departments of physical science. Nevertheless they have a feature in common. Both theories establish the existence of unsuspected critical values for certain mechanical concepts. In the theory of relativity, the concept is *velocity* and the critical value is c. In the quantum theory, the mechanical concept is *the change in the action* and the critical value is h. Moreover, the difference between the relativity theory and classical science is that the former proves the critical value to·be finite though exceedingly large, whereas classical science had assumed it to be infinitely great. The significant velocity is thus transferred by the theory of relativity from the infinitely great to the large but finite. Similarly, if we contrast classical science with the quantum theory, we note that in classical science, where action was assumed to vary continuously, the changes in the action of a system could be infinitely small. The quantum theory, on the other hand, restricts the changes in the action to the magnitudes $h, 2h, 3h \ldots$, where h is small but finite. Consequently, according to the quantum theory, if a change occurs, it is finite—never infinitely small. Thus, both the theory of relativity and the quantum theory displace our attention from the infinite (whether great, or small) to the finite (great or small).

A further point of resemblance between the two theories is the way they both tend to merge into classical science when appropriate limiting conditions are realized. The theory of relativity merges into classical science when the velocities of interest are small in comparison with the velocity of light. For this reason no discrepancies will be observed between the classical and the relativistic theories as long as the velocities are low.

Now a similar situation holds in the quantum theory. To understand how it arises, let us consider a mechanical structure formed of an electron constrained to move along a fixed circle. Circular motion being a' form of accelerated motion, the electron will radiate energy;* its velocity will decrease and so will its action. According to the.quantum theory, each successive drop in the value of the action is measured by nh, where n is a positive integer. In the particular mechanical system here considered, the drops will always be equal to h. The net result is that the motion of the electron will decrease by jerks. This conclusion is, of course, in contradiction with classical science, which would have required the motion to decrease continuously.

But suppose now that the mechanical system is of average proportions, so that its total store of action is very great, say 1,000,000 h. As contrasted with this enormous store of action, the successive drops h will be relatively negligible and will entail only imperceptible decreases in the motion of the system. Consequently, in spite of the fundamental jerkiness of the motion, we shall have the impression of a continuous slowing down, and no departure from the expectations of the classical theory will be observed. A similar impression of continuity is experienced when we watch the hands of a clock; they appear to be moving continuously whereas in all truth they are advancing by jerks.

From the previous illustration we may infer that quantum science passes over into classical science when the quantum drops in the action are small in comparison with the total action present. This important fact forms the basis of Bohr's celebrated *Correspondence Principle.* The conditions under which the foregoing passage takes place are therefore referred to as the "limiting conditions of the correspondence principle."

These limiting conditions thus play with respect to the quantum theory the same rôle as the limiting conditions of low velocities in the special theory of relativity. From this standpoint both theories prove to be mere

---

* We are following the classical presentation.

refinements of classical science. There is, however, a difference in the practical importance of the two classes of refinements. Phenomena in which velocities approximate that of light are not of a nature that can be observed on the commonplace level of experience. And so, in the interpretation of commonplace phenomena the theory of relativity may be ignored. But a similar conclusion would not always be correct in the case of the quantum theory. The fact is that conditions in which the total amount of action is of the same order of magnitude as the unit h may sometimes arise on the commonplace level; and in such cases the quantum theory cannot be ignored. An example in point is afforded by the familiar phenomenon of incandescence; indeed it was to account for this phenomenon that Planck devised the quantum theory. Incandescence, however, is practically the only commonplace phenomenon whose interpretation requires the introduction of the quantum theory. In nearly all other phenomena of a familiar kind, the quantum theory may be disregarded. Where the quantum theory cannot be disregarded is in the investigation of the subatomic world and in the study of phenomena occurring under enormous pressures or at exceedingly low temperatures.

The quantum theory has furnished considerable information on the phenomenon of radiation. According to classical science, if an electrified particle, $e.g.$, an electron, executes a periodic motion with a definite frequency, it emits radiation of the same frequency, and the emission is continuous. But in the quantum theory these conclusions must be revised. Let us first consider the frequency of the emitted radiation. According to the quantum theory, radiation is emitted when the radiating system drops from one stable level to another. Such drops are called "quantum drops." The emitted radiation will have as frequency the ratio defined by the drop in the energy of the radiating system divided by the unit of action h. In general, the frequency of the radiation differs from the mechanical frequency of the motion, but it will tend to coincide with this frequency when the drop in the action is small in comparison with the total action. We are then under the limiting conditions of the correspondence principle and we see once more that under these limiting conditions the quantum theory passes over into the classical theory.

Next let us consider the process of emission. The jerkiness of the quantum drops entails discontinuities in the emission of radiation. Such discontinuities appear difficult to reconcile with Maxwell's equations, which require that radiation be formed of continuous electromagnetic waves. Several experiments, however, seem to corroborate the implications of the quantum theory and to show that in some circumstances, at least, radiation behaves as though it were formed of discrete corpuscles.

On the other hand, in many experiments it is the wave aspect of radiation which imposes itself. Radiation thus appears to present a dual aspect. The same dual aspect was subsequently revealed by matter; and so, the belief grew that some fundamental underlying principle must be at stake.

What we have said to this point concerns the original formulation of the quantum theory by Planck, Einstein, and Bohr. But it was soon found that the anticipations of the theory were often incorrect, and that complementary assumptions had to be introduced now and then. For instance, when we have determined the stable states of a system by the methods of the quantum theory, we should expect that all drops from a higher to a lower level would be possible. But the observation of spectra shows that the radiations which would result from certain drops never arise, and hence we must conclude that for some reason or other these drops do not occur. Furthermore, even a superficial examination of a spectrum shows that the spectral lines are unequally intense. To account for this fact by means of the quantum theory, we must assume that, in a system formed of a large number of radiating atoms, certain kinds of drops are more numerous than others. Such drops, in their aggregate, will contribute a larger amount of radiation each second, with the result that the corresponding spectral lines will be more intense. Since we know nothing of the mechanism which generates the drops, the situation we have postulated is best expressed by the statement that, for some reason or other, the various drops are unequally probable. Obviously, if we could calculate the probabilities of the different drops, we should be able to predict the relative intensities of the different spectral lines. Unfortunately the quantum theory in its original form, at least, affords no means of calculating the probabilities of the drops and so gives no information concerning the intensities. Nor is any information given on the polarization (direction of vibration) of the emitted radiations. Some progress is made when we invoke the correspondence principle, and thus rely on the classical theory to furnish the required information. But this appeal to the discarded classical doctrine is inconsistent; besides, the information obtained is usually only approximate. This unsatisfactory situation is characteristic of the entire quantum theory in its original form: classical mechanics and the quantum restrictions are applied simultaneously, sometimes the one, sometimes the other being in the ascendency.

A second difficulty which the quantum theory in its original form is unable to remove arises from systematic discrepancies between the anticipations of the theory and the results of experimental observation. The quantum theory assumes that the stationary states of a radiating system are determined by the values

$$(1) \qquad\qquad 0, \, h, \, 2h, \, 3h, \, \ldots \, nh \, \ldots$$

for the action. In many cases this assumption appears justified, but in other situations a more complicated sequence of values seems to be demanded by experimental observation. For instance with the linear oscillator, the values

$$(2) \qquad\qquad \frac{h}{2}, \, 3\frac{h}{2}, \, 5\frac{h}{2}, \, \ldots \, (2n+1)\frac{h}{2} \, \ldots$$

are found necessary. And in the case of the space rotator (a molecule in rotation), the values

$$(3) \qquad 0, \quad h\sqrt{1 \times 2}, \quad h\sqrt{2 \times 3}, \quad \ldots h\sqrt{n(n+1)} \, \ldots$$

seem to be required. Theory was unable to account for the values (2) and more especially (3). Although it is true that the sequence of values (1), accepted by the older quantum theory, had received no theoretical explanation, it could be accounted for by the hypothesis of the atomicity of action. But the sequence of values (2) and especially the sequence (3) conflict with this hypothesis. The difficulty we are referring to is often called the difficulty arising from the "half-quantum numbers." This appellation was given because the form (2), in which the action assumes some half-integral number of times the unit value h, was discovered before the more complicated form (3). In any case the assumption of the atomicity of action appears incorrect.

With the realization that the original quantum theory of Planck and of Bohr could constitute but the first step in the unravelling of quantum phenomena, attempts were made to obtain a more refined theory. The Matrix Method of Heisenberg, the Wave Mechanics of de Broglie and Schrödinger, and the Quantum Mechanics of Dirac were the outcome. It must not be thought that these newer theories conflict. Mathematically they are equivalent, and they only differ in the conceptual trappings with which they are clothed. Some investigators prefer one method, others another. The new methods yield the correct intensities and polarizations of the emitted radiations. They show why certain energy drops are forbidden. Furthermore, they give the correct values (1), (2), or (3) which experiment demands, according to circumstances, for the determination of the stationary energy levels. In addition they lead to new discoveries. Heisenberg's Principle of Uncertainty, in particular, is a direct consequence of any one of these new mathematical theories. Thanks to this principle, the difficulty in comprehending the dual nature of radiation and also of matter is in a large measure removed; for the principle enables us to understand how the uncertainty of our observations may cause discreteness to yield the impression of continuity. The principle of uncertainty furnishes a new interpretation of the magnitude h: it shows that this constant measures, so to speak, the extent of the uncertainty and thereby serves as a link between the corpuscular and the wave aspects of radiation and also of matter. If h were infinitesimal, as classical science supposed, the uncertainty relations would vanish and so would all quantum phenomena. We should then be in the limiting conditions of the correspondence principle.

The newer quantum theories are but refinements of the original quantum theory, and when the store of action is progressively increased, they tend to pass over into the older theory. The passage is particularly clear in wave mechanics when we consider the transition from wave-optics to ray-optics. Finally, when the action is greater still, all the quantum theories pass over into classical science. We have here an illustration of the progressive approximations introduced by successive physical theories. Many examples have been given in this book; the progression illustrated in the sequence, classical science, original quantum theory, new quantum theories, affords an additional example.

The quantum theory has gradually found its way into practically every province of physical science. Originally devised by Planck to account for the phenomenon of incandescence, it was soon extended by Einstein to the problem of the specific heats of solids and of gases at low temperatures, to the photo-electric effect and its converse (the production of X-rays by the sudden stoppage of swiftly moving particles), and to photo-chemical reactions. The discovery of collisions of the second kind and of their importance in certain chemical reactions was brought about through quantum considerations. Similarly the calculation of the chemical constants was found to involve the constant h, and hence the quantum theory. The Third Principle of Thermodynamics was seen to be closely connected with this theory. The quantum theory was applied by Bohr to account for the spectra of atoms. In connection with these studies, the spinning electron and the ''Exclusion Principle of Pauli'' were discovered. The kinetic theory of gases was modified so as to comply with the quantum theory. Thanks to the new gas theory, the unexplained behavior of the electron gas in a metal was accounted for by Sommerfeld; and the behavior of gases under enormous pressures (*e.g.*, in the interior of the stars) was better understood. Dirac, in his attempt to obtain a fusion of the relativity theory and of the quantum theory, was led to the conclusion that a positively charged electron should exist. It has since been discovered (the positron). More recently still, Maxwell's equations have been investigated with a view of rendering them consistent with the new ideas. Finally, to the developments of the quantum theory are due the attacks on the doctrine of rigorous determinism. In short, optics, mechanics, thermodynamics, chemistry, the statistical laws, and many others have one by one come under the sway of the quantum theory. Needless to say, our entire outlook on the physical world has been affected.

In the following chapters we shall attempt to give an idea of the methods followed in the elaboration of the new doctrines.

CHAPTER XXIV

PLANCK'S ORIGINAL QUANTUM THEORY

The name radiation is given to all electromagnetic waves. These waves have the common property of being propagated with the velocity c, *in vacuo*. The frequency of a radiation is defined by the number of vibrations it executes every second at a given point of space. The wave length is the distance at any instant between two consecutive crests. The higher the frequency, the shorter the wave length *in vacuo*. In the quantum theory, frequency is a more useful notion than wave length, and so we shall classify the various radiations according to their increasing frequencies. Ranging them in this order, we have the long radio waves, the short ones, the infra-red radiations, the visible ones from red to violet, the ultra-violet radiations, X-rays, γ-rays, and possibly others of higher frequency. From the impersonal point of view of the physicist, all these radiations form a common family and differ from one another no more than a billiard ball moving fast differs from one moving slowly. Only when we confuse our subjective impressions with the impersonal point of view and introduce the extraneous idea of "value," is there any essential difference between yellow light, which we see, and infra-red light, which is invisible.

All bodies, with the exception of the so-called perfectly reflecting ones, absorb a part at least of the radiation that may fall upon them. They also emit radiation when heated. A general law due to Kirchhoff states that, for a radiation of given frequency, the ratio of the emissive power to the absorption coefficient is the same for all substances; it depends solely on the frequency of the radiation, on its plane of polarization, and on the temperature.

From the standpoint of their radiating properties, all substances fall between two extreme limiting types. One extreme is represented by perfectly reflecting surfaces, which absorb no radiation and therefore (according to Kirchhoff's law) emit none when heated. The other extreme is illustrated by the so-called "perfectly black bodies," or "black bodies," for short. A black body absorbs all radiations and consequently, under the influence of heat, emits them all. A piece of soot is a fair illustration of a black body. In practice the two extreme limiting cases are never realized rigorously. A sheet of polished silver is not a perfectly reflecting

447

surface, for some radiations are not reflected; and a layer of soot is not a perfectly black body, for though it absorbs all the visible radiations, it does not absorb completely all the invisible ones. Nevertheless, though perfectly reflecting surfaces and perfectly black bodies are mere idealizations, it is often useful to consider them in theoretical discussion. For similar reasons we speak of perfectly elastic bodies in mechanics and of perfectly reversible changes in thermodynamics.

What we shall be concerned with in this chapter is the emission of radiation by a black body. The phenomenon is familiar enough, since it is illustrated very approximately when a piece of soot is heated and brought to the state of incandescence. Commonplace observation shows that when the temperature is progressively increased, the soot becomes luminous, passing from a dull red to dazzling whiteness. But such observations are merely qualitative and of little value. What we must do is analyze the light, decompose it into its various monochromatic radiations, and then determine the intensity of each monochromatic radiation when the temperature of the black body has one value or another. The formula which condenses the results obtained is called the empirical law of "black-body radiation." The aim of the theoretical physicist is to secure a theoretical basis for this empirical law and thereby construct a theory of black-body radiation. Despite the familiarity of the phenomenon of incandescence, the theory of black-body radiation has proved to be one of extreme difficulty. Entirely new concepts unknown to the more classical theories were found necessary, and it was the introduction of these new concepts that marked the start of the quantum theory.

An important advance was made when a means of obtaining black-body radiation was devised. As we have said, a piece of soot is not a perfect example of a black body, so that the radiation emitted by incandescent soot is not true black-body radiation. But a theoretical argument shows that the radiation contained in a heated enclosure (equilibrium radiation) is the same as black-body radiation. By taking advantage of this discovery, experimenters were able to secure accurate information on the intensities of the various radiations which would be emitted by a black body at one temperature or another. These preliminaries dispensed with, we pass to a study of radiation in an enclosure.

Radiation Gas—Let us suppose that a certain amount of light (*e.g.*, a mixture of yellow and green light) is introduced into an evacuated enclosure the walls of which are perfectly reflecting. Since the walls can neither absorb nor emit radiation, the composition of the radiation in the enclosure remains unchanged.* If, then, we open the enclosure after any period of time, however long, exactly the same radiation will be found present, though of course it will escape immediately. The imprisoned radiation behaves in many respects like a gas. It exerts a pressure against the walls of the enclosure, and work must be expended if we wish to compress it into a smaller volume, e.g., by means of a piston. Because of this similarity in the behaviors of radiation and of gases, the radiation in an enclosure is sometimes referred to as *radiation gas*.

Equilibrium Radiation—In the idealized experiment we have just been considering, we have purposely assumed that the enclosure contains no trace of matter susceptible of emitting and absorbing radiation; the walls of the enclosure, being perfectly reflecting, do not interfere with this assumption. We now suppose that, in an enclosure maintained at constant temperature, a piece of matter, e.g., a piece of soot, is introduced. The only stipulation is that the matter should be susceptible of emitting at least some amount of radiation of every frequency (soot satisfies this requirement). We need not inquire whether radiation is already contained in the enclosure when the matter is introduced, for, as we shall see, the final result is always the same. The matter will emit radiation and at the same time will absorb some. Exchanges will continue until the rates of emission and of absorption by the matter are equal; a balance is thus reached, and thenceforth the radiation within the enclosure remains unchanged in its composition. We have here a state of statistical equilibrium. The name *equilibrium radiation* is often given to the peculiar mixture of monochromatic radiations that will be found in the enclosure. This appellation is justified by the fact that the radiation and the matter are in equilibrium.†

Kirchhoff's law establishes several important points. It shows that, for a given temperature of the enclosure, the composition of the imprisoned equilibrium radiation is exactly the same regardless of the nature of the matter present. The only restriction imposed is the one previously stated; namely, the matter must be susceptible of emitting at least some radiation of each conceivable frequency. It can also be shown that the shape and size of the enclosure do not affect results. To submit to experimental

---

* It is probable that a change in the radiation would take place, but since the reasons that prompt this statement are connected with recent discoveries, we need not press the matter further at this point.

† When matter is present in the enclosure, the equilibrium between the radiation and the matter is similar to the equilibrium between a vapor and the solid phase of the same substance.

measurement the intensities of the various monochromatic radiations, we perforate one of the walls with a pin. Some of the radiation streams out and may be analyzed by suitable instruments. We assume that the loss of radiation through the pinhole is too small to affect the conditions within the enclosure, and hence that the sample analyzed gives the correct composition of the equilibrium radiation.

As we have mentioned earlier, the equilibrium radiation for any given temperature of the enclosure is exactly the same as the black-body radiation that would be emitted by a perfectly black body at the same temperature. The reason for this equivalence is easily understood. Any radiation which falls on the opening in the wall passes into the enclosure, is reflected from wall to wall, and does not emerge again. To all intents and purposes the radiation is totally absorbed, just as it would be were it to fall on a perfectly black body. The advantages of substituting the heated enclosure for a black body are numerous. In the first place, we have said that no perfectly black body can be found, whereas the heated enclosure simulates this ideal existent. In the second place, the enclosure is more easily maintained at a stated temperature than is a piece of soot. Finally, since the radiation in the enclosure is a manifestation of an equilibrium condition rather than of an individual process, we may to a large extent ignore the mechanism by which the radiation is emitted from the matter and yet submit the problem of its composition to theoretical treatment.

Experimental Results—Before we consider the derivation of the correct law of equilibrium radiation by means of theoretical arguments, let us examine the general results established by experiment. An enclosure was heated to one temperature or another, and the radiation streaming from the aperture was analyzed. Experimenters found that the intensity i_ν of the radiation of frequency ν was not affected by the shape of the enclosure or by the material of which the enclosure was made; the intensity was found to depend solely on the frequency ν and on the temperature T of the enclosure. This dependency is expressed mathematically by the equation

(1) $i_\nu = F(\nu, T)$,

where $F(\nu, T)$ is some unknown function of the frequency and of the temperature.

To obtain the empirical law of equilibrium radiation we should have to determine the exact form of this function from direct measurements of intensity, frequency, and temperature. However, for reasons which will now be explained, the law of radiation was not discovered in this way. Human measurements being necessarily imperfect, the measurements performed by different experimenters (or by the same experimenter at different times) do not usually agree: slight discrepancies may be expected. As a result, slightly different forms are suggested for the same empirical law; and we cannot be certain that any one of the laws suggested is rigorously correct. Fortunately, in many cases, the various tentative laws differ so little from some law which is mathematically simple that this simple law is accepted as the correct one.* Of course, no modern physicist believes that the laws of physics are necessarily simple; nevertheless, whenever the correct form of a law is in doubt, we are obviously justified in selecting, at least tentatively, the simplest law that is consistent with the results of measurement. Thus, the criterion of simplicity provides a means of agreeing on the law which will be recognized as correct—at least provisionally. Now this criterion of simplicity fails us in connection with the radiation law, for no simple law is even vaguely compatible with the experimental measurements. As a result, the different experimenters suggested different more or less complicated radiation laws and were unable to agree on the correct form of the function $F(\nu, T)$ in (1).

Although the correct radiation law could not be determined empirically, the knowledge that the law must be of the general form (1) was soon supplemented by further information. Thus Stefan, as a result of his measurements, showed that the intensity of the sum total of all the radiations emitted was proportional to T^4, where T is the absolute temperature of the enclosure. This relation, which the correct radiation law must satisfy, is called Stefan's law.

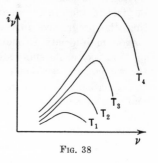

FIG. 38

A third important feature which must be consistent with the radiation law was likewise furnished by direct measurement; it is exhibited in the graph. The different graph lines show how the intensity of a radiation varies with its frequency for different temperatures. The higher the temperature, the greater the elevation of the graph line. Quite generally we may assume that a continuous range of frequencies (visible or not) is present, however low the temperature. The intensities of the

---

* Illustrations of this remark are afforded by Kepler's laws and by the gas laws.

visible radiations will be small unless the temperature exceeds 600° Centigrade; consequently, light will not be perceptible to the eye until this temperature is reached. We also see from the graph that, at any given temperature, one of the radiations is more intense than any of the others. Furthermore, when the temperature is increased, all the intensities increase, while the radiation of maximum intensity is displaced progressively towards the higher frequencies. Measurement shows that the frequency of the radiation of maximum intensity is directly proportional to the absolute temperature. The name "displacement law" is given to this fact of observation. Setting ν_{max} for the frequency of maximum intensity, and T for the absolute temperature, we may express the displacement law mathematically by

$$(2) \qquad \nu_{max} = \text{constant} \times T.$$

The constant factor of proportionality is determined by measurement.

The well-known phenomenon of incandescence affords an illustration of the displacement law. When a piece of soot is heated, we find that, at temperatures below 600° Centigrade the soot remains non-luminous, although radiation of the infra-red type is being emitted in perceptible amounts; the frequency of maximum intensity is now in the infra-red. When the temperature is progressively increased beyond 500°, visible radiations begin to appear; first red ones, then yellow, green, blue, and violet ones, etc.; the intensities of all these radiations increase with the temperature. When the temperature attains 1200° all the variously colored radiations are present in sufficient amounts for their superposition to yield to the eye the impression of dazzling whiteness. At the same time the progressive increase in temperature causes the radiation of maximum intensity to move towards the higher frequencies. If our piece of soot could stand still higher temperatures without passing into the vapor state, we should find that, at a temperature of about 3400°, the radiation of maximum intensity would emerge from the infra-red and pass into the visible part of the spectrum. As the temperature was increased still further the same gradual displacement of the frequency of maximum intensity would continue; this frequency would eventually pass into the ultra-violet, moving towards the X-rays.

Stefan's law or the displacement law may be utilized to determine the temperature of the sun. Let us consider, for example, the utilization of the displacement law. The frequency of the radiation of maximum intensity emitted by the sun can be determined from an analysis of the sun's light. If, then, we assume that the sun radiates like a black body (an assumption which is only approximately correct), we may deduce from the displacement law (2) the temperature of the sun's surface layers.

Let us summarize the results obtained thus far. Direct measurements show that the radiation law must satisfy the three following requirements: it must have the general form (1), and it must be consistent with Stefan's law and with the displacement law. On the other hand direct measurements cannot establish the exact form of the radiation law.

We are thus prompted to inquire whether theoretical considerations will be any more successful in furnishing the radiation law. As we shall see presently, the correct radiation law cannot be derived from the classical theories. Nevertheless, classical considerations, based on the laws of thermodynamics, do confirm our former empirical findings by showing that the radiation law should satisfy the three requirements listed in the previous paragraph. Furthermore the methods of thermodynamics enable us to particularize somewhat the form of the arbitrary function $F(\nu, T)$, which appears in (1). This last discovery was made by Wien when he sought to account theoretically for the displacement law.

Wien's Relation—Wien showed that the displacement law was suggested by thermodynamical considerations. We say *suggested*, because the thermodynamical theory, though compatible with the facts of observation, does not prove that they are inevitable. Wien approached the problem in an indirect way and obtained a result of major importance, of which the displacement law is, so to speak, a mere residue. Prior to Wien's investigations, theory was unable to assign any definite form to the unknown function $F(\nu, T)$ which appears in (1). Wien, thanks to an ingenious thermodynamical argument, proved that this function must necessarily fulfill certain requirements. In particular, the function $F(\nu, T)$ must be of the general form.

$$\frac{\nu^3}{c^2} f\left(\frac{\nu}{T}\right),$$

where $f\left(\dfrac{\nu}{T}\right)$ is an unknown function of the ratio $\dfrac{\nu}{T}$, and c is the velocity of light *in vacuo*. Substituting this expression for $F(\nu, T)$ in (1), we get for the law of radiation

$$(3) \qquad\qquad i_\nu = \frac{\nu^3}{c^2} f\left(\frac{\nu}{T}\right).$$

This important result is insufficient, of course, to furnish the radiation law, for the function $f\left(\dfrac{\nu}{T}\right)$ is still unknown. It might be any function of $\dfrac{\nu}{T}$, e.g., $\sqrt{\dfrac{\nu}{T}}$, or $e^{\frac{\nu}{T}}$, etc. Nevertheless the form of the unknown function $F(\nu, T)$ is narrowed down by Wien's discovery. Not all the conceivable laws which satisfy Wien's relation (3) entail a radiation of maximum frequency and lead to the displacement law; however, the relation (3) is not inconsistent with the displacement law, for we may imagine laws which satisfy the relation (3) and also the displacement law. This is what we meant when we said that Wien's investigations suggested the displacement law, though they did not necessitate it.

The theoretical results we have discussed exhaust all the information on the law of radiation that can be gathered when we restrict ourselves to the general methods of thermodynamics. To proceed further, hypothetical occurrences of one type or another must be postulated. We have here an illustration of the limited power of thermodynamical considerations, and we see that the progress of science would soon cease if all spirit of speculation were banned.

Rayleigh's Erroneous Law of Radiation—By basing his demonstration on the classical laws of electromagnetism and by introducing certain assumptions, Rayleigh was able to obtain a definite law of radiation. It bears his name. Jeans subsequently simplified the demonstration. We shall discuss, however, an entirely different derivation of the same law, given by Planck.

We have said that equilibrium radiation will be present in the enclosure when the emission and the absorption of radiation by the matter proceed at the same rate. We also mentioned that Kirchhoff's law shows that the precise nature of the matter in the enclosure is of no importance (subject to the reservations mentioned previously). Consequently, we are running no risk when, for reasons of mathematical simplicity, we utilize some schematic model to represent the emission and absorption of radiation by matter. Planck assumed that the matter was represented by a number of oscillators formed of electrons vibrating to and fro along straight lines. Each oscillator was characterized by a definite frequency, and it could emit and absorb only that radiation which had this frequency.

Suppose now that the state of statistical equilibrium is attained as between the oscillators and the radiation of the enclosure. Each oscillator will be emitting and absorbing radiation, and hence emitting and absorbing energy. Consequently the energy E of an oscillator will fluctuate in value, just as occurred for the molecules of a gas. At any given instant, therefore, the various oscillators having the same frequency ν will have different energies. We represent by $U(\nu, T)$ the average value of the energies of all the oscillators ν in the state of statistical equilibrium when the temperature of the enclosure is T. Equivalently we may suppose that $U(\nu, T)$ is the average value of the energy of one oscillator over a long period of time.

The continuous emission of radiation by oscillators had been studied previously by Lorentz on the basis of classical electrodynamics, and Planck accepted his results. Planck was then led to a relation between the intensity of the radiation of frequency ν present in the enclosure, and the average energy of any one of the oscillators vibrating with this frequency. The relation in question is

(4) $$i_\nu = \frac{\nu^2}{c^2} U(\nu, T),$$

where i_ν is the intensity of the radiation of frequency ν. The law of radiation will therefore be secured if we can calculate the average energy $U(\nu, T)$ of an oscillator.

The classical statistical theory furnishes the expression of $U(\nu, T)$. Thus let us consider the aggregate of oscillators which have some common frequency. The oscillators exchange energy by emitting and absorbing radiation of this frequency; they are therefore in a state of statistical equilibrium *inter se*. Consequently, the theorem of equipartition should apply as it does for colliding gas molecules, and each degree of freedom should have, on an average, a quantity of energy

(5) $$\frac{kT}{2},$$

where k is the gas constant. Our oscillators, being linear, have only one degree of freedom (in the mechanical sense); but, as mentioned on page 422, where we discussed the specific heats of solid bodies, we must double the number of degrees of freedom on account of the potential energy of the oscillators. The net result is that each oscillator must be regarded as having two degrees of freedom, and hence the average energy of each oscillator will be

(6) $$U = kT.*$$

---

* This expression shows that, according to classical theory, the average energy U of an oscillator depends on the temperature but not on the oscillator's frequency.

Substituting this expression in (4), we obtain Rayleigh's law of radiation, *i.e.*,

$$(7) \qquad\qquad i_\nu = \frac{\nu^2}{c^2} kT.$$

Rayleigh's law, though it satisfies the requirements of Wien's relation (3),* does not account for the displacement law. It indicates that the higher the frequency, the greater the intensity; as a result, the radiation of infinite frequency should be the most intense. This conclusion is in crass contradiction with observation. Moreover, Rayleigh's law is incompatible with Stefan's law. Finally, since the oscillators are assumed to have all possible frequencies, there must be an infinite number of oscillators. But then, if all of them are to share in the partition of energy, the energy must itself be infinite; and this is absurd.

To obtain correct results, it seems necessary to assume that the oscillators of higher frequency cannot participate in the partition of the energy. Under such conditions the total energy may be finite, the higher frequencies fail to appear, and a frequency of maximum intensity arise. An equivalent way of expressing the foregoing assumption is to say that the degrees of freedom of the oscillators of high frequency must be inoperative. We are here faced once again with the same peculiar situation noted in connection with the specific heats; and so we are led to suspect that this situation, occurring as it does in widely different phenomena, must connote some general law as yet unknown. An additional point of similarity between the present situation and the one noted for the specific heats is that Rayleigh's law tends to give the correct experimental values for the intensities when the temperature of the enclosure is high (and also when the frequencies considered are small). The law becomes increasingly incorrect when low temperatures are considered. Rayleigh's law, though obviously incorrect, has nevertheless played an important part in the development of the quantum theory. As Lorentz and Poincaré showed, in whatever way we proceed without introducing assumptions that would conflict with classical electromagnetics and the law of equipartition, Rayleigh's law is the inevitable outcome. Hence it seems certain that the rejection of Rayleigh's law will entail the rejection of classical theory, or at least will necessitate some important change in it.

Wien's Erroneous Law of Radiation—We mentioned that the complicated nature of the radiation law precluded it from being established

---

* In Wien's relation (3) we have but to set $f\left(\dfrac{\nu}{T}\right) = k\dfrac{T}{\nu}$ for the unknown function, and Rayleigh's law is obtained.

by direct measurements, and that for this reason the different experimenters were unable to agree on the exact form of the law. Wien suggested an empirical formula for the radiation law, and at one time this formula appeared to be in remarkable agreement with experimental measurements; later, however, discrepancies were detected. Wien's radiation law, which must not be confused with his relation (3), satisfies the requirements of the formula (3) and of the displacement law; it is also in harmony with Stefan's law. Wien's radiation law is, so to speak, the antithesis of Rayleigh's law, for whereas Rayleigh's law becomes correct at high temperatures and for low frequencies, Wien's radiation law becomes increasingly correct at low temperatures and for high frequencies. Wien first, and later Planck, sought to give this law a theoretical foundation, but the assumptions introduced appeared arbitrary. In any case the law is not in perfect accord with experiment so that we need not consider it further.

Planck's Law of Radiation—Inasmuch as both Rayleigh's and Wien's laws of radiation, though incorrect, appear to express facts correctly at opposite limits of temperature and frequency, we may presume that the correct law must have an intermediary form, passing over into Rayleigh's when T is large and ν small, and into Wien's when the reverse situation is contemplated. Planck, guided by these considerations, devised a new theory of radiation which he called the "Quantum Theory." From this theory Planck was able to derive a radiation law which satisfied Wien's relation (3), the displacement law, and Stefan's law, and which was in excellent agreement with experimental measurements at all temperatures. Planck's celebrated law is expressed by

$$(8) \qquad i_\nu = \frac{h\nu^3}{c^2} \; \frac{1}{e^{\frac{h\nu}{kT}} - 1}, \; *$$

where k is the gas constant, and h a new constant the value of which is exceedingly small and is approximately

$$h = 6,5 \times 10^{-27} \; \text{(erg. sec.)}.$$

We must now examine how Planck derived the radiation law (8) from his theory. Planck followed the same course he had pursued pre-

---

* We may derive Planck's law from Wien's relation (3) by setting

$$f\left(\frac{\nu}{T}\right) = \frac{h}{e^{\frac{h\nu}{kT}} - 1}.$$

viously when he derived Rayleigh's radiation law. As before, the matter in the enclosure was represented by a large number of linear oscillators, each one of which vibrated with a definite frequency and could emit and absorb only those radiations having that frequency. Planck, however, avoided being led to Rayleigh's erroneous law by introducing an innovation which we shall examine presently. In order to appreciate the nature of this innovation, let us first recall the results which Planck had obtained on the basis of the classical theory.

According to Planck's classical theory the energy E of an oscillator, in the state of statistical equilibrium, fluctuated continuously, as a result of the continuous emission and absorption of energy. The symbol $U(\nu, T)$ was taken to represent the average energy of an oscillator ν when the temperature of the enclosure was T. The classical treatment furnished the relation (4) between the average energy $U(\nu, T)$ and the intensity i_ν of the corresponding radiation. The classical expression of U, deduced from the theorem of equipartition is kT; and when this value was inserted in (4), Rayleigh's law was the result.

Since Rayleigh's law is incorrect, it is certain that something must be wrong in the classical derivation. On the other hand the classical treatment cannot be entirely wrong, for we know that Rayleigh's law tends to express conditions correctly at high temperatures. Planck therefore accepted the general procedure of the classical theory, and confined himself to determining at what point this theory needed modification. In this connection he assumed that the classical relation (4) was correct, and that the mistake in the classical treatment arose from its assigning the equipartition value, kT, to the average energy U of an oscillator. The problem confronting Planck was thus to obtain a revised expression for U.

Now the equipartition value for U appears to be unavoidable so long as we assume that the energy of an oscillator varies continuously during the exchange of radiation. Inasmuch as this continuous variation in the energy results from the continuity which the classical theory ascribed to the emission and absorption of radiation by the oscillators, Planck concluded that the classical assumption of continuous emission and absorption would have to be abandoned. In his first theory (which is the only one we shall consider here) he postulated that the emission and the absorption processes both occur discontinuously. This revolutionary postulate constitutes the essence of the quantum theory. The precise discontinuities postulated by Planck will be more easily understood if we examine the problem from the standpoint of action rather than of energy.

When the vibrating electron of an oscillator describes a small interval ds, the Maupertuis action over this interval is the product of the length of the interval and the momentum mv of the electron. The action over a back-and-forth vibration is obtained by integrating the elementary amounts of action over the entire vibration. In the case of an oscillator vibrating with energy E, the action for a complete vibration is found to be

$$(9) \qquad \text{action} = \frac{E}{\nu}.$$

According to classical theory, where emission and absorption are continuous, the action will fluctuate continuously, along with the energy E. But if we accept Planck's postulate of the discontinuity of emission and absorption, a discontinuous set of values will be obtained for the energy and hence for the action. Planck then particularized his postulate of discontinuity by making the following assumption:

The action of an oscillator can have only the values
$$(10) \qquad 0, h, 2h, 3h, \ldots\ldots nh, \ldots\ ;$$
and changes in the action can occur only by the amount h.*

A comparison of (9) and (10) shows that according to Planck's assumptions the total energy of an oscillator of frequency ν can have only the values

$$(11) \qquad 0,\ h\nu,\ 2h\nu,\ 3h\nu \ldots\ldots nh\nu \ldots,$$

and that when energy is radiated or absorbed, it will always be in parcels $h\nu$.

If now we calculate the average energy $U(\nu, T)$ of an oscillator in the state of statistical equilibrium, we find

$$(12) \qquad U(\nu, T) = \frac{h\nu}{e^{\frac{h\nu}{kT}} - 1},$$

instead of (6) as was required by the classical theorem of equipartition. If we substitute the expression (12) for the mean energy in the relation (4), Planck's formula (8) is obtained.

---

* We have modified Planck's theory by assuming that the action of an oscillator can change only by the amount h. Planck did not make this restriction; he supposed that changes nh could occur, where n is any positive integer. But the restriction in the text is required by the Correspondence Principle and by the subsequent development of the theory.

We may easily verify that Planck's law (8) degenerates into Rayleigh's law (7) when $\dfrac{h\nu}{kT}$ is small, and hence when the temperature T is high and the frequency ν is low. Now $\dfrac{h\nu}{kT}$ would also be small if ν and T were arbitrary and the value of h happened to be infinitesimal. We conclude that, if in our universe the magnitude h were infinitesimal, Planck's law would degenerate into Rayleigh's law for all values of the frequency and of the temperature. Rayleigh's classical law would thus be correct in such a world. We are here faced with a situation that will present itself repeatedly in the quantum theory. It may be stated:

In a universe where h would decrease progressively in value, passing from its actual finite value to an infinitesimal one, quantum phenomena would progressively be obliterated, and the quantum theory would pass over into classical science.*

In short, the essential change in our outlook brought about by the quantum theory is due to the discovery that the unit h is finite and not infinitesimal. Also, we perceive the similarity between the rôle played by h (finite and not infinitesimal) and the rôle played by the invariant velocity c (finite and not infinitely great) in the special theory of relativity. With a decrease in the value of h, the quantum theory would pass over into classical science; and, with an increase in the value of the invariant velocity c, the theory of relativity would likewise pass over into classical science.

Let us also observe that Planck's law passes over into Wien's radiation law when $\dfrac{\nu}{T}$ or $\dfrac{h\nu}{kT}$ is high. Consequently Wien's law would be correct in a world where the value of h was very great. The passage of Planck's law into Wien's is of less interest, so we shall not discuss it further.

---

* This statement may be too general. The statement appears to be borne out in connection with the law of radiation and also, as will be seen later, in many other cases. But we must remember that in the present stage of scientific development we are far from being acquainted with the various implications of the quantum h. It is conceivable that in a world where h was infinitesimal, life would be impossible. It is also to be presumed that the finite value of h is in some way connected with the very existence of such stable structures as electrons; and if this be so, the gradual vanishing of h might lead not only to the disappearance of quantum phenomena, but also to a physical universe which would differ entirely from the one in which we live. However this may be, the statement made in the text contains at least an element of truth, and the general law which it illustrates has been of considerable heuristic value in the development of the quantum theory.

Without going into mathematical detail, we may understand in a rough way how a lowering of the temperature causes the degrees of freedom of the oscillators of higher frequency to become inoperative. Thus we have seen that the total energy of an oscillator of frequency ν can have as value only some integral multiple of the quantity $h\nu$. Our oscillator can therefore absorb no amount of energy smaller than $h\nu$. This quantity increases with the frequency ν of the oscillator. Now, in our enclosure at temperature T, there is only a limited amount of energy. If, then, the frequency ν of an oscillator is so great that the minimum amount $h\nu$ of the energy it can absorb is greater than the total amount of energy to be distributed, this oscillator will necessarily be sacrificed in the distribution of the energy and will be as good as non-existent. We thus understand why the oscillators of extremely high frequency cannot share in the distribution of energy. But, of course, the exact manner in which the energy will be distributed, *i.e.*, the most probable manner, can be determined only through mathematical calculation. When the calculations are performed, Planck's law is obtained.

The foregoing analysis shows that Planck's principal innovation consents in the assumption that the "action" of the emitting and absorbing oscillators (which represent the matter in the enclosure) can change only by the amount h. Had Planck resorted to more complicated mechanisms than oscillators he would have had to assume that the action might change by nh, where n is any integer.* But, in either case, the significant fact is that the action can vary only by discrete quantities and not continuously. From this standpoint and on the basis of our present knowledge, we are justified in viewing h as the atom of action.

The atomicity of action entails a discontinuity in the emission and absorption of radiation-energy by matter. Thus radiation of frequency ν can be emitted only in bundles, or quanta, of energy $h\nu$. The notion of the atomicity of action is questionable in view of subsequent developments; but to speak of atoms of energy would be still more misleading. In the first place, the bundles of energy vary in magnitude according to the value of ν. Thus bundles $h\nu$ for violet light are nearly twice as energetic as for red light. So there are as many different kinds of bundles, or quanta, of energy as there are different frequencies, *i.e.*, a continuous infinity. Here is already a first difference between the bundles of energy and the atom of action, h.

A further difference is that the bundles of energy may be increased or decreased at will and are in no sense permanent. For instance, let

---

* The reasons for this statement receive their explanation when we consider the quantum theory of the atom.

462   PLANCK'S ORIGINAL THEORY

us suppose that green light is reflected against an advancing or a retreating mirror. We know that the reflected light may be violet or red. The reflected quanta of energy are then respectively more energetic or less energetic than the incident ones. In the former case, energy has been communicated to the incident quanta at the expense of the energy of the advancing mirror; in the latter case, energy has been withdrawn from the incident quanta and ceded to the mirror. The situation is much the same as when a tennis ball rebounds from an advancing, or from a retreating racket. And just as in this latter case the element of atomicity is represented by the ball and not by the energy it carries, so also with our radiation quanta it is the radiation, and not the energy, which in certain cases at least may be regarded as corpuscular.

At first the reaction of physicists to Planck's solution of the problem of equilibrium radiation was lukewarm. Against it there seemed to be weighty evidence. In the first place, we have assumed that each oscillator of frequency ν reacts, or exchanges energy, only with the radiation of the same frequency ν present in the enclosure. Under these conditions, each category of oscillators and the corresponding radiations are isolated, so to speak, from all other oscillators and radiations. Yet, when we calculate the mean energy of an oscillator, and therefore partition the total energy in the enclosure among the various oscillators, we are implicitly assuming that the oscillators are in statistical equilibrium *inter se* and are thus able to exchange energy. It seems necessary to make this assumption, and it is not difficult to understand how the exchanges in energy may occur. The exchanges may be due to gas molecules or electrons circulating within the enclosure, colliding with the oscillators and thereby conveying energy from one oscillator to another. We may also suppose that the molecules and electrons deviate the radiations and modify their frequencies; and that this process will bring about the exchange of energy between the various radiations, and thence between the oscillators. Finally, we may suppose that the different oscillators collide, and that an energy exchange occurs directly in this way. But the trouble is that if we take these subsidiary phenomena into account and treat them in the *classical* manner, we obtain Rayleigh's law, and not Planck's. If, then, Planck's law is correct, we must assume that none of the subsidiary energy interchanges we have just mentioned can proceed according to classical ideas. Quantum phenomena cannot therefore be restricted solely to the emission and absorption of radiation by matter, and so the whole of classical physics seems endangered. This conclusion in itself is not an argument against Planck's theory, but it does show that this theory entails more revolutionary consequences than might at first be supposed.

A second difficulty that confronts Planck's theory is exhibited by the following question: If radiation is emitted from matter in bundles, what becomes of Maxwell's theory with its *continuous* electromagnetic waves and fields? Planck himself adopted a cautious attitude. At the Solvay Congress of 1911, devoted to the discussion of the Quantum Theory, he stated:

"When the emission by an oscillator occurs in 'quanta,' Maxwell's equations retain their validity in surrounding space, but only at a sufficient distance from the oscillator . . . they must be modified inside the oscillator and in its immediate vicinity."

According to the views expressed in this passage the quanta of radiation, after leaving the oscillator, will merge, giving place to a continuous field. But Planck's semi-classical attitude was resisted by Einstein. Indeed the real revolutionary in this early stage of the quantum theory was not Planck but Einstein. From the outset he supported the idea that radiation not only is emitted in bundles but that it remains so constituted, at least until it enters into contact with matter.

The Specific Heat of Solids—The first physical phenomenon, which was not directly related with radiation but which nevertheless lent support to the quantum theory, was the general decrease in the specific heats of solids at low temperatures. When we discussed the specific heats of solids (see page 422) we saw that, if we assume the validity of the theorem of equipartition for the oscillating systems composing the crystal, calculation shows that the specific heat always has the same constant value, $3R$, at all temperatures and for all substances. This result agrees with the empirical law of Dulong and Petit at ordinary temperatures, but the agreement ceases at low temperatures. Experiment shows that as we approach the absolute zero the specific heat decreases continually. According to the third law of thermodynamics the specific heat should tend to zero. This falling off may be accounted for if we assume that as the temperature is lowered, the various degrees of freedom of the vibrating systems become inoperative. Einstein sensed that this phenomenon might be interpreted by the quantum theory.

The explanation is very similar to the one mentioned in connection with the radiation law. The vibrating systems in the crystal may be assimilated to so many oscillators vibrating in space with the same frequency ν. An oscillator vibrating in space, and hence in three directions, is equivalent to three linear oscillators, so that in effect we may view our crystal as composed of an aggregate of linear oscillators of the same frequency ν. These oscillators, according to our former conclusions, can assume only the energy values

$$0, \ h\nu, \ 2h\nu, \ 3h\nu \ \ldots nh\nu * \ \ldots$$

Consequently, if the temperature, or energy content, of the crystal is small, there may not be enough energy present to satisfy the demands of some of the oscillators. In this event some of the oscillators will receive no energy and will be inoperative. The specific heat of the crystal will thus tend towards zero when the absolute zero of temperature is neared. Qualitatively, therefore, the quantum theory accounts for the falling off in the specific heats of crystals and is thus in agreement with the third law of thermodynamics.

Einstein, in his original treatment, assumed that the crystal was formed of space-oscillators all of which had the same frequency. His theory was subsequently refined by Debye, Born, and Karman. Debye, for instance assumed a whole spectrum of frequencies of vibration instead of a single one, and he calculated the frequencies from the elastic properties of the body. Thus in Debye's theory we must consider oscillators having various frequencies instead of confining ourselves to oscillators having the same frequency. Except for this difference the general treatment is the same; but the quantitative results obtained by Debye are in better agreement with experiment.

The Specific Heats of Gases—According to the classical kinetic theory of gases, the specific heats of all gases should be $3R$, where R is the gas constant. Inasmuch as this conclusion conflicts with experiment, Boltzmann was led to suppose that, for some reason or other, the spherical molecules of a monatomic gas cannot rotate and can assume only translational motions; that the dumbbell-shaped molecules of a diatomic gas can rotate, but not along the line of the knobs; and that the molecules of irregular shape of a polyatomic gas may rotate in any way. With these restrictions imposed, the classical kinetic theory leads to the values

$$\frac{3}{2}R ; \quad \frac{5}{2}R ; \quad \frac{6}{2}R = 3R,$$

---

* Wave mechanics requires the values

$$\frac{h\nu}{2}, \ 3 \ \frac{h\nu}{2}, \ 5 \ \frac{h\nu}{2} \ \ldots (2n+1)\frac{h\nu}{2}\ldots$$

for the specific heats of the three kinds of gases; and these magnitudes are in agreement with experimental measurements at ordinary temperatures. But at extremely low temperatures, discrepancies appear: the specific heats of the diatomic and polyatomic gases decrease to the value $\frac{3}{2}R$, and thus assume the value that holds for a monoatomic gas. This falling off may be accounted for by supposing that when the temperature is lowered, the molecules one by one cease to rotate, so that their rotational degrees of freedom become inoperative. Einstein and Stern were the first to investigate the matter.*

At the time Einstein was initiating his investigations, in 1913, there was some doubt as to the correct method of applying the quantum theory to rotating bodies. Up till then, only oscillators had been considered; and Einstein's deduction was not rigorous. But the problem was solved by Ehrenfest soon after, and we shall follow his procedure. Ehrenfest extends to a rotating body Planck's postulate for the oscillators. He assumes that the action can take only the values

$$0, h, 2h, 3h \ldots nh \dagger \ldots$$

Now if we consider a molecule rotating about some axis, and if we call I the moment of inertia of the molecule about this axis, mechanical considerations show that the action and the energy E are connected by the formula

$$E = \frac{1}{8\pi^2 I} \times (\text{action})^2.$$

The values of the energy for the stable rotationary states are then secured when we substitute in this expression the permissible values for the action. We thus obtain as possible values for the energy,

$$E = 0, \quad \frac{h^2}{8\pi^2 I}, \quad \frac{4h^2}{8\pi^2 I}, \quad \ldots \quad \frac{n^2 h^2}{8\pi^2 I}, \quad \ldots$$

The smallest amount of energy our molecule can absorb is seen to be $\frac{h^2}{8\pi^2 I}$; and we note that this quantity is the smaller, the greater the value of the moment of inertia I.

---

* Nernst had already pointed out in 1911 that quantum effects should be expected in the rotation of diatomic and polyatomic molecules, but he did not investigate the problem theoretically.

† Wave mechanics shows that the correct values are

$$0, \quad h\sqrt{1 \times 2}, \quad h\sqrt{2 \times 3}, \ldots \ldots \quad h\sqrt{n(n+1)}, \ldots \ldots$$

466 PLANCK'S ORIGINAL THEORY

The value of I depends on the mass, size, and shape of the molecule and also on the axis of rotation. For a diatomic molecule, I is much smaller when the axis of rotation is the line of the knobs than when it is perpendicular to this line. The smallest amount of energy that can set a diatomic molecule into rotation is thus considerably greater when the rotation is around the line of the knobs than when it is around any perpendicular axis. In other words, the most greedy rotational degree of freedom of a diatomic molecule is the one connected with the rotation around the line of the knobs. Now, according to the usual arguments of the quantum theory, when the temperature of a diatomic gas is not extremely high and hence when the energy at our disposal is not very great, the first degrees of freedom that will be sacrificed in the partitioning of the energy are the more greedy ones. These more greedy degrees of freedom will thus become inoperative, and as a result the diatomic molecules will not rotate about the lines of the knobs though they may rotate about perpendicular axes. This is precisely the situation which Boltzmann assumed for a diatomic gas at ordinary temperatures when he wished to account for the specific heat $\frac{5}{2}R$. We conclude that the quantum theory justifies Boltzmann's assumption. At the same time his unsatisfactory hypothesis of perfectly smooth molecules becomes unnecessary.

Suppose now the temperature of the gas is high. Since more energy is at our disposal, there may be enough energy to distribute among the formerly inoperative degrees of freedom, with the result that the diatomic molecules may rotate in all ways. The specific heat of a diatomic gas should thus become $3R$ at high temperatures and hence be the same as for a polyatomic gas. These expectations are in agreement with experiment. At still higher temperatures there may be enough energy to satisfy the demands of other degrees of freedom, namely, those connected with the vibrations of the knobs in a dumb-bell molecule. The knobs will then start to vibrate and the specific heat of the gas will be increased further in consequence.

The reason why the spherical molecules of a monoatomic gas do not rotate at ordinary temperatures may be understood by similar arguments. Thus the moments of inertia of a tiny sphere being extremely small, the minimum amount of energy which a rotational degree of freedom can absorb is extremely large. At ordinary temperatures there is presumably not enough energy present to satisfy the rotational degrees of freedom; very much higher temperatures would be required before the spherical molecules could be set into rotation.

Finally, we examine the behavior which we may expect for diatomic and polyatomic molecules at extremely low temperatures, near the absolute zero. All the rotational degrees of freedom will be frozen in, and so the molecules will cease to rotate. According to our present theory, however, the molecules will still be in motion since they will be undergoing translations. For this reason the specific heat will be non-vanishing and will have the value $\frac{3}{2}R$, just as occurs for monoatomic gases at ordinary temperatures. Yet if we accept the inferences drawn from the third principle of thermodynamics, according to which the specific heat of a perfect gas should vanish at the absolute zero (degeneration of gases), we must suppose that the translational degrees of freedom will also eventually become inoperative. But the quantum theory, at its stage of development considered here, affords no means of introducing quantum restrictions on the translational movements of the molecules, with the result that gas degeneration cannot be accounted for. We shall see later that this difficulty is overcome in the New Statistics.

Photons—Planck's quantum theory raised at the outset the problem of the nature of radiation. Is it continuous and wave-like, or is it corpuscular? Planck himself believed that, only in the immediate vicinity of his oscillators would Maxwell's equations be at fault, and that elsewhere they would resume their validity. Planck's belief implied that radiation was continuous at points not in the immediate vicinity of an emitting system.

Einstein, however, subjected Planck's law to a series of theoretical tests and came to the conclusion that, everywhere throughout the enclosure, radiation must have a dual nature, part wave and part corpuscle. His investigations concerned the phenomenon of fluctuations, which we mentioned in connection with the molecules of the kinetic theory. Fluctuations in the density of the radiation occur incessantly during the state of statistical equilibrium. From Planck's law we may determine the fluctuations. Einstein found that the fluctuations should be composed of two parts: one part was consistent with the assumption that radiation is formed of waves; the other part required that radiation be corpuscular and that radiation of frequency ν be constituted of corpuscles having the energy $h\nu$. If, in place of Planck's law, we take Rayleigh's, we obtain only the wave part of the fluctuations. If we take Wien's law of radiation, the corpuscular aspect of the radiation is the only one in evidence. This illustrates a fact we already know; namely, Planck's law is a compromise between Rayleigh's and Wien's. Inasmuch as Planck's law passes over into Rayleigh's classical law when we imagine that h becomes smaller and smaller, we may view the corpuscular aspect of radiation as a quantum manifestation.

Einstein's theoretical deductions suggesting the dual nature of radiation soon received support in the experimental field. Experiment showed that on certain occasions radiation behaves as though it were wave-like, on others as though it were corpuscular. The corpuscles of radiation, which seem to betray themselves in certain phenomena, have since been named "photons." *In vacuo* they are assumed to be moving with the velocity c. A photon associated with a radiation of frequency ν is called a photon of frequency ν; its energy is $h\nu$ and its momentum (*in vacuo*) $\dfrac{h\nu}{c}$.

The Photo-Electric Effect—In support of his theoretical investigations, Einstein drew attention to the photo-electric effect. This effect, discovered accidentally by Hertz and studied by Lenard, consists in the emission of electrons from a strip of metal exposed to ultra-violet light. We must assume that the energy of the incident waves of light is communicated in some way to the electrons circulating in the metal, and that the electrons are thereby ejected. Quantitative measurements show that the velocity with which the electrons are ejected is independent of the intensity of the incident light, but increases with the frequency of the light. The phenomenon is not easy to account for on the basis of the wave theory of light, for according to this theory the energy of an incident wave increases with its intensity; and so a more intense light of the same frequency should generate greater velocities for the ejected electrons—and this is contrary to observation. Furthermore, according to the wave-theory the energy is distributed continuously over the surface of the wave front, and under these conditions it is surprising to find that the electrons are ejected only at discrete points.

But if a corpuscular theory is adopted for light, the phenomenon becomes comprehensible. The radiation corpuscles, or photons, fall in showers at random on the metal, and only at their points of impact need we expect electrons to be ejected. In particular, a photon of frequency ν has energy $h\nu$, and Einstein assumes that when the photon collides with the metal, its energy is transferred to an electron and appears in the form of kinetic energy. If we neglect the refinements of the theory of relativity, the kinetic energy of an electron of velocity v and of mass m is $\frac{1}{2}mv^2$, and the equation of the photo-electric effect becomes

$$\tfrac{1}{2}mv^2 = h\nu.^*$$

As a matter of fact, some of the energy of the photon is expended in tearing away the electron from the surface of the metal, and so the velocity with which the electron is ejected will be somewhat less than is indicated by this formula.

We may readily understand why the intensity of the incident light does not affect the electron's velocity. When the light is more intense, the energy conveyed by each one of the photons is no greater; it is only the number of photons that is increased. The effect of increasing the intensity will merely be to increase the number of electrons ejected. To increase the velocity with which these electrons are hurled out from the metal, we must increase the energy of the individual photons and hence the frequency of the incident light. When the frequency of the incident light falls below a certain value, the energy of the photons will be too small to tear the electrons from the metal; for this reason the photo-electric effect will not arise for visible light. It will be extremely violent if X-rays are used. Indeed, with X-rays, not only the superficial electrons, but also those present in the deep interior of the atoms may be torn out and hurled into space.

The converse of the photo-electric effect may also arise. It will occur when an electron moving at extremely high speed is suddenly stopped by impact against a metal. The kinetic energy of the electron is then destroyed and a photon of the same energy is emitted. By utilizing this phenomenon, physicists generate X-rays in the laboratory.

The Compton Effect—An effect which also suggests a corpuscular theory of light is the Compton effect. Compton's experiment consists essentially in directing radiation of high frequency against free electrons. The incident radiation is more or less deflected by the electrons and it decreases in frequency. At the same time, some of the electrons are thrown back as though they had been submitted to an impact. The interpretation of this effect by means of photons is simple. Thus we assume that the radiation is corpuscular, so that radiation of frequency ν moving *in vacuo* is identified with a shower of photons moving with velocity c; each photon has energy $h\nu$ and momentum $\dfrac{h\nu}{c}$. When radia-

---

* Since mv and ν can be measured, the formula of the photo-electric effect affords a means of determining the value of Planck's constant h.

tion of frequency v is directed against free electrons, collisions occur between the photons and the electrons. Each collision may be viewed as similar to one between two perfectly elastic billiard balls. By an application of the mechanical laws of the conservation of energy and of momentum, a relationship may be established connecting the three following magnitudes: the loss in energy of the rebounding photon; the direction of its deflection arising from the collision; and the direction of recoil of the electron. All three of these magnitudes can be measured experimentally. In particular, the loss in the energy of the photon exhibits itself in a lowering of the photon's frequency and hence in a lowering of the frequency of the radiation. Thanks to these measurements, the theory just sketched for the Compton effect may be tested. The agreement is satisfactory.

This theory of the Compton effect has since been refined by the application of wave mechanics and of Heisenberg's principle of uncertainty. Nevertheless the crude mechanical explanation on the basis of the analogy of billiard balls gives a good approximation, and we see from this example that even in the remote levels of the microscopic, far different from the level of commonplace experience, mechanical analogies may still retain a certain measure of validity.

The Fundamental Law of Photo-Chemistry—Some chemical reactions and dissociations which cannot occur under ordinary conditions take place when light falls on the chemical substances. Such reactions are called photo-chemical. If a photo-chemical reaction proceeds when light of a definite frequency falls on the substances, we usually find that, with a change in the frequency of the light, the reaction ceases. The frequency of the radiation is thus characteristic of the reaction. For instance, silver chloride, used in the photographic plate, dissociates into silver and chlorine (or at least the bonds between the atoms are loosened) when light ranging from the green to the ultra-violet and beyond is applied. Red light produces little effect. Similarly the chlorophyllian action of plants requires red light (as is apparent from the reflection of green light by the leaves).

An elementary interpretation of photo-chemical reactions based on the quantum theory was given in 1905 by Einstein and by Stern. In the particular case where the reaction is a dissociation, Einstein's law states:

If a molecule dissociates under the action of radiation of frequency v, the dissociation will absorb a quantum hv of energy.

The connection of this law with the quantum theory resides in the fact that a molecule cannot absorb less than a quantum of energy hv. We may suppose that a photon, having a quantum of energy $h\nu$, collides with the molecule and surrenders its energy to the molecule.

As an illustration let us consider the dissociation of gaseous hydrogen bromide HBr into hydrogen and bromine. At ordinary temperatures and pressures the free energy of the compound is less than the sum of the free energies of the component elements. According to the laws of thermochemistry a spontaneous dissociation is thus impossible. But if radiation of appropriate frequency ν falls on HBr gas, a quantum $h\nu$ is absorbed by a molecule, and the dissociation takes place. The free energy required to bring about the loosening of the atomic bonds is thus furnished by the quantum of radiation energy. Of course, in many cases, the initial action of the light may merely set free certain atoms, which in turn combine with others. In such cases the action of the light is more indirect. Further information is yielded on these chemical problems by the curious phenomenon of collisions of the second kind. See Chapter XXVIII.

The various applications of Planck's ideas, which we have mentioned in the course of this chapter, show that Planck's quantum theory cannot be regarded as a mere makeshift devised for the sole purpose of interpreting the law of equilibrium radiation. We must recognize therefore that the quantum theory has uncovered a new world of physical occurrences, a world formerly unsuspected. As we proceed, we shall find that one group of phenomena after another will reveal itself as controlled by quantum laws.

CHAPTER XXV

GENERALITIES ON THE ATOM

THE possibility of disrupting molecules and of separating them into atoms has been known for many years; it forms the basis of chemistry. But the destructibility and evolution of atoms has forced itself upon science only in recent times. So long as atoms were regarded as permanent and indestructible, the problem of accounting for their stability did not present itself. The atoms simply *were* and science had to accept them as fundamental units from which to build complicated structures. But when this permanency was found to be illusory, the reason for the extraordinary stability of the atoms became an urgent problem.

Here a statement of general policy may be mentioned. When we attempt to construct a theory so as to account for some new phenomenon, we seek to interpret the unknown in terms of the known or at least of the supposedly known. Accordingly, the first theories of the atom were based on the classical laws of mechanics or of electromagnetism. Of course, the critic is justified in questioning our right to apply the classical laws, especially so when we apply these laws to phenomena occurring in a microscopic world about which practically nothing is known. But this critical attitude, unless supplemented by some constructive idea, would be worse than useless, and, if applied consistently, would stifle the progress of science. The history of science shows that there is but one way to proceed: We extend the classical conceptions to some model of the phenomenon of interest, and then determine whether or not we are led to predictions that are verified by experiment. If, however much we vary the model, our predictions remain in conflict with experiment, we shall be assured that the classical laws are inadequate; it will then be time enough to discard these laws and to decide in what respects they must be amended. This is the course that has always been followed in theoretical physics, and we shall find that the theories of the atom are no exception to the rule.

In our attempt to devise a model for an entity as minute and as elusive as the atom, we must necessarily be guided by very indirect sources of information. The most important of these has been the study of the radiations which the atoms emit under suitable conditions. By means of resolving instruments these radiations may be separated, and they then appear as distinct luminous lines (spectral lines). But the atomic spectra are not the only clues at our disposal. Chemical properties, the color of substances containing the atoms of interest, electrical and magnetic properties, ionizing potentials, radioactive phenomena, and many others have all played a part in guiding physicists towards a better understanding of the atom.

The First Atomic Models —The first atomic model was devised by Lord Kelvin. At the time he was pursuing his investigations, the most conspicuous property credited to atoms was their stability; and the main purpose of Kelvin's model was to account for this stability. Now, the theoretical investigations of Helmholtz in hydrodynamics had established the peculiar stability of vortex motions, and so Kelvin availed himself of this discovery and assumed that an atom was a vortex in the ether. Kelvin's investigations have today but a historical interest. The next attempt was due to J. J. Thomson, though his model seems to have been suggested by Kelvin. Thomson's investigations were carried out in 1904, when the existence of the electron (corpuscle of negative electricity) had been established and its presence in the interior of atoms recognized. Inasmuch as atoms appeared electrically neutral, Thomson supposed that the negative electrons in the atom were counteracted by an appropriate amount of positive electricity. Now although negative electricity was known to be corpuscular, Thomson believed that positive electricity was a continuous fluid; and his model of the atom was represented by a sphere of uniform positive electrification in which electrons were embedded. The electrons placed themselves along rings and revolved about the centre of the sphere. Owing to their accelerated motion, they radiated electromagnetic energy (light) in accordance with the laws of electrodynamics. The conditions of stability were also investigated by Thomson. This conception of the atom was soon abandoned as a result of Rutherford's discoveries.

Rutherford's Atom—Rutherford, in 1911, submitted a thin foil of gold to a bombardment of those positively charged particles which are hurled out from atoms during radioactive explosions. The particles to which we refer have since been called "alpha particles"; the electrical charge of an alpha particle is twice that of an electron, and of course the sign of the electrification is reversed; the mass of an alpha particle is much greater than that of the electron. Rutherford, on bombarding the foil of gold with a stream of alpha particles moving at extremely high speeds, found that the particles usually passed straight through the foil without deflection, but that on rare occasions large deflections arose. Since each individual electron contained in an atom of gold had far too small a mass to be responsible for the deviations of the massive alpha particles, the large deviations that occasionally occurred were ascribed to the repulsive actions of massive positive charges present in the atoms. The rarity of large deviations suggested that in each atom the positive charge was concentrated in a small volume (called the nucleus), presumably situated at the centre of the atom. According to these ideas, an alpha particle, on passing through the metal foil, would be deviated perceptibly only when it passed sufficiently near to a nucleus; in all other cases it would proceed without perceptible deviation. Quantitative measurements of the deviations in a series of experiments enabled Rutherford to calculate the net positive electrical charges of the nuclei of various atoms (the nuclear charges) and also the approximate diameters of these nuclei.

Subsequent discovery proved that positive electricity was corpuscular in structure; its corpuscles were called *protons*. Though protons have the same charge as the electrons (but of opposite sign), their mass is some 1830 times greater. For this reason, practically the entire mass of the atom was thought to be due to the protons contained in its nucleus; and since the masses of the various atoms were known, physicists computed on this basis the numbers of protons present in the various nuclei. Now experiment showed that, except in the case of the hydrogen atom, the nuclear charge of an atom always fell below the total charge of the nuclear protons. This situation was accounted for by the assumption that a nucleus (other than that of hydrogen) contained electrons as well as protons, the charges of the former counteracting those of some of the latter. Furthermore, since an atom was electrically neutral as a whole, additional electrons were assumed to be circling around the nucleus under the pull of its electrostatic attraction.

The final picture of the atom as suggested by Rutherford is very similar to a miniature solar system: the massive nucleus formed of protons and of electrons, with a predominance of the former, plays the part of the sun; and the remaining electrons which circle around this central body behave like the planets. The nucleus occupies only a small part of the atom, and so the atom is represented by a practically empty region of space. The hydrogen atom which is the simplest of all atoms is formed by a single proton around which a single electron is circling. As for the alpha particle, to which reference was made in Rutherford's experiments, it is the nucleus of the helium atom and is a highly stable configuration, having a net positive charge of two units. Since it has a mass of four units, it was believed to be formed of four protons and two electrons. In the higher atoms the nuclei were assumed to contain larger numbers of protons and electrons.

Today, as a result of recent discoveries, a different conception of the nucleus is entertained, and many of the conclusions which we have just mentioned, have been revised. For the present, however, we shall accept the earlier conception of the atomic structure.

We have seen that according to this conception all atoms are formed of protons and electrons exclusively. Since the mass of an electron may be disregarded in comparison with that of a proton, the masses of the various atoms should be integral numbers of times the mass of a proton. This expectation, however, is not always borne out by experiment. For instance, the helium atom has as nucleus an alpha particle and so contains four protons, and the hydrogen atom contains a single proton; hence we should expect a helium atom to be exactly four times more massive than a hydrogen one. But such is not the case: the mass of the helium atom is slightly less than the required amount. The special theory of relativity affords an explanation of the discrepancy. According to the relativity theory the mass of a particle at rest is equal to its energy of constitution divided by c^2 (where c is the velocity of light). Now, if the protons and electrons in the helium nucleus, or alpha particle, are closely packed, the total potential energy of the helium nucleus will be reduced and hence the mass will be less than the sum of the constituent protons considered in isolation. Thus, the discrepancy is accounted for. But in some cases the atomic mass of an element appeared to differ considerably from a whole multiple of the mass of a proton; and in such cases the discrepancy could no longer be attributed to a relativity effect. Chlorine is an example in point; its atomic mass appears to be $35\frac{1}{2}$ times the mass of a proton. Subsequent investigation showed, however, that the chlorine gas obtained by ordinary chemical processes is in reality a mixture of two different chlorine gases having as atomic masses 35 and 37, respectively. In short, a chlorine atom of mass $35\frac{1}{2}$ is a myth, and the earlier belief in the existence of such an atom was due solely to the error of confusing a mixture with a pure substance. Thus, our assumption that all atoms contain an integral number of protons is not endangered by the apparent discrepancy noted in connection with the chlorine atom. Incidentally, the existence of atoms having the same chemical properties but different masses is of frequent occurrence. These various forms of the same element have been called *isotopes* by Soddy.

Rutherford's atom was imagined so as to account for his experiments, but these experiments elucidate only some of the features of the atomic structure. Other features must also be taken into consideration. In the first place, atoms are exceedingly stable structures, and any successful model of the atom must give an account of this fact. A rigorous proof of the stability is obviously out of the question, for it involves the problem of n-bodies. It is true that the stability of the solar system may be established over a long interval of time; but, in view of the tremendous speeds that must be credited to the planetary electrons, a long period of time for the solar system would correspond to a fraction of a second for an atom. For these reasons we cannot (apart from exceptional cases) obtain any rigorous proof of the stability of Rutherford's atom. Besides, Rutherford's model does not account for the observed distribution of the spectral lines. Nevertheless, we shall see that when supplementary assumptions are introduced, Rutherford's atom is taken over by Bohr.

Mendeleeff's Table—We shall advance a step further in our search for a satisfactory model of the atom by following momentarily a different train of inquiry. Many years ago Mendeleeff had the idea of ordering the atoms according to their increasing atomic weights, or masses. In so doing he noticed that atoms with similar chemical properties (such as fluorine, chlorine, bromine, iodine) recurred periodically at more or less regular intervals. To bring these chemical similarities into prominence, Mendeleeff cut into a number of sections the single line along which he had originally ordered the atoms. He placed the sections one below the other in such a way that atoms with similar chemical properties were situated in the same vertical column. This classification of the elements is known as Mendeleeff's Table, or the "Periodic Table." * Mendeleeff observed that some of the partitions in his table were left unoccupied. He attributed these vacancies to the existence of undiscovered elements. Subsequent investigations have justified his surmise, for today, with the discovery of new elements, all the vacant partitions have been filled.

Mendeleeff's classification of the elements is not entirely satisfactory. For instance, in his table, potassium precedes argon, and as a result potassium is situated in the same vertical column as neon, and argon in the same column as sodium. But to be in accord with the similarity of chemical properties, we should have to exchange the positions of these two elements. The same situation occurs for iodine and tellurium. We conclude that an ordering of the elements in accordance with the increasing atomic weights is incompatible with an ordering which stresses chemical similarities; one of the two methods of ordering must be sacrificed. This difficulty is present in all classifications and is not peculiar

---

* The name Periodic Table was given to Mendeleeff's classification because of the periodically recurring chemical properties of the atoms in the table. As we shall see presently, however, this name is now reserved for Moseley's table.

to Mendeleeff's. But a serious difficulty beset Mendeleeff's classification when isotopes were discovered by Soddy. Isotopes (to which reference has already been made in the case of chlorine) are atoms which are chemically indistinguishable, yet differ in their atomic weights. Ever-increasing numbers of them are being isolated. Their discovery necessarily entails considerable ambiguity in any rigid classification based on atomic weights.

Moseley's Discovery—The Atomic Number —Moseley's experimental investigations bear on the spectral lines emitted by the atoms. And this leads to a digression on the general characteristics of spectra. When a solid is heated, it emits radiation, and the frequencies of the radiations form a continuous aggregate. Matter in the form of a vapor or of a gas may also emit radiation. A commonplace illustration is afforded by the luminous gases in the so-called neon tubes, where the gases neon, argon, or helium are made to glow by electrical means. An important difference between the spectra of substances in the solid and gaseous forms is that, in the latter, the frequencies of the radiations emitted form a discontinuous and not a continuous spectrum (though a faint continuous spectrum is also present). For instance, if we analyze by means of a spectroscope the light generated by a tube of glowing neon, we find bright lines separated by wide dark intervals. The lines are not equally intense. With neon the greater number of the visible lines are red, and to this circumstance is due the red glow of the neon tubes; lines of other colors, however, are also present. In the theoretical investigation of the atom it is the line spectra due to incandescent gases, and not the continuous ones produced by incandescent solids, that are of particular interest. The former spectra are ascribed to internal changes occurring within the atoms; whereas, with the latter, the closeness in the packing of the molecules complicates the situation.

The converse of emission is absorption. If white light is transmitted through a gas, the gas (in accordance with Kirchhoff's law) absorbs precisely those incident radiations which it might emit.* The spectrum of the transmitted light is called the "absorption spectrum." It exhibits the same similarity to the spectra thus far considered (the emission spectra) as does a photograph to its negative. The absorption spectrum of a gas is thus composed of a discrete set of dark lines which intersect a brilliant band of color ranging from red to violet. An absorption spectrum is observed when the light of the sun is analyzed; for above the brilliant photosphere which radiates almost like a black body, lie layers of cooler gases that absorb some of the radiations emitted by the photo-

---

* The exceptions to this law are clarified by Bohr's theory of the atom.

sphere. The dark lines are called the Fraunhofer lines, and their distribution enables us to determine the nature of the absorbing gases in the sun's outer atmosphere.

The fact that atoms are associated with spectra which characterize them is an old discovery; it forms the basis of spectral analysis. We know, however, that scientific advance occurs only when our observations assume a precise quantitative form, for then alone is it possible to narrow down possibilities. The first advance in the study of spectra was made when the frequencies of the radiations emitted by the various atoms were measured with precision. Considerable difficulty attends such investigations. In the first place, many of the radiations are so faint that they are scarcely perceptible, and furthermore, the majority of them do not happen to fall within the range visible to the human eye. In such cases indirect methods of measurement (*e.g.* photography) must be applied.

The radiations emitted by the hydrogen atom and by the lighter atoms comprise visible radiations and also infra-red and ultra-violet ones, whose frequencies are respectively smaller and greater than those of the visible radiations. In the heavier atoms additional radiations having the frequencies of X-rays (several thousand times greater than the ultra-violet frequencies) may also occur. The spectral lines produced by these X-ray radiations constitute the X-ray spectra of the heavier atoms. Spectroscopists have grouped the lines of the X-ray spectra into families, or *series,* to which the names K-series, L-series, M-series, and so on have been given, in the order of decreasing frequencies. The different lines pertaining to the same series are sometimes designated by the letters a, β, γ. In particular the line of lowest frequency of the K series is represented by $K\alpha$. These preliminary notions will suffice to clarify the nature of Moseley's discoveries.

Moseley, having measured the X-ray frequencies of the various atoms, constructed a table of the elements, ordering them in such a way that the frequency ν_{Ka} of the Ka line increased from one element to the next. The frequencies of the other X-ray lines were then found to increase likewise. But the important point established by Moseley was that these successive increases in frequency did not take place in haphazard fashion but appeared to form more or less regular step-like progressions. In particular the progression of the Ka line was found by Moseley to be expressed by the following simple empirical formula:

$$\nu_{Ka} = \frac{3}{4}(Z-1)^2 \times \mathrm{Ry}, \quad i.e., \quad (Z-1)^2 \times \mathrm{Ry} \times \left(\frac{1}{1^2} - \frac{1}{2^2}\right).$$

In this formula Ry represents a constant which was already known to spectroscopists owing to its appearance in the formulae for the lines of the hydrogen spectrum. As for Z it is a positive integer which increases by one unit each time we pass from one element to the next in Moseley's table. Similar formulae containing Ry and Z were devised by Moseley for some of the other X-ray lines.

In 1913, however, when Moseley constructed his table, the number Z was occasionally found to change by more than one unit from one element to the next, so that gaps appeared. Moseley attributed these unexpected gaps to the existence of missing elements. His views have since been vindicated, for today all the missing elements have been discovered, and the frequencies of their X-ray lines verify Moseley's empirical formulae. Today, therefore, Moseley's table is complete; it extends from hydrogen * $(Z = 1)$ to uranium $(Z = 92)$, and beyond.

The value of Z being characteristic of an element and indicating its position in the table, Moseley named it the *atomic number*. We may therefore say that Moseley's X-ray classification is one based on the increasing values of the atomic number.

It is of interest to compare Moseley's X-ray classification with Mendeleeff's earlier one based on the increasing atomic weights. The two classifications coincide except in two or three cases. For example, in Mendeleeff's table, potassium preceded argon, and iodine preceded tellurium; Moseley's table transposes the positions of these elements two by two. Now we mentioned earlier that the foregoing transpositions were precisely those that were required in Mendeleeff's table to bring it into agreement with the classification based on the periodic recurrence of chemical similarities. We conclude that Moseley's X-ray classification is in full agreement with these recurring similarities, instead of with the increasing atomic weights.

Moseley's formulae for the X-ray lines have been of great assistance in facilitating the search for the missing elements. In this connection we recall that Mendeleeff's table, like Moseley's, suggested the existence of missing elements. But Mendeleeff's investigations afforded no means of recognizing these elements when they happened to be present in a

---

* It would appear from our explanations that Moseley's formula and his classification can apply only to the heavier atoms, which emit X-ray spectra. But this conclusion would not be correct, for the lines of the lighter atoms may be regarded as the analogues of the X-ray lines of the heavier atoms. For example, in the hydrogen atom the line of lowest frequency of the Lyman series (an ultra-violet line) is, in effect, the analogue of a $K\alpha$ line. Moseley's formula and his classification may thus be extended to all the atoms. In the case of hydrogen, however, which is the first atom in Moseley's classification (*i.e.*, $Z = 1$), we must make a slight change in his formula: we must replace $Z - 1$ by Z. In point of fact, as will be seen in Chapter XXVII, Moseley's formula is only approximate in any case; one reason for this lack of accuracy is that the $K\alpha$ line is found to be a doublet, not a single line.

chemical preparation. Moseley's formulae, on the other hand, by predicting the frequencies which the X-ray lines of the missing elements should have, facilitated the detection and identification even of minute traces of these elements by the ultra-sensitive methods of spectral analysis.

From a theoretical standpoint the most valuable part of Moseley's contributions was his discovery of the atomic number. This integer, which regularly increased by one unit from one element to the next, naturally suggested itself as having some fundamental significance, and its existence implied an underlying scheme of great regularity and simplicity in Nature's process of atom building. It should be noted that the regularity revealed in the atomic numbers is far more striking than that foreshadowed by the more or less capricious increases in the atomic weights. For this reason Moseley's table, based on the atomic numbers, has completely superseded Mendeleeff's earlier table.

The next problem was to determine the physical significance of the atomic number. Moseley surmised that it defined the net positive charge (in units e) of the atom's nucleus, and so he inferred that the positive charges of the nuclei belonging to the successive atoms of his table would differ by one unit e. This surmise is accepted today as correct. The nature of the isotopes is clarified by Moseley's ideas. The fact is that the isotopes of the same element all exhibit the same spectrum (except for minor details); consequently all such isotopes must have the same atomic number and therefore the same nuclear charge. Finally, let us observe that since an atom is electrically neutral as a whole (unless it is ionized), the total charge of its planetary electrons must be the same in absolute value as the charge of its nucleus. We conclude that in a neutral atom the number of planetary electrons is equal to the atomic number.

We mentioned on page 475 that the earlier conception of the atom's nucleus had been revised as a result of new discoveries. This revision was precipitated by the discovery of the "neutron," a particle which has the same mass as the proton but which carries no charge. According to the modern view, the nucleus of an atom, other than that of ordinary hydrogen (which is assumed to be a proton, as before) is formed of neutrons and protons instead of containing electrons and protons, as was formerly supposed. The alpha particle, for example, which in the earlier theory was viewed as formed of four protons and two electrons, is now assumed to contain two protons and two neutrons. The new arrangement, like the earlier one, accounts for the mass and charge of the alpha particle. As for isotopes, which in the earlier theory were thought to differ from one another in the number of pairs of electrons and protons in their nuclei, they are now believed to differ in the number of their nuclear neutrons.

Radioactive atoms are atoms whose nuclei are unstable; the nuclei explode, throwing off alpha particles, or else protons, or electrons, accompanied by an emission of γ-rays (which are X-rays of very high frequency). In some cases positrons (positive electrons) are expelled. When, following a radioactive explosion, the net positive charge of the nucleus is altered and hence the atomic number is changed, the atom of a different element is formed and transmutation occurs. Alterations in the constitution of the nucleus, with its attendant transmutations, may also be brought about artificially, *e.g.*, by submitting atoms to the bombardment of neutrons or alpha particles.

The new conceptions of the atom do not affect many of our former conclusions. As before, the net positive charge of the nucleus determines the atomic number; and also, as before, the atom is assimilated to a miniature solar system; *viz.*, a central nucleus around which planetary electrons (equal in number to the atomic number) are circling. Bohr's theory and the wave theories of the atom, which will be examined in this book, were developed before the newer conceptions of the nucleus had been advanced. In these theories, however, the behavior of the planetary electrons is the only subject of study. All that is considered of the nucleus is its charge (atomic number) and its mass; its constitution is disregarded entirely. For this reason the recent discoveries in nuclear physics do not impose any revision of the results established by the earlier theories of the atom.

Ritz's Combination Principle—Moseley's discovery of the regularities in the X-ray spectra was made in 1912. But many years earlier, in 1885, Balmer had discovered regularities in the spacings of some of the visible spectral lines of hydrogen—these are the lines of the so-called Balmer series. Then, in 1908, Ritz found that the spectral lines of many atoms were distributed according to a simple mathematical scheme, and he expressed the belief that this scheme would be found to be valid in all cases. Subsequent research appears to have confirmed Ritz's views. His scheme, called "Ritz's Combination Principle," has been of considerable assistance in the unravelling of spectra. For the present we are not attempting to justify Ritz's principle on theoretical grounds; we accept it as empirical. We shall now describe it in connection with the hydrogen spectrum.

Let us suppose that an infinite sequence of numerical terms is jotted down, *e.g.*,

$$T_1, T_2, T_3, \ldots \ldots T_n, \ldots \ldots T_\infty.$$

The terms are assumed ordered according to decreasing magnitudes. We then find that if appropriate numerical values are given to these various terms, the frequencies of all the radiations that hydrogen can emit are defined by the differences between the terms. The sequence of the numerical terms is called the Ritz sequence for the atom in question. The observation of spectra shows that, for hydrogen, the Ritz terms must have the values

$$\frac{Ry}{1^2}, \frac{Ry}{2^2}, \frac{Ry}{3^2}, \ldots \frac{Ry}{n^2}, \ldots \frac{Ry}{\infty^2}, \textit{i.e.,}\ 0,$$

where Ry is the Rydberg constant, which we have already encountered in Moseley's investigations. Its value is approximately

$$Ry = 3.27 \times 10^{15}\ \text{sec.}^{-1}.$$

All the frequencies ν which hydrogen can emit are thus, according to Ritz, comprised in the general formula,

$$\nu = \frac{Ry}{p^2} - \frac{Ry}{n^2} = Ry\left(\frac{1}{p^2} - \frac{1}{n^2}\right),$$

where p and n are any two integers. It is usual to subtract a smaller term from a larger one so as to obtain a positive value for the frequency. This requires that the integer p be smaller than the integer n. But the point is not of particular importance, and we may obviate the restriction by taking the absolute value of the difference.

With hydrogen, four spectral series are known: they are the Lyman, the Balmer, the Paschen, and the Brackett series, named after their discoverers. In particular, we obtain the frequencies of the lines of the Lyman series by setting $p = 1$ in our previous formula and by giving all integral values to n. The line of highest frequency of this series is called the "limiting line of the series"; its frequency is obtained by taking $n = \infty$, and is therefore

$$\nu = Ry.$$

This frequency is the highest one that hydrogen can emit (in the line spectrum). The Balmer, Paschen, and Brackett series are obtained in the same way but by setting $p = 2$ or 3 or 4. As for n, it must always be a positive integer greater than p.

Each different atom has its corresponding Ritz array; and the concern of experimenters is to determine in each case the exact numerical values of the Ritz terms. The application of Ritz's principle is particularly simple in the case of hydrogen, for not only is every frequency that hydrogen radiates defined by the difference of two of the Ritz terms (as required by the Ritz rule), but in addition the difference of any two of the Ritz terms defines a frequency which is actually observed in the hydrogen spectrum or which is assumed at least to be possible. We may express the situation by saying that all the theoretically possible Ritz frequencies coincide with actually observed ones. With the higher atoms, things are not so simple. As before, in a given atom we find that all the observed frequencies are equal to the differences between some of the Ritz terms of the array corresponding to the atom of interest.* But, as against this, not all differences between the terms define frequencies that have been observed. In other words, all the actually observed frequencies are comprised among the theoretical ones, but the converse is not true. The spectra of the higher atoms are much more complicated than that of hydrogen : their spectral lines are classed into series, or families, called the principal, diffuse, sharp, and fundamental series; sometimes there are also others. In such cases it is convenient to write the Ritz terms in superposed lines. The frequencies actually observed are then given by some of the differences between the terms situated in two adjacent lines.

The Classical Interpretation of Radiation—Up to this point we have neglected to consider how an atom was expected to emit radiation. We must remedy this omission for two reasons. In the first place, only when we take into consideration the classical understanding of radiation emission, can we appreciate the difficulty of obtaining an atomic model that will account for the spectral lines and for their distribution. Also, only in this way can we understand the necessity for the modifications introduced by Bohr. Finally, the classical theory of radiation, though discarded in Bohr's atom, is nevertheless an essential adjunct to Bohr's theory; it still plays a fundamental rôle owing to its intrusion in the Correspondence Principle. This will be understood in the following chapter.

When we apply the laws of electromagnetism to an electron moving in a Galilean frame, we find that no electromagnetic energy will be radiated by the electron unless its motion is accelerated. We recall that the motion of a particle is accelerated whenever the velocity changes either in

---

* This statement is at present a mere assumption, for there are many lines in spectra which have not yet been classified.

direction or in magnitude. Circular motion in particular, even with uniform speed, is an example of accelerated motion.

The motions we shall consider are the so-called periodic motions. One of the simplest is harmonic motion along a straight line. It may be defined as follows: A point P describes a circle with uniform velocity; the projection Q of this point P on any diameter of the circle then exe-

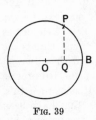

FIG. 39

cutes a to-and-fro periodic motion about the centre O; this motion is called *simple harmonic*. The number of complete vibrations a second is called the *frequency* of the periodic motion, and the maximum elongation OB is its amplitude. If we call x the elongation OQ of the vibrating particle Q from the centre of oscillation O at any instant t, and if we denote the frequency by ν and the amplitude by A, the harmonic motion is expressed by

(1) $x = A \cos 2\pi\nu t$.

The vibrating system here described is called a "linear harmonic oscillator." Planck's oscillators are of this type.

The application of classical electromagnetism to an electron vibrating harmonically proves that a monochromatic electromagnetic wave will be emitted continuously, speeding away with the velocity of light. The frequency of the radiation is the same as the mechanical frequency of the oscillation; and the intensity of the radiation is proportional to the square of the amplitude A.* Finally, the vibration of the electric vector in the propagated wave is parallel to the line of oscillation of the electron; the radiation is said to be *plane polarized*. Since the vibrating electron is radiating energy, it should eventually come to rest. But we may assume that, owing to some restoring mechanism, energy is furnished as fast as it is being lost, so that the motion of the electron is prevented from dying down.

A harmonic motion of the electron yields, as we have seen, a radiation of one single frequency (monochromatic). But we must now examine

---

* The intensity of the radiation emitted by the oscillator is by definition the total energy carried by the radiation and transmitted every second through a closed surface (*e.g.*, a sphere) surrounding the oscillator. It is measured by

$$\frac{16\pi^4\nu^4}{3c^3}e^2A^2,$$

where e is the charge of the oscillating electron, A the amplitude of the oscillation, and c the velocity of light.

more complicated periodic motions of the electron. Suppose that the electron vibrates in some arbitrary continuous way over a straight-line segment. We shall assume that the motion is periodic and hence that it repeats after each complete vibration. The number of complete vibrations performed every second is called the frequency ν, as before. Fourier's discoveries, mentioned in Chapter XIV, show that this motion may be viewed as due to the superposition of a finite, but usually infinite, number of harmonic vibrations of frequencies

$$(2) \qquad \nu, 2\nu, 3\nu, \ldots \ldots n\nu, \ldots \ldots,$$

and of amplitudes which are to be calculated. We shall denote these amplitudes by

$$(3) \qquad A_1, A_2, A_3, \ldots A_n \ldots.$$

The sum of partial harmonic vibrations thus obtained illustrates a Fourier series. It will be noted that the harmonic vibration of lowest frequency has the same frequency ν as the given non-harmonic vibration. This frequency is called the *fundamental frequency* of the non-harmonic vibration; the other frequencies, which are integral multiples of the first, are called the *harmonics* of the fundamental frequency. As for the amplitudes (3) of the component vibrations, they can be obtained only by calculation and will vary in value according to the nature of the vibrational motion which we have split up. Some of these amplitudes may be zero, and in this event the corresponding vibrations do not arise. In the particular case, where the given periodic motion of frequency ν is harmonic, all the amplitudes vanish, except the first one.

To revert to the problem of radiation, classical electromagnetism shows that we may treat the periodic motion of the electron as though it were equivalent to the superposed motions of an infinite number of harmonic oscillators vibrating with the frequencies (2) and having the amplitudes (3). The net result is that our present vibrating system will emit, at one and the same time, radiations of all the frequencies listed in (2), and that the intensity of each one of these radiations will be proportional to the square of the corresponding amplitude. For instance the second harmonic, of frequency 3ν, will have an intensity proportional to $A_3{}^2$. If the Fourier decomposition yields a zero value for the amplitude A_3, the radiation of frequency 3ν will not be emitted. We may also mention that all these radiations will be plane polarized, with their electrical vibrations parallel to the line of oscillation of the electron. The emission of sound waves by a vibrating harp string is a phenomenon of practically the same kind as the one just considered. Here also, if the vibration of the harp string is harmonic, only one musical note is emitted (the fundamental note). And if the periodic motion of the harp string is more complex, we may decompose it into its Fourier components, each one of which will yield a musical note. The string will then emit simultaneously the fundamental note and any number of others, namely, the several different harmonics of the fundamental note.

We must now examine what will happen when the electron describes a curve, *e.g.*, an ellipse. We assume, as before, that the motion is periodic. The frequency is defined by the number of revolutions performed by the electron every second. Here also a distinction between harmonic and non-harmonic motions arises. Suppose that, when the motion of the electron P along the ellipse is projected on the principal axes OA and OB, harmonic motions are obtained for the points of projection Q and Q'. In this case the motion along the ellipse is likewise called harmonic. The two linear harmonic vibrations of the projected points Q and Q' have the same frequency ν as has the elliptical motion of the point P. We may now view the elliptical motion as due to the simultaneous occurrence of the two simple harmonic motions along the perpendicular axes. These two harmonic motions will give rise to radiations of frequency ν, and the two radiations when combined will yield the radiations of frequency ν generated by the elliptical motion. The intensity of this latter radiation will be the sum of the intensities of the two partial radiations. As for the polarization of this radiation, it will be of the type called elliptical. This means that the electric vector in the radiated wave rotates perpendicularly to the direction of advance while remaining parallel to the rotating segment OP in the figure. If the curve followed by the electron is a circle instead of an ellipse, the light will be circularly polarized; and if the motion in the circle is uniform, only one frequency will be emitted.

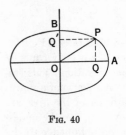

FIG. 40

It is easy to pass to cases where the periodic motion in the ellipse is not harmonic or where the plane curve described periodically by the electron is more complicated. We have but to decompose the motion into two motions along perpendicular axes. These two linear motions will be periodic, though no longer harmonic. But we may apply the Fourier decomposition to each one of them, and, as before, we shall obtain a superposition of harmonic vibrations yielding radiations of frequencies ν, 2ν, 3ν, ..., with corresponding intensities. Combining the radiations of the same frequency which are derived from the rectangular decomposition along the two directions, we obtain the total radiation (of this frequency) due to the electron's motion along the plane curve. These considerations may also be extended to the case where the electron is pursuing a periodic motion in space instead of in a plane. In short, the point to keep in mind is that according to classical conceptions a single electron describing a path with periodic motion may generate a number of different electromagnetic waves having different frequencies, and that all the waves are emitted simultaneously. Furthermore, the frequencies, intensities, and polarizations of the radiations may be computed from the motion of the electron. We shall see in the next chapter that Bohr's theory, though it gives an entirely different rule for computing the frequencies of the emitted radiations, is obliged nevertheless to rely on the classical rules in order to predict the intensities and the polarizations.

CHAPTER XXVI

BOHR'S ATOM

BOHR's atom is an adaptation of Rutherford's. We shall consider first the atom of hydrogen. It is the simplest of the atoms and was the first to be studied by Bohr. According to Rutherford's ideas the hydrogen atom is composed of a single proton around which revolves a single electron. The attraction exerted on the electron by the proton varies in accordance with Coulomb's law of the inverse square—the same law as Newton's, except that the force is electrostatic instead of gravitational. The orbits will necessarily be conics, whether ellipses, parabolae, hyperbolae. We shall exclude from consideration the open orbits (parabolae and hyperbolae); because, if the electron were to follow an open orbit, it would leave the atom for good, and we should no longer be dealing with a hydrogen atom.

The system proton-electron forms an isolated and conservative dynamical system, and according to the general principles of mechanics its total energy remains constant. In the present case this total energy is represented by the sum of the kinetic energy of the proton, the kinetic energy of the electron, and the mutual potential energy. First, let us consider the kinetic energy. The principles of mechanics require that the centre of gravity of our dynamical system remain fixed in the Galilean frame in which it is initially at rest. Inasmuch as the mass of the proton is very much greater than that of the electron, the centre of the proton will at all instants coincide approximately with the fixed centre of gravity. Thus the proton will scarcely move, and we may neglect its contribution to the kinetic energy of the system. When this approximation is made, the total energy is equal to the sum of the kinetic and potential energies of the electron alone; and the total energy thus defined will remain constant during the motion. The kinetic energy of the electron is $\frac{1}{2}\mu v^2$, where μ is the electron's mass, and v its velocity.

We now pass to the definition of the electron's potential energy. We shall adopt the convention that this potential energy vanishes when the electron is situated at infinity. The value of the potential energy when the electron is at a current point P then becomes determinate; it is defined by the following magnitude: the negative value of the work we should have to expend against the force of attraction in order to

488

remove the electron from the point P to a point at infinity. This definition shows that the potential energy is always negative when the electron is situated at any finite distance from the nucleus.

The total energy of our dynamical system can be shown to have a negative value whenever the orbit described by the electron is elliptical. For a parabolic orbit the total energy is zero, and for a hyperbolic one it is positive. Since the value of the potential energy at any point is determined only when its value at some particular point is stipulated, no special significance should be attached to the possible negative values of the potential energy and hence to negative values of the total energy. For instance, the negative value of the electron's energy on an elliptical orbit as contrasted with its positive value on a hyperbolic orbit merely means that the energy is less on the elliptical orbit than on the hyperbolic one. The advantage of agreeing, as we have done, on a vanishing value for the potential energy at infinity is that the various kinds of orbits are now classified immediately according to the sign of the total energy.

Let us suppose the electron is describing some particular elliptical orbit. With this orbit is associated a definite energy. But the total energy by itself does not determine the orbit uniquely, for an infinite number of other elliptical orbits varying in shape and in size have exactly the same total energy. It can be shown, however, that all such orbits have a common geometrical property: *viz.*, their major axes have the same length. In particular, if we consider a circular orbit of given energy and any number of elliptical orbits associated with the same energy, the diameter of the circle and the major axes of the various ellipses will all be equal in length. A further common property of all these orbits of equal energy follows directly from Kepler's third law of planetary motions; namely, any one of these orbits will be described in the same time. In other words, the frequency of the motion will be the same over any one of the orbits of equal energy. All these conclusions remain valid when Bohr's atom is considered, for insofar as the motion of the electron on an orbit is concerned, Bohr's theory accepts classical mechanics.

We must now examine what kinds of radiation an electron moving in an orbit should emit, according to the classical laws of radiation mentioned in Chapter XXV. If the orbit is circular, only one radiation will be emitted namely, that which has the same frequency as the electronic motion. The radiation will be circularly polarized. But suppose that the motion is elliptical. It will never be harmonic, for a harmonic motion would arise only if the attractive force of the nucleus were proportional to the distance, whereas here we are dealing with a Newtonian attraction. Since the motion though periodic is not harmonic, more than one radiation

will be emitted.* The emitted radiations will have as frequencies the
fundamental frequency ν of the periodic motion and its harmonic
frequencies 2ν, 3ν, ... $n\nu$, ... The intensities of these radiations
can be derived from the Fourier series which express the electron's motion; they will depend on the ellipse followed. Owing to the loss in
energy that accompanies the radiation, the electron will fall progressively
to lower orbits of smaller total energy; and, in so doing, it will revolve
faster and faster around the nucleus—a situation that may be understood
when we note that the planet Mercury revolves more rapidly round the
sun than does the Earth. The fundamental frequency ν of the motion
and all its harmonics will thus increase in value, and so the radiations
emitted by the electron will likewise increase in frequency. Since these
increases will be extremely rapid, the spectral lines should appear broad-
ened and blurred.

These classical expectations conflict with observation. The classical
frequencies are incorrect; and the spectral lines are not blurred, they
are sharp, well defined. The latter discrepancy may be remedied if we
assume that energy is communicated in some way to the system as fast
as it is being lost by radiation; the frequencies would then remain fixed
and the spectral lines be well defined. But it is unnecessary to gloss over
details, because the major discrepancy concerns the exact frequencies
radiated. The frequencies actually observed in the hydrogen spectrum
are not the successive integral multiples of some fundamental frequency,
as would be required by the classical theory. Ritz made an ingenious
attempt to account for the observed frequencies. He assumed that tiny
bar magnets which he called "magnetons" were present inside the nucleus,
and that they could adjust themselves in various ways. According to
the number and the arrangements of these magnets, different types of
orbits would arise for the electron, and Ritz hoped by this means to
account for the spectral lines. Quite aside from the artificial nature of
Ritz's assumptions, the magnetons he postulated were never isolated and
so his theory was discarded.

Bohr's Atom—Bohr accepts Rutherford's model of the hydrogen
atom; namely, a central proton around which a single electron circles.
But Bohr introduces additional assumptions:

(A) The electron can move only along privileged ellipses.
These are determined by the condition that, over them, the total

---

* In fact, the decomposition of the elliptical motion by means of Fourier series
shows that all the harmonics of the fundamental frequency will be present. Consequently, according to the classical theory, all the harmonic frequencies (an infinite
number) should be radiated.

Maupertuisian action of the electron must be an integral multiple of Planck's constant h.

(B) No radiation will occur when the electron circulates along a permissible orbit. Since no energy is lost by the atom under these conditions, its state is said to be "stable," or "stationary." The permissible orbits are therefore also referred to as stable or stationary orbits.

(C) The electron may drop spontaneously and spasmodically from a permissible orbit of higher energy to one of lower energy. The energy, thus liberated from the atom, will appear as a radiation of perfectly definite frequency. Conversely, if radiation is directed against an atom, or if the atom is submitted to collisions, it may absorb energy, and its electron is then jerked up to an orbit of higher energy.

(D) The frequency ν of the radiation emitted when the electron drops from an orbit of total energy E_1 to one of total energy E_2 is defined by Bohr's *frequency condition*, viz.,

$$(1) \qquad \nu = \frac{E_1 - E_2}{h}, \; i.e., \; \frac{\text{drop in energy}}{h}.$$

We may note that in our present treatment, where the motion of the proton nucleus is disregarded, the total energy of the electron moving in an orbit is also the total energy of the atom. Instead, then, of saying that the electron is in an orbit of energy E, we may equivalently say that the atom has the energy E, or is situated in the energy level E. Bohr's assumptions appear strange; yet they are natural generalizations of those introduced by Planck and by Einstein in connection with the earlier development of the quantum theory. This point will be better understood when we analyze Bohr's assumptions in detail.

Bohr's Assumption (A)—This assumption is in crass contradiction with classical mechanics, for according to classical (or relativistic) mechanics the electron might describe any one of the elliptical orbits: none would be excluded as impossible. Assumption (A) is called the *quantizing condition*. As we shall now see, it is a mere generalization of the restriction postulated by Planck for his linear oscillators.

We recall that when a mass μ is moving with velocity v over a tiny stretch ds, the Maupertuisian action over this stretch is defined by the product $\mu v ds$. The Maupertuisian action over a finite distance is obtained by integrating such elementary amounts of action over the entire path described. We may therefore calculate the Maupertuisian action which is associated with an electron's to-and-fro vibration in an oscillator or with an electron's motion around an elliptical orbit in Bohr's atom. The quantizing condition adopted by Planck for his oscillators was mentioned in Chapter XXIV. It requires that the action over a complete oscillation should have a value determined by nh, where n is any positive integer. Bohr's quantizing condition is exactly the same, except that now the periodic motions are along ellipses.

If we call J the value of the Maupertuisian action of the electron over an ellipse, the privileged ellipses will be determined by the condition

$$(2) \qquad\qquad J = nh \qquad\qquad (n = 1, 2, 3, \ldots).$$

The integer n is called the *quantum number*. According to the value given to n, we have the first, second, or third stable orbit, and so on. Along the first orbit the total action is h, along the second $2h$, and along the nth it is nh. The formula (2) gives the mathematical expression of Bohr's quantizing condition.

To each value ascribed to J corresponds a value of the total energy E, so that orbits associated with the same value of the action are automatically associated with the same value of the total energy. An orbit is thus equally well determined by the action or by the energy with which it is associated. The relationship between the action and the energy is easily derived from classical mechanics; it is given by

$$(3) \qquad\qquad E = -\frac{2\pi^2 \mu e^4}{J^2},$$

where μ and e are the mass and charge of the electron, respectively. The stable *energy levels* are obtained when we substitute in this expression of the energy the permissible values of the action J. We thus find for the stable energy levels of the hydrogen atom the values

$$(4) \qquad\qquad E = -\frac{2\pi^2 \mu e^4}{n^2 h^2}.$$

The lowest energy level corresponds to the orbit which is nearest the proton-nucleus; we obtain it by setting $n = 1$ in the previous expression. Successive higher integral values ascribed to n give the successive higher energy levels. Two points may be noted.

Firstly: The formulæ (2) and (4) express exactly the same quantizing condition; the only difference is that in (2) the stable energy levels are determined by the action, whereas in (4) they are determined by the energy. The greater simplicity of the formula (2) shows that, in the present formulation of the quantum theory, action is more fundamental than energy.

Secondly: For a given value of the quantum number n, the formula (4) defines a corresponding value of the energy and hence defines an energy level, namely the nth energy level. The energy level is thus well determined, but we have seen that an infinite number of different orbits are associated with the same value of the energy. The quantizing condition does not therefore pick out any particular one of these orbits of equal energy; and so, when the system is in the nth energy level, the electron may, for all we know, be following one or another of the orbits associated with this level. In short, each energy level may be viewed as represented by any one of a large number of equivalent orbits.

Bohr's Assumption (B)—According to this assumption no radiation occurs when the electron is describing a quantized orbit. The classical laws of electrodynamics are thereby violated, for, the motion of the electron being accelerated, these laws require that radiation be emitted. It is not the first time, however, that physicists have been led to question the validity of the classical laws in similar situations. For example, some years before Bohr developed his theory, Langevin had ascribed permanent magnetism to the circlings of electrons in a metal. And in order to account for the permanence of the magnetism, Langevin made the same assumption that was subsequently made by Bohr; namely, the circling electrons did not radiate energy.

If we connect Langevin's idea with Bohr's model of the atom, we must expect that a hydrogen atom with its circling electron should be equivalent to a little magnet. The magnetic properties of the atom will, however, be discussed on a later page, and we shall not dwell on them here.

Bohr's condition (B) also postulates the stability of the quantized motions. But we must first understand clearly in what sense the word "stability" is used. As will be explained presently, an electron revolving in a stable orbit does not remain there indefinitely. Unless the orbit be the one of lowest energy, the electron soon drops to a lower orbit, emitting radiation in the process. It would thus be incorrect to suppose that the stability of the privileged electronic motions implies their permanency. All that is meant by the word *stability* is that the privileged orbits constitute possible resting places for the electron, and that the electron may circulate for a certain length of time (extremely short) on one of these orbits before it falls to a lower one.

The nature of the stability is best understood when we examine under what conditions the electron will be jerked up to a higher level. Let us suppose that the atom is in a state characterized by the value $3h$ for the action J. If light falls on the atom, the light may yield part (or all) of its energy to the electron, thereby causing the atom to pass from the state $3h$ to a higher state, say $4h$. Similar changes will occur in the state of the atom if it is subjected to a violent collision; the energy surrendered by the colliding body throws the atom into a higher state. But only when the energy of the incident light is sufficiently great or the collision is sufficiently violent, will the atom be raised to a higher state. For weak collisions in particular, the state of the atom will remain unaffected, so that in such cases the state of the atom shows stability.

These considerations enable us to formulate in another way the stability of the privileged states of the atom. Thus according to Planck's assumptions, light of high energy is light of high frequency. Hence light of high energy generates a force which varies with extreme rapidity. Similarly a violent impact is necessarily more sudden than a weak one. We may therefore say that the privileged states remain unchanged,* and hence manifest stability, when the atom is subjected to weak forces which vary slowly. This stability breaks down under the action of strong forces applied violently.

Adiabatic Invariants—At first sight, we might suppose that the type of stability just discussed is a characteristic of Bohr's postulates and was unknown in classical mechanics. But this is not so. All periodic mechanical systems, in which no quantum assumptions need be introduced, exhibit peculiar magnitudes which remain unchanged in value when the system is submitted to slowly varying forces. Ehrenfest calls these semipermanent magnitudes "adiabatic invariants." Suppose, for instance, that while a pendulum is swinging, we shorten its length gradually. We may do this by displacing slowly downwards a small ring which embraces the string of the pendulum near the point of suspension. To lower the ring we must expend work against the tension of the strong, thereby increasing the total energy of the pendulum. But though the energy increases, the Maupertuisian action remains unchanged in value. This action affords an example of an adiabatic invariant. Similarly in our mechanical model of the hydrogen atom, the Maupertuisian action J of the electron can be shown to be an adiabatic invariant. As a result, if the atom is submitted to forces which are slowly varying, the action J will remain unchanged and the atom will manifest stability.

Thus, quite independently of Bohr's postulates, the existence of stable mechanical motions in an atom must be expected on the basis of classical

---

* Spontaneous changes to lower levels are disregarded in this discussion.

considerations. In this respect the only novelty introduced by Bohr is that *not all* values of the action are stable, but only the quantized ones

$$J = nh \qquad (n = 1, 2, 3 \ldots).$$

Bohr's Assumption (C)—Since, according to Bohr, radiation occurs only as a result of a drop of the electron from a higher to a lower energy level, we can expect radiation only when the electron is not already in the lowest orbit. Equivalently we may say that the atom must be in one of the higher energy states. Such states are called *excited states*; the higher the energy of the orbit in which the electron is moving, the greater the "excitation." When, as a result of its successive drops, the electron has reached the lowest level, all further radiation ceases; the atom is then said to be *unexcited*. Now, the spectrum of an incandescent gas exhibits spectral lines which endure. It is therefore necessary to suppose that the gas atoms which have radiated and have fallen into the unexcited state are subsequently re-excited so that the drops may be repeated. We may readily understand whence the restoring energy is derived in practice. At the high temperature of the incandescent gas, the collisions between molecules are (according to the kinetic theory) sufficiently violent to restore the excited states of the atoms. In this way, so long as the temperature of the gas is maintained, radiation may continue indefinitely.

These considerations bring to light an important difference between the classical and the quantum conception of radiation. The classical theory ascribes the radiations to the mechanical motion of the electron; it also assumes that the frequencies of the radiations are the frequencies into which the periodic motion of the electron may be decomposed by Fourier's method. Hence, according to classical conceptions, the electron may emit different frequencies simultaneously. A single hydrogen atom should thus be susceptible of emitting all the frequencies at the same time. But in Bohr's theory, the situation is entirely different, for owing to the discrete succession of energy drops, the atom can emit only one frequency at a time, not two or more different ones. The reason we observe in practice a large number of lines simultaneously in a spectrum is that we are dealing with the large aggregate of radiating atoms present in an incandescent gas. The simultaneous exhibition of different spectral lines is thus a cumulative effect of an orchestra of atoms, as it were.

Bohr's Assumption (D)—Here Bohr postulates the frequency of the radiation which will be generated by a drop from a higher energy level E_1 to a lower one E_2. The frequency emitted is given by the drop in the value of the energy, divided by h. The new rule is revolutionary from the classical standpoint, for, in contradistinction to classical ideas, the frequency of the radiation now has no connection with the mechanical frequency of the electron on its orbit. Bohr's rule is, however, a mere generalization of Einstein's rule mentioned in connection with the photoelectric effect. We recall that, according to Einstein, when a fast-moving electron is suddenly stopped, it yields its energy (here kinetic) to a photon. The frequency of vibration of the radiation associated with the photon is then defined by

$$\nu = \frac{\text{kinetic energy of electron}}{h}.$$

The frequency is thus equal to the drop in the energy of the electron divided by h; and this is also Bohr's rule. But Bohr's rule is more general, for in the hydrogen atom the energy lost by the electron during a drop is the *total* energy (kinetic + potential), not the kinetic energy alone. Furthermore, the electron does not lose all its energy but only a part. Thus, Einstein's rule appears as a particular case of Bohr's.

Verifications of Bohr's Theory—We have now to investigate whether Bohr's theory can account for the precise spectral frequencies which experiment reveals for hydrogen. We saw in the previous chapter that all these frequencies are contained in the empirical formula

$$(5) \qquad \nu = Ry\left(\frac{1}{p^2} - \frac{1}{n^2}\right),$$

where Ry is Rydberg's constant and where p and n are any positive integers. We assume $p < n$, so as to obtain positive values for the frequencies. In particular, if we set $p = 1$ and give all positive integral values to n, we get the lines of the so-called Lyman series; the line of highest frequency, or the limiting line, is obtained when we set $n = \infty$. We then get $\nu = Ry$, which is the line of highest frequency that the hydrogen atom can emit (in the line spectrum). The Balmer series is obtained when we set $p = 2$ and give all higher values to n. The Paschen series arises when $p = 3$, and the Brackett series when $p = 4$. A still higher series, in which $p = 5$, has also been observed.

According to Bohr's theory the radiated frequencies produced by a drop from the nth to the pth energy level are, by (4),

$$(6)\ \nu = \frac{E_n - E_p}{h} = -\frac{2\pi^2\mu e^4}{h^3}\left(\frac{1}{n^2} - \frac{1}{p^2}\right) = \frac{2\pi^2\mu e^4}{h^3}\left(\frac{1}{p^2} - \frac{1}{n^2}\right).$$

The two formulae (5) and (6) will yield exactly the same frequencies provided the numerical value of the Rydberg constant Ry, as determined by experiment, and that of the expression

(7)
$$\frac{2\pi^2\mu e^4}{h^3}$$

are the same.

Now the value of the Rydberg constant as determined by empirical measurements on the spectra is

$$Ry = 3.27 \times 10^{15} \text{ sec.}^{-1}.$$

The same value is found when in the expression (7) we replace e, μ, and h (*i.e.*, the charge and mass of the electron and Planck's constant h) by their experimental values.* Consequently, we are certain that Bohr's theory gives the correct distribution of the spectral lines.

Aside from its success in accounting for the hydrogen spectrum, we see that Bohr's theory has also given an interpretation of Rydberg's optical constant. The value of Ry is now known to be defined by (7), a formula in which the universal constants e, μ, and h alone appear. This formula is of extreme importance, for, just as the chemist endeavors to show that all substances can be derived from the various combinations of a limited number of elementary ones, so is it the aim of the theoretical physicist to express the various constants of physics in terms of a limited few.

Bohr's theory of the hydrogen atom also accounts for the continuous spectrum which covers the line spectrum and prolongs it in the direction

---

* Bohr's theory may be refined if we take into consideration the wobbling motion of the proton nucleus. We have disregarded this motion in the text because, in view of the larger mass of the proton as compared with that of the electron, the refinement would scarcely modify our results. If, however, we wish to take the refinement into consideration, the motion of the electron round the proton will not be quite the same, and the energy levels will be slightly different. The corrections in the values of the energy levels, in those of the frequencies radiated, and in the expression of the Rydberg constant are secured if we replace in the formulae (4), (6), and (7), the mass μ of the electron by

$$\frac{\mu}{1 + \dfrac{\mu}{M}},$$

where M is the mass of the proton. Since $M = 1830\mu$ the effect of the correction will merely be to multiply our former numerical values by $\dfrac{1830}{1831}$, and will therefore be very minute.

of the higher frequencies. We may explain the presence of the continuous spectrum by taking into consideration the hyperbolic orbits which an
electron sometimes describes. Such orbits are not periodic; and since the
quantum theory imposes no restrictions on non-periodic motions, no
quantization is imposed on the hyperbolic orbits; all such orbits are possible. Let us suppose, then, that a hydrogen atom has lost its single
electron: the atom is now said to be *ionized*; all that remains of it is the
proton. Of course, no radiation can be emitted since there is no electron
to drop. Suppose, however, that a vagabond electron comes into proximity with the proton and, under the proton's attraction, describes a hyperbolic orbit. (In astronomy we have the simile of a comet following a
hyperbola under the gravitational attraction of the sun.) A drop can
now occur from the more energetic hyperbolic orbit to one of the less
energetic elliptical orbits. Bohr's formula for the frequency emitted is

$$\nu = \frac{E_H - E_e}{h},$$

where E_H and E_e are the energies of the hyperbolic and the elliptical
orbits. Inasmuch as E_H may have any one of the values of a continuous
set, the same will be true of the emitted frequency; a continuous spectrum
is thus accounted for. Drops between hyperbolic orbits may also arise.

The Spectrum of Ionized Helium—The atom of helium is formed
of a nucleus consisting of an alpha particle around which two planetary
electrons are revolving. All attempts to investigate the helium atom by
Bohr's original method were failures, and only after the hypothesis of the
spinning electron was advanced was any progress made. Even then,
many obscure points remained. Not until Heisenberg investigated the
problem by the methods of the new quantum mechanics was the riddle
of the helium spectrum elucidated.

Despite the difficulty of accounting for the spectrum of neutral helium,
the problem of the once-ionized helium atom is simple. In the ionized
atom one of the two planetary electrons is removed, and we are left with
one electron revolving around an alpha particle. The situation is thus
the same as in the hydrogen atom, except that the net positive charge of
the nucleus is double and hence the attractive force it exerts on the
planetary electron is twice as great. In view of this similarity the ionized
helium atom is said to be hydrogen-like; and we may expect its spectrum
to resemble that of hydrogen and to differ considerably from the spectrum
of neutral helium. Bohr's theory requires that the energy levels and the
frequencies of the spectral lines of ionized helium should be approximately
four times as high as those of hydrogen, and that some of the lines of

ionized helium should occupy very nearly the same positions as some of the hydrogen lines.* When Bohr first applied his theory to the ionized helium atom, some of the spectral lines which he attributed to this atom had been observed in the laboratory and also in the spectra of certain stars. But the prevalent opinion was that these lines belonged to a new hydrogen series. Bohr disputed this belief, asserting that ionized helium, not hydrogen, was responsible for the lines. Finally, the matter was put to a test when the spectrum of ionized helium was carefully studied in the laboratory by A. Fowler. Bohr's contention proved correct. This incident was regarded at the time as affording a powerful argument in favor of Bohr's theory.

The atom following helium is lithium, with three planetary electrons. When two of these electrons are removed, and hence when the lithium atom is twice ionized, we again obtain a hydrogen-like atom. The nuclear charge, or atomic number, of lithium is 3, in place of 2 for helium and 1, for hydrogen; but aside from this difference the situation is the same as in once-ionized helium or in hydrogen, and the spectrum may be predicted by Bohr's method. Still higher hydrogen-like atoms are exemplified by beryllium ionized three times, boron ionized four times, and so on.

The Correspondence Principle—We have now reached a point where we may discuss the famous "Correspondence Principle," which is so important both from a philosophical and a practical standpoint. In Bohr's theory the spectral lines of hydrogen, for example, are produced by the various atoms dropping from higher to lower energy levels. We assume that, as a result of collisions, energy is continually imparted to the atoms, exciting them afresh, so that the drops may be repeated. If all the drops were equally probable, they should on an average occur equally often and the various spectral lines should have more or less the same intensity. But this is not what is observed. Instead we find that some lines are very much fainter than others, and that, in many atoms, lines which we should expect (from our knowledge of the distribution of the energy levels) do not occur at all.† We are thus compelled to recognize that all drops are not equally probable, and that some of them are so improbable that they never take place. These are the *forbidden drops*, as they are sometimes called.

---

* The lines in question would actually coincide, were it not for the larger mass of the helium nucleus. This point becomes clear when the motion of the nucleus is taken into account.

† Hydrogen is an exception in that all the theoretically possible drops are believed to be possible in practice.

Bohr's theory affords no means of calculating the probabilities of the
various drops and is thus unable to furnish any information on the relative intensities of the various spectral lines. In particular, it cannot predict the "forbidden drops." Finally, Bohr's theory has nothing to say
on the polarizations of the radiations. In this respect classical science is
superior in principle, for it professes to give not only the frequencies
of the radiations, but also their intensities and polarizations. It is true
that the classical frequencies usually conflict with experiment, but the
classical theory does at least make an attempt to answer our questions,
and it thereby gives us something to work on. To remedy these defects in
his theory, Bohr introduced his correspondence principle. This principle
may be divided into two parts.

1. Bohr shows that the radiation frequencies derived from his frequency condition tend to merge into the mechanical frequencies of the
electron's motion, and hence into the radiation frequencies predicted by
classical science, whenever the drops in the energy (or in the action)
are small and the levels between which the drops occur are high. We
shall refer to these restricted situations, in which the Bohr and the classical frequencies tend to coincide, as the *limiting conditions of the correspondence principle*. Accordingly, we may say: Under the limiting conditions of the correspondence principle, the quantum theory passes over
into classical theory. This first part of the correspondence principle
results immediately from mathematical analysis; it involves no hypothetical assumptions.

2. The second part of the correspondence principle is speculative; it
concerns the intensities and the polarizations, which Bohr's theory, as
such, is unable to calculate. Bohr circumvents this deficiency in his
theory by postulating that even when the limiting conditions of the correspondence principle are not realized, the intensities and polarizations of
the radiations (in contradistinction to their frequencies) are given
correctly by classical theory.

Let us examine more closely the first part of the correspondence principle. We mentioned in Chapter XXV that when an electron is performing a periodic motion with a mechanical frequency ν_0, the motion
may be decomposed into a number of superposed harmonic vibrations of
mechanical frequencies

(8) $\nu_0, 2\nu_0, 3\nu_0, \ldots \, n\nu_0, \ldots$

and of amplitudes

(9) $A_1, A_2, A_3, \ldots A_n, \ldots$

These amplitudes can be calculated mathematically from the known motion of the electron. According to the classical theory of radiation, the electron will radiate just as would an aggregate of electrons vibrating simultaneously, each electron of the aggregate vibrating with one of the frequencies (8) and with the corresponding amplitude (9). The frequencies of the radiations emitted will be the mechanical frequencies (8), and the intensities will be proportional to the squares of the corresponding mechanical amplitudes (9). Finally, the polarizations of the radiations are determined by the nature of the mechanical vibrations. Let us apply these classical considerations to an electron describing an ellipse in Bohr's hydrogen atom. We assume that the ellipse is one of the quantized ellipses of Bohr's theory, *e.g.*, the ellipse characterized by some integral value for the quantum number n. Since the motion of the electron in the ellipse is known, the frequencies that should be radiated are obtained from the classical theory. These frequencies exhibit, of course, the general relationships expressed in the sequence (8).

Suppose now that we consider the hydrogen atom in the same state as previously but that we apply Bohr's quantum principles to determine the radiated frequencies. The electron, initially in the nth orbit, drops to a lower orbit. If we call E_n, E_{n-1}, E_{n-2} . . . the energies of the nth and of the successively lower orbits, the Bohr frequencies are given by

$$(10) \qquad \frac{E_n - E_{n-1}}{h}, \qquad \frac{E_n - E_{n-2}}{h}, \qquad \frac{E_n - E_{n-3}}{h}, \ldots$$

In general, these Bohr frequencies will differ from the classical, or mechanical, frequencies (8). Nevertheless, if the energy level E_n is high, it can be shown that the values of the Bohr frequencies (10) tend to coincide with the classical ones ν_0, $2\nu_0$, $3\nu_0$ . . . . . This discovery constitutes the first part of the correspondence principle. The approximation holds best for the lower frequencies, which result from small energy drops (*e.g.*, the ones written in (10)), and will tend to become less accurate when the higher frequencies are considered. We may summarize our findings by saying that when an electron drops from a high energy level to the one immediately below, or to the second lower one, or to the third lower one, and so on, the corresponding Bohr frequencies coincide, respectively, with the fundamental mechanical frequency ν_0 of the motion on the higher orbit, or with its successive harmonics $2\nu_0$, $3\nu_0$ . . . . . This rule holds, however, only for the relatively small drops. All these results are obtained mathematically without the introduction of additional assumptions, and furthermore they are general and do not hold solely for the hydrogen atom.

A simple graphical representation of these results may be given. Thus let us consider a hydrogen atom, formed as usual of a proton-nucleus around which an electron is circling. We first disregard Bohr's quantum restrictions and assume, in accordance with classical ideas, that the planetary electron may be describing any elliptical orbit. Formula (3) expresses the general relation between the energy E of the electron and the action J associated with the electronic orbit. Let us take J and E as Cartesian coordinates, and plot the curve defined by the relation (3); this curve is represented by the dotted line in the figure. We now mark off any distance Oa along the horizontal axis, and through the point a we draw the vertical, which intersects the curve at a point E. The lengths Oa and aE then measure corresponding values for the action and the energy of the electron (as required by the classical formula (3)).

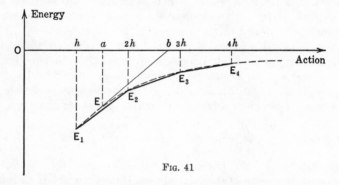

Fig. 41

The frequency of the electron's motion along its orbit may also be represented in the graph. We draw the tangent Eb to the curve at E. The trigonometric tangent of the angle made by the line Eb and the horizontal axis Ob may then be shown to measure the mechanical frequency of the electron. For simplicity, we shall say that the "slope" of the tangent Eb represents this frequency, so that the more nearly vertical the tangent Eb, the higher the frequency.

Now, according to classical theory, the frequency ν_0 of the electron's motion and the various harmonics of this frequency (i.e., $2\nu_0$, $3\nu_0$,....) are also the frequencies of the radiations which the electron will emit. We shall confine ourselves, however, to the radiation of lowest frequency, i.e., to the one whose frequency is the fundamental frequency ν_0 of the electron's motion. As we mentioned previously, this frequency is represented by the slope of the tangent Eb in the figure.

Thus far we have considered the classical theory. We now take Bohr's theory into account. The relation (3) between the action and the energy holds as before, but in Bohr's theory the only possible orbits of the electron are those for which the action J has any one of the values nh ($n = 1, 2, 3 \ldots$). Let us, then, plot these privileged values of the action along the horizontal axis. The verticals drawn through the points thus obtained intersect the dotted curve at points $E_1, E_2, E_3 \ldots$ The distances of these points of intersection from the horizontal axis measure the energy values corresponding to the privileged values of the action. Such energy values are the energy levels of the atom. The lowest of the energy levels is E_1, the next lowest is E_2, and so on. In short, whereas in the classical treatment any point on the dotted curve could define a possible energy state of the atom, we must now suppose that only such points as E_1, E_2, E_3, etc., are possible.

Next, we pass to the frequencies that may be radiated according to Bohr. The frequency condition shows that when the electron drops from an energy level to the one immediately below it, say from E_3 to E_2, the frequency of the radiated light is defined by the slope of the chord E_2E_3. On the other hand, if the classical theory were accepted, the electron circling with energy E_3 would emit a radiation whose frequency was defined by the slope of the tangent to the dotted curve at E_3. Obviously the slopes of the tangent at E_3 and of the chord E_2E_3 are not the same. We therefore conclude that the classical frequency and the corresponding Bohr frequency will be different. But we observe that the curve tends to become straight as it proceeds to the right, i.e., for high values of the energy. Consequently, for any two points, E_n and E_{n-1}, which are situated at a sufficiently great distance to the right, the chord $E_{n-1}E_n$ and the tangent to the curve at E_n will tend to coincide; hence the classical frequency and the Bohr frequency will tend to the same value.

Thus we have the following general result: If the energy of an orbit is high, the Bohr frequency emitted when the electron drops from this orbit to the one immediately below will tend to coincide with the fundamental frequency ν_0 of the electron's motion along the orbit, and hence with the frequency of the radiation that would be expected in the classical theory. This conclusion is rendered intuitive by the graph. To establish the other results mentioned in the first part of the correspondence principle, we should have to show that the larger drops yield frequencies which coincide with the harmonics of the electron's motion. We shall, however, accept this correlation without proof because its demonstration requires mathematical analysis.

The graphical method of representation has the advantage of bringing out another point. Let us imagine that the universe be so constituted that Planck's constant h is many times smaller than it is in our actual universe. The points E_1, E_2, E_3, E_4, which define the energy levels, would then be more closely packed, though of course they would still lie on the same curve. The closer packing would imply that the energy levels differed by smaller amounts. Now, we see that if the energy level E_2, just below the level E_3, differed but little from E_3, the segment E_2E_3 would tend to coincide with the tangent at E_3. Consequently, even for the relatively low energy level E_3, the classical and the Bohr frequencies would tend to coincide; and we should not have to pass to the high energy levels in order to bring about the merging of the two frequencies. In short, the classical and the quantum theories would tend to become indistinguishable, not only for the high energy levels, but also under all conditions. A very general principle is here illustrated; it often recurs in the quantum theory and was noted on previous occasions. We may express it by saying: If in our universe the constant h were to be infinitesimal instead of merely small, the quantum theory would pass over into classical theory; the latter would then be correct.*

We now examine the second part of Bohr's correspondence principle. In contradistinction to the first part, which is a mathematical necessity, the second part of the correspondence principle is a sheer assumption. We have seen that Bohr's theory does not even attempt to predict the intensities and polarizations of the radiations whereas the classical treatment professes to give full information. Now common experience shows that, under the limiting conditions of high energy values and small drops, not only the frequencies radiated but also the intensities and polarizations of the radiations are predicted correctly by the classical methods. To remedy the deficiency in his theory, Bohr therefore suggests that the intensities and polarizations be calculated classically in all cases. It must be admitted that this extrapolation of the classical methods appears gratuitous. We know that the extrapolation would lead to incorrect results if it were applied to the radiated frequencies. Why, then, should it be permissible in the case of the intensities and polarizations? No answer can be given to this question,† and so the extrapolation has no theoretical justification. Be this as it may, the extrapolation appears justified on empirical grounds and it has been of considerable use in the development of Bohr's quantum theory.

The correspondence principle readily furnishes the polarizations of the emitted radiations, but it is less successful in determining the intensities; some measure of vagueness is usually unavoidable. This vagueness is due to the peculiarities ascribed to the radiation process in Bohr's

---

* See the note page 460.

† At least not at the present stage of the discussion.

theory. Thus let us assume that an electron is describing one of the Bohr orbits. We apply Bohr's frequency condition, and the frequency of the radiation is obtained. We wish to utilize the correspondence principle to determine the intensity of the radiation. The principle tells us that we must refer to the classical theory and derive the intensity from the motion of the electron on the orbit. But in Bohr's theory a difficulty arises at this point; for there are two orbits, the initial one and the one to which the electron drops; and we have no means of deciding which of the two orbits should be taken. Two different values are thus suggested for the intensity. Usually a mean value, corresponding to an intermediary orbit, is selected. Obviously, however, the procedure is arbitrary.

Nevertheless there is one important case in which the correspondence principle furnishes definite results. It occurs when the decomposition of the classical motion fails to yield some of the harmonic motions. For example, suppose that in the mechanical motion of frequency ν_0, the amplitudes of the harmonics $2\nu_0$ and $4\nu_0$ are both zero. This circumstance implies that in the classical treatment the radiations of frequencies $2\nu_0$ and $4\nu_0$ will have zero intensity and hence will not be radiated at all. Now, under the limiting conditions, where the frequencies of Bohr and of classical theory merge, the classical frequencies $2\nu_0$ and $4\nu_0$ are also those which, in Bohr's theory, would be generated when the electron drops from a high energy level E_n to the lower but one (*i.e.*, E_{n-2}) and to the lower but three (*i.e.*, E_{n-4}). And since these frequencies do not arise in the classical treatment, the correspondence principle informs us that neither can they arise in Bohr's theory. Accordingly the two drops listed above cannot occur. Thus, the correspondence principle enables us to determine the forbidden drops in Bohr's atom or indeed in any quantized mechanical system.*

The usefulness of the correspondence principle is well illustrated in the plane rotator. This is the name given to the system formed of a charged body, such as an electron, revolving in a fixed circle around a fixed centre. Bohr's quantizing condition gives for the energy levels the values

$$0, \quad \frac{h^2}{8\pi^2 I}, \quad \frac{2^2 h^2}{8\pi^2 I}, \quad \frac{3^2 h^2}{8\pi^2 I}, \quad \cdots \quad \frac{n^2 h^2}{8\pi^2 I} \cdots \cdots,$$

where I is the moment of inertia of the system. If we were ignorant of the correspondence principle, we should assume that all drops between the energy levels were possible. But if we apply the correspondence prin-

---

* In the next chapter we shall see that some forbidden drops cannot be predicted from the correspondence principle.

ciple, we find that some drops are impossible. To understand how the correspondence principle operates in the present case, we must first treat the problem classically. The electron is describing a circle with uniform motion. The frequency of the motion is, say, ν_0; and the motion being uniform, no harmonic motions of higher frequency occur in the Fourier decomposition. The classical theory thus requires that the rotator should emit a single radiation frequency ν_0, i.e., the fundamental mechanical frequency. In addition, the radiation should be circularly polarized. We now apply the correspondence principle. Since according to the classical theory the rotator can emit only the fundamental frequency ν_0, we must assume, in accordance with our previous explanations, that in the quantum treatment the only possible energy drops are those between consecutive levels. All other drops are forbidden. Furthermore, the correspondence principle enables us to predict that the radiation will be circularly polarized.

Let us verify that these permissible quantum drops will, under the limiting conditions of the correspondence principle, yield the classical frequency. According to Bohr's frequency condition, the frequency radiated during a drop from the nth to the $(n-1)$th energy level is

$$\nu = \frac{1}{h}\left[\frac{n^2h^2}{8\pi^2 I} - \frac{(n-1)^2h^2}{8\pi^2 I}\right] = \frac{h}{4\pi^2 I}\left(n - \frac{1}{2}\right).$$

On the other hand, the classical frequency of the rotator, when its energy is that of the nth energy level, is found to be $\dfrac{h}{4\pi^2 I}\,n$. The two frequencies are not the same. But if the energy, and hence n, is high, the relative discrepancy between the two frequencies tends to vanish, so that the requirements of the correspondence principle are satisfied. We may note that the two frequencies would also tend to coincide if the value of h were to decrease.

As a second illustration of the correspondence principle, we consider the linear harmonic oscillator, i.e., a Planck oscillator. It is constituted by an electron vibrating back and forth with simple harmonic motion along a straight line. If we call ν_0 the mechanical frequency of the motion, the quantizing conditions show that the energy levels are

$$0, \; h\nu_0, \; 2h\nu_0, \; \ldots \; nh\nu_0, \; \ldots \; *$$

---

* The correct energy levels, as derived from wave mechanics or the matrix method, are

$$\frac{h\nu_0}{2}, \quad \frac{h\nu_0}{2} + h\nu_0, \quad \frac{h\nu_0}{2} + 2h\nu_0, \ldots\ldots \frac{h\nu_0}{2} + nh\nu_0, \ldots\ldots$$

But this change does not affect the conclusions stated in the text.

The theoretically possible radiated frequencies, corresponding to all conceivable drops, are contained in the general formula

$$\nu = \frac{nh\nu_0 - ph\nu_0}{h} = (n - p)\nu_0,$$

where n and p are any two positive integers or zero. We assume $n > p$. We must, however, apply the correspondence principle so as to determine which drops (if any) are impossible. We first note that the mechanical motion is a simple harmonic motion of frequency ν_0, and that a simple harmonic motion cannot be decomposed into a Fourier series. Consequently, from the classical standpoint, the fundamental mechanical frequency ν_0 is the only one that can be radiated. According to the correspondence principle this circumstance implies that the theoretically possible Bohr frequencies can be radiated only when $n - p$ has the value unity, and hence only when the energy drops take place between consecutive levels. All other drops are forbidden. We conclude that the quantum treatment supplemented by the correspondence principle permits only the frequency ν_0 to be radiated by the harmonic oscillator.

The result we have just obtained is important. It shows that the quantum frequency radiated by the oscillator is always exactly the same as its mechanical frequency ν_0, and is therefore the frequency which should be radiated according to classical ideas. This situation is exceptional, for usually it is only in the case of high energy levels that the quantum and classical frequencies tend to coincide (*e.g.*, in the rotator). In a sense this peculiar feature of the harmonic oscillator is fortunate, because Planck, when he laid the foundations of the quantum theory, assumed that an oscillator present in the heated enclosure radiated light of the same frequency as the motion of the vibrating electron. Planck, at the time, knew nothing of Bohr's frequency condition and his assumption was based solely on classical considerations. We now see that in the general case Planck's identification of the quantum frequencies with the mechanical or classical ones would have been untenable, except under the limiting conditions of the correspondence principle. Fortunately for the development of the quantum theory, the linear oscillator is an exception, and so Planck's assumption, though founded on erroneous premises, was correct.

In the hydrogen atom the correspondence principle shows that all drops may occur. Consequently there is nothing to change in our previous presentation of the hydrogen atom. But we may note that if the electron in the hydrogen atom described only circular orbits, the situation would be the same as in the rotator, and the correspondence principle would exclude all drops between non-consecutive orbits. This restriction would render impossible the production of the majority of the spectral lines of hydrogen. The fact that many of the other mathematically possible spectral lines of hydrogen have been observed, shows that the electron's orbits cannot always be circles; elliptical orbits must also be assumed.

The various examples we have given demonstrate the importance of the correspondence principle in the application of Bohr's theory. Nevertheless this combination of the semi-classical correspondence principle with Bohr's revolutionary quantum theory is an extremely unsatisfying feature of the theory. A properly conceived quantum theory should furnish physical predictions by its own methods without having to resort to the help of discarded classical conceptions. Objections of this sort were among those which prompted the formulation of the newer quantum theories. In particular, the inability of Bohr's theory, as such, to calculate the intensities of the radiations was, according to to Born, one of the controlling motives which prompted Heisenberg and himself to develop the Matrix Method.

The Significance of the Correspondence Principle —The classical theory of radiation furnishes correct predictions whenever we are operating under the limiting conditions of high energy values and of small quantum drops. An illustration is afforded by the emission of radio waves by an oscillating electromagnetic system. Here the emitted frequency coincides with the frequency of the alternating current oscillating within the system, and so the classical theory is in no need of revision. Inasmuch as the limiting conditions occur in a large number of cases, even in many laboratory experiments, we may well understand why the classical theory was accepted for so many years.

Closely allied with the range of validity of the classical theory of radiation is the problem of mechanistic conceptions in general. We recall that, according to the classical theory, the frequencies of the radiations are those of the electron's mechanical motion, the intensities are proportional to the squares of the mechanical amplitudes, and the polarizations are determined by the shape of the path followed by the electron. Bohr's theory on the other hand, though it utilizes mechanical concepts as a scaffolding, subsequently rejects these concepts when it interprets the phenomenon of radiation. The frequencies radiated are determined by Bohr's frequency condition, and the various intensities of the different radiations are assumed to be proportional to the probabilities of the different quantum drops. Thus the merging of Bohr's theory with classical science, under the limiting conditions of the correspondence principle, implies that, as these conditions are approximated, the mechanical categories, frequency of motion, amplitude of motion, and direction of motion, become more and more adapted to the interpretation of radiation phenomena. Since, in the main, the earlier experiments occurred under the limiting conditions, mechanistic interpretations were able to account for the phenomena observed and so no particular objection to their introduction was voiced. Suspicions arose when the field theories were investigated, but it was only after the more remote subatomic realms had been explored that mechanical categories were recognized as inadequate for the majority of physical processes. In short, the discarding of mechanistic interpretations was due to the pressure of facts, not to a change in the psychology of the theoretical physicists.

It is misleading, therefore, to say that mechanism was a wrong departure; rather was it a necessary consequence of the too limited range of phenomena known to the earlier investigators. Indeed, as mentioned in the earlier part of this book, the meaning of the word mechanism needs revision, for today, since mechanics has been revised, there is no reason to associate mechanism with a form of mechanics which is recognized as a mere approximation.

When we recall that the limiting conditions, under which the classical mechanical theory of radiation gives correct conclusions, would always be realized in a universe where the constant h was infinitesimal, we see that the breakdown of mechanism, in the outworn sense, can be ascribed to the existence of Planck's constant h, finite and not infinitesimal. In particular, if we should imagine a progressive decrease in the value of h, the classical theory of radiation and its mechanical categories would express results with increasing correctness.*

Sommerfeld's Discovery of the Fine Structure of the Hydrogen Lines—Sommerfeld remarked that the electron in the hydrogen atom must be circling at tremendous speed and that under these conditions the variation of its mass with its velocity, as required by the theory of relativity, cannot be disregarded. The relativistic refinement does not affect the force exerted by the proton on the electron, for this force is electrostatic and is expressed by the same law of the inverse square in the theory of relativity as in classical theory. We are not dealing here with a gravitational force, and so Einstein's modification of Newton's law does not enter into consideration. But although the force acting on the electron will be the same as heretofore assumed, the variation in the electron's mass will modify the ordinary elliptical orbit.

---

* See note page 460.

Calculation shows that the orbit and the motion are the same as they would be if the electron's mass were to remain constant (as in classical mechanics) while the central force attracting the electron to the nucleus were increased by the addition of a small force varying as the cube of the distance. The exact numerical value of the change in the force is of no interest here. All we need note is that the relativistic orbit and motion may be deduced if we assume that classical mechanics holds but that a slightly different attractive force is acting on the electron. This equivalent way of stating the problem enables us to anticipate the form of the orbit. The fact is that the law of the inverse square is, among the laws of central forces, one of the very few which can yield closed curves (ellipses) for the circling body.* Any slight departure from this law, such as the one we have just been led to consider, will give rise to an open orbit differing slightly from an ellipse. Mathematical analysis shows that the orbit will be rosette-like and that it will not be a closed curve.† Its general aspect is shown in the figure. The orbit lies in a plane and is contained between two concentric circles. The nature of the motion is easily understood when we say that the electron revolves in an ellipse as before but that, while the first motion is taking place, the ellipse as a whole moves in its plane, precessing very slowly around the nucleus. The orbit and motion are of the kind illustrated by the planet Mercury: a progressive advance occurs in the position of the perihelion (or rather of the perinucleon in the case of an electron and a proton-nucleus).

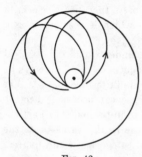

Fig. 42

Two periodicities are connected with the motion: the first is defined by the motion of the electron in the ellipse, and the second by the revolution of the ellipse as a whole. These two periodicities are independent. By "independent" we mean that when the one is given, the other is not determined thereby. The existence of two independent periodicities is also revealed graphically by the shape of the orbit: the orbit does not close and it covers an area.

According to classical (and also relativistic) mechanics, one of those stable mechanical magnitudes called adiabatic invariants is always asso-

---

* The elastic force (*i.e.*, the central force proportional to the distance) is the only other central force (depending solely on the distance) which yields closed orbits; these also are ellipses.

† The orbit may close accidentally.

ciated with each independent partial motion of a periodic nature. In Bohr's treatment of the hydrogen atom there was only one independent periodicity in the motion and consequently only one adiabatic invariant, which we called J. But in Sommerfeld's treatment, where there are two independent periodicities, there will be two adiabatic invariants J_1 and J_2. The first of these adiabatic invariants is associated with the periodic motion in the ellipse, and the second with the rotation of the ellipse as a whole; these adiabatic invariants measure the Maupertuisian actions of the two partial motions, respectively.

In classical (and also in relativistic) mechanics the adiabatic invariants may have any values within certain continuous ranges, and according to their values the motion has one aspect or another. The first adiabatic invariant J_1 is the one that we called J in Bohr's more elementary treatment; the greater its value, the greater the swing of the orbit. The mechanical significance of the second adiabatic invariant J_2 will be considered presently. Suffice it to say that J_2 controls the eccentricity of the revolving ellipse and also its rate of revolution, or of precession, around the nucleus. The greater the value of J_2, the smaller the eccentricity (*i.e.*, the less pronounced the flattening) of the revolving ellipse, and also the slower its speed of revolution. Furthermore, J_2 can never exceed J_1 in value. If $J_2 = J_1$ the orbit is circular. If J_2 vanishes, the electron will oscillate through the nucleus—a situation which may be discarded as physically impossible. Consequently, the value of J_2 will always be some positive number comprised between 0 and the value of J_1.

We next consider the energy. The energy of a periodic (or multiply periodic) system is determined by the values of all the independent adiabatic invariants. Thus in Bohr's treatment the energy was determined by the value of the single adiabatic invariant J, whereas in Sommerfeld's treatment, where there are two adiabatic invariants J_1 and J_2, the energy will be determined by the values of both J_1 and J_2.

What we have said up to this point is a mere consequence of the ordinary mechanical laws; the quantum theory has not been considered. But we know that a characteristic of Bohr's theory is to place quantum restrictions on mechanical motions which manifest periodicity. Since in Sommerfeld's treatment the motion of the electron has two independent periodicities, the application of Bohr's method requires that both these periodic motions be quantized. The quantization is most easily obtained through the medium of the two adiabatic invariants, J_1 and J_2.

We recall that in the absence of quantum restrictions the adiabatic invariants J_1 and J_2 could have any values within appropriate ranges. Bohr's quantum restrictions now require that only a discrete set of values within these ranges be possible. In particular the value of either adiabatic invariant must be a multiple of Planck's constant h: Our present quantum restrictions are thus expressed by

$$(11) \quad \begin{cases} J_1 = nh & (n = 1, 2, 3 \ldots .) \\ J_2 = kh & (k = 1, 2, 3, \ldots n), \end{cases}$$

where n and k are two positive integers. The numbers n and k are called the quantum numbers of the problem. It will be noted that the quantum number k cannot exceed the value of n, since J_2 cannot exceed J_1. Similarly k cannot take the value zero, since, for the reasons stated previously, J_2 cannot vanish.

Let us revert to the energy. We have said that it depends on J_1 and J_2. Now since J_1 and J_2 must be quantized according to (11), we conclude that the energy can have only a discrete set of values; these values define the energy levels. Instead of saying that the energy depends on J_1 and J_2, we may equivalently say that it depends on the values of the quantum numbers n and k, which define J_1 and J_2. The difference between the present situation and the one encountered in the ordinary treatment of the hydrogen atom is that now an energy level is no longer determined by a single quantum number n; two quantum numbers are required for its specification. For instance, if $n = 3$, then k may have the values 3, 2, or 1; so that three energy levels will be associated with the same value of n. The separations between the energy levels that correspond to the same value of n but to different values of k are relatively small; and we may regard the various levels which differ solely in the value of k as forming so many sublevels of the main level determined by n. For this reason the quantum number n is called the *main quantum number*. The net result is that when we follow Sommerfeld by introducing the theory of relativity into the study of Bohr's hydrogen atom, Bohr's original energy levels are usually split into sublevels. There is only one exception; it occurs for the lowest energy level, which remains single as before. Our analysis shows that these complications are due to the presence of two quantum numbers and hence to the existence of two independent periodicities in the mechanical motion.

Owing to the increase in the number of energy levels, a larger variety of drops will be possible and consequently a larger number of spectral lines may be expected. To determine the modifications that will arise in the spectrum of hydrogen, let us examine what will happen to the lines of the Balmer series. In Bohr's original scheme these lines were generated by the drops to the level $n = 2$ from all levels $n > 2$. But with our relativistic refinements, the Balmer lines arise from the drops between the following sublevels:

$$(n > 2; \; k = 1, \, 2, \, 3 \ldots . n) \longrightarrow (n = 2; \; k = 1 \text{ or } 2).$$

A larger number of different mathematically possible energy drops is thus suggested. For instance, if we consider a drop from the level $n = 3$ to the level $n = 2$, we see that in place of the single drop considered by Bohr there are now six different ones; these drops differ according to the particular sublevel, $k = 1$ or 2, to which the electron falls and also according to the sublevel whence the fall occurs.

The number of lines to be expected is, however, restricted, when the correspondence principle is applied. Thus, when we calculate the frequencies that should be radiated according to the classical theory and apply the rules outlined in connection with the correspondence principle, we find that, during a drop, the quantum number k must always change by ± 1. When account is taken of this restriction, the possible drops from given sublevels $n > 2$ to the sublevels ($n = 2$, $k = 1$ or 2) are reduced to the following three:

$$(n > 2; \, k = 1) \longrightarrow (n = 2; \, k = 2)$$

$$(n > 2; \, k = 2) \longrightarrow (n = 2; \, k = 1)$$

$$(n > 2; \, k = 3) \longrightarrow (n = 2; \, k = 2).$$

Let us consider one of these triplets of drops, say the triplet corresponding to drops from the level $n = 3$ to the level $n = 2$. Since sublevels $(n, \, k)$ which have the same value for n and different values for k differ but slightly in energy, the three drops we have considered will likewise differ but slightly. The three spectral lines generated by these drops will therefore be exceedingly close to one another, so much so that it may be difficult to distinguish them. The net result is that each Balmer line, believed by Bohr to be single, must be formed of three closely packed fine lines, or "components." This structure of the lines, predicted by Sommerfeld, is called the "fine structure" of the Balmer lines of hydrogen.

We may verify, by taking into account the restrictions in the permissible changes of the quantum number k, that the lines of the Lyman series will not be split, and that only the lines of the higher series (Balmer, Paschen, and Brackett) will exhibit the fine structure. Furthermore, the number of components to be expected increases with the order of the series considered: each Balmer line is formed of three fine lines; each Paschen line of five; each Brackett line of seven. The theory just developed applies to all atoms of the same type as hydrogen, *i.e.*, to the hydrogen-like atoms, such as once-ionized helium, twice-ionized lithium, and so on. But the fine structure should be easier to detect in the case of these higher atoms, because theory shows that the separation of the components increases with the nuclear charge and hence with the atomic number.

Sommerfeld's theory prompted spectroscopists to examine the hydrogen and hydrogen-like spectra more carefully. The first investigations that were then performed appeared to corroborate Sommerfeld's theory; and at the time, these corroborations were viewed as lending powerful support to Bohr's theory, to the theory of relativity, and to the validity of the correspondence principle. But subsequent measurements revealed discrepancies. We may understand their nature by reverting to the Balmer lines. Sommerfeld's theory, as we have said, requires that each line of the Balmer series should be formed of a number of closely packed lines. The correspondence principle then excludes all but three of the lines, with the result that each Balmer line should be a triplet. But more recent spectroscopic analysis has shown that five lines, and not three, are present. Inasmuch as the two additional lines were found to occupy the positions of two of the forbidden lines ruled out by the correspondence principle, doubt was at one time cast on the validity of the principle. Today the situation is interpreted in a totally different way. Sommerfeld's relativistic refinements and the correspondence principle are retained, but the situation is complicated by the introduction of an additional hypothesis—that of the spinning electron. We shall not consider the matter further at this stage; it will be resumed in the next chapter.

The Physical Significance of the Quantum Number k—We have seen that Sommerfeld's treatment of the hydrogen atom introduces a second adiabatic invariant J_2 and a second quantum number k. We now propose to show that simple mechanical interpretations can be given to these two magnitudes. The hydrogen atom constitutes an isolated mechanical system (as long as it is not radiating). Consequently, in accordance with the laws of mechanics, its total angular momentum remains constant in magnitude and in direction. Since we are viewing the nucleus as fixed and non-rotating, the total angular momentum of our hydrogen atom becomes that of the electron. The angular momentum (with respect to the nucleus) of the electron along its orbit is

$$\mu r_0 v_0,$$

where μ is the mass of the electron, r_0 is the distance from the nucleus to the perinucleon, and v_0 is the electron's velocity at the instant it passes the perinucleon. The value written above is, then, also the value of the angular momentum at any other instant.

Obviously, the magnitude of the angular momentum would be exactly the same if the motion of the electron were reversed or if the orbit were situated in some other plane differently inclined. The bare magnitude of the angular momentum gives us therefore no means of determining the situation. But we may obviate this ambiguity if we represent the angular momentum by a vector, *i.e.*, an arrow. The length attributed to the vector measures the magnitude of the angular momentum. The vector is then placed perpendicularly to the plane of the orbit which the electron is describing, with the result that the direction of the vector determines the orientation of this plane. Finally, the sense in which the electron is describing its orbit is indicated by making the vector point in one sense or the other in accordance with the following convention:

If we imagine that we are standing at the nucleus on the plane of the orbit and find that, on looking down, the electron is circling in a counterclockwise direction, we shall place the vector so that it points from our feet to our head. If the circulation is clockwise, the vector must be made to point in the opposite sense.

Thanks to this convention, the angular-momentum vector determines at one and the same time the magnitude of the angular momentum, the plane of the orbit, and the sense of circulation of the electron. The principle of mechanics which asserts that the angular momentum is conserved implies that the vector remains fixed during the motion. In particular, the permanency in the vector's direction shows that the orbit must lie in a fixed plane; and the permanency in the vector's magnitude entails the law of areas (which Kepler discovered by observing the planetary motions). Remembering these general considerations, let us revert to Sommerfeld's hydrogen atom.

We recall that Sommerfeld's adiabatic invariant J_2 was connected with the presence of a second periodicity in the electron's motion. This adiabatic invariant, like the first, measures a certain action pertaining to the motion. But we may also prove that it measures 2π times the constant angular momentum of the circling electron. If, then, we represent the magnitude of this angular momentum by K, Sommerfeld's second quantizing condition becomes

$$J_2 = 2\pi K = kh \qquad\qquad (k = 1, 2, 3 \ldots n),$$

whence

$$K = k\,\frac{h}{2\pi} \qquad\qquad (k = 1, 2, 3 \ldots n).$$

From this formula we see that Sommerfeld's second quantum condition
can be expressed by the statement that, in the stable orbits, the value of
the angular momentum must be some multiple k of the quantity $\dfrac{h}{2\pi}$.*

Because of the frequent occurrence of the magnitude $\dfrac{h}{2\pi}$ in Bohr's
theory, it is often taken to represent one "unit of angular momentum."
We may say, therefore, that the quantum number k represents the number of units of angular momentum which the electron may have in a
stable state of the atom.

The foregoing analysis reveals an important difference between
Sommerfeld's quantizing conditions and the simpler one used by Bohr.
In Bohr's non-relativistic treatment, there was no need to quantize the
angular momentum, and in Bohr's nth energy level any value of the
angular momentum, compatible with the energy of this level, was possible.
The introduction of relativity considerations by Sommerfeld has thus
restricted the possible orbits of the nth energy level by restricting the
possible values of the angular momentum.

Bohr's Magneton —In the hydrogen atom the negatively charged
electron is circling around the nucleus. Since this motion is exceedingly
rapid, its physical effects will be very similar to those developed by an
electric current of suitable intensity describing the same orbit. The laws
of electromagnetism show that such a current will develop a magnetic
field which is the same as would be developed by a very thin and flat
magnet limited by the orbit. Consequently, we may expect the atom to
manifest the properties of a tiny magnet.

A magnet is characterized by its *magnetic moment,* a concept which
we shall now investigate. Consider a bar magnet. In a simplified way, we
may represent it by two magnetic poles situated near the opposite ends of
the bar, the two poles being of opposite sign but equal in absolute value.
The magnetic moment M of the magnet is, then, defined by the numerical
value of the positive pole multiplied by the distance separating the two
poles (approximately the length of the magnet). This definition gives the
bare magnitude of the magnetic moment; but, as in the case of an angular

---

* As will appear later, it is not $k\dfrac{h}{2\pi}$ which measures the angular momentum of the
circling electron; it is $(k-1)\dfrac{h}{2\pi}$, or $l\dfrac{h}{2\pi}$ (setting $l = k - 1$). The significance of
this change cannot be understood at this point; it receives an interpretation only when
the methods of wave mechanics are applied.

momentum, we must also define its direction by a vector.* The convention usually adopted is that the vector points from the south to the north pole. For reasons of convenience which will appear presently, we shall make it point in the opposite direction, *i.e.*, from the north to the south pole.

We now return to the hydrogen atom. A calculation based on classical considerations shows that when the electron (or atom) has an angular momentum K, the magnetic moment M generated by the circling of the electron has a magnitude

$$(12) \qquad M = \frac{e}{2\mu c} K,$$

where e is the charge of the electron, μ its mass, and c the velocity of light.† Also, the vector of the magnetic moment is perpendicular to the plane of the orbit (like the vector of the angular momentum) and points in the same direction as the angular-momentum vector. Both vectors will be assumed to pass through the centre of the atom, *i.e.*, through the proton-nucleus. The line and the direction defined by the angular-momentum vector is called the *atomic axis*, because our present angular-momentum vector, which is that of the circling electron, is also the angular momentum of the entire atom. We may therefore say that the magnetic-moment and the angular-momentum vectors both point along the atomic axis.

If we take into account Sommerfeld's relativistic refinement, the value of the angular momentum K is always k times the unit of angular momentum $\frac{h}{2\pi}$ (where $k = 1, 2, \ldots n$). Substituting this value for K in formula (12), we obtain

$$(13) \qquad M = \frac{e}{2\mu c} k \frac{h}{2\pi} = k \frac{eh}{4\pi\mu c} \qquad (k = 1, 2, 3 \ldots n).$$

In the lowest orbit we have $n = 1$, and hence k has the value 1; the magnetic moment is then

$$(14) \qquad M = \frac{eh}{4\pi\mu c}.$$

This magnitude is called *Bohr's magneton*. It defines, in a certain sense, a unit of magnetic moment, and, as will be noted, it involves Planck's

---

* When we are concerned solely with the magnitude of the magnetic moment and not with its orientation, we express the magnetic moment by M. When the direction is taken into consideration, we write $\overrightarrow{M}$. This rule will apply to all vectors.

† The electric charge e is measured in electrostatic units, and the magnetic moment M in electromagnetic units.

constant h. The formula (13) shows quite generally that if the electron
in the hydrogen atom has k units of angular momentum $\dfrac{h}{2\pi}$, the magnetic
moment of the atom in this state will be k magnetons. Such at least are
the conclusions to which we are led in the present state of the theory.*

Thus, in Bohr's theory, magnetism is interpreted as a macroscopic
manifestation of hidden electric occurrences. This view is not new; it
was first suggested by Ampère, and was assumed explicitly by Lorentz
in his theory of electrons and by Langevin in his theory of magnetism.
The novelty of Bohr's treatment is that, in his formula for the magneton,
Planck's constant h appears. Incidentally the great importance of this
constant once again asserts itself.

Space-Quantization—When a top which is not spinning is set with
its point against the ground, it falls on its side owing to the pull of
gravity. But if the top is spinning rapidly, it does not fall; instead, its
axis precesses with uniform motion and describes a cone about the direction of the force, *i.e.*, about the vertical. We have here an illustration
of a gyroscopic motion in which the force of gravity seems to be defeated.
A similar situation arises if, in place of a top in a gravitational field, we
consider a bar magnet suspended at its centre of gravity in a magnetic
field. Let us suppose that we have a vertical magnetic field $\vec{H}$ directed
upwards. In such a field the positive, or north, pole of a magnet will be
attracted upwards in the direction of the field, while the negative, or
south, pole will be attracted in the opposite direction. If, therefore, a
bar magnet is suspended at its centre of gravity and is abandoned at
some slanting inclination with respect to the field, it will start to oscillate
back-and-forth indefinitely about the direction of the field. Of course,
if there is friction, the oscillations will soon cease, and the magnet will
point its north pole in the direction of the field. This is the familiar
phenomenon illustrated in the magnetic compass.

The potential energy of the magnet, due to the action of the field, is
a minimum when the north pole points in the direction of the field. And
since we are defining the direction of the magnetic-moment vector of the

---

* Later we shall see that the correct value of the electron's angular momentum
is not $k\dfrac{h}{2\pi}$, but $(k-1)\dfrac{h}{2\pi}$. According to this change the magnetic moment should
be $(k-1)$ magnetons. As a result, in the unexcited state, where $k=1$, the hydrogen
atom should have no magnetic moment; but this expectation is not borne out by
experiment. The anomaly is accounted for by the hypothesis of the spinning electron,
a hypothesis which will be discussed in the next chapter.

magnet as the direction from the north to the south pole, we conclude that the potential energy is a minimum when the magnetic-moment points in the direction opposite to the field, and is a maximum when it points in the same direction as the field. The value of the potential energy is determined by the product of the intensity of the field and the projection of the magnetic moment on the direction of the field. Thus, if θ is the angle the magnetic-moment vector $\vec{M}$ makes with the vector defining the direction of the field $\vec{H}$, the potential energy is

(15) $MH\cos\theta.$

As occurs in all definitions of the potential energy, the position of zero potential energy is determined by convention. In the foregoing definition the position of zero potential energy is attained when $\cos\theta = 0$, *i.e.*, $\theta = \dfrac{\pi}{2}$, and hence when the magnet is set perpendicularly to the field.

We now suppose that the bar magnet is spinning about its axis. An angular momentum is generated about the axis, and a new phenomenon occurs. If we assume that the axis of the spinning magnet is released from a position of rest, we find that the axis precesses with uniform speed around the direction of the field, while the initial angle between the axis and the field remains constant. We have here the exact analogue of the spinning top in a gravitational field.

Instead of considering a bar magnet we may consider a hydrogen atom. We have seen that a hydrogen atom is analogous to a tiny magnet *in rapid rotation*; for the circling of the electron generates at one and the same time the angular momentum and the magnetic moment. Both vectors are perpendicular to the plane of the orbit and point in the same direction—the direction of the atomic axis. A precessional motion of the atomic axis around the direction of the field will therefore take place, and the axis will retain a constant inclination θ with respect to the field. Larmor was the first to calculate the frequency of the precessional motion (number of precessions or revolutions, per second). This Larmor frequency, as it is called, is represented by ω_L, and its value is

(16) $$\omega_L = \frac{eH}{4\pi\mu c}.$$

Here e is the charge of the electron (in electrostatic units), μ its mass, and H the intensity of the magnetic field in electromagnetic units. The Larmor frequency increases with the intensity of the field. The direction of the Larmor precession about the magnetic field $\vec{H}$ is such that, if we look along the lines of force, the axis of the atom will appear to be precessing in a clockwise direction.

The resultant motion of the electron in the hydrogen atom under the action of the magnetic field is now clear: the electron describes an ellipse (or rather the more complicated rosette curve) in the plane perpendicular to the atom's axis, and, while this first motion is proceeding, the plane precesses with a constant and slow uniform speed about the direction of the field (the inclination of the plane relative to the direction of the field remaining unchanged). It is this latter slow precessional motion which represents the Larmor precession. Thus, in addition to the two periodicities of the electron's motion in its plane (as in the relativistic atom in the absence of a field), we now have a third periodic motion superposed on the other two (due to the precessional motion of the orbital plane, or of the atomic axis). These three periodicities are independent in the sense explained previously.

The existence of three independent periodicities in the motion shows that, in addition to the two adiabatic invariants J_1 and J_2 noted in Sommerfeld's atom (no field), there will be a third adiabatic invariant J_3 independent of the former two. Furthermore, since according to mechanics the total energy of the atom involves all the independent adiabatic invariants and is determined when their values are specified, we conclude that the total energy E of the hydrogen atom in a magnetic field will be a function of J_1, J_2, and J_3.

We explained in connection with Sommerfeld's atom that the adiabatic invariant J_2 is equal to 2π times the angular momentum K of the revolving electron. This adiabatic invariant retains the same significance when the magnetic field is applied, and its numerical value is not affected by a slow application of the field. As for the third adiabatic invariant J_3, required by the new periodicity in the motion, its value is 2π times the projection of the electron's angular momentum onto the direction of the field. Calling K the magnitude of the angular momentum and therefore $K\cos\theta$ its projection on the direction of the field, we may set

(17) $$J_3 = 2\pi K \cos\theta = J_2 \cos\theta.$$

The following argument, which is sufficiently rigorous for our present purpose, will enable us to verify that the total energy of the atom in the magnetic field is determined by the values of the three adiabatic invariants J_1, J_2, J_3. Let us suppose that at the initial instant there is no magnetic field; the axis of the hydrogen atom may then be pointing in any direction. If we adopt the relativistic treatment for the atom, we know that the energy will be defined in terms of the two adiabatic invariants J_1 and J_2. We call $E_0(J_1, J_2)$ the value of this energy. We now suppose that a magnetic field H is applied. Since we are viewing our atom as equivalent to a little magnet of moment M, the presence of the field will contribute the potential energy $MH\cos\theta$ to the atom (see (15)), where θ is the angle between the direction of the field and that of the magnetic moment, or atomic axis. The total energy E of the atom in the field will thus be the sum of the energy in the absence of a field and the added potential energy. Thus

$$(18) \qquad E = E_0(J_1, J_2) + MH\cos\theta.$$

If in this formula we replace the magnetic moment M by its value (12), we obtain

$$(19) \qquad E = E_0(J_1, J_2) + \frac{eKH\cos\theta}{2\mu c}.$$

In view of (17), the formula (19) may also be written

$$(20) \qquad E = E_0(J_1, J_2) + \frac{eH}{4\pi\mu c}J_3.$$

This formula shows that, when the three adiabatic invariants J_1, J_2, and J_3 are specified, the total energy of the hydrogen atom in a given magnetic field of intensity H is determined. Thus far we have treated the problem classically.

We wish now to examine the restrictions which will be imposed on the motion by the quantum theory. In the classical treatment the adiabatic invariants, and J_3 in particular, could have all values compatible with the system. The same was therefore true for the ratio $\frac{J_3}{J_2}$ and hence for $\cos\theta$. This fact implied that all inclinations of the atomic axis with respect to the direction of the field were permissible. But if we introduce quantum restrictions, all our adiabatic invariants must be quantized; and in addition to the restrictions (11), *i.e.*,

$$(21) \qquad \begin{aligned} J_1 &= nh & (n = 1, 2, 3 \ldots) \\ J_2 &= kh & (k = 1, 2, 3 \ldots n), \end{aligned}$$

we must set the third quantum restriction

$$(22) \qquad J_3 = mh,$$

where m is a new quantum number which must be some integer. Inasmuch as $\frac{J_3}{J_2} = \cos\theta$ (see 17), we must have

$$(23) \qquad \frac{m}{k} = \cos\theta.$$

The only values a cosine can assume are contained between -1 and $+1$. Consequently, for a given value of the positive integer k, the new quantum number m must assume one of the $2k + 1$ integral values

$$(24) \quad -k, -k+1, -k+2, \ldots -1, 0, 1, \ldots k-1, k.$$

Reverting to (23) we see that only a limited number of orientations θ are possible for the atomic axis in a magnetic field; all other orientations are impossible according to the quantum theory. This limited number of possible orientations for the atomic axis illustrates the quantum manifestation called *space-quantization* by Sommerfeld.

An important feature of space-quantization is that it can arise only when there is a privileged direction in space, which has a physical significance and with respect to which the inclination of the atomic axis can be measured. In the problem we have discussed, the magnetic field furnishes this direction. In free space, where there is no privileged direction, space-quantization does not come into consideration and so the atomic axis may point in any direction.

The Normal Zeeman Effect—Lorentz, on the basis of his electronic theory, predicted that if a radiating atom were placed in a magnetic field, each spectral line which the atom emitted in the absence of the field would be split into three components. He also predicted that if we observed one of these triplets from a direction at right angles to the field, the radiations corresponding to the three lines of the triplet would appear plane polarized in well-defined directions; and that if the observation were made in the direction of the field, only the two outer lines of the triplet would be seen, the corresponding radiations being circularly polarized in opposite senses.

This phenomenon, predicted by Lorentz, is named after Zeeman, who established it experimentally. It is called the *normal Zeeman effect*. According to Lorentz's theory, it should manifest itself in all atoms and not only in the hydrogen atom. But experimenters soon found that with the higher atoms a more complicated splitting of the spectral lines usually occurred. The name *anomalous Zeeman effect* was given to this more complicated phenomenon.* The normal Zeeman effect thus appears to be the exception; and in the light of these discoveries, the

---

* Recent investigation has shown that even hydrogen manifests the anomalous, and not the normal, effect, as had formerly been assumed. The past misconceptions on this score have been traced to the too high intensity of the magnetic field used in the experiments; a strong magnetic field generates a secondary effect called the Paschen-Back effect, which has the same appearance as the normal Zeeman effect.

terminology should logically be reversed: the normal effect should be called anomalous, and vice versa. However, the original terminology is always retained, and so we shall adhere to it.

Lorentz's theory of the normal Zeeman effect was developed subsequently by Larmor. Larmor, before the advent of the quantum theory, showed that the classical theory accounts for the normal Zeeman effect. His treatment is easily understood when we first assume that there is no magnetic field. The electron in the hydrogen atom is then describing some ellipse situated in a fixed plane. The motion of the electron may be decomposed by the Fourier method into a number of superposed harmonic vibrations. Let ν_0 be the frequency of any one of these. According to the classical treatment, where the mechanical and the radiated frequencies are the same, the mechanical frequency ν_0 of the electron should also be the frequency of one of the radiations emitted by the atom in the absence of a field. We know of course that the classical theory is unable to give the correct frequencies. But this fact is unimportant from our present standpoint, because, in the Zeeman effect, our concern is not with the exact frequencies of the hydrogen lines, but only with the manner in which spectral lines in general are split under the action of a magnetic field.

Let us, then, suppose that a magnetic field is applied. The motion of the electron is now complicated by the precession of the atomic axis, with the Larmor frequency ω_L (see (16)), about the direction of the field. As a result of this precession two additional mechanical frequencies, $\nu_0 \pm \omega_L$,* will be superposed on the original frequency ν_0 of the electron's motion. Hence, the frequency ν_0 will be radiated as before, but there will also be two other radiated frequencies of magnitudes $\nu_0 \pm \omega_L$. In place of a single spectral line we shall have three: the original line and two others equally spaced to the right and to the left. When account is taken of the numerical value of the Larmor frequency ω_L (which measures the separation between the two displaced lines and the central one), the exact Zeeman split disclosed by experiment is accounted for. The polarizations may also be determined classically.

In view of the known deficiencies of the classical conceptions in so many cases, we cannot accept the classical theory of the Zeeman effect, and we must now attempt to interpret it by means of Bohr's theory and its quantum postulates. We have seen that, when a hydrogen atom

---

* The motion of the electron may be decomposed into a harmonic vibration of frequency ν_0 along the direction of the field, and into two additional vibrations $\nu_0 \pm \omega_L$, which are represented by circular motions in opposite directions occurring in a plane perpendicular to the direction of the field.

situated in an energy level (n, k) is placed in a magnetic field of intensity H, its total energy is increased by the presence of the field. If we call $E_0(n, k)$ the energy of the atom in the absence of a field, the formula (20), in which J_3 is replaced by its permissible values mh, becomes

$$(25) \qquad E = E_0(n, k) + \frac{eHh}{4\pi\mu c}m,$$

where the values of the quantum number m are restricted to integers extending from $-k$ to $+k$. Each sublevel (n, k), which exists in the absence of a field, is thus subdivided into $2k + 1$ sub-sublevels (n, k, m), which we define by giving to the quantum number m all permissible values (*i.e.*, all integral values between $-k$ and $+k$). This splitting of the former sublevels into more numerous sub-sublevels entails a correspondingly increased variety of quantum drops, and a larger number of spectral lines may be expected. But the application of the semi-classical correspondence principle, which throws us back on the electron's motion, shows that the possible drops are considerably restricted by the following requirements:

During a quantum change, either the quantum number m must not change at all (n or k will then change), or else it must change by $+1$ or by -1.

Under these conditions it is easy to see that each drop which could have occurred in the absence of the field will be supplemented, thanks to the presence of the field, by two other possible drops differing slightly from it. As an example, consider the spectral line which arises in the absence of a field when the drop from $(n = 4, k = 3)$ to $(n = 2, k = 2)$ takes place. If we operate in a magnetic field, m may also suffer a change by passing, let us suppose, from m_1 to m_2. If we apply Bohr's frequency condition and utilize the value (25) of the energy, we see that the frequency of the spectral line that is now obtained differs from the former one by

$$(26) \qquad \frac{eH}{4\pi\mu c} \frac{hm_1 - hm_2}{h}, \quad i.e., \quad \frac{eH}{4\pi\mu c}(m_1 - m_2).$$

And since $m_1 - m_2$ must be 0 or $+1$ or -1, the new spectral line will have the same frequency as the original line or else its frequency will exceed the original frequency or fall below it by the amount

$$(27) \qquad \frac{eH}{4\pi\mu c} \; i.e., \text{ the Larmor frequency } \omega_L.$$

We thus obtain three possible lines, and one of the three new lines coincides with the original one. Incidentally we may note that (27) also measures the Larmor frequency (16), so that our present results coincide with those obtained in the classical treatment of the Zeeman effect.

In practice we observe a large number of atoms simultaneously, and since, for a given drop, the quantum number m can behave in three permissible ways, we may assume that the three different situations will occur for the various atoms. For this reason the three lines of the triplet will be observed simultaneously. Finally, the polarizations of the rays are obtained from the classical treatment by an application of the correspondence principle, and the results are in agreement with observation. Although Bohr's interpretation of the normal Zeeman effect cannot be regarded as a distinct triumph for his theory since the effect is also accounted for by classical methods, it is satisfying to know that the normal effect does not conflict with the theory.*

A point that may seem strange is that the classical theory should have been able to account for the normal Zeeman effect which after all, is a quantum phenomenon. However, this point is clarified when we examine the expression (27) for the shifts in the frequency. In this expression Planck's constant h does not appear at all; it is cancelled when (26) is derived from (25). Had the constant h been present in our formula, the classical theory, which ignores h, would have been helpless. The success of the classical theory in the present case is thus accidental.

The Stark Effect—The Stark effect consists in a more or less complicated splitting of each of the spectral lines when an electric field is applied. The interpretation of this effect by Bohr's theory is a marked success for the theory, because classical methods proved ineffective. We shall not discuss the Stark effect further. Suffice it to say that, under the influence of the electric field, an additional periodicity appears in the electron's motion, and that the number of energy levels and attendant spectral lines is thereby increased.

The Various Kinds of Degeneracy—In Bohr's theory the only motions that are subjected to quantum restrictions are those that mani-

---

* We mentioned in a previous note that recent experiment has shown that hydrogen exhibits the anomalous effect. It would thus appear that Bohr's theory, which predicts the normal effect, must be incorrect. In point of fact, however, it is not only Bohr's theory in its present form which is incorrect; our model of the hydrogen atom also needs revision. In the next chapter we shall see that the circling electron is in a state of spin and that complications arise therefrom.

fest periodicities. In celestial mechanics such motions are called "conditioned periodic" or "multiply periodic." They are not usually periodic, but they include periodic motions as a special case. We shall here be concerned only with very particular illustrations, and to simplify the exposition we shall confine our attention to the motion of a single particle.

A particle (point-mass) moving in space has three degrees of freedom, for three coordinates are required to define its position in a given frame; and the motion of the particle in space is determined when we specify how the three coordinates vary with time. We shall suppose that the three coordinates are Cartesian coordinates x, y, z. Let us project the motion of the particle onto the three coordinate axes Ox, Oy, Oz. We obtain three component motions along these axes. The motion of the particle is called conditioned periodic when each one of the three component motions is periodic. Let us call ν_1, ν_2, and ν_3 the respective frequencies of the three component periodic motions. Here several situations may arise.

Firstly, it may happen that no one of the ratios of the three frequencies, taken two by two, is a rational number, *i.e.*, a common fraction (or an integer). The three periodicities of the component motions are then said to be independent. Despite the periodic nature of the component motions, the resultant motion is not periodic in this case. The particle describes a curve which never closes and which fills a volume. The conditioned-periodic motion is said to be *non-degenerate*.

Secondly, one among the ratios of the three frequencies may be a rational number, though the other ratios are irrational. Only two among the three periodicities are now independent. The resultant motion in space is still aperiodic. The particle describes a curve which never closes but which, this time, covers an area instead of filling a volume. The conditioned-periodic system is said to be *once-degenerate*.

Finally, if all the ratios of the three frequencies are rational numbers, there is only one independent periodicity. The motion in space is itself periodic, and under these conditions the conditioned-periodic motion becomes periodic. The path in space is now a closed curve, and the motion is said to be *totally degenerate*.

In the study of the hydrogen atom, we have encountered all three kinds of motion. In Bohr's treatment of the atom, the path of the electron is an ellipse which is described with a periodic motion. There is but one independent periodicity and the motion is completely degenerate. In Sommerfeld's relativistic treatment, the motion has two independent periodicities, represented by the periodic motion along the ellipse and the precession of the ellipse as a whole around the nucleus. The motion is thus once-degenerate, and the path followed by the electron covers an area. Finally, when a magnetic field is applied, a third independent periodicity appears; it is exhibited by a precession of the orbital plane round the direction of the field. The motion is now non-degenerate, and the path described by the electron fills a volume.

Calculation shows that when a motion which is degenerate is submitted to a perturbing force, its degeneracy is decreased. For example, we have seen that the relativistic refinements, when superimposed on Bohr's treatment, are equivalent to a small modification in the Newtonian force of attraction exerted by the proton-nucleus on the circling electron. The relativistic refinements thus impose a small perturbation on the original motion, and, for this reason, the motion formerly totally degenerate becomes once-degenerate. A second type of perturbation is introduced when a magnetic field is applied (Zeeman effect); the motion has its degeneracy reduced still further and becomes non-degenerate. In the Stark effect, a perturbation is created by means of an electric field, and the degeneracy is likewise completely removed.

We wish now to examine the bearing of this general discussion on the quantum theory. We recall that with each independent periodicity of the motion is associated one adiabatic invariant, and that the total energy of the system depends on the numerical values of all the adiabatic invariants present. In the quantum theory the adiabatic invariants may assume only discrete sets of values, i.e., they are quantized, and the allotment of the quantized values introduces the corresponding quantum numbers. There are thus as many quantum numbers as there are adiabatic invariants or, what amounts to the same, as there are independent periodicities. Since the various energy levels are determined by the different sets of integral values that may be given to the quantum numbers, there will be a larger variety of distinct energy levels, and consequently of spectral lines, when the supply of quantum numbers increases. Collecting these results, we conclude that as the order of the degeneracy is decreased (and therefore the number of adiabatic invariants and of quantum numbers is increased), the spectrum will betray a richer variety of spectral lines. For this reason, a perturbation imposed on an atom increases the complexity of its spectrum.

These conclusions are readily verified in the hydrogen atom. Thus, when the degeneracy of Bohr's original model is decreased, first by the relativistic refinements and then by the application of a magnetic field, the spectrum predicted by the theory progressively increases in complexity.

An alternative method of quantizing the orbits is sometimes advantageous because it leads to a new interpretation of degeneracy which proves valuable in Schrödinger's wave theory of the atom. Consider Bohr's hydrogen atom when the relativistic refinements are disregarded and no magnetic field is applied. We shall refer to this atom as the "undisturbed atom." We saw that in the undisturbed atom the major axes of the elliptical orbits were quantized by the introduction of the quantum number n, but that the eccentricities and orientations of the orbits were left arbitrary. Thus the energy states E_n of the atom were determined by the single quantum number n. When the relativistic refinements were introduced, the eccentricities of the orbits were quantized by means of the quantum number k, though the orientations of the orbits still remained arbitrary. The quantizing of the eccentricities entailed the splitting of the energy levels E_n into sublevels $E_{n,k}$. Finally, when a magnetic field was applied, the orientations of the orbital planes were submitted to space-quantization, and the third quantum number m was introduced in consequence. The sublevels $E_{n,k}$ of the relativistic atom were thus split into sub-sublevels $E_{n,k,m}$.

In the alternative method of quantization, the orbits must always be submitted to the three quantizations, n, k, m (or to equivalent ones in the case of k and m). Consequently, even in the undisturbed atom the three quantum numbers will appear. This implies that the energy states of the undisturbed atom will be represented not by levels E_n, but by sub-sublevels $E_{n,k,m}$. The new procedure does not entail any change in the spectra, for, in the undisturbed atom, all those sub-sublevels $E_{n,k,m}$, which are associated with the same value of n, will have the same energy and hence will coalesce into the single level E_n. Similarly in the relativistic atom, those sub-sublevels, which differ only in the value of m, coalesce into the corresponding sublevels $E_{n,k}$ of Sommerfeld.

An obvious objection to our present retention of seemingly irrelevant quantum numbers is apparent in connection with the quantum number m. This number determines the inclinations of the various orbital planes with respect to the direction of the magnetic field. But then m must become meaningless when no magnetic field is applied. How, then, can it affect the quantization of the undisturbed atom? This difficulty is mitigated to a certain extent when we observe that $2\pi m$ is an adiabatic invariant in the presence of a magnetic field, so that when the field is progressively reduced to the vanishing point, m still retains its original value and the orbit its original inclination. At all events we shall accept the new method of quantization for the purpose of discussion.

According to our present views the hydrogen atom is degenerate whenever more than one sub-sublevel is associated with each energy state. Degeneracy thus consists in a coalescing of some or all of the sub-sublevels. In the undisturbed atom all the sub-sublevels $E_{n,k,m}$, pertaining to the same value of n, coalesce, and hence the degeneracy is total. The relativistic refinements, which are equivalent to a perturbation, remove this degeneracy in part, for in the relativistic atom coalescence endures only for those sub-sublevels which are associated with the same values of both n and k. Finally, the perturbation imposed by a magnetic field removes the degeneracy entirely: all the sub-sublevels now become distinct. From this account we see that a perturbation does not create sub-sublevels; it merely separates some or all of those which already existed but which happened to coalesce.

The process whereby a perturbation removes the degeneracy is readily understood. The perturbation modifies the energies of all the orbits, but not necessarily to the same extent. If, then, a perturbation is applied to a degenerate atom, two orbits that had the same energy may assume different energies, with the result that two sub-sublevels that coalesced may now be separated and become distinct. If all the sub-sublevels are thus separated, the degeneracy is removed entirely and a further perturbation cannot cause any additional separation.

The considerations we have just developed in connection with the hydrogen atom are general. In all cases, whether we be dealing with sub-sublevels or with levels of a still lower order, degeneracy implies coalescence, and a perturbation, by destroying the coalescence, removes the degeneracy.

For the present, however, Bohr's simpler method of quantization will suffice for our purpose. Accordingly, we shall mention the existence of the sub-sublevels (n, k, m) of the hydrogen atom only when they are distinct.

The Higher Atoms.—We explained in the last chapter that, according to the views of Rutherford and of Moseley, an atom of atomic number N is formed of a nucleus of positive electric charge N around which N planetary electrons are circling. When all the electrons except one are removed from a higher atom, we obtain a so-called hydrogen-like atom, this name being given on account of the similarity of an atom of this sort with the atom of hydrogen. Examples of hydrogen-like atoms are afforded by the once-ionized helium atom and the twice-ionized lithium atom. The hydrogen atom and the hydrogen-like atoms are the simplest of the atoms, and it is with these that Bohr's theory registered its greatest success. Considerable progress was also made by Bohr in unravelling the complexities of other atoms. But owing to the increase in the number of planetary electrons, the difficulties were very much greater. If we attempted to operate as we did in the case of the hydrogen atom, we should first have to determine the motions of the electrons in the absence of restrictive quantum conditions, and then establish the nature of these quantum restrictions. But such attempts would be futile, for even the first step in our program would be impossible since it would throw us back on the unsolved problem of n-Bodies in celestial mechanics.

In view of the difficulties besetting a direct approach, indirect methods were utilized by Bohr. Every possible clue was scanned. The spectroscopic evidence, the chemical properties, the color of salts, ionization potentials, X-ray spectra, magnetic properties, and many others played a part in determining the models of the various atoms and the distributions of their planetary electrons. The progress was piecemeal, and in many cases results originally believed correct were subsequently found to need revision in the light of new discoveries. If we were to state the final results of Bohr's theory in its completed form without mentioning the intermediary steps, we should convey a totally wrong impression of the methods of theoretical physics. Besides, the final conclusions would appear extravagant and gratuitous. We shall therefore examine the successive models suggested for the atoms and explain, as we proceed, in what respects and for what reasons they were revised. In the present chapter the first of the models devised by Bohr is the only one we shall discuss. In the next chapter the completed model will be investigated.

Bohr's treatment of the higher atoms consists in imposing quantum restrictions on the motions of the planetary electrons. For the present we shall defer consideration of the manner in which these restrictions are formulated. Suffice it to say that the N electrons of an atom of atomic number N are assumed to be circling along prescribed orbits which are determined by means of quantum numbers. Let us consider two electrons describing two different orbits in a given atom. The total energy of the first electron may be less than that of the second. In this event the first orbit is said to be of lower energy than the second. The lower the energy of the orbit on which an electron is circling, the greater the amount of work we must expend to tear the electron from the atom. Orbits of lower energy pass within shorter distances of the nucleus and are usually enveloped by the orbits of higher energy. We may therefore refer to the orbits of lower energy as the inner orbits, and to those of higher energy as the outer orbits. For similar reasons the electrons describing the respective orbits are often called inner and outer electrons. An orbit can contain only one electron, or else none. When an atom is unexcited, its electrons are moving in the lowest orbits available, so that the innermost orbits will be the first ones filled.

Next, let us consider the problem of radiation. In the hydrogen atom, where there is only one electron, radiation occurs when this electron drops from a higher to a lower level. But in the higher atoms there are always two or more electrons, and the problem of radiation emission must be investigated further. Bohr assumes that, in the general case, when a higher atom is excited and is thus in a position to radiate, only one of the electrons is removed from its normal orbit and is raised to an unoccupied orbit of higher energy. The subsequent drops of this electron to the lower unoccupied orbits then give rise to radiation. The electron which is raised during the process of excitation is the one which, in the unexcited state, is describing the outermost of the occupied orbits, *i.e.*, the orbit of highest energy. This electron is called the *optical electron*. Being less firmly bound to the atom than are any of the others, it can more easily be removed to a higher level. As for the inner electrons, they circulate in their orbits, and since all lower orbits are occupied, these inner electrons cannot drop and so cannot radiate. When there are two or more outer electrons, one alone of these outer electrons is usually assumed to play the part of the optical electron. In some cases, however, it would seem that two electrons may act as optical electrons. In other words, two outer electrons may be displaced to higher levels and then fall back simultaneously. But we must not suppose that two different radiations are emitted thereby. Both drops add up, as it were, producing the same effect as a single drop from a higher energy level. The evidence in favor of this assumption is that, in some cases, the frequency radiated is greater than it would be if a single electron were to drop from infinity; and since the drop from infinity represents the highest energy drop possible when one electron alone is involved, we must suppose that the energies of the two falling electrons have been added together.

The major problem confronting Bohr was to decide in which precise orbits the electrons of a specific atom would be circling. Here, he was guided by a number of clues. Some of these will be mentioned in the remaining pages of this chapter.

Ionized Atoms—An atom is said to be once-ionized when one of its planetary electrons is removed completely; a once-ionized atom has therefore a net positive charge of one unit. An atom which is not

ionized will be called a *neutral* atom. In a once-ionized atom the missing
electron is always the electron that is least firmly bound, *i.e.*, the optical
electron. If, from a neutral atom, we remove an inner electron instead
of the optical one, an outer electron will drop immediately into the
vacated inner orbit; and the final result will be the same as if we had
removed the optical electron. Thus, the same once-ionized atom is obtained regardless of the particular electron that is withdrawn from
the atom. A definite amount of work, depending on the kind of atom
considered, must be expended to withdraw the optical electron from a
neutral atom. This work can be measured and may be expressed by a
difference in electric potential, called the *ionization potential*. The once-
ionized atom, like the neutral atom, becomes excited when its outermost
electron is displaced to a higher orbit; the drops of this electron then
generate radiation. This electron is thus the optical electron of the
once-ionized atom.

Having obtained a once-ionized atom, we may ionize it afresh by
removing its optical electron. We thus obtain a twice-ionized atom, the
net electric charge of which is +2. The second ionization will be more
difficult to perform because the new optical electron which must be
removed, though attracted by the same nucleus as the optical electron
in the neutral atom, is repelled by a smaller number of planetary electrons (on account of the first ionization). Still higher ionizations may
be considered; and for the reasons just stated, each successive ionization
will require the expenditure of a greater amount of work.

The spectrum of a neutral atom differs entirely from that of the
corresponding once-ionized atom. An illustration of this diversity in
the spectra was mentioned in connection with the neutral and the once-
ionized helium atoms. The spectra of the variously ionized atoms of
the same element also differ entirely from one another. Since the higher
the degree of ionization of a given atom, the more firmly bound is its
optical electron, the line of highest frequency (which is emitted when
the optical electron falls to the lowest available orbit from infinity) will
increase in frequency with the degree of the ionization. Consequently,
the spectral lines of the successively ionized atoms of the same element
will extend more and more towards the X-ray frequencies. High temperature, entailing as it does violent collisions between molecules and
atoms, facilitates ionization; and this explains why the spectrum of
an atom excited in the electric spark (spark-spectrum) differs from the
spectrum of the same atom excited in the electric arc (arc-spectrum).
In the former case we observe the spectrum of the ionized atom, whereas
in the latter case the spectrum of the neutral atom is obtained. Low

pressure also favors ionization (or at least prolongs the duration of the ionized state) ; and the cumulative influence of high temperature and low pressure in certain stars is responsible for the highly ionized states of some of their atoms.

An important clue to the distributions of the electrons in the atoms is afforded by a comparison of the spectrum of an ionized atom and of the neutral atom which immediately precedes it in Moseley's classification by atomic numbers. The once-ionized atom of atomic number N has a spectrum similar to that of the neutral atom of atomic number $N - 1$. Also, the twice-ionized atom of atomic number N and the neutral atom of atomic number $N - 2$ have analogous spectra. For example, we may verify that the once- and the twice-ionized sodium atoms (atomic number 11) have spectra that are analogous, respectively, to those of neutral neon (atomic number 10) and neutral fluorine (atomic number 9). The rule appears to be general. It thus seems permissible to suppose that when an atom is once-ionized, the arrangement of its remaining planetary electrons is the same as that of the neutral atom immediately preceding it. We may also suppose that when the electron is restored to the ionized atom, the underlying electronic orbits will not be affected by its return. This view leads to the assumption that the electronic arrangements in two successive atoms differ merely by the fact that the atom of higher atomic number has one additional electron situated in the orbit of lowest energy available.

A second important clue to the arrangements of the planetary electrons is found in the chemical properties. Among the elements are certain gases called the *rare gases*. These are helium (atomic number 2), neon (atomic number 10), argon (18), krypton (36), xenon (54), radon (86). With the exception of the last, the rare gases are present in the atmosphere. Argon is fairly plentiful, but the other rare gases are present only in minute concentrations. In this sense they are rare, and from this peculiarity their name was derived. The rare gases are chemically inert, and only with extreme difficulty can they be made to enter into combination with atoms of the same family as themselves or with totally different ones. For this reason the rare gases are often called *inert gases*. Indeed, whereas two atoms of hydrogen combine to form a molecule (the same is also true for oxygen and nitrogen), the atoms of a rare gas do not combine, and hence the rare gases are monatomic. Exceedingly powerful agents are required to ionize (tear an electron away from) an atom of an inert gas. This fact leads to the suspicion that the planetary electrons of an inert-gas atom must form a dynamical structure of great stability, which cannot easily be disturbed. It is plausible to associate stability with symmetry, and we are thus led to suppose that, in the atom of an inert gas, the outermost electrons are describing orbits of the same .type and therefore orbits of the same energy. The assumption that different orbits may have the same energy illustrates a situation which we did not have to countenance in the hydrogen atom, because in this atom only one electron was present. Orbits of the same energy are sometimes said to form a group of orbits. Accordingly we may say that the outermost electrons of an inert-gas atom fill the orbits of a group; this group of orbits thus contains its full quota of electrons—a kind of saturation is realized. The situation is often expressed by the statement that an inert gas exhibits a closed electronic configuration.

These assumptions appear to be supported by a large number of additional facts. For instance, argon is an inert gas. The atom immediately following it is potassium. If, as we have supposed, the additional electron of potassium revolves in an orbit situated outside the underlying closed configuration of argon, we may expect it to be more loosely bound than the remaining electrons, for it is further removed from the nucleus and does not form a part of a closed underlying group of electrons. It should therefore be relatively easy to effect a first ionization of the potassium atom by removing this loosely bound outer electron. Our expectation is verified. Furthermore, if Bohr's ideas are correct the once-ionized potassium atom will be limited by a stable group of electrons, and it should be as difficult to ionize it once again as it is to ionize the inert gas argon. This expectation is also verified.

The investigation of atomic volumes seems to corroborate these ideas further. Thus, since we are assuming that neutral potassium has an electronic configuration which differs from that of argon by the presence of an extra electron circling outside the closed structure of argon, we may reasonably suppose that the volume of the neutral potassium atom should be larger than that of the argon atom. On the other hand, a once-ionized potassium atom should have approximately the same volume as the atom of argon. Indirect experiment seems to verify this conclusion.

Chemical properties also show that we are on the right track. According to our assumptions, the atom of chlorine, which immediately precedes argon, should exhibit an outer group of orbits that lacks one electron to be filled completely. But then a chlorine atom should combine readily with an atom like potassium, which has one dangling electron that it parts with easily. This expectation is verified, for chlorine and potassium combine violently. Similar considerations apply to atoms such as calcium and sulphur. The atomic numbers of calcium and sulphur are 20 and 16 respectively, and the atomic number of the inert gas argon is 18. Bohr's scheme requires that calcium should have two outer electrons circling around a closed configuration similar to that of the atom of argon, and that sulphur should lack two electrons for this closed configuration to be realized. Calcium and argon should therefore combine readily for the same reasons as before, namely, calcium has two electrons which it is willing to lose whereas sulphur is eager to acquire two electrons. For the same reasons also, calcium should combine with two atoms of chlorine, each chlorine atom taking up one of the two loosely bound electrons of calcium. Inasmuch as these chemical reactions are known to occur, they afford added confirmation of Bohr's electronic arrangements.

The phenomenon of electrolysis also justifies Bohr's views. When an electric current is transmitted through a solution of potassium chloride, the potassium atoms move in the direction of the positive current, proving themselves thereby to be positively electrified; the chlorine atoms move in the opposite direction. Quantitative measurements show that the potassium atoms must have one unit of positive charge, and the chlorine atoms one unit of negative charge. Bohr's conception of the atoms accounts readily for these facts. We have but to assume that when chlorine and potassium combine to form potassium chloride, the chlorine atom takes so firm a hold on the loosely bound electron of potassium that, even when the two atoms are separated again by the electric current, the electron remains attached to the chlorine atom and thereby gives it a unit negative charge.

A name that arises frequently in the discussion of the structure of atoms is the word *core*. In the case of a non-ionized atom the core refers to that part of the atom which remains when the optical electron is removed; it is thus the once-ionized atom. Lacking as it does a single electron to secure electric neutrality, the core has a net charge which is always one positive unit charge, just as occurs for the proton-nucleus of hydrogen. We may therefore say that the optical electron describes its orbit under the attraction of the core. In the atom of an alkali metal, such as sodium, the optical electron is revolving around the closed stable electronic configuration which is characteristic of the inert gas immediately preceding the alkali metal. In the case of sodium this inert gas is neon. Consequently, the core of the sodium atom is represented by the closed electronic configuration of neon; but of course the nucleus at the centre of the core is not the same as the nucleus of neon, for its electric charge (atomic number) is greater by one unit. Since the cores of the alkali-metal atoms have closed electronic configurations, they are compact units and are not fringed with loosely bound exterior electrons. In this respect they have a greater resemblance with the proton-nucleus of hydrogen than have the cores of the other atoms. For this reason the atom of an alkali metal, represented as it is by an optical electron circling under the attraction of the core, has a certain resemblance with the atom of hydrogen, in which an electron is revolving under the attraction of the proton-nucleus. We should therefore expect the spectra of the alkali metals to resemble that of hydrogen. Now, this expectation is verified; at least to the extent that, of all the higher atoms, those of the alkali metals yield spectra which differ least from the hydrogen spectrum. We may regard these spectral similarities as affording additional support to Bohr's views.

Bohr's method of distributing the electrons in the atoms is further corroborated by the spectra of the rare gases. If Bohr's ideas are correct, we should expect that an atom containing two or more outer electrons would have a more complicated spectrum than an alkali-metal atom. The possibility of simultaneous drops occurring for the several outer electrons in an excited atom would in itself justify this expectation. The atom of a rare gas, in particular, owing to its large number of outer electrons, should give rise to a most complicated spectrum. This surmise is verified. Indeed, even the simplest of all the rare-gas atoms, namely, neutral helium, exhibits a spectrum which Bohr's theory was unable to interpret until the hypothesis of the spinning electron was introduced.

Let us consider the orbits in the higher atoms. In the relativistic treatment of the hydrogen atom we noted that the orbits of the optical electron were no longer ellipses (as in the classical treatment), but rosette orbits. Precisely the same general type of rosette orbit occurs for all the electrons of the higher atoms, and to some extent the relativistic refinements are responsible for the more complicated motions. But in the higher atoms other factors, the importance of which far exceeds the relativistic influences, also concur towards yielding the rosette-like motions, and for this reason we may forego the application of the theory of relativity.

We shall first examine the possible orbits of the optical electron. Some of these may pass through the core, whereas others may circle around the core without entering it. We shall suppose that the optical electron is describing the former type of orbit. Now so long as the optical electron is well outside the core, it will be attracted by the core as a whole and hence will move as it would if the core were replaced by a proton. The orbit outside the core will thus be elliptical. But when the optical electron penetrates inside the core, it passes between the nucleus and some of the core-electrons; the nucleus, being no longer screened by the core-electrons, exerts a powerful attraction on the optical electron, an attraction which will considerably exceed the attraction a proton would exert. Inside the core, the optical electron will thus swing round the nucleus with greater rapidity than it would if the nucleus were a proton; and so the electron will be deviated from the elliptical orbit which it was originally following and will describe a slightly displaced ellipse. Each time the optical electron penetrates the core afresh, the same displacement will be repeated. The electron's motion is therefore represented by a motion along an ellipse which is precessing around the nucleus. A rosette-like orbit will thus arise, and the departure from the elliptical orbit will be more pronounced for those orbits which penetrate more deeply into the core. If the eccentricity of the ellipse is small, so that the orbit is more or less circular, or if the ellipse is large, the optical electron will not penetrate into the core and the orbit will then remain elliptical, no rosette motion occurring. In the majority of cases, the departure of the orbit from the elliptical form will be much more accentuated than it was in Sommerfeld's relativistic hydrogen atom.

In this discussion we have restricted our attention to the optical electron, but the conclusions we have obtained may be extended to each one of the core-electrons. Each core-electron will therefore be describing a rosette-like orbit.

We must now consider the quantum restrictions which Bohr imposes on the motions. The rules of quantization for the present rosette motions are the same as those described in connection with Sommerfeld's hydrogen atom. Let us concentrate on any one of the electrons in our higher atom. Its motion exhibits two periodicities, which arise, respectively, from the motion in the ellipse and from the precession of the ellipse as a whole. Two quantum numbers, n and k, are therefore introduced as before. We recall that n may be any positive integer 1, 2, 3, . . . and that the possible values of k associated with a given value of n are the values 1, 2, 3, . . . n. The value of n defines the main energy level, whereas that of k defines the sublevel of the main energy level; and both numbers together determine the energy of the electron on the orbit. When $k = n$, the orbit is a circle, and the rosette orbit no longer arises. The smaller the value of k, the more eccentric the rotating ellipse and the more the orbit penetrates into the core; the more pronounced therefore is the rosette appearance of the orbit. The value $k = 0$ is excluded as impossible for physical reasons, because it would indicate that the optical electron is oscillating back and forth through the nucleus. Of course in addition to the orbits that the various electrons are describing in the unexcited atom, there are higher orbits which are usually empty but which may receive the optical electron when the atom is excited. These higher orbits (called virtual orbits) are also defined by means of the quantum numbers n and k.

Before proceeding, we must mention a difficulty. According to Bohr the quantum numbers, n and k, of each electron must be integers; and, except in the case of the optical electron when the atom radiates, these numbers must not change as the electrons describe their orbits. Bohr calls this postulate "the postulate of the invariance and of the permanence of the quantum numbers n and k." From the physical standpoint this postulate implies that the energy of each electron remains constant during the motion. That Bohr's postulate is a necessary adjunct to his theory is obvious when we apply it to the optical electron in an excited orbit. If, contrary to Bohr's postulate, the energy of the electron were to vary during the motion, the loss in energy that the electron would sustain when it dropped to the basic orbit would vary from one instant to another, and the radiation emitted would have a variable frequency. Since this conclusion is incompatible with the sharply defined frequencies of the spectral lines, we see that Bohr's postulate must be accepted in the case of the optical electron. A similar reasoning applies to the core-electrons when we consider the generation of the X-ray spectra. But Bohr's postulate is in conflict with the laws of mechanics. Thus, when the atom is not radiating and is therefore a conservative isolated system, the mechanical laws state that the total energy of the atom remains constant, but that owing to the mutual electronic actions the energy of each individual electron varies, energy exchanges taking place with the other electrons. (For example, in the solar system, energy is continuously exchanged among the various planets.) Since Bohr's postulate is incompatible with the ordinary laws of mechanics, we must recognize that some new quantum manifestation is involved.

We mention this point to show that, in Bohr's atom, not only are the principles of mechanics violated in the initial postulates of the theory, but they are also violated practically at each step. Many other violations of the principles of mechanics will be recorded as we proceed.

The Spectral Series of Sodium—Whatever may be the defects of the present quasi-mechanical model of the atom, this model is the one which was accepted in the earlier stages of Bohr's theory. Let us investigate it more fully in connection with the atom of sodium, which, in view of its similarity to the hydrogen atom, is one of the simplest of the higher atoms.

The systematization of spectra on the part of the spectroscopists has shown that the spectral lines of sodium may be grouped into four main series. They are called the *Principal,* the *Sharp,* the *Diffuse* and the *Fundamental* series. Other less conspicuous series have also been detected, but we shall confine our attention to the more important ones just listed. The Fundamental series bears in the spacing of its lines a close resemblance to the Paschen series of hydrogen; whereas the Diffuse, the Sharp, and the Principal series differ progressively from the hydrogen series. In common with the general characteristics of all spectral series, the spacings between the lines of a given series decrease as we proceed towards the ultra-violet; and the extreme line of a series is called the "limit" of the series. In the case of sodium, the limit of the Principal series is in the ultra-violet, and the limits of the other series are displaced towards the lower (infra-red) end of the spectrum. It remains to be seen whether these spectral series can be interpreted on the basis of Bohr's model.

The atom of sodium has the atomic number 11 and therefore contains eleven planetary electrons circling around the nucleus. We shall assume for the present, without further explanation, that ten of the electrons fill completely all the orbits ($n = 1$; $k = 1$) and ($n = 2$; $k = 1$ and 2). Consequently, the lowest orbit available for the last electron (the optical electron) is one of the more loosely bound orbits associated with the quantum number $n = 3$. We shall show in due course that the lowest available orbit is one of the orbits ($n = 3$; $k = 1$). The ten electrons present in the two lower levels are the core-electrons, and in the phenomenon of radiation they play no direct part; the drops of the optical electron alone generate radiation. When the atom is excited, the optical electron is removed to some higher orbit, and it then falls back to a lower orbit emitting radiation. The fall to a lower orbit may occur either in a single drop or by progressive stages. In the present discussion, however, we shall be concerned only with the direct drops to the lowest sublevels ($n = 3$; $k = 1, 2, 3$). If we remember that, according to the correspondence principle, the quantum number k must always change by $+1$ or -1 during a drop, we may easily predict the drops that can arise.

   The following table gives a list of the possible drops from the higher
sublevels on the left to the lower ones on the right.

$(n \geqslant 3; k = 2)$ to $(n = 3; k = 1)$—Principal series

$(n > 3; \; k = 2)$ to $(n = 3; k = 3)$—? (unknown)

(28) $(n > 3; k = 1)$ to $(n = 3; k = 2)$—Sharp series

$(n \geqslant 3; k = 3)$ to $(n = 3; k = 2)$—Diffuse series

$(n \geqslant 4; k = 4)$ to $(n = 3; k = 3)$—Fundamental series.

In the table, we have connected the various families of drops with the
different spectral series; the justification for these connections will be
explained presently. If, for the time being, we assume the correctness
of the table, we see that the succession of lines of the principal series
arises when the optical electron drops to an orbit $(n = 3; k = 1)$ from the
succesively higher orbits determined by the fixed value $k = 2$ and by
increasing values for n $(n \geqslant 3)$. The highest possible drop pertaining to
this series occurs when the electron falls from infinity to the basic orbit
$(n = 3; \; k = 1)$; and this drop will therefore yield the line of highest
frequency of the principal series (i.e., the limit of the series). The
smallest drop pertaining to this series is realized when the electron falls
from an orbit $(n = 3; k = 2)$ to an orbit $(n = 3; k = 1)$; we then get the
line of lowest frequency of the principal series. This line is the well-
known yellow line emitted by incandescent sodium vapor. In point of
fact the yellow line is a doublet, so that two slightly different drops
should be possible; but this complication need not detain us for the
present; we shall revert to it when we discuss the theory of the multiplet
lines. For future reference let us note that, according to the table (28),
it is the change in the value of k that characterizes the spectral series;
changes in the upper quantum number n merely give the different lines
of the same series.
   We now examine some of the clues which guided Bohr in associating
the various drops with the respective spectral series of sodium (as indicated in the table (28)). In the hydrogen atom, the mechanical model
is so simple that there is no difficulty in calculating the energies of the
various quantized orbits, or sublevels. Knowing the energy values of
the orbits, we have but to apply Bohr's frequency condition, and the
numerical values of the corresponding radiations are obtained. It is
then an easy matter to establish the correspondence between the drops
and the spectral lines. But in the sodium atom, the mechanical model
is far more complicated, especially so since the electron may pass through

the core during its motion and be subjected to perturbations which we cannot calculate with precision. Various clues, however, allow us to be reasonably certain of the correctness of the table (28).

In the first place we mentioned that, of all the spectral series of sodium, the fundamental series is the one which bears the closest resemblance to a hydrogen series. We may therefore expect the fundamental lines to be generated by the drops between those orbits which differ least from the hydrogen ones. In the sodium atom we shall obviously approximate the conditions existing in the hydrogen atom if we consider the orbits which do not pass through the core. For such orbits, the core as a whole will exert very nearly the same electrical attraction as does the single proton of hydrogen. Now, the orbits that stretch around the core without entering it (*i.e.*, the non-penetrating orbits) are orbits of small eccentricity. Such orbits are circular or very nearly so, and hence are characterized by large values of the quantum number k. If we revert to the table (28), we perceive the justification for identifying the drops at the bottom of the table with those responsible for the lines of the fundamental series. Furthermore, since the principal series is the one that differs most from a hydrogen series, we are led to connect the drops between the most penetrating orbits with the spectral lines of the principal series. The most penetrating orbits are characterized by the smallest values of the quantum number k, and so the general scheme of the table (28) appears to be justified.

Let us determine the basic orbit of the optical electron, *i.e.*, the orbit which the electron will be describing in the unexcited atom. The basic orbit will necessarily be the orbit of lowest energy which is not already occupied by a core-electron. Calculation, based on Bohr's mechanical model, indicates that the basic orbit will be the orbit $(n = 3; k = 1)$, a fact we might have anticipated since the orbit $(n = 3; k = 1)$ is associated with the smallest possible value for n and is highly eccentric (small value of k) and hence firmly bound. Other methods of investigation which utilize certain empirical formulae or the phenomenon of resonance radiation would lead to the same conclusion.

A matter of interest is to establish the energy of the basic orbit, *i.e.*, the energy of the optical electron when it is describing that orbit. Calculation involves too many approximations to furnish accurate information, and so we must rely on other methods. One way of determining the energy of the basic orbit would be to measure the limiting frequency of the series which is generated by the drops of the electron to the orbit $(n = 3; k = 1)$, *i.e.*, the limiting frequency of the principal series. Thus if ν were the limiting frequency, $-h\nu$ would be the energy of the basic orbit. Limiting frequencies cannot, however, be measured directly because the drops from infinity, which would generate them, occur too seldom. But we may compute the energy of the basic orbit indirectly by determining the ionization potential of the atom. We recall that the ionization potential measures the work we should have to expend in order to remove the optical electron from the basic orbit and displace it to a point at infinity (in practice to a point at a small distance from the centre of the atom). This ionization potential measures the energy of the basic orbit (with its sign reversed), and since the ionization potential can be determined, the energy of the basic orbit is obtained.

Having obtained the energy value of the basic orbit $(n = 3; k = 1)$, we may derive the energy values of higher orbits from spectroscopic measurements. The lines of the principal series, for instance, are given by the drops

$$(n \geqslant 3; k = 2) \longrightarrow (n = 3; k = 1);$$

and by ascribing to n in the upper level the successive integral values 3, 4, 5 - - , we obtain the drops corresponding to the successive lines of increasing frequency. For example, the line of second lowest frequency results from the drop

$$(n = 4; k = 2) \longrightarrow (n = 3; k = 1).$$

According to Bohr's frequency condition, the product $h\nu$ (where ν is the frequency of the radiation emitted) gives the energy radiated by the drop. Since the frequency of the radiation may be measured and since the energy of the lower of the two orbits is already known, the energy of the higher orbit $(n = 4; k = 2)$ can be determined. By applying in this way Bohr's frequency condition to the direct spectroscopic measurements of the radiated frequencies, we may determine one after another the energies of all the orbits of the sodium atom.

The illustrations we have given indicate the type of evidence that has been utilized in connecting the various quantum jumps with the observed spectral lines. Many other clues might be mentioned, but for lack of space we shall not discuss them here. Suffice it to say that they all appear to corroborate the general arrangement of the table (28).

Atom Building—We are already familiar with the hydrogen atom. When the relativistic requirements are applied, the orbits are determined by the two quantum numbers, n and k. Since in the lowest orbit, n has the value 1, and since k can assume only the values 1, 2, 3 . . . n, we conclude that only one type of orbit is associated with the value $n = 1$; namely, the circular orbit $(n = 1; k = 1)$. When the atom is unexcited, it is in this orbit that the electron is revolving. In the hydrogen atom, owing to the presence of a single electron, we have no means of deciding whether the orbit $(n = 1; k = 1)$ is unique or whether other orbits associated with the same quantum numbers may not also exist. The study of the higher atoms shows that there must be two such orbits, although only one of them comes into consideration in the hydrogen atom.

The next atom is helium of atomic number 2. It is formed of an alpha particle as nucleus and of two planetary electrons. We wish to determine in what orbits these electrons will be moving. According to Bohr's ideas the addition of a second electron in no wise disturbs the quantum numbers allocated to the first electron, so that one of the two electrons of helium will necessarily be in the orbit $(n = 1; k = 1)$, just as in the case of the hydrogen atom. To determine the orbit of the second electron we note that since helium is one of the rare gases, its electronic configuration must exhibit a closed structure. This fact implies that both electrons must be moving in orbits which have the same energy and therefore the same quantum numbers. Consequently, the two planetary electrons of neutral helium must both be moving in orbits $(n = 1; k = 1)$. In addition, since the structure of helium is closed, we must assume that there are no further orbits $(n = 1; k = 1)$, which subsequent electrons in the higher atoms might occupy. Hence the two orbits just mentioned must form what we have called a group of orbits.

A further vindication of these conclusions is obtained when we consider the atom of atomic number 3, *i.e.*, the lithium atom. Inasmuch as its nucleus has a three-fold net positive charge, there must be three planetary electrons. Since, with helium, the orbits of the innermost group are assumed to have received their full quota of electrons, we may expect the third electron of lithium to be situated in an orbit of the next group, while the two first electrons fill the two innermost orbits $(n = 1; k = 1)$. Other considerations lead to the same view. Thus lithium is easily ionized; hence its third electron must be moving in an outer orbit; this electron cannot therefore be situated in the same group of orbits $(n = 1; k = 1)$ as the other two electrons. Furthermore, lithium is an alkali metal, and, as such, exhibits a spectrum similar to that of hydrogen. This circumstance suggests the presence of the third, or optical, electron in an outer orbit encircling a closed configuration. The foregoing clues and many others indicate that, in the unexcited atom, the third electron of lithium must be moving in an orbit connected with the value 2 for the main quantum number n.

But there are two kinds of orbits associated with $n = 2$; these are the elliptical orbit $(n = 2; k = 1)$ and the circular one $(n = 2; k = 2)$. The problem is to decide in which of these two kinds of orbits the third electron will be moving. We know of course that the third electron will set itself in the lowest, or the most closely bound, of the two orbits, and calculation indicates that this orbit will be the elliptical one. But because of our ignorance of so many factors, we must not place too much reliance on·calculations, especially so, since they are but approximations at best. Fortunately we have other means of clarifying the situation. The general spectroscopic evidence and the precise measurement of the spectral frequencies, utilized in the same way as was explained in connection with the atom of sodium, show that the elliptical orbit $(n = 2; k = 1)$ is indeed the one of lower energy, and that the third electron of lithium must be situated in this orbit when the atom is unexcited. The electronic configuration of neutral lithium is thus represented by the two core-electrons in the orbits $(n = 1; k = 1)$ and by the optical electron in an orbit $(n = 2; k = 1)$.

Following lithium, come the atoms of beryllium, boron, carbon, nitrogen, oxygen, fluorine, and neon. The latter of atomic number 10 is a rare gas and must exhibit a closed electronic structure. Inasmuch as neon is the first of the rare gases after helium, and since with helium the closed structure involves the orbits $(n = 1; k = 1)$, we must assume that, in the closed structure of neon, one or both of the groups of orbits associated with the quantum number $n = 2$ will be filled. Various clues show that both groups of orbits $(n = 2; k = 1$ and $2)$ have their full quota of electrons; in other words, each orbit belonging to either one of these two groups contains one electron. Since neon contains ten planetary electrons and helium two, the two groups of orbits $(n = 2; k = 1, 2)$ will comprise eight orbits in all. A decision on the correct number of orbits in each one of these two groups would be speculative at this stage. But several considerations, involving subsequent discoveries (notably the X-ray spectra and Pauli's exclusion principle), lead to the conclusion that a group of orbits (n, k) contains $2(2k - 1)$ orbits, so that there must be two orbits $(n = 2; k = 1)$ and six orbits $(n = 2; k = 2)$. We may therefore suppose that, for beryllium, the fourth electron joins the third in an elliptical orbit $(n = 2; k = 1)$, and that the six following electrons which are introduced in succession as we pass from boron to neon are situated in circular orbits $(n = 2; k = 2)$. The closed structure of neon is thus represented by two electrons in the first group of orbits, which contains the two circular orbits $(n = 1; k = 1)$; by two more in the second group, which contains two elliptical orbits $(n = 2; k = 1)$;

and by six in the third, or outermost, group, which comprises six circular orbits $(n = 2; k = 2)$. All three of these groups of orbits are filled with their quotas of electrons.

After neon comes the alkali metal sodium of atomic number 11. Its optical electron must start the formation of a new group exterior to the closed structure of neon. The electron must therefore be in the level $n = 3$. This level contains three groups of orbits $(n = 3; k = 1, 2, 3)$; and the added electron will be situated in an orbit belonging to the lowest of these three groups. We have already seen that for several reasons, notably on account of the spectroscopic evidence, the orbits $(n = 3, k = 1)$ are the lowest. The orbit and the energy level of the newly added optical electron is thus known.

Following sodium come the seven elements, magnesium, aluminum, silicon, phosphorus, sulphur, chlorine, and argon. . The last is a rare gas, and with it a closed stable formation must be attained. The allocation of the electrons is here more difficult. We might naturally expect that the three groups of orbits of the level, $n = 3$, would be filled when argon is reached. But this is not the case. Only the two lower groups $(n = 3; k = 1$ and 2) are filled (with two and six electrons respectively). The orbits of the third group $(n = 3; k = 3)$ are unoccupied. We might suppose that these unoccupied orbits were the lowest of the unoccupied ones, and we should therefore conclude that in the next atom, which is potassium, the added electron would be moving in one of these orbits. Yet this is not so, for various clues show that the highly eccentric orbits $(n = 4; k = 1)$ are of lower energy than the circular ones $(n = 3; k = 3)$. This result is not altogether surprising when we consider that a highly eccentric orbit, passing as it does very near the nucleus, may quite well be more firmly bound than a circular orbit of lower main quantum number n. At all events, the spectroscopic evidence shows that, as also occurs for lithium and sodium, the optical electron of potassium must be in an orbit associated with the quantum number $k = 1$. And this leaves no other choice than to place the electron in one of the orbits $(n = 4; k = 1)$. We have here an illustration of a new group of orbits being filled even though the orbits of the underlying group $(n = 3, k = 3)$ are unoccupied. Following potassium comes calcium (atomic number 20). In the case of calcium the new electron is situated in the same group $(n = 4, k = 1)$ as the optical electron of potassium; the orbits $(n = 3, k = 3)$ still remain unoccupied.

The Interpolated Elements—The next nine elements following calcium are scandium, titanium, vanadium, chromium, manganese, iron, cobalt, nickel, and copper. With scandium the added electron is situated in one of the skipped orbits $(n = 3; k = 3)$, and so this group of orbits begins to be filled. We conclude that, when two electrons are already present in orbits $(n = 4, k = 1)$, the orbits of lowest energy available are those of type $(n = 3; k = 3)$. The filling of these orbits continues as we pass from scandium (atomic number 21) to copper (atomic number 29). In the last atom one of the two electrons formerly in the orbits $(n = 4, k = 1)$ has passed into an orbit $(n = 3, k = 3)$, and, owing to this occurrence and to the adjunction of eight electrons as we pass from scandium to copper, there are ten electrons in the group of orbits $(n = 3, k = 3)$. This group is then filled.

The peculiarity in the construction of the atoms from scandium to copper is that, with them, a level $n = 3$ is being filled while electrons are already present in a level $n = 4$. We seem to be filling the orbits backwards, reversing what would appear to be the normal order. For this reason, these elements are called "interpolated elements." The filling of an inner group (when an outer one has already been started) is peculiar to Bohr's theory of the atom and is not encountered in the static atom of the chemists. At first sight this situation appears strange and needs justification before it can be accepted. But as a matter of fact there is strong evidence to show that Bohr was on the right track. In the first place, the elements from scandium to copper exhibit ambiguous chemical valencies. Secondly, they yield colored salts (with the exception of scandium). Thirdly, they are paramagnetic instead of being diamagnetic like the majority of the other elements. Finally, their X-ray spectra exhibit a departure from the regularities disclosed in Moseley's law. All these facts suggest that some new structural feature is involved.

Let us first examine the bearing of the ambiguous chemical valencies on the problem of the interpolated elements. An atom is said to have a positive valency N if it combines with N atoms of chlorine. It is said to have a negative valency N if it combines with N atoms of hydrogen. Chemical valencies are intimately related with the arrangements of the electrons in the atoms, and indeed valency considerations were among the numerous clues utilized by Bohr in his atom building. For our present purpose we may restrict ourselves to an example bearing on positive valencies. Chlorine, as we know, has seven electrons in its uncompleted group, and so one additional electron is required to fill all the orbits of the outlying group and thereby to complete a closed configuration. An atom which combines with one atom of chlorine (i.e., an atom having a unit positive valency) must therefore have one exterior electron which it can cede easily. Sodium, Na, is of this type, and as a result an atom of sodium combines readily with an atom of chlorine, Cl, to form NaCl.

The close connection which appears between the electronic arrangement and the valency of an atom suggests that, when the valency of an atom is ambiguous (owing to its ability to combine in different proportions with the same element), the ambiguity may be ascribed to a breakdown in the simple arrangement of the electrons in the atom. Now, the elements from scandium to copper have ambiguous valencies, and so it is plausible to suppose that in these elements the arrangements of the electrons are more confused. Bohr's conception of interpolated elements is thus rendered plausible by a certain measure of qualitative evidence.

We have also mentioned that the salts of the interpolated elements exhibit color and that this coloration supports Bohr's electronic structure for these elements. Here a short digression is necessary. A translucent body which manifests color is a body which absorbs some of the visible radiations that fall upon it while allowing others to pass. Iron sulphate, for instance, since it appears green, must absorb the complementary radiations to green, namely, the red ones. According to Bohr's theory, a radiation is absorbed by an atom' when the energy of the radiation is communicated to the atom and excites it by displacing its optical electron to a higher level. In the case of a solid, however, the removal of the electron to an outer unoccupied orbit cannot occur, for a removal of this sort would swell the volume of the atom and would be resisted by the proximity of the other atoms of the solid. This crude mechanical interpretation must, of course, be viewed only as schematic, but it suffices to account for the phenomenon we are about to discuss. Guided by it, we must assume that, when the atom of an element responsible for color is excited by the absorption of radiation, the removal of the optical electron to a higher level must be such as to produce no increase in the volume of the atom. Consequently, the electron must be jerked up to some sublevel which already contains one or more electrons but which is susceptible of receiving yet another.

Bohr's interpretation of the interpolated elements appears to be compatible with this situation. Thus in the case of iron, there are two electrons in the orbits ($n = 4$; $k = 1$) and six in the orbits ($n = 3$; $k = 3$). The orbits of the latter group are of higher energy. Since all of them are not filled with electrons, we may presume that when the atom of iron present in the salt is excited, an electron is jerked up from an orbit ($n = 4$; $k = 1$) to one of the unoccupied orbits ($n = 3$; $k = 3$). This transition is not accompanied by an increase in the atom's volume, so that the previous requirement is satisfied. Furthermore, the difference in the energy between orbits $(n = 4; k = 1)$ and $(n = 3; k = 3)$ cannot be very great, and hence the light that will be absorbed may very well be among the visible frequencies rather than in the ultra-violet. The coloration of the salt is thus accounted for, at least qualitatively.

The third argument, adduced from the paramagnetic properties of the interpolated elements, need not detain us. The fact is that the magnetic properties of solids are ascribed not solely to the peculiarities of the individual atoms, but also to the groupings of the atoms into clusters. Nevertheless we may presume that the destruction of the symmetry in the electronic arrangements, which characterizes the atoms of the interpolated elements, should exert some effect on solid bars made of these elements, e.g., on iron bars.

The evidence afforded by the X-ray spectra in favor of Bohr's conception of the interpolated elements is, however, more convincing. The X-ray spectra are intimately connected with the distribution of the inner electrons. Hence the sudden irregularity in this distribution which, according to Bohr, occurs when we pass from calcium to the first of the interpolated elements (scandium) would account for the experimentally verified irregularity in the disposition of the X-ray lines.

Let us now proceed with the building up of the following atoms. Our scheme of atom building indicates that copper (atomic number 29) is formed of one outer electron circling around completed inner shells. The atom of copper is thus built on the same plan as that of an alkali metal, i.e., a single electron circling round a core which has a closed configuration. Consequently, we should expect some similarity between the spectrum of copper and those of the alkali metals. These anticipations are verified by observation.

With zinc, the element following copper, a further electron is added to the outer shell $(n = 4; k = 1)$, yielding an atom analogous in its structure to magnesium and calcium. The spectral similarities anticipated are again observed. When we reach the atomic number 36, we have the inert gas krypton, with its closed configuration represented by two electrons in the group of orbits $(n = 4; k = 1)$ and by six electrons in the group $(n = 4; k = 2)$. The higher orbits $(n = 4; k = 3$ and $4)$ are unoccupied. With the next element, rubidium (atomic number 37), we encounter the same situation as in the case of potassium. The new electron, instead of being placed in an orbit $(n = 4; k = 3)$, goes into an orbit $(n = 5; k = 1)$; the two groups of orbits $(n = 4; k = 3$ and $4)$ are skipped. Next to rubidium comes strontium for which a second electron is added to the outer group of orbits $(n = 5; k = 1)$. But after strontium a new series of interpolated elements occurs. The electrons are placed in one of the skipped groups of orbits, namely, in the group ($n = 4$; $k = 3$), and when palladium (atomic number 46) is reached, the group ($n = 4$; $k = 3$) is filled with ten electrons. The group ($n = 4$; $k = 4$) still remains unoccupied. These interpolated elements manifest the same ambiguity of valencies, coloration of salts, and paramagnetic properties mentioned previously.

From palladium to lanthanum the electrons are placed in the outer levels $n = 5$ and $n = 6$; the orbits ($n = 4$; $k = 4$) still remain empty. But immediately following lanthanum, beginning with cerium and extending to lutecium, another series of interpolated elements arises. In this interpolated series, called the series of the "rare earths," the orbits ($n = 4$; $k = 4$) are progressively filled with fourteen electrons; the series is terminated by the element lutecium (atomic number 71).

At the time Bohr was developing his theory, the element following lutecium was undiscovered. But Bohr, on the strength of his electronic arrangements, asserted that in this unknown element the added electron would have to be placed in the level $n = 5$. This would imply that the unknown element would have an electronic configuration analogous to that of zirconium. Inasmuch as elements that are chemically analogous are usually found in the same minerals, Bohr claimed that the missing element should be sought in minerals containing zirconium. It was indeed in such ores that the missing element, called hafnium, was detected by Hevesy.

We need not pursue further this process of atom building. All we need mention is that the procedure continues till we reach uranium of atomic number 92. In that atom the electrons are distributed among the seven levels $n = 1, 2, 3, 4, 5, 6, 7$. But all the orbits in the levels $n = 5$ and $n = 6$ are not filled, so that if the process of atom building could be continued, more interpolated elements should arise. Uranium is the last of the atoms known at the present time, but no obvious theoretical reason opposes the existence of still higher atoms.[*] We must remember, however that our present method of atom building concerns only the arrangements of the planetary electrons around the various nuclei and affords no information on the constitution of the nuclei themselves. For this reason the assurance that atoms higher than uranium are possible under ordinary conditions must await a better understanding of nuclear physics or else the direct discovery of such atoms by experiment.

---

[*]Since the first edition of this book appeared, atoms higher than uranium have been produced artificially, viz. the atoms of neptunium, plutonium, and still others.

CHAPTER XXVII

BOHR'S ATOM (*Continued*)

The Multiplet Lines—On closer examination we find that by far the larger number of spectral lines are not single but are formed of two or more lines. Such groups of lines are called "multiplets" and are qualified by the names doublets, triplets, quartets, . . . according to the number of lines in the group. The single lines, which also occur in certain spectra, are called "singlets"; they must not be confused with the individual lines of a multiplet group.

If we start with the atom of an alkali metal, such as potassium, and then consider the consecutive atoms of increasing atomic number, we find that the parity of the multiplet lines changes for each successive atom. The following list illustrates these remarks.

Potassium yields doublets

calcium—singlets and triplets

scandium—doublets and quartets

titanium—singlets, triplets, and quintets

vanadium—doublets, quartets, and sextets

chromium—triplets, quintets, and septets

manganese—quartets, sextets, and octets

iron—triplets, quintets, and septets

cobalt—doublets, quartets, and sextets

nickel—singlets, triplets, and quintets

copper—doublets and quartets

zinc—singlets and triplets.

Quite generally, all the atoms of the alkali metals, *e.g.*, lithium, sodium, potassium, are associated with doublets, whereas those of the alkaline earths, *e.g.*, beryllium, magnesium, calcium, emit singlets and triplets. The foregoing illustrations indicate that an atom is associated either with even or with odd multiplets and cannot be associated with both. To avoid misconceptions on the subject of the multiplets, we must mention that when, for instance, we speak of sodium as associated with a spectrum of doublets, the stress is placed not on the spectral lines, but on the energy sublevels (n, k) of the atom. The lines themselves may exhibit higher multiplicities, but the sublevels (n, k) are never split into more than two sub-sublevels; hence their multiplicity can never exceed two. These points will be elucidated as we proceed.

In the last chapter we stated that the once-ionized and the twice-ionized atoms of atomic number N have spectra similar to those of the neutral atoms of atomic numbers $N - 1$ and $N - 2$, respectively. This rule, which results from the sameness in the electronic configurations of the atoms considered, extends also to the details of the multiplet structure. For example, once-ionized calcium has a spectrum of doublets like neutral potassium. Thus, the origin of the multiplet lines must be sought in the peculiarities of the electronic distributions.

To interpret the multiplet structures, we must revert to the general considerations developed in the last chapter. We saw that, in the hydrogen atom (when the relativistic refinements were taken into account) and quite generally in the higher atoms, the energy values of the orbits were determined by two quantum numbers, n and k. We also found that orbits (n, k) of the same energy appeared in groups. For example, there were two orbits $(n = 3; k = 1)$ and six orbits $(n = 3; k = 2)$; and in the sodium atom, the yellow line of the principal series was generated when the optical electron dropped from any one of the six orbits of the latter group to either one of the two orbits of the first group. But we know that this yellow line, formerly treated as single, is in reality a doublet formed of two yellow lines very close together. Consequently, two different energy drops must exist in place of the single drop heretofore assumed. Our original interpretation must therefore be refined, and we must recognize that the two orbits $(n = 3; k = 1)$ or else the six orbits $(n = 3; k = 2)$ do not have the same energy. Further research showed that the second solution was the correct one.

The group of six orbits $(n = 3; k = 2)$ will thus be divided into two kinds of orbits differing in energy, and the one or the other of the two yellow lines of the sodium doublet will be generated according to whether the electron in its drop to an orbit $(n = 3; k = 1)$ falls from the one or the other of the two kinds of orbits $(n = 3; k = 2)$. In practice, both yellow lines are observed simultaneously, because the spectrum is generated by a large number of excited atoms undergoing energy drops independently of one another.

In order to differentiate the two kinds of orbits comprised in the original group $(n = 3; k = 2)$, Sommerfeld introduced a qualifying number j. In his original scheme he represented the orbits of the lower and of the higher kind, respectively, by $(n = 3; k = 2; j = 1)$ and by $(n = 3; k = 2; j = 2)$. Since the two orbits of the basic group $(n = 3; k = 1)$ were credited with the same energy, there was no immediate necessity to introduce the number j in connection with them. But for the sake of consistency Sommerfeld retained the number j; and both orbits of the lower group $(n = 3; k = 1)$ were represented by the notation $(n = 3; k = 1; j = 1)$.

From this account we see that the multiplet structure of the spectral lines indicates that, in many cases, the energy levels of the higher atoms must be formed of distinct sub-sublevels (n, l, j), and not of the simpler sublevels (n, k), as was formerly supposed.

We must now examine the significance of Sommerfeld's new number j. Our former quantum numbers n and k were derived when we applied Bohr's postulates to a mechanical model. But in our present study of the multiplets we have no mechanical model (as yet) at our disposal, and no definite significance can be attributed to the number j; nor is there any means of deciding what numerical values should be assigned to j in various situations. The introduction of this number is suggested solely by the direct observation of the multiplet structure. Yet, although the physical significance of j is still obscure, we have every reason to suppose that it is some new quantum number; and in view of our past experience with quantum numbers we may presume that its possible values should always be integers. The exact numerical values of j are of no importance for the present, and so there is no objection to the values we ascribed to j in the foregoing illustrations.

Our present aim is then to split the various sublevels (n, k) of the sodium atom into sub-sublevels (n, k, j) in such a way that the numbers of lines observed in the various spectral series of sodium will be accounted for. After considerable labor it was thought that the following scheme would secure correct results: With the exception of the sublevels of type $(n \geqslant 3; k = 1)$, each one of the unoccupied orbits, or sublevels, of sodium is split into two sub-sublevels. The notational scheme originally suggested by Sommerfeld consisted in representing the sub-sublevels which were single by $(n \geqslant 3; k = 1; j = 1)$, and the other sub-sublevels which occurred in pairs by $(n \geqslant 3; k > 1; j = k - 1 \text{ or } k)$.

It is instructive to determine the allotments of quantum numbers which, according to this scheme, define the drops responsible for the lines of the various spectral series. The table (28) of the last chapter shows that the lines of the principal series of sodium are associated with drops in which k changes from the value 2 to the value 1. Consequently, when the new quantum number j is taken into account, the lines of the principal series result from the drops

(34)
$$(n \geqslant 3; k = 2; j = 1)$$
$$\searrow$$
$$(n = 3; k = 1; j = 1).$$
$$(n \geqslant 3; k = 2; j = 2) \nearrow$$

When some fixed value is attributed to n in the two sub-sublevels on the left-hand side, we obtain the two drops which give the two lines of the same doublet. In particular, if we set $n = 3$, we have the two lines of the doublet of lowest energy, *i.e.*, the well-known yellow doublet. Similarly, bearing in mind that the lines of the sharp series of sodium are associated with a change in the value of k from 1 to 2 (see (28)), we obtain for the drops corresponding to the lines of this series the pairs

$$\nearrow (n = 3; k = 2; j = 1)$$
$$(n > 3; k = 1; j = 1)$$
$$\searrow (n = 3; k = 2; j = 2).$$

The lines of the sharp series should thus also be doublets—a fact verified by observation.

A further point is brought out when we consider the lines of the diffuse series. The table (28) shows that these lines are generated by drops in which k changes from 3 to 2. Sommerfeld's scheme indicates that these drops are of type

(35) $(n \geqslant 3; k = 3; j = 3 \text{ or } 2) \longrightarrow (n = 3; k = 2; j = 2 \text{ or } 1)$.

When we give some fixed value to the integer n on the left-hand side, four drops to the two lower sub-sublevels are seen to be possible, and consequently the diffuse series of sodium should be formed of quartets. Observation, however, reveals only three lines in each multiplet (such a triplet of lines is often called a doublet with a satellite). The discrepancy is overcome by placing restrictions on the permissible changes in the value of j during a drop. In this connection the following rule was adopted by Sommerfeld: The integer j need not change in value during a drop; but, if it does change, it can change only by ± 1. Thanks to this restriction, or selection rule, one of the four drops expressed in (35) is impossible (*i.e.*, the drop in which j would change from 3 to 1) and so a triplet structure is in order for the lines of the diffuse series.

Incidentally the presence of more than two lines in some of the multiplets of sodium illustrates the warning we mentioned earlier in this chapter. We said that the association of sodium with doublets arises not so much from the structure of the spectral lines as from the manner in which the sublevels (n, k) are split into sub-sublevels. In the sodium atom some of the sublevels (n, k), remain simple (*i.e.*, those associated with the quantum number, $k = 1$), but all the other sublevels are double. The number 2 thus measures the greatest number of sub-sublevels (n, k, j) into which any sublevel (n, k) is split; and it is this feature that is responsible for the association of sodium with doublets. This maximum number 2, which arises in the case of sodium, measures what is called the *maximum order of multiplicity* for the sodium atom. Quite generally, the kind of multiplets associated with any atom is determined by the maximum order of multiplicity for that atom.

Next we consider atoms, like magnesium, which can emit singlets and also triplets. When singlets are emitted, the energy sublevels (n, k) remain single, as has hitherto been supposed; but when triplets are emitted, we shall have to assume that a different situation arises and that the energy sublevels (n, k) are split into triplets of sub-sublevels (n, k, j). In view of this alteration, we are justified in assuming that some important difference must distinguish the magnesium atom when it is in a position to emit singlets or to emit triplets. For the present this assumption is little better than a guess, but we shall find that it is corroborated in the sequel.

Let us, then, consider an atom of magnesium in the state conducive to the emission of triplet lines. The basic sublevel for the principal series is found to be the energy sublevel $(n = 4; k = 1)$.* The lines of the principal series are generated when drops to this lowest sublevel occur from the higher sublevels $(n \geqslant 4; k = 2)$. But since the lines are triplets, we must assume that one or the other of these types of sublevels is split into triple sub-sublevels. As was the case for sodium, the higher sublevels have this peculiarity. We are thus led to introduce the quantum number j and to subdivide the sublevels $(n \geqslant 4; k = 2)$ into triplets of type $(n \geqslant 4; k = 2; j = 0, 1, 2)$. All the other sublevels (n, k), in which $k > 2$, will likewise be split into triplets; and appropriate integral values for j will be chosen for the purpose. As before, multiplets of order higher than three may appear among the spectral lines, but the maximum multiplicity of the energy levels will be 3. We have said that the atom of magnesium may also emit singlets, and that when it is in a state conducive to the emission of singlets, none of the energy sublevels (n, k) is split. For the sake of consistency, however, the quantum number j is still introduced,

---

* The sublevel $(n = 3, k = 1)$ is also available, but when the atom is in a position to emit triplet lines, the basic sublevel is the one mentioned in the text. Pauli's Exclusion Principle furnishes the reason for this unexpected situation. We shall examine it on page 581 when we consider the similar case of the helium atom.

but now only one value of j is associated with each value of k in the sublevels (n, k).

To obtain the correct number of spectral lines in all cases, empirical selection rules must be imposed. We mentioned some of these in connection with the doublets. But here an additional rule must be postulated. It is to the effect that, if $j = 0$ in the higher sub-sublevel, it must always change to the value $+1$ during a drop. We may summarize all the selection rules by saying:

1. k must always change by ± 1 during a drop.
2. If $j > 0$ before the drop occurs, it must change by ± 1 or else must remain fixed during a drop.
3. If $j = 0$ before the drop occurs, it must change to $+1$ as a result of the drop.

The first of these selection rules was established on the basis of our former atomic model by appealing to the correspondence principle. But the two other rules affecting j have as yet no theoretical basis and must be viewed as empirical. Only when we have devised an appropriate mechanical model for the atom, shall we be in a position to give mechanical significance to the third quantum number j; and only then, by an application of the correspondence principle to the mechanical model, shall we justify the selection rules for j. In the following pages the first tentative model which was suggested for this purpose will be examined.

Landé's Core + Electron Model of the Atom—To obtain a theoretical interpretation of the multiplet structure of the spectral lines, we shall follow Bohr and represent the atom by means of a mechanical model. By pursuing this course we are not necessarily endorsing the thesis that the internal atomic processes can be represented in terms of the mechanical categories. Indeed, we already have every reason to believe that a mechanical representation is impossible, for even in the simpler cases discussed in the last chapter, we saw that the laws of mechanics were constantly violated. The advantage of imagining a mechanical model, if only as a working hypothesis, lies in a different direction: the model enables us to connect different facts and features which might otherwise exhibit no connection. Even at a later date when the model is discarded, we realize that it has played a useful part, precisely because it has drawn attention to hidden relations. At all events, Landé sought to devise a model of the atom which would account for the multiplet structure. He was guided by the following considerations:

Quantum numbers in Bohr's theory are associated with periodic motions. In the mechanical model discussed in the last chapter two independent periodicities were found for the motion of the optical electron.

These were the periodic motion of the electron in an ellipse and the
precession of the elliptical orbit as a whole in its plane around the
nucleus. From these two periodicities the two quantum numbers n and k
were derived. But the introduction of a third quantum number j (which
seems to be required by the multiplet structure) necessitates a third independent periodicity. This periodicity would be generated if, for example, the entire orbital plane of the optical electron were to precess
under a fixed inclination around some direction. The motion of the
optical electron would then be non-degenerate. A precession of this sort
occurred when we submitted our former model of the atom to the action
of a magnetic field: the Larmor precession then took place around the
direction of the magnetic field; a third quantum number m was introduced thereby; and the splitting of the spectral lines in the normal
Zeeman effect was accounted for.

The multiplet lines, however, contrary to the Zeeman lines, occur in
the absence of an impressed field, and we must seek elsewhere the origin
of the third periodicity. Landé traced it to the magnetic field developed
internally in the atom by the circling of the core-electrons; it is this
magnetic field which, according to Landé, secures the precession of the
optical electron's orbital plane. From Landé's standpoint the multiplet
lines thus result from a kind of internal Zeeman effect.

The model suggested by Landé to account for the third periodicity
involves some rather delicate points. We shall therefore preface our
description of his model with a few mechanical considerations. Suppose
we have an axially symmetrical top spinning around its axis. The spin
is represented by a vector placed along the axis; and, according to

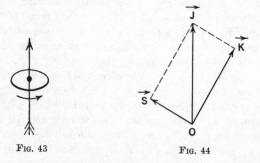

Fig. 43   Fig. 44

the usual conventions, it is set in such a way that if we look along the
vector from its tip to its tail, we shall see the top spinning in a counterclockwise direction (Figure 43). Our spinning top has an angular momen-

tum which is also represented by a vector situated along the axis. This vector points in the same direction as the first one, and its length is equal to the numerical value of the angular momentum of the top. If we call I the moment of inertia of the top about its axis, and ω the angular velocity of the spin, then the angular momentum is $I\omega$.

Let us now suppose that we have two axially symmetrical spinning tops, which need not be of the same size and need not be spinning at the same rate. We assume that the axis of either top is fixed to the same point O, but that the axes are not rigidly connected, so that they are free to pivot independently of each other about this fixed point. The first top, because of its spin, has an angular momentum which we represent in direction and in length by the vector $\overrightarrow{OS}$ set along the axis of spin (Figure 44). We proceed in the same way with the second top of angular momentum $\overrightarrow{OK}$. We shall suppose that no force of gravity is acting on the tops; hence the system may be viewed as floating in space. The two tops will then continue to spin with their axes remaining fixed in space, and aside from the spins no motion will occur.

Next we assume that the extremities of the two tops attract each other; for instance, we may suppose that an elastic band is placed over the tips of the two axes, drawing them together. The general principles of mechanics show that, under the action of the rubber band, the axes of the two tops cannot merely be drawn closer the one to the other without any further change occurring. This point is readily understood when we note that if, under the action of the rubber band, the relative inclination of the two axes of spin decreased and no other change took place, the total angular momentum of the system (which is defined by the diagonal $\overrightarrow{OJ}$ of the parallelogram of sides $\overrightarrow{OS}$ and $\overrightarrow{OK}$) would necessarily increase. But then the total angular momentum of the isolated system would not remain constant; and this would violate the laws of mechanics. We must suppose therefore that when the axes of the two tops are drawn closer by the rubber band, some counterbalancing angular momentum will arise offsetting the former increase. Conceivably, the spins of the two tops might slow down, or the axes of the two tops might precess in such a way as to ensure a decrease in the total angular momentum.

At all events it is unnecessary to resort to guesses, for the correct motion can be predicted through the medium of mathematical analysis. Accordingly, we shall merely state results: The two tops are first drawn slightly closer by the rubber band, the spins of the tops remaining practically unaffected. Then the two axes $\overrightarrow{OS}$ and $\overrightarrow{OK}$ will start precessing (as though rigidly connected) around $\overrightarrow{OJ}$ (*i.e.*, around the total-angular-momentum vector) which remains fixed in space; the precession will appear clockwise to an observer viewing the situation from the tip of the vector $\overrightarrow{OJ}$. This precessional motion offsets the increase in the angular momentum due to the approach of the two axes, and in this way the total angular momentum remains constant. We may also show that if the pull of the elastic band were increased, the precessional motion would be more rapid. If, instead of a rubber band, we placed a compressed spring between the two extremities of the tops, so that a repulsion would ensue, the precession would take place in the opposite direction.

At first sight it seems difficult to understand why the axes of the tops are not brought into coincidence under the action of the rubber band. The situation is no more mysterious, however, than that illustrated in a top spinning on the ground. Here also, so long as the top is spinning, the gravitational pull does not cause the top's axis to assume a horizontal position, it merely causes the axis to execute a precessional motion about the vertical. The peculiar gyroscopic motion of the two tops connected by a rubber band would also occur if each top were replaced by a bar magnet spinning along its length; the attraction of the opposite poles of the two magnets would here take the place of the action of the elastic band. We now propose to show that similar gyroscopic motions should arise in Landé's model of the atom.

In Landé's model the atom is represented, as before, by a large number of electrons circling around the nucleus, the most loosely bound of these electrons being the optical electron. According to the laws of mechanics the various electrons should exchange energy and angular momentum with one another during the motion, so that the energies of the core and of the optical electron should vary. In our treatment of the atomic model in the previous chapter we mentioned, however, that, owing to certain obscure quantum manifestations, no energy exchanges arise. Consequently, the energy and the angular momentum of the core will remain fixed, and so will the energy and the angular momentum of the optical electron (provided the atom is not radiating).

Now, the optical electron will have an angular momentum due to its circling, and as a result a magnetic field will arise. We may therefore liken the circling electron to a bar magnet which is spinning about the line of its length set perpendicularly to the orbital plane. Also the core may have angular momentum. This will be the case if there is a predominance of core-electrons circling in the same direction; the circling of the core-electrons as a whole will then develop a magnetic moment, and the core will behave like a bar magnet spinning about the axis of its angular momentum. The net result is that the atom may be likened to a system of two spinning bar magnets the axes of which are fixed to a common point at the centre of the atom. Since the opposite poles of the two spinning magnets exert attractive forces on each other, our model will be similar mechanically to our former system of two spinning tops connected by an elastic band; and the same gyroscopic motion will occur.

The motion of the optical electron in the atom is now readily understood. Thus let us revert to the figure of the two spinning tops (Figure 44), and let $\vec{OK}$, $\vec{OS}$, and $\vec{OJ}$ represent the angular momenta of the optical electron, of the core, and of the total atom. For short we shall refer to these vectors as $\vec{K}$, $\vec{S}$, $\vec{J}$. The vector $\vec{K}$ is perpendicular to the orbital plane of the optical electron. During the motion the vectors $\vec{S}$ and $\vec{K}$ will precess uniformly about the vector $\vec{J}$ of the resultant angular momentum, and it will be as though the three vectors $\vec{S}$, $\vec{K}$, and $\vec{J}$ formed a rigid framework pivoting about the diagonal $\vec{J}$. Thus the motion of the optical electron may be depicted as follows: the electron describes an ellipse in a plane; this ellipse is precessing in the plane with a constant angular motion about the fixed nucleus, and hence a rosette orbit results for the electron in the plane; finally this orbital plane, rigidly connected to the axis of the core, is precessing with uniform motion about the atom's resultant angular momentum $\vec{J}$ fixed in space. The three periodicities we have sought to detect in the motion of the optical electron are thus accounted for.

In what follows we shall represent the bare magnitudes (regardless of direction) of the three angular-momentum vectors $\vec{K}$, $\vec{S}$, $\vec{J}$ by K, S, J (i.e., without arrows). Since K, S, and J represent angular momenta, they must be positive numbers or else must vanish. The possibility of K vanishing may, however, be disregarded, for as we explained in the last chapter, the vanishing of K would imply that the electron was oscillating through the nucleus—a physical impossibility. On the other hand there is no reason why S and J should not vanish. Collecting these results, we see that K will always be positive, and S and J may be positive or else vanish. Further restrictions are placed on the angular momenta K, S, and J as a result of the following considerations.

Since the vector $\vec{J}$ is the diagonal of the parallelogram constructed on the two vectors $\vec{S}$ and $\vec{K}$, its magnitude J will depend not only on the magnitudes S and K of the two vectors $\vec{S}$ and $\vec{K}$, but also on the angle between them. The greatest possible value of J is attained when $\vec{S}$ and $\vec{K}$ point in the same direction. In this case the parallelogram collapses, and J (*i.e.*, the magnitude of $\vec{J}$) is equal to the sum of the magnitudes S and K of $\vec{S}$ and $\vec{K}$. Physically, this situation will occur when the angular momenta of the core and of the optical electron point in the same direction.

The smallest value of J is attained when $\vec{S}$ and $\vec{K}$ point in opposite directions. Its magnitude is then $K - S$ or $S - K$ (whichever of the two has a positive value). All possible situations are thus represented by the inequalities

$$(36) \qquad K + S \geqslant J \geqslant |K - S|.*$$

Up to this point we have introduced no quantum restrictions; we must now remedy this deficiency. The principles of mechanics show that the angular momenta S, K, J, when multiplied by 2π are adiabatic invariants which define certain Maupertuisian actions. Consequently the quantum restrictions consist in specifying that $2\pi S$, $2\pi K$, and $2\pi J$ must be integral multiples of h. We therefore set

$$(37) \qquad 2\pi S = sh; \quad 2\pi K = kh; \quad 2\pi J = jh;$$

where the quantum numbers s, k, and j are integers. The remarks we made on the permissible classical values of S, K, and J show that the quantum numbers s and j may have all positive integral values and may also vanish, whereas k must always be a positive integer.

The restrictions (37) may be written in the following equivalent form:

$$(38) \qquad S = s\frac{h}{2\pi}; \quad K = k\frac{h}{2\pi}; \quad J = j\frac{h}{2\pi};$$

and in this form, each of the three angular momenta is seen to be some integral multiple of the unit of angular momentum $\frac{h}{2\pi}$. The quantum condition affecting K is already known to us; it expresses the quantizing of the angular momentum of the optical electron. The first and last quantum conditions, referring to the core and to the total angular momentum, are new. We must also remember that the quantum condition which introduces the quantum number n is still in force; it fixes the major axis of the elliptical orbit. We do not write it here because it

---

* The symbol | | means the absolute value, *i.e.*, the positive value.

introduces no new element into the situation. Landé expressed the hope
that the quantum number j, associated with the total angular momentum
of the atom (and indirectly with the third periodicity in the electron's
motion), would be none other than the additional quantum number that
had been introduced by Sommerfeld to account for the spectra of multiplets. As for the quantum number s, we shall find that it serves to
represent a particular kind of state of the atom.

The relations (36) which held in the classical treatment are still valid
when the quantum restrictions are imposed. If, then, in (36) we introduce the expressions (38) for S, K, and J, we get a relation connecting
quantum numbers, namely,

$$(39) \qquad k + s \geqq j \geqq |k - s| .$$

Let us consider the inequalities (39). Since s, k, and j are restricted to integral values, we see that when s and k are specified, j cannot

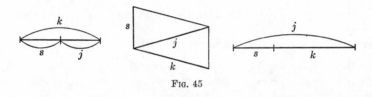

Fɪɢ. 45

have all values compatible with (39) but only those which are integers.
Suppose $s = 1$ and $k = 2$. Then, according to (39), j can have only the
values 1, 2, or 3. The possible configurations of the parallelogram are
thus considerably restricted; the three possible configurations are exhibited in Figure 45.

Our analysis shows that, as a result of the quantizing conditions, not
only are the magnitudes of the angular-momenta $\vec{S}$, $\vec{K}$ and $\vec{J}$ restricted
to certain discrete values, but also that the relative inclinations of these
vectors cannot be arbitrary. A kind of space-quantization, similar to
that discussed in the previous chapter, is thus revealed. The principal
difference between the two kinds of space-quantization is that, in the
present case, the quantization is automatic and does not arise from the
action of a magnetic field applied externally; it now occurs as a result
of the internal magnetic field developed by the rotating core.

Landé's mechanical model involves four periodicities, i.e., the three
periodicities of the optical electron and the periodicity exhibited by the
rotation of the core on its axis. To the first three periodicities correspond
the three quantum numbers n, k, j, and to the periodicity connected with

the core corresponds the new quantum number s. Since the empirical
evidence derived from the observation of multiplets seems to require
only the three quantum numbers n, k, j, and hence only three periodicities,
it would appear that the quantum number s and the fourth periodicity are
redundant. But this is not so for a number of reasons.

In the first place, we have seen that the same atom may emit different
kinds of spectra. (Calcium, for example, has a spectrum of singlets, but
it also has a spectrum of triplets.) Since the distribution of the sub-
sublevels must be different in the various cases, we must qualify an atom
further and must specify whether the atom is in a state conducive to the
emission of one kind of multiplet or of another. As we shall see, the
office of the quantum number s will be to define the state of the atom in
the sense just explained.

Other ways of clarifying the significance of the quantum number s
may be considered. Let us suppose, for instance, that s is zero; the core
then has no angular momentum, and as a result the angular momentum
J, or $j\dfrac{h}{2\pi}$, of the entire atom coincides with the angular momentum K,

or $k\dfrac{h}{2\pi}$, of the optical electron on its orbit. In this case the quantum
numbers j and k are the same. Since a separation in the values of k and
j occurs when s is non-vanishing (*i.e.*, when the core has angular momentum), we conclude that the angular momentum of the core serves
to create a difference between k and j and thereby gives significance to
the quantum number j. In short, were it not for the angular momentum
of the core, the third quantum number j, required for the interpretation
of the multiplet structures, would not be available.

Landé's mechanical model supplemented by the correspondence principle is able to account for Sommerfeld's empirical selection rules. The
proof is obtained by calculating the frequencies of the radiations that
should be emitted by the optical electron in Landé's model when the
classical laws of radiation are accepted. Thus we know that the frequencies radiated under the classical treatment are the mechanical frequencies
of the motion. Suppose, then, we have calculated the classical frequencies.
If now we apply the correspondence principle and thereby rule out the
quantum drops which cannot occur, Sommerfeld's selection rules for the
quantum number j, and also the older selection rules for the quantum
number k, are obtained.

We have seen that Landé's mechanical model justifies the introduction
of Sommerfeld's quantum number j and that it leads to the correct selection rules. But we cannot expect to calculate from this model (or indeed
from any other mechanical model) the exact energy values of the

various sub-sublevels and hence the exact frequencies that should be radiated. This would be to expect too much, for, as we have already seen in connection with the simpler model discussed in the last chapter, many of the principles of mechanics are violated in the quantum theory. And besides, the difficulty of the mathematical calculations would compel us to resort to approximations in any case. All that we may demand of Landé's model is that, when the quantum restrictions are taken into consideration, it should account for the various multiplicities of the sub-sublevels which the spectroscopic observation requires. In order to deduce these multiplicities from Landé's mechanical model we must assign definite angular momenta to the core and to the optical electron in every case. Now the angular momentum of the optical electron, whether in the basic or in an excited orbit, is known. Thus, if the optical electron is in an orbit (n, k), its angular momentum is $k\dfrac{h}{2\pi}$. But to determine the angular momentum of the core is not so easy. The core contains a large number of electrons, and up to the present we know nothing of the orientations of the planes in which the core-electrons are circling; nor do we know in what directions the circulations are proceeding.

Some insight into the matter is obtained when we consider the atom of an inert gas, e.g., neon. This atom contains an even number of electrons, and in view of its great stability we may suppose that its electronic motions are highly symmetric. In particular, we may assume that the orientations of the planes of the various electrons exhibit a symmetric pattern and that the electrons are circling two by two in opposite directions. If this be so, the atom of neon should have no net angular momentum and hence no magnetic moment (magnetic moment and angular momentum being necessarily associated when circling electrons are considered). These plausible expectations appear to be verified by experiment; for neon gas is diamagnetic, and Langevin's theory of magnetism shows that diamagnetism and the absence of a magnetic moment are closely connected. Inasmuch as the atom of an inert gas has the same electronic structure as the core of an alkali-metal atom, we infer that the core of an alkali-metal atom has no angular momentum. The angular momentum of an alkali-metal atom, e.g., sodium, may therefore be determined, and since this atom has the electronic configuration of the core of the magnesium atom, the angular momentum of this latter core is also known. Inferences of the foregoing type often lead to incorrect results, but at the present stage of the discussion they constitute our only means of guessing at the angular momenta of the various cores.

Before we examine whether Landé's model can account for the various multiplets, let us consider the numerical schemes which will express the partitioning of sub-sublevels required by the multiplet structure. We shall accept, at least provisionally, the inequalities (39), suggested by Landé's model, and we shall also take into consideration the restrictions placed on the values of the integers l, j, and s.

We first consider triplets. To obtain triplets, we must set $s = 1$ in the inequalities (39); for since k is always a positive integer and j is a positive integer or zero, we thereby obtain three possible values for j, namely,

$$k - 1, \quad k, \quad k + 1.$$

This result implies that each sublevel (n, k) must be formed of three distinct sublevels (n, k, j).

We have seen, however, that the triplet structure, though requiring that most of the sublevels be triple, necessitates that those sublevels in which k has the value unity should remain single. We conclude that the inequalities (39) are unable to express all the peculiarities of the triplet structure. This result should not surprise us, for the inequalities (39) were derived from a mechanical model, and we know that the macroscopic mechanical laws are no longer applicable inside the atom.

The Quantum Number l—We may overcome the discrepancy noted in the previous paragraph by resorting to an artifice. We agree that, whenever in the numbering of our levels a numerical value is assigned to k, we must change this numerical value, reducing it by one unit. Thus a sublevel $(n; k = 1)$ will be written $(n; k = 0)$. To avoid confusion, we introduce a new quantum notation. We replace k by the quantum number $l = k - 1$. Our sublevels (n, k) will henceforth be represented by (n, l). Also, the inequalities (39) will be replaced by

(40) $$l + s \geqslant j \geqslant |l - s|.$$

Finally, since our former k could have integral values only between 1 and n, the values that l may assume are 0, 1, 2, . . . $(n - 1)$.

Let us verify that, thanks to this artifice, the sublevels of magnesium associated with $k = 1$, *i.e.*, $l = 0$, will be single, whereas those connected with $k > 1$, *i.e.*, $l > 0$, will be triple. This point is immediately proved from (40), in which s is given the value 1 as before, and l is first given the value zero and then any positive integral value. If we set $s = 1$ and $l = 0$ we get

(41) $$1 \geqslant j \geqslant 1; \text{ whence } j = 1.$$

There is thus only one value for j and hence the sublevels $(n; l = 0)$ are single as required. If now we set $s = 1$ but give some positive integral value to l, e.g., $l = 1$, we get from (40)

$$(42) \qquad\qquad 2 \geqslant j \geqslant 0; \text{ whence } j = 0 \text{ or } 1 \text{ or } 2.$$

Three values are obtained for j, so that each sublevel $(n; l > 0)$ furnishes three sub-sublevels (n, l, j). Everything thus appears to be in order.

Next, we examine how singlets will be accounted for. We may verify that, when we set $s = 0$ in the inequalities (40), all the sublevels (n, k), or (n, l) with our new notation, will remain single. Thus when we set $s = 0$, the inequalities (40) become

$$(43) \qquad\qquad l \geqslant j \geqslant l;$$

and we conclude that j has always the same value as l, and hence has only one value.

More generally, we find that when s is given the values 0, 1, 2, 3, . . . , we may account for singlets, triplets, quintets, and septets . . . , respectively.

All the multiplets considered so far are of odd parity, and it would thus appear that multiplets of even parity cannot arise. How then are we to explain the doublets, quartets, . . . , which are revealed in some of the spectra? To account for doublets, we must introduce a further change in our numerical scheme. For instance, if we suppose that, contrary to our former beliefs, the quantum numbers s and j may assume half-integral values, multiplets of even parity can be accounted for.

Thus let us set $s = \frac{1}{2}$ in the inequalities (40), and let us suppose that j can take only half-integral values, though l assumes integral values as before. We get

$$(44) \qquad\qquad l + \frac{1}{2} \geqslant j \geqslant \left| l - \frac{1}{2} \right|.$$

The implications of these inequalities are rendered more explicit in the following scheme.

$$(45) \qquad \text{If } l = 0, \quad \text{then} \quad j = \frac{1}{2},$$

$$\text{and if } l > 0, \quad \text{then} \quad j = l - \frac{1}{2} \text{ or } l + \frac{1}{2}.$$

Thus, all the sublevels $(n; l = 0)$ are single and are of type $\left(n; l = 0; j = \frac{1}{2} \right)$; and all the other sublevels are double, i.e., $\left(n; l; j = l - \frac{1}{2} \text{ or } l + \frac{1}{2} \right)$. In short, the correct divisions of the levels of sodium, as required by the spectroscopic evidence are accounted for.

We may verify that in every case the maximum order of multiplicity (*i.e.*, the maximum number of sub-sublevels (n, l, j) into which a sublevel (n, l) is divided) is given by $2s + 1$. According to whether s has the value 0 or $\dfrac{1}{2}$ or 1 or $\dfrac{3}{2}$. . . . . , the value of the maximum order of multiplicity is 1, 2, 3, 4. . . . . The dependence of the order of multiplicity of the spectrum on the numerical value of s is thus manifest.

Difficulties Encountered in the Mechanical Model—The half-integral values credited to s and j and the change from the quantum number k to the quantum number l have no other justification than to permit the devising of a numerical scheme for the various multiplets. We must therefore inquire whether these modifications are consistent with the requirements of our mechanical model. We first consider the half-integral values of s and of j.

The quantum numbers s, j, and also l, when multiplied by h, define the permissible values of the adiabatic invariants connected with the core, the total atom, and the circling optical electron. Now a fundamental tenet of Bohr's theory has been that adiabatic invariants may be only whole multiples of the quantum of action h. The half-integral values we have just ascribed under certain conditions to the two first adiabatic invariants violate this basic tenet. However, since the original tenets of the quantum theory were adopted so as to account for particular occurrences, we cannot object to their being modified when more complicated phenomena are considered.

But the change from the quantum number k to the quantum number l confronts us with a new difficulty. The angular momentum of the optical electron in Landé's model was $K = k\dfrac{h}{2\pi}$. The change from the quantum number k to the quantum number l thus implies that, henceforth, this angular momentum must be replaced by

$$(46) \qquad L = l\frac{h}{2\pi} = (k-1)\frac{h}{2\pi} = K - \frac{h}{2\pi}.$$

The new definition requires that the angular momentum of the optical electron should fall below its former value by one unit $\dfrac{h}{2\pi}$. On the other hand, the angular momentum S of the core is still expressed by $S = s\dfrac{h}{2\pi}$; it is subjected to no revision, except that s may now assume half-integral values. The total angular momentum $\vec{J}$, however, since it is now the resultant of $\vec{L}$ and $\vec{S}$, not of $\vec{K}$ and $\vec{S}$, has neither the same magnitude nor the same direction as before.

Now, this change in the value of the optical electron's angular momentum cannot be understood from the standpoint of mechanics. Thus far, it appears as a mere artifice devised to account for the multiplet structure, and its deeper significance escapes us entirely. The strange consequences to which it leads in the field of mechanics may be illustrated in the case of the hydrogen atom. Let us consider a hydrogen atom in the basic state. The basic orbit is the circle $(n = 1; k = 1)$, so that, according to mechanics, the electron has one unit $\frac{h}{2\pi}$ of angular momentum. But our revised ideas compel us to say that this basic orbit must be written $(n = 1; l = 0)$, so that, as a result of $l = 0$, the circling electron has no angular momentum at all. From the standpoint of mechanics this is incomprehensible, for an electron moving in a circle must have angular momentum; if it had no angular momentum, it could not describe a circle or an ellipse but would merely oscillate through the nucleus. We have here another illustration of the difficulties which attend any attempt to devise a mechanical model for the atom. Inasmuch as the change from k to l is required in many other cases, we must recognize that in the subatomic world the ordinary concepts of mechanics lose their validity.

We now inquire whether Landé's mechanical model can account for the various multiplets when all the modifications previously noted are introduced. As an example we consider the case of the magnesium atom. This atom differs primarily from the atom of neon in that its nucleus is not the same. In particular, the magnesium nucleus has a positive charge exceeding that of the nucleus of neon by two units. For the present, however, we are concerned solely with the electronic configurations, and from this standpoint the atom of magnesium may be represented by the closed structure of neon around which two electrons are circling. One of these two outer electrons is the optical electron of the magnesium atom, and so the core of this atom is represented by the closed electronic structure of neon and by one outer electron. We have already observed that the closed electronic configurations of the inert gases have no net angular momentum. Hence the core of the magnesium atom must have the angular momentum of the single outer electron. Now, the spectroscopic evidence shows that this outer electron is in an orbit $(n = 3; k = 1)$, or $(n = 3; l = 0)$ with our revised notation. The quantum value, $l = 0$, indicates that the outer electron has no angular momentum and hence that the core of the magnesium atom likewise has none. Accordingly the value, $s = 0$, must be assigned to the core. We conclude that the magnesium atom should emit a spectrum of singlets.

But the difficulty is that magnesium may also be in the triplet state and emit triplets. To account for triplets we must assume that s now has the value unity, and therefore that the core of magnesium has one unit of angular momentum, $\dfrac{h}{2\pi}$. The core's outer electron must therefore also have one unit of angular momentum, in spite of the fact that the spectroscopic evidence shows that it is circling in the same orbit ($n = 3$; $l = 0$) as before, and hence in an orbit for which there should be no angular momentum. This difficulty which appears insuperable from the standpoint of Landé's mechanical model will receive a satisfactory explanation when the hypothesis of the spinning electron is considered.

A similar difficulty confronts us when we wish to account for the spectra of doublets. Thus, the sodium atom emits doublets. Its core has the same electronic configuration as the atom of neon and hence should have no angular momentum. We should therefore expect $s = 0$. But, to account for doublets, we must set $s = \frac{1}{2}$ and ascribe a half-unit of angular momentum to the core. This difficulty will likewise be removed when the hypothesis of the spinning electron is considered.

The inequalities (40) have proved fundamental in the interpretation of atomic spectra, and their importance is in no wise dependent on our ability to justify them by means of a model. For future reference we have summarized their implications in the following table.

$$(47) \begin{cases} \text{singlets;} \quad s = 0; \quad l = j = 0, 1, 2, 3 \ldots (n-1). \\[2mm] \text{doublets;} \, s = \dfrac{1}{2}; \, l = 0, 1, 2, 3 \ldots (n-1) \\[2mm] \qquad\qquad j = l \pm \dfrac{1}{2} \qquad\qquad (\text{if } l > 0) \\[2mm] \qquad\qquad j = \dfrac{1}{2} \qquad\qquad\quad\; (\text{if } l = 0) \\[2mm] \text{triplets;} \quad s = 1; \quad l = 0, 1, 2, 3 \ldots (n-1) \\[2mm] \qquad\qquad j = l \pm 1, l \qquad\quad (\text{if } l > 0) \\[2mm] \qquad\qquad j = 1 \qquad\qquad\quad\;\; (\text{if } l = 0) \\[2mm] \text{quartets;} \, s = \dfrac{3}{2}; \, l = 0, 1, 2 \ldots (n-1) \\[2mm] \qquad\qquad j = l \pm \dfrac{3}{2}, l \pm \dfrac{1}{2} \qquad (\text{if } l > 1) \\[2mm] \qquad\qquad j = \dfrac{1}{2}, \dfrac{3}{2}, \dfrac{5}{2} \qquad\quad (\text{if } l = 1) \\[2mm] \qquad\qquad j = \dfrac{3}{2} \qquad\qquad\quad\;\; (\text{if } l = 0). \end{cases}$$

We have omitted to consider quintets and higher multiplets, but there is no difficulty in continuing the table so as to cover all cases.

When we consider the drops which generate the spectral lines, we must take into account the selection rules that restrict the permissible changes of the quantum numbers l and j. These rules were given on a previous page for k and for j, and since $l = k - 1$, the same rules will hold for l and for j. For future reference we shall restate these selection rules in terms of the quantum numbers l and j:

1. l must always change by ± 1 during a drop.
2. If $j > 0$ before a drop occurs, j must change by ± 1, or else it must remain fixed.
· 3. If $j = 0$ before a drop occurs, j must change to the value $+1$.

The selection rules we have stated show that a drop in which l does not change, or changes by ± 2, is impossible. Similarly a drop in which j changes by 2 is excluded. But as will be explained in the next chapter, drops which violate the selection rules are sometimes observed. Accordingly, we shall be on safer ground if we say that the drops excluded by the selection rules are improbable though not impossible.

There is also a third selection rule which we have applied implicitly. It concerns the quantum number s. Heretofore we have assumed that when an atom can emit different kinds of multiplets, its state (determined by the value of s) does not change during a drop: the quantum number s remains fixed. But further investigation shows that in certain cases the value of s does change. The lines which result from such drops are called "intercombination lines"; one of them is conspicuous in the spectrum of mercury. Of course, such lines do not belong to the spectra we have been discussing.

Landé's mechanical model gives no information on the manner in which s may change. Some insight into the matter was obtained when the spinning electron was introduced, but it was only after the discovery of the newer methods of quantum mechanics (e.g., wave mechanics) that a satisfactory solution was given.

The Anomalous Zeeman Effect—In the last chapter we mentioned that if an atom emitting radiation is submitted to the action of a magnetic field, each spectral line is split into three or more lines, the separations of which increase with the intensity of the field. This general phenomenon is called the Zeeman effect. But there are two kinds of Zeeman effects: the normal effect and the anomalous one. In the normal

effect each spectral line is decomposed into three lines (called components), whereas in the anomalous effect more than three components usually
appear.   Even when the anomalous effect yields only three components,
the separations are not the same as in the normal effect, so that the two
Zeeman effects appear to be entirely different.   The anomalous effect is
generated whenever the decomposed line belongs to a group of multiplets;
the normal effect occurs only when the line is a singlet.

We saw that Bohr's model of the hydrogen atom accounted for the
normal Zeeman effect, which hydrogen was believed to exhibit.[*]   The
next step was therefore to interpret the anomalous effect disclosed in the
case of the multiplet lines of the higher atoms.   But here several difficulties and complications beset Bohr's semi-mechanical theory.   So· as to
simplify the presentation we shall dwell only on the more essential points.

The first attempt to explain the anomalous Zeeman effect was undertaken by Landé, who utilized in this connection the core + electron model
of the higher atoms.   In this model, the core, having the angular momentum $S = s\dfrac{h}{2\pi}$, should have a magnetic moment of s magnetons—at
least this is what we should expect if we were to be guided by the classical
relation between magnetic moment and angular momentum, $viz.$,

$$(48) \qquad \frac{\text{magnetic moment (in magnetons)}}{\text{angular momentum}\left(\text{in units }\dfrac{h}{2\pi}\right)} = 1.\dagger$$

Similarly the optical electron and the total atom, having angular mo-
menta $L = l\dfrac{h}{2\pi}$ and $J = j\dfrac{h}{2\pi}$ respectively, should have magnetic moments of l and of j magnetons.

If the core + electron model is placed in a magnetic field, the angular
momenta $\vec{S}$ and $\vec{L}$ of the core and of the optical electron will precess as
before around the atomic axis.   At the same time the atomic axis $\vec{J}$, which
remained fixed in the absence of a magnetic field, will now precess at
some constant inclination around the direction of· the field.   Owing to

---

[*] Subsequent observations of the hydrogen spectrum have shown that hydrogen
exhibits the anomalous effect, not the normal one.   This feature will be discussed at
the end of the chapter.

† This relation is derived immediately from (13), page 517.

space-quantization, only a certain discrete set of inclinations will be possible for the atomic axis. These quantized inclinations are determined by the permissible values of the quantum number m (the same number we encountered in the normal Zeeman effect). In the present case we must suppose that there are $2j + 1$ quantized inclinations and hence $2j + 1$ possible values for m, viz.,

$$m = -j, -j+1, -j+2, \ldots \ldots j-2, j-1, j.$$

Since j may be an integer or a half integer, we see that m may itself be an integer or a half integer. Thus, half-integral values appear once more in Bohr's theory. The quantum number m also measures, in magnetons, the projection of the total magnetic moment of the atom on the direction of the magnetic field.

During the process of emission, abrupt changes in the inclination of the atomic axis may occur, with the result that different Zeeman components will replace the single line which would be emitted in the absence of a magnetic field. However, since the selection rules require that m should be restricted to changes ± 1 or else remain unchanged, our present model predicts only three Zeeman components, and hence the normal Zeeman effect. The increased complication of the core + electron model over the simple hydrogen model is thus of no avail in accounting for the anomalous Zeeman effect.

In view of this setback, Landé modified the core + electron model. He still retained the characteristic structure of a core around which the optical electron is revolving, but he assumed that the projection of the total magnetic moment of the atom onto the direction of the field (when the atom was in a state determined by the numbers s, l, j) was not m magnetons but was gm magnetons, where

$$(49) \qquad g = 1 + \frac{j(j+1) + s(s+1) - l(l+1)}{2j(j+1)}.$$

As a result of this change, Landé showed that the correct number of Zeeman components, and also their separations, could be predicted in every case. The normal effect for singlet lines and the various forms of the anomalous effect for multiplet lines were thus accounted for simultaneously. The modification introduced by Landé does not affect many of our former results: as before, the atomic axis precesses around the field and can assume only $2j + 1$ quantized inclinations.

Landé's unexpected value for the magnetic moment may in turn be explained if the following assumptions are made:

1. We must suppose that the angular momenta of the optical electron, core, and total atom, respectively, are

$$(50) \qquad \sqrt{l(l+1)}\frac{h}{2\pi}, \qquad \sqrt{s(s+1)}\frac{h}{2\pi}, \qquad \sqrt{j(j+1)}\frac{h}{2\pi},$$

instead of

$$(51) \qquad l\frac{h}{2\pi}, \qquad s\frac{h}{2\pi}, \qquad j\frac{h}{2\pi}.$$

2. The ratio of magnetic moment to angular momentum retains the classical value (48) in the case of the optical electron, but it must have *twice* the classical value in the case of the core. It then follows that the value of this ratio in the case of the total atom will be g times the classical value.

It must be emphasized that Landé's revisions are as mysterious as the anomalous Zeeman effect he was seeking to explain. Especially is this true of the values (50) for the angular momenta. One of the fundamental tenets of Bohr's theory has been that the successive quantized values of an angular momentum should differ by $\frac{h}{2\pi}$ and this rule was observed even after half-quantum numbers had been introduced. But Landé's expressions (50) for the angular momenta show that Bohr's tenet is disregarded, and that an unexpected progression of values is required. There thus appeared to be no comprehensive scheme underlying Bohr's theory. As Andrade expressed the situation: "The immediate problem is to find cookery receipts, rather than physical principles." *

The Paschen-Back Effect—The Zeeman effect occurs only when the magnetic field is not too intense—at least this is so in the case of the anomalous effect. When the field is very strong, a new effect called the Paschen-Back effect comes into existence. We may readily understand its nature. Let us suppose that our atom emits multiplets, and that a relatively weak magnetic field is applied. Each line of a multiplet will undergo the anomalous Zeeman decomposition, and when the field is progressively increased the separations of the Zeeman components in-

* Andrade. The Structure of the Atom. Third Edition. Harcourt, Brace and Co., New York, 1926; p. 553.

crease in proportion. But when the intensity of the field reaches a certain point, we find that all the Zeeman components pertaining to the same multiplet suddenly coalesce into three lines. The separations and polarizations of these three lines are precisely those of the normal Zeeman triplet which would be generated by the normal Zeeman decomposition of a singlet line. This transformation of a multiplet into a Zeeman triplet under the action of a strong magnetic field constitutes the Paschen-Back effect.

The Paschen-Back effect may be accounted for by means of the core + electron model—at least with the same measure of success as the Zeeman effect. We recall that, in the Zeeman effect, the angular-momentum vectors $\vec{S}$ and $\vec{L}$ of the core and of the optical electron were compounded into the resultant vector $\vec{J}$, and that this vector was then acted upon by the magnetic field and submitted to space-quantization. To interpret the Paschen-Back effect, we must suppose that when the intensity of the field reaches the point where this effect is about to take place, the coupling of the vectors $\vec{S}$ and $\vec{L}$ suddenly snaps, being no longer strong enough to resist the disruptive action of the field. Thus the vector $\vec{J}$ ceases to be formed, and the vectors $\vec{S}$ and $\vec{L}$ are now acted upon directly by the magnetic field, precessing independently of each other around the field's direction, under quantized inclinations. When suitable selection rules are imposed on the new quantum numbers that appear in this treatment, the Paschen-Back effect can be accounted for. It remains to be said that the sudden snap in the coupling of the vectors $\vec{S}$ and $\vec{L}$ illustrates the sudden breakage of some non-mechanical constraint, obviously of a quantum nature.

Pauli's Objections to the Core + Electron Model—Pauli investigated the core + electron model by means of the mechanics of relativity. He observed that the relativistic variations in the masses of the core-electrons, caused by their motions, would increase in importance as the heavier atoms were considered. He showed that under these conditions the ratio of the core's magnetic moment to its angular momentum would necessarily vary with a change in the atomic number, so that Landé's g-formula would be subjected to a progressive change when we passed from one atom to the next in Moseley's table. This would imply that the separations of the Zeeman lines should be dependent on the atomic number of the atom of interest. Inasmuch as this expectation is not borne out by experiment, Pauli concluded that the core + electron model was incompatible with the theory of relativity and should be rejected in consequence. His suggestion was that a division of the atom into a core and an optical electron should be abandoned in favor of some scheme which would place all the electrons on the same footing. Such a revision would, of course, compel us to devise some new interpretation for the quantum number s, formerly attributed to the angular momentum of the core. We shall see that the study of the X-ray spectra appeared to confirm Pauli's views.

The X-ray Spectra—Thus far we have been concerned with the spectral lines generated by the drops of the optical electron (optical spectra). But spectral lines are also generated when one of the core-electrons is removed. Some other electron then drops into the vacated orbit, and radiation is emitted. We thus obtain a new kind of spectrum, called an X-ray spectrum on account of the X-ray frequencies of its lines.

The examination of X-ray spectra has shown that they are analogous to the optical spectra of doublets, such as are emitted by sodium and potassium. According to table (47), doublets in the optical spectra are associated with the quantum number $s = \frac{1}{2}$ and with corresponding sets of values for l and j. We must expect therefore that all X-ray spectra will be connected in some way with these quantum numbers. However, it is certain that the X-ray quantum numbers cannot have the same physical significance as those which intrude in the optical spectra. This point is made clear when we consider the number s. In the optical spectra the value of s varies from one atom to another and may also have different values for the same atom. But in the X-ray spectra, s must always retain the value $\frac{1}{2}$. Obviously the number s cannot refer to the same situation in both cases. Similar arguments apply to l and j. In short, the study of the X-ray spectra indicates the existence of energy levels involving quantum numbers l, j, s which are not the same as the numbers encountered in the optical spectra. In addition the fourth quantum number n must be considered in the X-ray spectra just as it was in the optical ones. The problem is thus to give physical significance to the X-ray quantum numbers, n, l, j and $s = \frac{1}{2}$.

Now we recall that in Bohr's process of atom building, two quantum numbers were attributed to each electron. We called these numbers n and k (or n and l with the revised notation). It would have been preferable, however, to call these numbers n_i and l_i, the index i meaning that the numbers are associated with the ith electron. We shall follow this practice here, and shall assume that these two numbers n_i and l_i are the two X-ray quantum numbers n and l. In view of this association of the quantum numbers n_i and l_i with the individual electrons, physicists suspected that the other X-ray quantum numbers, j and $s = \frac{1}{2}$, should also be allocated to the electrons. This would imply that the ith electron was connected with the four numbers n_i, l_i, j_i, and $s_i = \frac{1}{2}$.

We propose to show that the numbers l_i, j_i, s_i are not truly independent numbers. In order to establish this result, we first recall that l_i, j_i, and s_i must satisfy the relations of doublets exhibited by l, j, and $s = \frac{1}{2}$ in the table (47). We conclude that the relationships of l_i, j_i, and s_i are the same as those of l, j, and s. Now only two of the numbers l, j, and s (i.e., the magnitudes of our former vectors $\vec{l}$, $\vec{j}$, and $\vec{s}$) are independent, because $\vec{j}$ was obtained by compounding $\vec{l}$ and $\vec{s}$, so that when two of the vectors were specified, the third was determined. We may infer therefore that only two of the electronic numbers l_i, j_i, and s_i are independent. As a result each electron is associated in effect with only three *independent* quantum numbers, e.g., n_i, l_i, and j_i.

Now we know that an atom which is not submitted to a perturbation is degenerate, its state being then determined by the three independent quantum numbers n, l, and j. The degeneracy may be removed by the application of a magnetic field, and when this is done, a fourth quantum number, m, is introduced. In view of the fact that we have already allotted the analogues of n, l, and j to each electron, it appears plausible to suppose that when the degeneracy of an atom is removed by a magnetic field, each electron should receive a fourth independent quantum number m_i, analogues to m. Each electron would thus be associated with a quadruplet of independent quantum numbers $(n_i,\ l_i,\ j_i,\ m_i)$. Pauli defended this view. The principal objection to it was that independent quantum numbers were associated with independent periodicities in the motion, and that an electron, treated as a point-mass, could have only three independent periodicities, since it had only three degrees of freedom. This difficulty was removed by the hypothesis of the spinning electron.

The Spinning Electron—So as to overcome the objection just noted, Uhlenbeck and Goudsmit supposed that an electron does not merely undergo displacements but can also spin about an axis. The required fourth degree of freedom is thus accounted for.

Uhlenbeck and Goudsmit's hypothesis is then as follows:

1. Each electron in an atom is spinning with a half unit of angular momentum, $i.e.$, $\frac{1}{2}\frac{h}{2\pi}$. The quantum number $s_i = \frac{1}{2}$, which we wished to attach to the electron, merely represents the magnitude of the angular momentum of spin $\left(\text{in units } \frac{h}{2\pi}\right)$.

2. Each electron has, as a result of its spin, a magnetic moment of one magneton. It will be observed that the ratio of magnetic moment to angular momentum has here twice the classical value (48), just as was postulated by Landé for the core in the core + electron model. The reason for this assumption will be understood presently.

We now examine the changes that the spinning-electron hypothesis will entail in our earlier outlook. The quantum numbers s_i, l_i, j_i are the magnitudes of the angular-momentum vectors $\vec{s_i}$, $\vec{l_i}$, $\vec{j_i}$ of the ith electron. The vector $\vec{s_i}$ is the spin vector; $\vec{l_i}$ is the angular-momentum vector due to the circling of the electron on its orbit; and $\vec{j_i}$, obtained by compounding $\vec{s_i}$ and $\vec{l_i}$, is the total angular-momentum vector of the ith electron, due to its spin and to its circling $\left(\text{all these angular momenta are expressed in units } \frac{h}{2\pi}\right)$. The number n_i is the main quantum number of the orbit the electron is describing.* The fourth independent quantum number m_i will be examined when we consider Pauli's exclusion principle.

We next consider the connections between these quantum numbers and those of the table (47), which refer to the atom as a whole. The spin vectors $\vec{s_i}$ of the various electrons will set themselves parallel and antiparallel, owing to the mutual magnetic actions of the spinning electrons. The resultant vector $\vec{s}$ will thus have one of the values $0, \frac{1}{2}, 1, \frac{3}{2}, 2 \ldots$. We have here the interpretation of our former quantum number s of (47), which Landé attributed to the angular momentum of the core. Similarly, the resultant of the vectors $\vec{l_i}$ is the vector $\vec{l}$, which was formerly thought to represent the angular momentum of the optical electron. The resultant of $\vec{s}$ and $\vec{l}$ will be $\vec{j}$, and will therefore define, as before, the total angular momentum of the atom and hence the direction

* The permissible values of the quantum numbers l_i and j_i are the same as those of l and j in the table (47) when $s = \frac{1}{2}$.

of the atomic axis. As for the number n, it is the number n_i of the optical electron.

The name *kernel* is given to the closed electronic configuration which underlies the exterior (or valency) electrons. We must assume that the electrons of the kernel are spinning and also revolving two by two in opposite directions. Consequently the resultants $\vec{s}$ and $\vec{l}$ for the kernel both vanish, so that the kernel has no angular momentum and therefore no magnetic moment. These assumptions account for the absence of magnetic moment manifested by the inert gases, whose atoms necessarily reduce to a kernel. A further consequence is that in an atom which is not an inert gas, the $\vec{s}$ and the $\vec{l}$ of the total atom will be the resultants of spin and of circulation of the valency electrons alone.

Our understanding of the multiplet structures is considerably simplified by the hypothesis of the spinning electron. For example, in the sodium atom, there is but one electron (the optical electron) circling around the kernel. The s of the atom is thus merely the $s_i = \frac{1}{2}$ of the optical electron. When the atom is excited, the electron is raised to a higher orbit and may be spinning and circling in the same or in opposite senses. The energy of the atom is not quite the same in the two cases, and so when the electron drops back to the lowest orbit, one or the other of the two lines of the doublet is generated. The doublet structure is thus due to the electron spin. Let us also consider the case of the magnesium atom. Here there are two valency electrons, which may be spinning in the same or in opposite directions. In the former case, the resultant of the spins yields $s = 1$, and the atom is in the triplet state; in the latter case the spins cancel, so that $s = 0$, and the atom is in the singlet state. Thus the difficulty we encountered formerly, when we wished to understand how the same core should sometimes have one unit of angular momentum and at other times none, is obviated entirely.

The spinning-electron hypothesis supplies us with a new model of the atom. The earlier division of the atom into two parts, the core and the optical electron, no longer holds. All the electrons now contribute jointly to the numerical values of the quantum numbers s and l.

The new model does not change appreciably our understanding of the Zeeman effects. As before, the atomic axis $\vec{j}$ (now the resultant of the spin and circulation vectors of all the electrons) precesses around the field and can assume only $2j + 1$ quantized inclinations determined by the permissible values of the quantum number m. Landé's unexpected value for the ratio of the core's magnetic moment to its angular momentum is merely transferred from the core to each individual electron. His

g-formula (49) and the expressions (50) are retained, but their significance remains as mysterious as before. One of the triumphs of the new quantum mechanics has been to account for these expressions.

To account for the Paschen-Back effect, we must assume that the strong magnetic field overcomes the coupling between the total spin vector $\vec{s}$ and the total circulation vector $\vec{l}$. These two vectors are no longer compounded into a vector $\vec{j}$, and the magnetic field now acts on them directly and individually, causing them to precess independently of each other under quantized inclinations around the direction of the field. Incidentally we see that the way in which the electronic vectors are coupled may vary from one situation to another.

Pauli's Exclusion Principle —Let us consider the fourth independent quantum number m_i which we have said to be associated with each electron when the degeneracy of the atom is removed by a magnetic field. This fourth number did not appear in the treatment of the Zeeman effect or of the Paschen-Back effect, because the methods of coupling the electronic vectors, utilized to interpret these effects, were not of the kind that reveals the number m_i. We observed, however, that the coupling of vectors may be performed in various ways according to conditions. We shall suppose therefore that the electronic vectors $\vec{l_i}$ and $\vec{s_i}$ are coupled as we assumed originally, i.e., $\vec{l_i}$ and $\vec{s_i}$ are compounded and furnish $\vec{j_i}$, which represents the total angular momentum (in units $\frac{h}{2\pi}$) of the ith electron. If now a magnetic field is applied the vector $\vec{j_i}$ will undergo space-quantization and will precess around the magnetic field. The quantized inclinations of $\vec{j_i}$ are determined by its projections m_i on the direction of the field. We thus obtain the $2j_i + 1$ quantized values

$$(52) \quad m_i = -j_i, -j_i + 1, -j_i + 2, \ldots \ldots j_i - 2, j_i - 1, j_i.$$

In short, when the degeneracy is removed by the magnetic field, each electron will be associated with a quadruplet of quantum numbers (n_i, l_i, j_i, m_i), these numbers serving to define the corresponding sub-sub-sublevels of the electrons.

Let us recapitulate the various values which the quantum numbers n_i, l_i, j_i, m_i may assume. The number n_i may assume all positive integral values. The numbers l_i and j_i may take all the values which were allowed to l and j in the table (47) when s was supposed to have the value $\frac{1}{2}$. Consequently l_i may have only the values 0, 1, 2, . . . $(n_i - 1)$; j_i may have either one of the two values $l_i \pm \frac{1}{2}$, except when $l_i = 0$. In this last case the value of j_i is $\frac{1}{2}$. As for m_i its permissible values are defined by (52).

The possible quadruplets of numbers in the case of $n_i = 1, 2, 3,$ are listed in the table (53).

$$(53) \begin{cases} n_i = 1; \quad l_i = 0; \quad j_i = \frac{1}{2}; \quad m_i = \pm \frac{1}{2} \\[2mm] n_i = 2; \begin{cases} l_i = 0; \quad j_i = \frac{1}{2}; \quad m_i = \pm \frac{1}{2} \\[2mm] l_i = 1; \begin{cases} j_i = \frac{1}{2}; \quad m_i = \pm \frac{1}{2} \\[2mm] j_i = \frac{3}{2}; \quad m_i = \pm \frac{3}{2}, \pm \frac{1}{2} \end{cases} \end{cases} \\[8mm] n_i = 3 \begin{cases} l_i = 0; \quad j_i = \frac{1}{2}; \quad m_i = \pm \frac{1}{2} \\[2mm] l_i = 1; \begin{cases} j_i = \frac{1}{2}; \quad m_i = \pm \frac{1}{2} \\[2mm] j_i = \frac{3}{2}; \quad m_i = \pm \frac{3}{2}, \pm \frac{1}{2} \end{cases} \\[4mm] l_i = 2; \begin{cases} j_i = \frac{3}{2}; \quad m_i = \pm \frac{3}{2}, \pm \frac{1}{2} \\[2mm] j_i = \frac{5}{2}; \quad m_i = \pm \frac{5}{2}, \pm \frac{3}{2}, \pm \frac{1}{2}. \end{cases} \end{cases} \end{cases}$$

Each one of these quadruplets of values (n_i, l_i, j_i, m_i) defines a sub-sub-sublevel available to the electrons of an atom. There is no difficulty in continuing the table for higher values of n_i.

It should be noted that our table affords no indication of the number of electrons which may occupy any given sub-sub-sublevel. But Pauli, guided by several clues on the atomic structures, surmised that each sub-sub-sublevel would contain one electron or else none. In other words no two electrons could occupy the same sub-sub-sublevel. It would be as though the presence of one electron in a sub-sub-sublevel excluded the presence of another; whence the name "Exclusion Principle" often given to Pauli's principle.

Pauli's principle is most simply expressed in terms of the quantum numbers. It then assumes the following form:

No two electrons in the same atom can have the same quadruplet of quantum numbers.*

Our statement of Pauli's principle involves four independent electronic quantum numbers, and hence assumes that the atom is in a non-degenerate condition. In the foregoing treatment we removed the degeneracy from the atom by the application of a magnetic field, and in this way the fourth quantum number m_i was added to the three existing ones, n_i, l_i, j_i. But suppose we withdraw the magnetic field. In this case m_i loses all physical significance, and so we might believe that Pauli's principle could no longer be applied.

This difficulty may be obviated, however, if we follow the general procedure of retaining irrelevant quantum numbers that was outlined on page 528 in connection with Bohr's original model of the atom. Thus we shall suppose that, even in the absence of a magnetic field, the fourth quantum number m_i is retained and credited with the values indicated in the table (53). The electrons will then be associated with exactly the same values of m_i as before, the only difference being that all sub-sub-sublevels (n_i, l_i, j_i, m_i), in which m_i alone differs in value, will coalesce. The seemingly meaningless number m_i may be viewed in this case as having a potential significance, in the sense that it indicates how the various electronic axes will orientate themselves if a magnetic field is applied. In view of the permissibility of retaining the number m_i even when the atom is degenerate, we conclude that Pauli's principle may be applied without change to the degenerate atom.

It is of interest to determine how many electrons in the same atom can occupy the same sub-sublevels (n_i, l_i, j_i), or sublevels (n_i, l_i), or levels (n_i). Pauli's principle, taken in conjunction with the table (53) supplies the answer. Thus we may verify that not more than $2j_i + 1$ electrons in an atom can have the same n_i, l_i, and j_i; not more than

---

* The four quantum numbers may be selected in a different way. Thus let us view n_i, l_i, and $s_i = \frac{1}{2}$ as the independent quantum numbers, and let us suppose that an exceedingly strong magnetic field, far stronger than in the case of the Paschen-Back effect, is applied. We may suppose that this field will disrupt all couplings between the electronic vectors. In this case the vectors $\vec{l_i}$ and $\vec{s_i}$ will be acted on directly by the field and will undergo space-quantization. Their projections m_{l_i} and m_{s_i} (on the direction of the field) will thus assume only quantized values. A table similar to (53) may be constructed for the four quantum numbers n_i, l_i, m_{l_i}, m_{s_i}, and Pauli's principle may be expressed in terms of the new numbers.

$2(2l_i + 1)$ can have the same n_i and l_i; and not more than $2n_i{}^2$ can have the same n_i.

The exclusion principle has many applications. For instance, when Bohr first attempted to build up the higher atoms by adding successive electrons, there was considerable doubt as to the number of electrons that could be placed in any particular group of orbits (n_i, l_i). Pauli's principle yields this number, i.e., $2(2l_i + 1)$,* and it thus enabled Bohr to rectify earlier mistakes.

A second application of the exclusion principle concerns the unexcited helium atom. This atom can emit singlets or triplets. In the triplet state we have, $s = 1$, so that both electrons must be spinning in the same sense. But then, according to Pauli's principle, both electrons cannot be situated on the lowest orbit $n_i = 1$; for if they were, they would be associated with the same quadruplet of quantum numbers. We must suppose therefore that one of the two electrons is on a higher orbit, $n_i = 2$. This situation does not occur when the atom is in the singlet state because in that case we have $s = 0$, and the two electrons, spinning as they are in opposite senses, may both revolve on the lowest orbit.

The Revised Model of the Hydrogen Atom—The hypothesis of the spinning electron compels us to revise Bohr's original model of the hydrogen atom. We must now suppose that this atom is formed of a proton around which an electron is not only revolving but is also spinning on its axis. This change in the model removes several difficulties which beset the earlier theory.

In the original model, after we had replaced the quantum number k by $l = k - 1$, we had to suppose that the electron, revolving on the lowest orbit, had no angular momentum and hence that the atom had no magnetic moment. But this conclusion was impossible in view of the known magnetic properties of atomic hydrogen. With the revised model the contradiction disappears: the spin of the electron generates the required magnetic moment (one magneton).

We next consider the bearing of the new model on the hydrogen spectrum. Bohr's original model indicated that the hydrogen lines should be singlets accompanied by fine lines (caused by the relativistic refinements). On the other hand, the electronic configurations of hydrogen and of an alkali metal (e.g. sodium) were known to be similar. In either atom one outer electron revolved around a positively charged centre (the proton for hydrogen, and the closed configuration of neon for sodium).

---

* With our earlier system of notation where the quantum number $k = l_i + 1$ was taken in place of l_i, the formula in the text would become $2(2k - 1)$. We mentioned this formula in the last chapter, page 544.

Now sodium exhibits a spectrum of doublets, and so in view of the similarity just mentioned, many physicists suspected that hydrogen likewise should emit a spectrum of this kind. But this suspicion did not appear to be verified, for no doublet structure was detected by the spectroscopists in the hydrogen spectrum; and besides, hydrogen seemed to exhibit the normal and not the anomalous Zeeman effect, a fact which implied that the spectrum was one of singlets.

The revised model, however, furnished positive assurance that the hydrogen spectrum must have a doublet structure; for in the new model, the quantum number s of the atom is merely the s_i of the electron and thus has the value $\frac{1}{2}$, characteristic of doublets.

To settle the matter, experimenters renewed their investigations on the Zeeman effect for hydrogen. Finally the anomalous effect was disclosed and the reason why the normal effect had seemed to occur in former experiments was explained: the former magnetic fields were too intense, with the result that the Paschen-Back effect, closely resembling the normal effect, had been generated unwittingly.

There still remained the necessity of detecting the doublet structure in the absence of a magnetic field. Here a short digression is necessary. In the last chapter we saw that Sommerfeld's relativistic treatment of Bohr's original hydrogen atom led him to predict that each supposedly single Balmer line should in reality be formed of several fine lines closely packed. Sommerfeld's theory required five fine lines, but since two of these were ruled out by the correspondence principle, only three lines should be observed. We mentioned that the first careful spectroscopic observations revealed the three fine lines predicted by Sommerfeld, but that subsequent observations disclosed also the two lines which he had excluded. Doubt was therefore cast on the correspondence principle.

At the time no one suggested that the five fine lines merely betrayed the doublet structure of the hydrogen spectrum. Indeed, had this suggestion been made, it would in all probability have been rejected for the following reason: If the hydrogen spectrum were to have a doublet structure, the separations of the three doublets connected with the same Balmer line would certainly be very small; consequently each Balmer line would appear to be formed of seven fine lines * instead of five, as observed

---

* Consider the Balmer line generated by the drop from the level $n = 3$ to the level $n = 2$. If the hydrogen spectrum is one of doublets, as we are now supposing, this line will in reality represent an agglomeration of seven lines generated by the following drops: There will be two drops in which l passes from 1 to 0; two in which l passes from 0 to 1; and three in which l changes from 2 to 1. These three groups of drops will yield respectively a principal doublet, a sharp doublet, and a diffuse doublet with a satellite, so that there will be seven lines in all.

in practice. Even the revised model of the hydrogen atom does not enable us to answer this objection with any assurance. Today, however, the objection has lost its force because Dirac's investigations, based on the new wave mechanics, have shown that although the hydrogen spectrum is one of doublets, four among the seven lines happen to coalesce in pairs, so that only five lines are distinct. We conclude therefore that the doublet structure of hydrogen, required by the revised model, has actually been observed.

A coalescence of spectral lines implies a coalescence of energy levels and hence represents a condition of degeneracy. The levels that coalesce in the hydrogen atom (and also in the hydrogen-like atoms) are the so-called "screening-doublet" levels. This coalescence must be regarded as accidental, for it does not take place in any of the other optical spectra of doublets or in the X-ray spectra. The manner in which the coalescence is brought about in the hydrogen atom is readily understood. Consider two orbits of unequal energy; and suppose that on the lower orbit the electron is spinning and circling in the same sense, whereas on the higher orbit the spin and the sense of circulation are opposite. In the former case the energy of the atom is increased by the electron's spin; in the latter it is decreased. Thus, the spins decrease the difference in the energies of the two orbits. In the hydrogen atom this decrease is just sufficient to ensure the coalescence of some of the levels.

One of the defects of Bohr's theory, even when the spinning electron is introduced, is that it does not furnish any assurance that the foregoing screening-doublet levels will actually coalesce. The reason for this failure is that Bohr's theory is based on mechanical models, which as we know, cannot be relied upon in the representation of sub-atomic phenomena; consequently the results of Bohr's theory usually have a qualitative rather than a quantitative value.

In our analysis we appear to have rejected Sommerfeld's relativistic treatment. But this impression would be too hasty. The relativistic refinements are indispensable in any case and Sommerfeld's error was merely that he applied his treatment to the wrong model of the hydrogen atom. We may also add that the appelation "fine structure of the hydrogen lines" is now inappropriate, for this name suggests some peculiar kind of structure, whereas the hydrogen spectrum differs from other spectra of doublets only in that the various lines are very much closer, and in some cases coincide.

CHAPTER XXVIII

APPLICATIONS OF BOHR'S THEORY

The original aim of Bohr's theory of the atom was to account for the atomic spectra. The reason why spectroscopic phenomena played so large a part in the elaboration of the theory is that spectral analysis provides one of the most conspicuous means of exploring the subatomic world; it permits, moreover, highly exact measurements, and exact quantitative information alone is of value in theoretical physics. The success of Bohr's theory in the interpretation of the atomic spectra has been recorded in the two preceding chapters. But the success of a theory in one particular field of investigation does not ensure its acceptance; and had Bohr's theory been in conflict with the discoveries made in the other departments of physics, it would have been rejected from the start or else profoundly modified. Attempts were made therefore to verify the implications of Bohr's theory in fields foreign to spectroscopy. In the present chapter we shall review some of the experiments devised.

One of the most significant of the new conceptions introduced by Bohr is that of the discrete energy levels of the atom. The evidence in favor of these levels is afforded, as we know, by the distribution of the lines in the spectral series. But we may establish the existence of the energy levels in a totally different way. We recall that, in Einstein's interpretation of the photo-electric effect, when radiation of frequency ν falls on a metallic plate which is electrified, a quantum of energy $h\nu$ of the incident light is transferred to an electron in the plate, and this electron is then expelled with a kinetic energy equal to that of the quantum $h\nu$ absorbed. In point of fact the kinetic energy of the photo-electron ejected always falls below the theoretical amount $h\nu$, but this circumstance is readily explained when we note that some of the energy must have been used up in tearing the electron away from the plate. Now a phenomenon very similar to the photo-electric effect occurs when atoms are submitted to the action of ultra-violet light. The more loosely bound optical electron is torn from the atom, and the atom becomes ionized. A certain amount of work is expended in the removal of the optical electron from its orbit to a point outside the atom. Consequently, if we represent this work by W, we have the equation

$$h\nu = \text{kinetic energy of electron} + W,$$

where $h\nu$ is the quantum of energy absorbed from the radiation. Since h is known, and the frequency ν of the incident radiation can be measured, and since the velocity (and hence the kinetic energy) of the electron as it leaves the atom can be determined, the foregoing equation enables us to obtain the value of W. Thus, the work which must be expended to ionize the atom can be measured with a high degree of accuracy, and as a result the total energy, $-W$, of the optical electron when it was describing its orbit inside the atom is known.

Suppose that, instead of submitting the atom to the action of ultra-violet light, we submit it to the more powerful action of X-rays. In place of the loosely bound optical electron one of the deeply embedded core-electrons may now be ejected, and by proceeding as before, we may determine the energy of the orbit on which this core-electron was moving. M. de Broglie undertook a large number of experiments of this sort and devised a means of measuring the kinetic energies of the ejected electrons. He found, in full agreement with Bohr's theory, that the various core-electrons were bound with energies forming a discrete set of values. In particular the sub-sublevels (n_i, l_i, j_i) enumerated in the table (52) of the last chapter were disclosed by this method.

Other experiments which demonstrate the existence of Bohr's energy levels have been devised. One of the first was conducted by Franck and by Hertz. The essence of this experiment may be understood from the following considerations. Suppose an electron moving at high speed collides with an atom assumed at rest. We shall further suppose that the atom is in its unexcited state. According to Bohr's theory the atom is in the lowest energy level, and the immediately higher level will differ in energy by a finite amount. The colliding electron has itself a certain energy, namely, its kinetic energy $\frac{1}{2}\mu v^2$, and in the process of collision part of this energy will be transferred to the atom. But this transferred energy may betray itself in two ways. It may impart a translational motion to the atom as a whole, or it may pass into the interior of the atom and excite it to a higher energy state. The second possibility cannot, of course, occur if the energy of the colliding electron is less than the energy required to excite the atom from its present level to the next higher one. Let us first assume that the energy of the colliding electron is too small to excite the atom. The collision will then be of the type which occurs when two perfectly elastic bodies meet, the transfer of energy being external, as it were. Since the mass of the atom is considerably greater than that of the electron, the impact will scarcely move the atom, and the colliding electron will rebound with practically no loss in its kinetic energy and hence in its velocity. In this case the collision of the electron with the atom is said to be "elastic". But suppose that the kinetic energy of the electron is greater than the amount required to excite the atom. The electron during the impact may then lose this required amount of energy, ceding it to the interior of the atom; and so the electron will rebound from the atom with a correspondingly reduced velocity. As for the atom, it will now be in an excited state. The collision in this case is said to be of the "inelastic" type, because the decrease in the velocity of rebound of the electron is of the same kind as arises when two inelastic balls collide, or when a billiard ball rebounds from an inelastic cushion.

Franck and Hertz sought to verify these anticipations of Bohr's theory by causing an electron moving at high speed to collide with an atom. In a series of experiments the electron was given increasing velocities, and hence kinetic energies. According to Bohr's theory we should expect that, for certain critical energies of the oncoming electron, a sudden decrease in the velocity of its rebound should arise. These sudden decreases should occur every time the kinetic energy of the electron was just sufficient to excite the atom from its lowest level to one of the higher levels. Franck and Hertz verified these expectations, and by measuring the energies which the electron lost as a result of its collisions, they were able to determine the differences in the energy values of the various atomic levels. The values of the energy levels derived from these experiments were then checked against the values derived from spectroscopic observations. For instance, let us suppose that the energy of the oncoming electron is just sufficient to excite the atom from its lowest level, of energy E_1, to the immediately higher level, of energy E_2. The colliding electron will lose its energy $E_2 - E_1$, and the atom will become excited. The atom is now in a position to radiate, and according to Bohr's frequency condition it will radiate a frequency

$$\nu = \frac{E_2 - E_1}{h}.$$

We may measure this frequency and determine therefrom the difference, $E_2 - E_1 = h\nu$, between the energies of the two levels; and we may then verify that this difference is precisely equal to the energy lost by the colliding electron. This experimental test proves the existence of the energy levels, and at the same time it proves the correctness of Bohr's frequency condition.

Application to the Kinetic Theory—In the Franck and Hertz experiments we have been concerned with collisions between atoms and electrons, but our former general conclusions apply also to collisions between atoms. Thus, if two unexcited atoms collide and if the kinetic energies are insufficient to excite one of them, the collision will be elastic. If, on the other hand, the kinetic energies of the colliding atoms are sufficiently high, the collision will be inelastic and one or both of the atoms may become excited. As we shall now see, the elastic collisions predicted from Bohr's theory explain away one of the difficulties noted in Boltzmann's kinetic theory of gases.

Boltzmann assumed that the collisions between the molecules were perfectly elastic, an assumption which implied that the sum of the kinetic energies of two colliding molecules was conserved in spite of the collision. In particular, no part of the energy was communicated to the internal degrees of freedom of the molecules or of the atom; these structures, insofar as exchanges of energy were concerned, thus behaved as would closed systems. The study of the specific heats of gases corroborates Boltzmann's assumption (at least at ordinary temperatures) for if the internal degrees of freedom were to participate in the partition of the energy, the specific heats would be considerably greater than they are actually found to be.

However necessary Boltzmann's assumption may have been in the construction of his theory, it seemed to express an isolated fact having no connection with any other physical phenomenon then known. But when account is taken of Bohr's theory, we realize that Boltzmann's assumption exhibits a general feature of the atomic and subatomic worlds. The fact is that the atoms and molecules in a gas at ordinary temperatures have kinetic energies which are insufficient to bring about an excitation of the atoms and the loosening of their molecular bonds. For this reason the collisions are elastic. Boltzmann's assumption thus exhibits a quantum manifestation.

At the same time Bohr's theory shows that if a gas is heated sufficiently, the energies of the collisions should become sufficiently high to excite the atoms, with the result that inelastic collisions should occur. The internal degrees of freedom of the atom will then participate in the partition of energy, and the specific heat will be increased. Also, the excited atoms will radiate and the gas become luminous. We have here the interpretation of incandescence due to heating. A gas may also become luminous at relatively low temperatures, even though the energies of the collisions are insufficient to excite the atoms. But in this case the excitation is due to other causes. For example, in a neon tube the excitation is due presumably to the impacts of electrons against the atoms; and in certain nebulae it is caused by the radiation emitted from neighboring stars.

Some of the other difficulties which beset Boltzmann's theory have already received an interpretation by means of the quantum theory (Chapter XXIV). In particular, quantum considerations have shown why spherical molecules cannot rotate at ordinary temperatures, and why polyatomic molecules cease to rotate at extremely low temperatures. There still remains the problem of gas degeneracy *i.e.*, the tendency of the specific heat to vanish entirely at the absolute zero. This vanishing is suggested by the third law of thermodynamics. To account for it on the basis of our present conceptions, we should have to suppose that at extremely low temperatures the translational degrees of freedom of the molecules also become inoperative. The quantum theory in its present form throws no light on this phenomenon. It will be cleared up only when we consider the New Statistics (Chapter XL).

Band Spectra—The existence of quantized rotations of molecules is demonstrated indirectly by the band spectra. Let us consider a hydrogen molecule. Its shape is presumably that of a dumbbell, and according to the quantum theory, at ordinary temperatures, it can rotate only about an axis perpendicular to the line of the knobs, and its permissible rotational frequencies are quantized. Each quantized rotational motion determines a stable energy level. The levels that are generated in this way are of course no longer energy levels of the intra-atomic variety, which play so large a part in the interpretation of atomic spectra; but they are energy levels, notwithstanding, and we explained in Chapter XXIV how they intrude in the quantum interpretation of the specific heats of gases. Suppose, then, that white light is directed against hydrogen molecules which are rotating with quantized energies. The atoms in the molecules will absorb some of the radiations and will become excited. But the molecules also will absorb certain radiations, their rotational energies increasing in consequence. Absorption lines corresponding to these two kinds of absorption will therefore be present when the transmitted light is analyzed. Since the successive energy levels of a rotating molecule usually differ only little in energy, the radiations absorbed by the molecules will be of low frequency, often in the infra-red. The small differences in energy between the successive levels will cause the absorption lines to lie very close together, and the lines will thus yield the appearance of continuous dark stretches. The name "molecular absorption bands" is given to these agglomerations of lines.

Measurements of the infra-red absorption lines have afforded means of computing the quantized rotational energies of the molecules and, incidentally, have thus served to vindicate Boltzmann's idea that the molecules are in rotation. We may note that if the hydrogen gas happened to be in the atomic state, the molecular absorption lines would not occur. A study of the absorption spectrum of a gas thus permits us to decide whether the gas is in the molecular or in the atomic state.

Collisions of the Second Kind—The inelastic collisions so far considered are called "collisions of the first kind." Klein and Rosseland have shown that collisions in which the transfer of energy takes place in the opposite direction should also be possible. In particular, when an electron collides with an excited atom, the energy of excitation should occasionally be transferred from the atom to the electron; the electron would thus acquire a higher kinetic energy and would rebound from the erstwhile excited atom with increased velocity. To this kind of collision the name "collision of the second kind" was given. Of course in a collision of the second kind the atom, which surrenders its internal energy of excitation to the electron and falls to the unexcited state, cannot in addition lose this internal energy in the ordinary way by the emission of radiation. The principle of conservation of energy makes this statement obvious. The atom will therefore fall to a lower level *without radiating*. If we compare an excited atom to a watch that has been wound up, it would be as though the energy stored in the spring could be communicated directly to an object colliding with the watch. After the collision, the spring would be unwound and slack, with its energy delivered to the colliding object. A simple theoretical reasoning shows why collisions of the second kind must be expected. In an enclosure containing matter and maintained at constant temperature, the atoms will emit and reabsorb radiation until the state of statistical equilibrium characteristic of the radiation (according to Planck's law) is attained. If the temperature is sufficiently high, some of the atoms in the enclosure will presumably be ionized, and free electrons will be present, rushing hither and thither and colliding with the atoms. These free electrons may be likened to the molecules of a gas, the "electron gas." The state of statistical equilibrium will then be represented by a condition of equilibrium between the atoms, the radiation, and also the electron gas.

Starting from this state of statistical equilibrium, in which the entropy is necessarily a maximum, we shall suppose that only collisions of the first kind can arise. If we assume that the temperature is high, the collisions between atoms and electrons will be sufficiently energetic to excite the atoms. But each time a collision of the first kind occurs, the electron loses energy (or at least receives none) and after a sufficient number of collisions the average energy of the electrons will be too low to cause any further excitation of the atoms. Part of the original energy of the electron gas will therefore have been lost and imparted to the atoms, and thence also to the radiation. Since the distribution of energy among the atoms, radiation, and electrons will no longer be what it was initially, the state of statistical equilibrium will have been departed from. We conclude that if only collisions of the first kind are possible, a spontaneous and permanent departure from the state of statistical equilibrium will arise. But this contradicts the law of entropy. To overcome the difficulty, we must suppose that there is some means whereby the atoms excited by electronic impacts can subsequently surrender their energy of excitation to other electrons with which they collide. In other words, the transfer of energy must be reversible. This assumption is equivalent to postulating collisions of the second kind.

Such collisions are of great importance in chemistry, for they throw a new light on those mysterious chemical reactions called *catalytic*. Some chemical reactions take place only when small traces of an apparently irrelevant substance, called a *catalyser*, are present. What the function of this seemingly irrelevant substance may be has always been obscure; the substance appears to remain totally unaffected by the reaction, as though it had remained neutral and indifferent throughout. Yet some important function it must certainly perform, for, when the substance is absent, no reaction occurs.

Many photo-chemical reactions and dissociations appear to require the presence of catalysers; these agents are called "photo-sensitisers." Only when they are present does the energy of the incident radiation bring about the reactions or the dissociations. The sensitizing action of dyes on silver chloride is an illustration; this action is utilized in the panchromatic plates of photography. Another example is furnished by the sensitizing action of mercury vapor in photo-chemical dissociations. Thus, when ultra-violet light falls on hydrogen gas which contains traces of mercury vapor, the hydrogen molecules are dissociated into atoms, whereas no dissociation occurs when all trace of mercury vapor is removed. Obviously the mercury vapor plays the part of a photo-sensitizer. Collisions of the second kind afford a simple interpretation of this phenomenon. We may assume that the mercury atoms are excited by the ultra-violet radiations, and that when a mercury atom collides with a hydrogen molecule, it surrenders not only the normal amount of energy to be expected from the impact, but also its internal energy of excitation (in accordance with the theory of collisions of the second kind). The net result is that the energy delivered to the hydrogen molecule is now sufficient to disrupt the molecule into its constituent atoms. Inasmuch as hydrogen gas in the atomic state is far more active chemically than in the molecular state, the dissociation we have just considered may generate secondary reactions which would otherwise have been impossible. For instance, if copper oxide is present, the hydrogen in the atomic state will react with it, at ordinary temperatures, yielding water and copper. In this way, thanks to the original dissociation and thus to the collisions of the second kind, subsequent chemical reactions may occur.

The catalytic action of mercury vapor, in the example discussed, is very special since it takes place only under the action of radiation, whereas the majority of catalytic actions proceed in the absence of radiation. Nevertheless the part played by collisions of the second kind in the foregoing illustration does entail some progress in our understanding of catalytic phenomena. When we realize that many of the chemical processes occurring in living tissues are presumably of a catalytic nature, we may well appreciate the important rôle which the quantum theory may play some day in accounting for certain obscure biochemical reactions.

As contrasted with sensitization, the opposite phenomenon of "inhibition" is sometimes encountered: traces of an apparently irrelevant substance may arrest the progress of a chemical reaction, or may prevent excited atoms from radiating. In this latter case the phenomenon is accounted for if we suppose that the presence of the substance favors the production of collisions of the second kind, as a result of which the excited atoms, which would normally emit radiation, fall back to the unexcited state without radiating.

The Duration of the Excited States —When an atom is excited by the removal of its optical electron to a higher orbit, it remains in this state of excitation for a certain interval of time before falling to a lower level. Experiment shows that, if at an initial instant a large number of atoms are in a given excited state, the drops to lower states appear to be governed by a law sometimes called the law of chance.* This is the same law that holds for radioactive explosions, and so we may illustrate its peculiarities in connection with these explosions. According to the law of chance, the rate of decay of the radioactive substance (*i.e.*, the rate at which the explosions are proceeding) is proportional to the number of unexploded atoms still present.† The ratio of the latter number to the rate of decay is called the *mean life* τ of the atom. The higher the value of the mean life, the longer will it take a large number of radio-active atoms to disintegrate to any appreciable extent. The value of the mean life differs considerably from one type of radioactive atom to another; it varies from a thousand million years to an incredibly small fraction of a second. An equivalent definition of the mean life is to say that it represents the interval of time that must elapse for the number of unexploded atoms to be reduced to $\frac{1}{e}$ times the initial number. Note that τ does not give the life of any individual atom, and that even when τ is very small and even after a protracted period of time, there may be

---

* The nature of the experiments whereby the results stated in the text were obtained is easily understood. Let us suppose that a large number of similar atoms in the same state of excitation are shot through a slit, with extremely high velocities in the same direction. A parallel beam of swiftly-moving atoms is thus formed. As the atoms proceed along the beam they lose their excitation, undergo energy drops, and emit radiation. If we assume that there is no restoring influence to re-excite those atoms which have lost their excitation, the atoms which have already radiated will be unable to do so a second time and will thus pursue their courses in the unexcited state.

Observation shows that the beam is highly luminous in the immediate neighborhood of the slit and decreases progressively in brightness at points further removed from the slit. Since the relative intensities of the radiation emitted from the various portions of the beam are assumed proportional to the numbers of atoms which are radiating as they cross these portions, we may deduce, from the observation of the intensities, the relative numbers of atoms that are experiencing energy drops in the various parts of the beam. Furthermore, knowing as we do the common velocity of the atoms, we know the time taken by any atom to pass from the slit to any given part of the beam. Consequently, we may deduce the percentage of atoms which remain excited for any given period of time.

Wien, who was the first to perform experiments of this type, utilized a beam of canal rays generated by means of a Crookes tube.

† If we call n_t the number of unexploded atoms present at time t, and n_0 their number at the initial instant, the law of radioactive explosions is expressed by

$$n_t = n_0 e^{-\frac{t}{\tau}},$$

where $e = 2, 87\ldots$, and τ is a magnitude which varies in value from one atom to another. This magnitude τ is called the ''mean life'' of the atom.

atoms that have not yet exploded. The actual explosions occur, so far as we know, by chance, and τ represents a mere average life. The question of deciding whether the explosions truly occur by chance, or whether they are governed by some rigorously deterministic scheme which we ignore, is left unanswered.

The experimental measures referred to in the note show that the mean life of many of the excited states must be exceedingly short, *i.e.*, about one hundred-millionth part of a second. This value differs, however, from one excited state to another in the same atom, and it also varies with the atom considered. Furthermore, let us make clear that the mean lives we are here discussing refer to situations in which the excited atoms are allowed to drop to lower levels without interference. In the phenomenon of phosphorescence, for example, the mean lives of the excited states are increased. In a phosphorescent substance the atoms are first excited by being exposed to the action of light; the excited atoms subsequently drop to lower levels and radiation is emitted. But owing to the presence of an agent which exerts a kind of inhibitive effect on the drops, the mean lives of the excited atoms are considerably extended. The drops are spread over a protracted period of time, and the phosphorescent body continues to glow even after it is placed in the dark and hence withdrawn from the exciting action of the white light. If the inhibitive agent is removed, the excited atoms fall back to the lower levels in the ordinary sudden way, and the prolonged luminescence is no longer observed. In the following paragraphs the mean lives of the excited states to which we shall refer will be those for excited atoms under ordinary conditions.

The shorter the mean life of a given excited state, the greater the probability that this atom will experience an energy drop within a given interval of time. Thus, the mean life and the probability of a drop are related magnitudes. Einstein, in his revised deduction of Planck's radiation law,* qualified the various probabilities still further.

Metastable States and Forbidden Drops—The basic state of any atom is perfectly stable, because no spontaneous departure from this state can arise. The excited states, on the other hand, are not stable in the sense just defined, for energy drops can occur spontaneously. We must remember, however, that as a result of the selection rules imposed by the correspondence principle on the quantum numbers l and j, some of the theoretically possible drops are forbidden. Consequently, if all

---

* See page 598.

the drops from some excited level happen to be forbidden, this level, though excited, will yet be stable. Stable excited levels must therefore be expected according to Bohr's theory; they are called *metastable levels*, and the corresponding states of the atom are called *metastable states*.

The foregoing metastable states result from the selection rules controlling the changes of l and j. But the observation of spectra shows that there is another selection rule affecting the transitions of the quantum number s; and this additional selection rule will likewise entail metastable states. The selection rule for s cannot, however, be derived from the correspondence principle. Our only reason for believing in its existence is that, in some situations, lines which would be generated by drops in which s changes in value are not detected, so that such drops appear to be forbidden. On other occasions, however, drops of this sort arise. Bohr's theory sheds no light on the selection rule for s, but we shall see in Chapter XXXV that wave mechanics clarifies the problem.

Illustrations of the different kinds of metastable states are furnished by the neutral helium atom. In the lowest energy state of this atom the two electrons are circling on the innermost orbit, $n_i = 1$, and hence they are both situated in the sub-sublevel ($n_i = 1$; $l_i = 0$; $j_i = \frac{1}{2}$). According to Pauli's exclusion principle, the two electrons must therefore be spinning in opposite senses, so that $s = 0$. The lowest state of the atom is thus a singlet state, and the quantum numbers associated with it are $s = 0$, $l = 0$, $j = 0$. The next lowest states of the helium atom are the basic triplet states connected with the quantum numbers $s = 1$, $l = 1$, $j = 0, 1, 2$. The lowest of the three is the first. In it, one of the electrons is circling in the orbit $n_i = 1$, while the second electron (the optical one) is in the excited orbit $n_i = 2$; * the spins of the electrons are now in the same direction. The study of the helium spectrum shows that intercombination lines generated by drops from any one of the three triplet levels to the basic singlet level do not occur; consequently these three levels are metastable. The two drops in which j changes from 2 to 0, or retains the value 0, are forbidden by the selection rule for j, so that the two corresponding metastable states issue from this selection rule. But the third drop (in which j and l both change from 1 to 0) is permitted by the selection rules for j and for l. The fact that this drop does not occur must be ascribed therefore to some selection rule which prevents s from changing in value. The corresponding metastable state is thus connected with this selection rule for s.

---

* See page 581.

The mercury atom also affords examples of metastable states. In the mercury atom two outer electrons are circling around an electronic structure formed of 78 inner electrons. The spins and the circlings of all the inner electrons are opposite two by two and thus neutralize one another. Consequently, the vectors $\vec{s}$ and $\vec{l}$ for the entire atom result from the spins and the circlings of the two outer electrons. If these two electrons are spinning in opposite senses, we have $s = 0$, and the corresponding level is one of singlets. If the spins have the same sense, we have $s = 1$, and the atom is in the triplet state.

When the atom is not excited, its energy state, which of course is the lowest possible, is found to be a singlet state; both electrons are in the sub-sublevel $\left(n_i = 6; l_i = 0; j_i = \dfrac{1}{2} \right)$ with their spins in opposite senses. The resultant quantum numbers for the entire atom are thus $s = 0$, $l = 0$, $j = 0$. Immediately above this basic state we have the triplet states defined by $s = 1$, $l = 1$, $j = 0, 1, 2$; the spins of the electrons are now in the same direction. In short the general situation is the same as in the helium atom.

Just as in the case of the helium atom, two of the drops are excluded by the selection rules for l and j, and hence two metastable levels are at once apparent. The third drop (in which l and j both change from 1 to 0) is permitted by these selection rules. In the helium atom this last drop did not occur, owing to some selection rule which prevented s from changing in value, but in the mercury atom the drop takes place frequently and furnishes one of the conspicuous ultra-violet lines of the mercury spectrum. Incidentally, we perceive that no uniform selection rule can be provided for the changes of s.

Let us examine the metastable states more fully. Suppose an atom is in a metastable state connected with the selection rules for l and j. Such metastable states, as we know, can be predicted from the correspondence principle. If we regard the correspondence principle as absolutely valid, we must assume that a drop cannot occur spontaneously. Nevertheless, the atom may be removed from the metastable state in two ways:

(a) It may be jerked up to a higher level either by the action of incident light or by a collision of the first kind. It will then be in a position to fall to lower levels and in particular to the basic unexcited level, provided it does not get sidetracked again into a metastable level. Inasmuch as the drops occur at random, we may assume that in many cases the drop to the basic orbit will take place.

(b) Another way in which the metastable state might be departed from would be under the action of a collision of the second kind. The atom would then surrender its energy to the impinging particle and fall to the basic level, without radiating. We might think that an occurrence of this kind would contradict the correspondence principle, which prohibits the drop in question. But this is not so, for the principle exercises no restrictions on drops that take place under the action of collisions of the second kind, or on jumps to higher levels which are generated by collisions of the first kind. Only spontaneous drops or else transitions taking place under the influence of radiation are controlled by the correspondence principle.*

Thus far we have argued as though the drops forbidden by the correspondence principle were truly impossible and hence the metastable states truly stable. But experiment has shown that many of the metastable states are not absolutely stable and that drops from them do occur. The absolute validity of the correspondence principle is thus disproved, and so it is plausible to suppose that no metastable state is truly stable. The various metastable states on which experiments have been performed appear to have a mean life of the order of $\frac{1}{1000}$th of a second. Though this is an exceedingly short period of time when judged by ordinary standards, it is nevertheless one hundred thousand times longer than the average life of many of the ordinary excited states.

Although the correspondence principle and the selection rules which it entails can no longer be viewed as absolutely valid, the conditions under which spectra are usually observed have conspired to make these rules yield accurate results in practice. First of all, since the average lives of the metastable states are so much longer than those of the ordinary excited states, the probability of a drop from a metastable state will be correspondingly smaller. Consequently, the intensities of the forbidden lines will be so low that the lines may be imperceptible. Furthermore, what appear to be accidental circumstances accentuate the difficulty of detecting the forbidden lines. The fact is that in practice, when we examine the spectra of incandescent gases and vapors, we are dealing with gases at normal pressures and at high or moderate temperatures. And in this case the kinetic theory of gases shows that the collisions between atoms will occur with such frequency that the majority of the atoms in metastable states will be subjected to impacts before they have had

---

* This may be understood when we note that in the deduction of the correspondence principle vibrationary motions play a prominent part.

time to radiate. Since the impacts may excite these atoms to higher levels or may diminish their energies (collisions of the second kind), the drops from the metastable states will be even more infrequent than they would normally be. Owing to these various causes, the correspondence principle gives reliable results in practice, and the most delicate experiments are required to expose its inaccuracy.

The foregoing discussion suggests that the most favorable conditions for observing the forbidden lines will be realized if we can observe the spectrum emitted by a gigantic cluster of excited atoms, because the intensities of the lines being cumulative, even faint lines may then become perceptible. Furthermore, so as to decrease the frequency and the violence of the collisions, which occur in gases at ordinary temperatures and pressures and which would disturb the metastable states, we must contrive to operate on a gas at extremely low temperature and pressure, and hence on a cold and highly rarefied gas. While all these conditions could scarcely be obtained in the laboratory, they happen to be realized in the gaseous nebulae. A nebula has an enormous mass, and its rarefaction is extremely high. In addition its temperature is exceedingly low; and the excitation of its atoms is due not to the effects of the collisions which accompany high temperatures, but to the radiations emitted by neighboring stars. Finally, the low pressure facilitates the excitation of the atoms.

Considerations of this sort led Bowen to an important discovery. Some of the gaseous nebulae emit spectral lines which cannot be matched in the laboratory. These lines were attributed tentatively to an unknown element called "nebulium." But Bowen, on calculating the forbidden lines for ionized oxygen and ionized nitrogen, found that some of them coincided with some of the lines ascribed to nebulium. Whence he concluded that the mysterious lines detected in the spectra of the nebulae were none other than the forbidden lines of ionized oxygen and ionized nitrogen; and that these elements, not nebulium, were present in the nebulae. As a result, the hypothesis of nebulium automatically became useless.

Einstein's Derivation of Planck's Radiation Law—Planck, in his deduction of the law of equilibrium radiation, treated the atoms of matter in a heated enclosure as though they were oscillators, and assumed that each oscillator was susceptible of emitting and of absorbing only one definite radiation. But, in view of the more elaborate conception of the atom furnished by Bohr's theory, it would be preferable to obtain Planck's law on the assumption that the emitting and absorbing systems are Bohr atoms, rather than Planck oscillators. These considerations prompted Einstein to revise Planck's treatment. We shall merely indicate Einstein's procedure.

When the radiation in an enclosure has attained the state of statistical equilibrium, the densities of the various radiations do not change (except for inevitable fluctuations). We must therefore suppose that the atoms of the enclosure emit and absorb radiation at the same rate. In particular the number of atoms in any specific energy state E_i must not vary.

Let A_{ij} denote the probability that any one of the atoms in the energy state E_i will drop to the lower state E_j within one second. The coefficients A_{ij} pertaining to the various energy drops, are called the "Einstein probability coefficients of spontaneous emission"; they are closely related with the mean lives of the atoms in the different excited states. If we could determine the numerical values of these coefficients for some specific kind of atom, we could deduce therefrom the relative intensities of the various lines in the spectra of this atom. Bohr's theory, however, furnishes no means of calculating these coefficients, and in Einstein's treatment their values are left blank.

We have also to consider the absorption process. In the more familiar process of absorption, the atom absorbs the energy of the radiation and is thereby excited to a higher level. We are then dealing with so-called "positive absorption." But in other cases the atom drops to a lower level under the influence of the radiation and thereby radiates energy. This second kind of absorption (which is in reality an emission of energy) is called "negative absorption."

Einstein therefore introduces probability coefficients B_{ji} and B_{ij} corresponding to these two kinds of absorption. After utilizing Bohr's frequency condition and also one of Boltzmann's statistical-equilibrium formulae so as to determine the relative numbers of atoms in the various excited states, Einstein specifies that the emission and the absorption of each individual radiation must proceed at the same rate (as is required in the state of statistical equilibrium). He then shows that Planck's radiation law will be obtained if the following relations hold between the probabilities:

$$\frac{A_{ij}}{B_{ij}} = \frac{8\pi}{c^3}\nu^3 h; \quad B_{ij} = B_{ji}.$$

It will be observed that ratios alone are obtained and that a third numerical relation would be necessary to assign definite numerical values to the probabilities. At all events, provided the relative values just written be assumed, Einstein's analysis yields Planck's law of radiation. At the same time, Einstein's treatment furnishes the above important relations between the probabilities—a result of considerable interest, since it is these probabilities which govern the intensities of spectral lines and the absorptive properties of gases.

Einstein's deduction of Planck's radiation formula is of historic importance because it marks the first systematic introduction of probability factors in the mathematics of the quantum theory. At the time Einstein developed the theory just outlined (in 1917), there was no incentive to view these probabilities as expressing anything deeper than our current ignorance of the underlying precise laws regulating the quantum transitions. The probability coefficients were viewed as mere expedients which might in theory be deduced from the requisite rigorous laws; their status was thus much the same as that of the probabilities introduced in the kinetic theory of gases and in other departments of physics. In particular, no suggestion was made that these probabilities betrayed any fundamental indeterminacy. In the following chapters we shall see that these original beliefs have had to be revised.

The Experiment of Stern and Gerlach—Bohr's theory was devised, in the main, for the purpose of interpreting the atomic spectra and the various changes which these spectra underwent when the atoms were submitted to the actions of magnetic fields (Zeeman effect). With a view of testing the correctness of Bohr's theory, physicists performed many experiments which had no direct connection with the atomic spectra. In the experiments of Franck and Hertz, electrons were made to collide with atoms, and the velocities with which the electrons rebounded were found to justify Bohr's conception of the discrete atomic energy levels. In the experiments of M. de Broglie, atoms were submitted to the action of X-rays, and the energies with which the core-electrons were ejected demonstrated the existence of sub-sublevels (n_i, l_i, j_i). But the energy levels are not the only characteristic features of Bohr's theory. To interpret the Zeeman effect, we have had to assume that the atoms may betray magnetic properties, and that the phenomenon of space-quantization occurs when a magnetic field is applied. Furthermore, Landé's splitting factor g must be introduced.

Let us recall the results obtained by Bohr and Landé. When an atom, in a state (l, j, s), is placed in a magnetic field, its axis can assume only $2j + 1$ different orientations (space-quantization). Furthermore, its magnetic moment is gj * magnetons, where

$$g = 1 + \frac{j(j+1) + s(s+1) - l(l+1)}{2j(j+1)}.$$

The experiments of Stern and Gerlach were devised to test these predictions of the Bohr-Landé theory. The essentials of these experiments are easily understood.

Suppose that a parallel beam of similar atoms, moving at high speed, is directed at right angles through a magnetic field. We may assume that before the atoms enter the field, their axes have random directions. The atoms now penetrate the field and while they are advancing, their axes precess at fixed inclinations about the direction of the field. If space-quantization is a myth, the inclinations of the atomic axes during the precessions will be the random ones that existed originally. But if space-quantization is a fact, the atomic axes, on entering the field, will suddenly assume the quantized inclinations and will then precess at these inclinations. Since we have no reason to suppose that one particular quantized inclination is more probable than any of the others, we may assume that each inclination is taken by approximately the same number of atoms. In order to ascertain whether space-quantization actually occurs, Stern and Gerlach utilized an inhomogeneous magnetic field. A field of this kind exerts different attractions and repulsions on the variously inclined atoms, so that the atoms will be deflected in different directions when they emerge from the field. If space-quantization is a fact, the field will split the beam into $2j + 1$ variously deflected beamlets, each beamlet containing those atoms whose inclinations are the same. If space-quantization does not exist, the field will merely cause the beam to spread out more or less uniformly, like a fan. Thus the existence of space-quantization may be tested.

When Stern and Gerlach performed this experiment with different atoms, they found that space-quantization was accurately verified; $2j + 1$ beamlets were observed in all cases. Furthermore, the experiment enabled them to test Landé's g-formula. The fact is that the deflection of any one of the beamlets by the magnetic field depends not only on the

---

* In reality gj magnetons represents the maximum value of the *projection* of the atom's magnetic moment onto the direction of the field. The magnetic moment itself is $g\sqrt{j(j+1)}$ magnetons. Since, however, in the Stern-Gerlach experiments it is the projection gj of the magnetic moment that is measured, physicists often refer to gj rather than to $g\sqrt{j(j+1)}$ magnetons as the magnetic moment in connection with these experiments.

inclination of the atoms of the beamlet, but also on the magnetic moments of these atoms. A simple formula connects the magnetic moment, the intensity of the field, and the time of transit through the field. Hence by measuring the deflections of the beamlets Stern and Gerlach deduced the magnetic moments of the atoms. They found, in full agreement with Landé's formula, that for an atom in a state qualified by the numbers s, l, j, the magnetic moment was gj magnetons. In this way indirect confirmation of Landé's theory of the Zeeman effects was obtained.

For example, neutral silver atoms are characterized by the quantum numbers ($s = \frac{1}{2}$; $l = 0$; $j = \frac{1}{2}$), and the value of g in this case is 2. Hence we should expect that a beam of neutral silver atoms would be split into two beamlets ($2j + 1 = 2$), and that the magnetic moment of these atoms would be one magneton ($gj = 1$). These expectations were confirmed in the Stern-Gerlach experiments.

Conclusions on Bohr's Theory—Bohr's theory borrows from Rutherford's model of the atom the idea of a central nucleus surrounded by planetary electrons. The constitution of the nucleus is disregarded; the sole reference to it is its net positive charge (atomic number) and its mass. The arrangements of the planetary electrons are the chief concern of the theory, and hence only the most superficial part of the atomic structure is investigated. The planetary electrons are responsible for the main features of the spectra, for the chemical properties of the atoms, and also for some of the magnetic ones. But it is the nucleus which determines the nature of the atom and informs us whether we are dealing with one element or with another. We mention these points so as to emphasize that, when we speak of Bohr's theory as a theory of the atom, we must be careful to note that the theory deals only with some of the features of atoms.

The most important contributions of the theory are the discovery of the energy levels and the formulation of the frequency condition

$$\nu = \frac{\text{drop in energy}}{h}.$$

Both of these contributions will be retained in the new quantum mechanics. We pointed out (see page 491) that the existence of peculiar quantized energy states violates the laws of classical and of relativistic mechanics. The frequency condition is also in open violation with the laws of classical electromagnetism. Indeed, it is even more revolutionary than the existence of privileged energy levels, for something analogous to the latter is found in the adiabatic invariants of classical mechanics. However this may be, we may grant that the energy levels and the frequency condition are expressions of quantum laws formerly unknown. We have then no alternative but to accept them, and no criticism can be directed against Bohr's theory on this score.

Nevertheless, on other scores the theory is open to serious objections. The first condition we must demand of any theory is that it be consistent, and that rules once formulated be applied throughout. This norm is certainly not satisfied in Bohr's atom. Thus we start with the quantum rule that all independent adiabatic invariants in our mechanical model must assume values which are multiples of the quantum of action h. And, yet, in the course of the theory we are compelled to modify this rule as and when needed, invoking sometimes half-quantum values and at other times still more complicated values, such as those of type $\sqrt{j(j+1)}\,h$, mentioned in connection with Landé's g-formula for the Zeeman effect. All these alterations, though necessitated by experimental results, have no theoretical basis. The change from the quantum number k to the quantum number $l = k - 1$ is also incomprehensible. Furthermore, the quantum rules explicitly expressed are not the only ones that play a part in Bohr's theory. One illustration among many is the assumption that the optical electron moves through the core without any change in its total energy. This assumption violates classical mechanics and to justify it we must invoke some new quantum form of stability only dimly understood. Pauli's exclusion principle is also utterly mysterious from our present standpoint. The spinning electron is introduced for empirical reasons, and however useful this new hypothesis, it is not even suggested by the initial premises of the theory.

Quite aside from these considerations, the spectra of hydrogen and of the hydrogen-like atoms are the only ones for which Bohr's theory gives a satisfactory interpretation; and, even here, the theory is unable to prove that the screening-doublet levels should coalesce, as the experimental evidence demands. The simplest neutral atom after hydrogen, $i.e.$, the helium atom, emits a spectrum certain features of which Bohr's theory is unable to anticipate. In those higher atoms, such as sodium and potassium, where more satisfactory results are obtained, it is qualitative information rather than quantitative numerical values that can be predicted.

The number and the disposition of the spectral lines are not, however, the only matters that must be considered; the intensities and polarizations of these lines are also important. Bohr's theory affords no means of predicting the intensities and the polarizations. The best it can do is to refer us to the correspondence principle, and thereby to classical mechanics and electromagnetism which it had discarded. Even so, the calculation of the intensities is far from precise. The forbidden energy drops, which must be assumed if the spectra are to be interpreted, are predicted to some extent by the correspondence principle, but the selection rules obtained affect only the quantum numbers l and j. The permissible changes in the value of the quantum number s are not deducible from the correspondence principle, and we cannot predict when s may or may not change during a drop. In the helium atom, for instance, s cannot change, whereas in the mercury atom it can.

As will be gathered from this rapid enumeration, Bohr's theory suffers from a large number of defects, and we may well wonder whether the root of the difficulties does not lie in our having attempted to secure a model of the atom based on too familiar concepts. The subsequent development of the quantum theory suggests that this is indeed the case, and we shall find, as we proceed, that familiar concepts will fade away gradually, giving place to mathematical symbolism. At the same time we shall find that the more highly developed theories, though differing conceptually from Bohr's, are in many respects but refinements of his theory, and that many of the results acquired by Bohr will be retained. From this standpoint, Bohr's theory of the atom may be viewed as a kind of half-way theory, linking the older concepts to the new ones and serving thereby as a scaffolding for the newer theories soon to appear. Physical science is once again following the same procedure of advancing by successive approximations.

The important position occupied by Bohr's theory in this gradual evolution justifies in our opinion the considerable amount of space we have devoted to it. In any case, it will always serve as a first approximation for those who wish to enter into a more detailed study of the atom.

CHAPTER XXIX

DE BROGLIE'S WAVE MECHANICS

So FAR as we are able to judge from L. de Broglie's earlier papers, he was led to his important discovery by the following considerations: In many experiments radiation seems to manifest the properties of waves, whereas in others it behaves as though it were corpuscular; in some mysterious way the wave aspect and the corpuscular aspect appear to be closely associated. De Broglie therefore suggested that this dualism detected in radiation might be symptomatic of some general condition holding also for matter.

In the case of radiation, the aspect first discovered was the wave aspect, exhibited in the numerous interference phenomena studied in optics. Only later in the photo-electric and in the Compton effects did the corpuscular aspect manifest itself. In the case of matter, the historical sequence was reversed, for it was the corpuscular aspect, revealed in the electrons and protons, which was discovered first; and when de Broglie initiated his theory the wave aspect of matter was unknown. The originality of de Broglie's idea is thus to assume that a wave aspect should also be exhibited by matter. In making this assumption, de Broglie was obviously taking a leap in the dark, but, as we shall see, subsequent events have vindicated his intuition. Had de Broglie limited himself to the bare expression of his ideas without coordinating them into a consistent mathematical theory, they would have received little attention—rather would they have been classed with those so frequently expressed by people whose main desire is to satisfy their personal whims or to strike a note of originality. And when the wave effects of matter were finally disclosed by experiment, de Broglie would not have been regarded as the originator of wave mechanics. But de Broglie, as we shall now explain, gave a mathematical formulation of the waves which he was postulating and showed how these waves should behave and what phenomena they should entail.

Let us first recall certain results mentioned in connection with radiation. Prior to the discovery of the quantum theory, radiation was regarded as a wave manifestation. In vacuo, the waves progressed with velocity c, and these waves were believed to transport energy and momentum, the density of the momentum being $\dfrac{1}{c}$ times that of the energy.

But when the quantum theory was advanced by Planck, the wave theory of light was amended, and a semi-corpuscular theory was proposed in its stead. In the revised theory corpuscles of light (photons) were supposed to accompany the waves. Einstein, in his analysis of the photo-electric effect, assumed that the photons had energy $h\nu$, where ν is the frequency of the light. He also attributed a certain momentum to each photon; and so as to retain the relationship between energy and momentum prescribed for the waves in the electromagnetic theory, Einstein supposed that photons, associated with radiation in vacuo, had a momentum $\dfrac{h\nu}{c}$ directed along the rays of the waves. The expression of the momentum may equivalently be written $\dfrac{h^{*}}{\lambda}$, or $h\sigma$, where λ and σ are the wave length and wave number of the wave. Einstein's results showed that in experiments where the corpuscular aspect of radiation impressed itself, the mechanical concepts of energy and momentum formerly ascribed to the waves should be transferred to the corpuscular photons. In such cases therefore the associated waves were characterized by their frequency and wave length and were in no wise the carriers of energy and momentum. At this juncture, de Broglie suggested that even when the corpuscular aspect of light was not apparent and the wave aspect alone was manifest, the energy and momentum of the radiation should still be attributed to the photons. In support of his contention de Broglie pointed out that the pressure of radiation could be ascribed to the impacts of the photons against an obstacle. These preliminaries enable us to understand de Broglie's theory of matter.

De Broglie, guided by the association of photons and radiation waves, postulated that a similar association should hold between material particles and some new kind of wave (a de Broglie wave), the underlying nature of which remained mysterious. De Broglie assumed therefore that a particle in motion was associated with a plane sinusoidal wave, and that the energy and momentum of the particle were connected with the frequency and wave length of the wave by the same relations that held for photons and waves of light. Thus if we designate the energy of a par-

---

* The relation connecting the frequency, wave length (or wave number) and wave velocity V of a wave is

$$\lambda\nu = \frac{\nu}{\sigma} = V.$$

Since in the present case $V = c$, we have $\dfrac{\nu}{c} = \dfrac{1}{\lambda} \equiv \sigma.$

ticle by W, and its momentum (a vector) by $\vec{p}$, the relations postulated by de Broglie were

$$(1) \qquad W = h\nu; \qquad \vec{p} = \frac{h\nu}{\vec{V}} = \frac{h}{\vec{\lambda}} = h\vec{\sigma},$$

where ν, V, λ, and σ are the frequency, wave velocity, wave length, and wave number of the associated wave. The last equality expresses the fact that the wave is advancing in the same direction as the particle.

De Broglie first examined the case where the particle was moving in free space (*i.e.*, in empty space in the absence of a field of force). The particle would then be following a straight path with constant speed v. If, then, in the general relations (1) we substitute the relativistic expressions of the particle's energy and momentum, we get

$$(2) \qquad W = \frac{m_0 c^2}{\sqrt{1 - \dfrac{v^2}{c^2}}} = h\nu; \qquad \vec{p} = \frac{m_0 \vec{v}}{\sqrt{1 - \dfrac{v^2}{c^2}}} = \frac{h\nu}{\vec{V}} = \frac{h}{\vec{\lambda}} = h\vec{\sigma}.$$

These relations connect the frequency, wave velocity, wave length, and wave number of the mysterious wave with the energy and momentum of a particle of rest mass m_0 moving with constant velocity v.[*]

Here a possible difficulty had to be considered. De Broglie viewed the relations (2) as the expressions of natural laws. But for de Broglie's views to be possible, these relations should endure regardless of the particular velocity v of the particle, and hence regardless of the particle's energy and momentum. If this condition were not satisfied, the relations (2) would represent mere accidental relationships holding only for a certain specified velocity of the particle.

Now the velocity of the particle will be changed automatically if we change the frame of reference in which this velocity is measured. On the other hand, according to the theory of relativity, a change in the frame of reference requires that we apply a Lorentz transformation. Consequently, in order to test de Broglie's relations (2) and justify their aptness to express natural laws, we must submit them to a Lorentz transformation and ascertain whether the changes thereby incurred by the magnitudes v, V, ν, λ, or σ are consistent with the maintenance of these relations. De Broglie applied this test and found that the relations endured, *i.e.*, he found that they were covariant under a Lorentz transformation. Of course, this circumstance in itself did not prove that

---

[*] The expression $\dfrac{m_0 c^2}{\sqrt{1 - \dfrac{v^2}{c^2}}}$ of the energy comprises the rest energy $m_0 c^2$ of the particle.

de Broglie's relations actually expressed laws of Nature; it merely proved that the laws suggested were not impossible.

Before considering the different aspects of de Broglie waves, we must recall certain results and definitions mentioned in Chapter XIX. We said that in a steady propagation of sinusoidal waves, the phase remains uniform over appropriate fixed surfaces, called "equiphase surfaces." More precisely, the phase at all points of any specified equiphase surface is the same at a given instant, though it varies rhythmically in the course of time. These equiphase surfaces form a simple infinity of surfaces placed one behind the other. The orthogonal trajectories of the family of equiphase surfaces constitute the mathematical rays, and when the conditions of ray-optics are satisfied, these mathematical rays are also the physical rays.

We also defined a "wave front" as a mobile surface over which the phase not only remains uniform but also retains a fixed value throughout time. These wave fronts progress along the rays with the wave velocity V at each point, and as a result of their motion they coincide in succession with one equiphase surface after another. A wave crest has a clear meaning in the case of ocean waves and of waves propagated along a stretched string, but it has no intuitive significance when we are dealing with waves in space. Nevertheless it is sometimes advantageous to speak of wave crests in connection with waves in space. A wave crest then refers to a wave front over which the disturbance is a maximum.

These preliminaries being recalled, we return to the de Broglie waves. The de Broglie wave associated with a particle moving in free space is assumed to be an infinitely extended plane wave of the sinusoidal kind. The equiphase surfaces of the wave will be fixed parallel planes, intersecting the straight trajectory of the particle at right angles. The wave fronts are likewise plane, and they advance with the wave velocity V, moving in the same direction as the particle. The wave velocity is derived immediately from the relations (2). Thus for a particle moving in free space with velocity v, the wave velocity of the associated de Broglie wave is

(3)
$$V = \frac{c^2}{v},$$

so that the wave crests move much faster than the particle—indeed their velocity is greater than that of light.

We also see that if the particle is at rest ($v = 0$), the wave velocity and the wave length both become infinite. In this case the sinusoidal wave degenerates into a standing wave represented by a simultaneous heaving and sinking of some existent in the space around the particle.

The frequency ν_0 of the standing wave is now $\nu_0 = \dfrac{m_0 c^2}{h}$; it is smaller than the frequency of the progressive wave associated with a moving particle.

A point that may seem strange is that the wave velocity V of the waves, *i.e.*, $\dfrac{c^2}{v}$ is greater than that of light. This feature would appear to place de Broglie's theory in conflict with the theory of relativity, which prohibits the existence of such velocities. But our fears would be unfounded, because the theory of relativity prohibits velocities greater than c only insofar as these velocities refer to the transfer of energy. The de Broglie waves, however, transport no energy, for we have seen that the energy is restricted to the particle. Thus there is no conflict with the theory of relativity on this score.

We now pass to de Broglie's subsequent investigations. Up to this point we have considered the wave associated with a particle moving in free space. The next step will be to assume that the particle is moving in a conservative field of force. The complication caused by the field of force does not affect de Broglie's fundamental relations (1), except that now the total energy W of the particle includes the potential energy E_{pot} due to the field of force. If we call E_{kin} the relativistic kinetic energy of the particle, the total energy W may be written

$$(4) \qquad W = \frac{m_0 c^2}{\sqrt{1 - \dfrac{v^2}{c^2}}} + E_{pot} = m_0 c^2 + E_{kin} + E_{pot}.$$

The momentum $\vec{p}$ of the particle has, however, the same expression as before, *viz.*,

$$\vec{p} = \frac{m_0 \vec{v}}{\sqrt{1 - \dfrac{v^2}{c^2}}}.$$

As before, the relations (2) connect the energy and momentum, W and $\vec{p}$, of the particle with the wave magnitudes: frequency ν, wave length λ and wave-number vector $\vec{\sigma}$ of the associated wave. Let us write these relations so that the wave magnitudes are expressed in terms of the mechanical ones. We have:

$$(5) \ \ \nu = \frac{W}{h}; \quad (6) \ \ \lambda = \frac{h}{p}; \quad (7) \ \ \vec{\sigma} = \frac{\vec{p}}{h}.$$

Also, utilizing $V = \lambda\nu$, we get for the wave velocity $\vec{V}$ of the wave

(8) $$\vec{V} = \frac{W}{\vec{p}}.$$

These relationships cover the cases where the particle is moving in a field of force or in free space. In a field of force, however, the frequency and the wave velocity will not have the values they had in free space for a particle moving with the same velocity. This is because the value of W is not the same in the two situations.

It must be borne in mind that the wave magnitudes, λ, $\vec{\sigma}$, $\vec{V}$ which depend on the variable momentum $\vec{p}$ of the particle, are defined only at the point where this momentum has a meaning at the instant considered, *i.e.*, at the point which the particle happens to occupy. Elsewhere, these wave magnitudes, and hence the wave itself, are left undetermined. It would appear therefore that the wave can be defined only at the point which the particle occupies at the instant considered.

This limitation may be overcome to a certain extent when we note that the position and momentum of the particle at any given instant determine its position and momentum at any subsequent (or prior) instant. Hence we may suppose that the wave magnitudes at a point P through which the particle has not yet passed are determined by the energy and momentum that the particle will have when it passes this point. Yet even when this assumption is made, the wave can be determined only at points on the particle's trajectory and not elsewhere. In free space we overcame this difficulty by supposing that the wave was plane and that the equiphase surfaces extended to spatial infinity. But in a field of force where the trajectory of the particle is usually a curve, we cannot proceed so simply.

We also see from (7) that the direction of advance of the wave, at any point on the particle's trajectory, coincides with the direction of the particle's motion. Consequently the trajectory of the particle will intersect the equiphase surfaces of the wave at right angles.

Next let us consider the frequency of the wave. Since in a conservative field of force the energy W of the particle is constant, the relation (5) shows that the frequency of the associated wave will always be the same. In the language of optics, the wave is monochromatic.

We are thus led to the following conclusions: A particle describing a trajectory in a conservative field of force is associated with a monochromatic de Broglie wave, the wave fronts of which advance while remaining orthogonal to the trajectory; the frequency, wave length, and wave velocity are determined by the formulae (5), (6), and (8); but nothing is known of these magnitudes, or of the shape of the equiphase surfaces, at points which are not in the immediate neighborhood of the trajectory.

Now we have seen in Chapter XIX that in a wave propagation the name "rays" is given to the orthogonal trajectories of the equiphase surfaces. By the de Broglie rays we shall therefore mean the orthogonal trajectories of the equiphase de Broglie surfaces. Since the equiphase surfaces of the wave associated with a particle in motion are orthogonal to the trajectory described by the particle, we conclude that the de Broglie ray and the particle's trajectory coincide. Furthermore, since the trajectory is usually a curve, the de Broglie ray will likewise be a curve.

The fundamental relations (5) and (6) enable us to obtain the analytical expression of the de Broglie wave associated with a particle which is moving with total energy W. Thus let us first consider any kind of sinusoidal wave, of frequency ν and of wave length λ, moving along a trajectory. We shall assume that at the instant $t = 0$ a wave crest is passing a given point O on the trajectory. In this event the wave is represented at any time t and at any point P on the trajectory by

$$(9) \qquad C \cos 2\pi \left(\nu t - \int_0^P \frac{dr}{\lambda} \right).$$

Here C is an amplitude which may be a constant or may vary from point to point along the trajectory. As for dr, it represents an infinitesimal segment of the trajectory. To obtain the de Broglie wave associated with a particle describing the trajectory with the total energy W, we must apply the formulae (5) and (6) to (9). This gives

$$(10) \qquad C \cos \frac{2\pi}{h} \left(Wt - \int_0^P p\,dr \right).$$

Now the integral defines the increase in the Maupertuisian action of the particle as it moves from the point O to the point P. If we assume that the action has the value zero at the point O, the integral will represent the action of the particle at the point P. Let us denote this action by S. The expression (10) of the de Broglie wave may then be written.

$$(11) \qquad C \cos \frac{2\pi}{h} (Wt - S).$$

This formula will be of use presently.

Thus far, when considering the de Broglie wave associated with a particle, we have found that the wave and its equiphase surfaces can be determined only in the immediate vicinity of the trajectory. Elsewhere, the wave appears to be meaningless or at least indeterminate. But we may complete the wave and define it throughout a volume of space. To do this, however, we shall find it necessary to consider a swarm of similar particles. We now propose to show how the wave associated with the swarm can be constructed.

The de Broglie Wave Associated with a Swarm of Particles—First let us recall certain explanations given previously. Any trajectory which a particle may describe with energy W in a given conservative field of force will be called a trajectory of energy W, or a trajectory W for short. Now in Chapter XIX we saw that if P is any point on a trajectory W, this point will also be situated on an infinite number of other trajectories W. We also saw that when the particle passes the point P, its momentum p will always have the same magnitude regardless of the particular trajectory W which the particle has followed in reaching P. Thus, when we confine our attention to particles moving with the same energy W, we may say that each point of space is connected with a definite value of the momentum p.

Consider now the de Broglie wave associated with a particle which is describing a trajectory W. De Broglie's relation (8) shows that the wave velocity V at any point on the trajectory will be inversely proportional to the momentum p of the particle at this point. On the other hand, in the case of wave motions generally, the wave velocity at a point is inversely proportional to the refractive index of the medium at this point. We conclude that, in any given conservative field of force, the refractive index of space for de Broglie waves associated with particles of energy W is proportional to the value of the momentum p.*

Now we saw in Chapter XIX, when discussing Hamilton's investigations on the principles of Fermat and of Maupertuis, that the proportion-

---

* Since in a field of force the value of p varies from point to point, we see that space will exhibit a variable refractive index for the de Broglie waves. This result becomes intuitive when we recall that in a field of force the de Broglie rays are usually curves. Nevertheless, at first sight it seems strange to find that empty space may have a variable refractive index, for this conclusion seems to be incompatible with the idea that space is homogeneous. We must remember, however, that if there were no field of force, the momentum p would have the same value at all points, and the refractive index would be constant. Hence it is the field of force in space, and its variable potential from place to place which supplies the heterogeneity necessary to account for the variability of the refractive index (see the note on page 614).

ality of the refractive index of a medium to the momenta of particles moving with energy W is precisely the relation that must hold if the rays of a wave propagated in the medium are to coincide with the trajectories W of the particles. Thus the curious situation commented upon by Hamilton, and formerly viewed as a mere mathematical curiosity, is now seen to be realized physically in the case of particles and their associated de Broglie waves.

The upshot of this discussion is that if any de Broglie wave of frequency $\nu = \dfrac{W}{h}$ is propagated through the field of force, the rays of the wave will coincide with trajectories W, and the wave fronts of the wave will advance normally to these trajectories.

The foregoing explanations enable us to construct an extended de Broglie wave which will be associated with a whole swarm of particles and will thus be defined over a wide region of space instead of along a single trajectory. Consider an arbitrary continuous surface Σ in the field of force, and let us suppose that at an initial instant this surface is one of the equiphase surfaces of a de Broglie wave of frequency $\nu = \dfrac{W}{h}$. The rays of this wave are the orthogonal trajectories of the family of equiphase surfaces, and hence start perpendicularly from the surface Σ. The wave fronts progress with the wave velocity $V = \dfrac{W}{p}$ at each point. Now we know from our previous explanations that the rays of the wave are also trajectories W. Consequently, if similar particles are thrown with energy W from the surface Σ along the normals to the surface, these particles will follow the rays. (At least, this will be so if we assume that the particles exert no mutual actions, so that each particle will follow a trajectory W as though the other particles were inexistent.)

If we consider only that portion of the extended wave which moves in the immediate vicinity of a trajectory, we obtain a limited wave. This limited wave is none other than the wave that we should have associated with the particle describing the trajectory before the extended wave was investigated. Hence we conclude that the extended wave represents a linking up of the individual, limited de Broglie waves which are associated with the respective particles. We may therefore view the extended wave as associated with the whole swarm of non-interacting particles.

To summarize: A swarm of similar non-interacting particles, projected normally with the same total energy W from the various points of an arbitrarily given continuous surface in a conservative field of force, may be associated with a monochromatic de Broglie wave of frequency $\nu = \dfrac{W}{h}$; the equiphase surfaces of the wave are orthogonal to the family of trajectories; the wave length and wave velocity from point to point are defined by the fundamental relations (6) and (8). In particular, if there is no field of force and the particles are describing parallel straight lines, the associated wave will be plane and its wave velocity will be $\dfrac{c^2}{v}$.

An alternative interpretation is possible. Having obtained the de Broglie wave and the equiphase surfaces associated with the swarm of particles, we may restrict our attention to one special particle of the swarm. It will still be permissible to associate the single particle with the same wave and hence with the same family of equiphase surfaces. However, many other de Broglie waves having differently situated equiphase surfaces might with equal justification be associated with the single particle. The fact is that a single particle describing a definite trajectory does not yield sufficient information to permit the construction of the equiphase surfaces, except in the immediate vicinity of the trajectory. To extend the surfaces, we must always consider a swarm of particles and hence an initial surface Σ from which the particles are thrown normally. If, then, we view our particle, moving as before, as belonging to a different swarm connected with a different surface Σ, we shall obtain a different wave and different equiphase surfaces. This shows that when we are dealing with a single particle, considerable arbitrariness is attached to the shapes of the equiphase surfaces.

The last remark may be illustrated very simply in the particular case where there is no field of force. We imagine a particle to be moving with energy W from a point P in a given direction. This particle may be supposed to belong to a swarm of non-interacting particles which are moving with energy W, and which start normally from the surface of a sphere Σ passing through the point P. On the other hand, our particle may also be supposed to belong to another swarm, the particles of which are thrown normally with energy W from a plane surface Σ' which is tangent to the sphere at the point P. The equiphase surfaces of the de Broglie wave are spheres in the former case and planes in the latter. If now we consider the original particle in isolation, we see that we may associate it at pleasure with one or the other of the two de Broglie waves. Thus, the particle moving in the same way in either case is associated with differently disposed waves.

The association of de Broglie waves with trajectories W allows us to translate a mechanical problem into an optical one, and vice versa. Suppose we wish to determine the trajectories W that start normally from a given surface Σ in a conservative field of force. We have here what appears to be a mechanical problem, and indeed we might solve the problem by the methods of mechanics, *e.g.*, by integrating the Hamilton-Jacobi partial differential equation. But we may also utilize the wave theory. Thus if we can determine the equiphase surfaces of the de Broglie wave which is associated with these trajectories W, the trajectories themselves will be obtained immediately, since they will be the rays, or orthogonal trajectories, of the equiphase surfaces. Let us, then, examine how the wave problem can be solved.

The first step will be to determine the refractive index of space for the waves of frequency $\nu = \dfrac{W}{h}$. Now de Broglie assumes, by analogy with the expression of the refractive index in the case of waves of light, that the refractive index μ of space for the de Broglie waves is defined by

$$(12) \qquad\qquad \mu = \frac{c}{V}.$$

If in (12) we replace the wave velocity V by its value defined by (8), we obtain the expression of μ in terms of the field of force and of the energy W of the associated particles.* Our problem is now to determine the equiphase surfaces of the wave which is transmitted through this medium, and for which Σ is an equiphase surface. If we can solve this problem, we obtain the rays (*i.e.*, the orthogonal trajectories of the equiphase surfaces), and hence the mechanical trajectories. In short,

---

* On performing the calculations, we find

$$(13) \qquad\qquad \mu = \frac{\sqrt{(W - E_{pot})^2 - m_0{}^2 c^4}}{W},$$

where W is the total energy with which the particles are thrown, and E_{pot} is the potential energy of a particle at the point where the refractive index is to be calculated. The preceding formula is the rigorous relativistic one. A less precise formula is

$$(14) \qquad\qquad \mu = \sqrt{\frac{2(E - E_{pot})}{m_0 c^2}},$$

where E is the classical total energy, in which the rest-energy $m_0 c^2$ is disregarded, *i.e.*,

$$(15) \qquad\qquad E = W - m_0 c^2,$$

$\dfrac{v^2}{c^2}$ being assumed very small compared with unity.

the mechanical problem of determining the trajectories in a given field of force, and the optical problem of obtaining the rays in a medium of given refractive index are seen to be equivalent.

De Broglie Wave Packets—Let us suppose that a de Broglie wave is associated with a family of trajectories, one of which is described by a particle. The wave fronts are normal to the trajectories and are advancing with a velocity $V = \dfrac{W}{p}$ at each point. Now, nothing in this wave picture permits us to single out the precise trajectory which the particle is describing, and still less does the picture afford any means of locating the particle at any particular instant of time. The only velocity suggested by the wave picture is the wave velocity V, and this velocity has nothing in common with the particle's velocity v. In free space, for instance, the value of V is $\dfrac{c^2}{v}$. Nevertheless we may modify the wave picture in such a way that it gives an approximate description of a definite particle moving along a definite trajectory.

To understand how this can be done, let us revert to the refractive properties of the space in which the de Broglie waves are moving. The refractive index for a wave of frequency ν is given by the formula (13) in which the energy W is replaced by $h\nu$. The refractive index of space is thus seen to depend on the frequency ν of the de Broglie waves. We conclude that space behaves not only as a refractive medium for de Broglie waves, but also as a dispersive medium. Formula (13) shows that the higher the frequency ν of the waves, the higher the refractive index; hence the dispersion is of the normal type.

In Chapter XIX we saw that, when optical waves differing but little in frequency and proceeding very nearly in the same direction are superposed, a wave packet may be formed. We also mentioned that if the conditions of the propagation are consistent with ray-optics, the packet will hold together, at least for a certain time. Furthermore, if the medium exhibits the normal type of dispersion, the velocity of the packet will fall below the wave velocities of the superposed waves. The conclusions here stated are not peculiar to optical waves and would hold for waves generally and in particular for de Broglie waves. As we shall now see, it is possible to construct a wave packet of de Broglie waves which moves along with the material particle and thereby serves to afford a representation of the mobile particle in the wave picture.

The wave packet, associated with a particle of energy W and momentum $\overrightarrow{p}$, results from the superposition of a large number of monochromatic de Broglie waves whose frequencies and wave numbers differ only by small amounts from the values $\nu = \dfrac{W}{h}$ and $\vec{\sigma} = \dfrac{\vec{p}}{h}$. The last equality indicates that the waves of the packet will advance more or less in the same direction as the particle. Since we are supposing that the conditions of the propagation are those compatible with ray-optics, a wave packet surrounding the particle can be constructed.

The next step is to verify that this wave packet will move along the particle's trajectory with the velocity v of the particle. According to the explanations given on page 296, if $\Delta\nu$ and $\Delta\sigma$ are the ranges in the frequencies and in the wave numbers of the individual waves which by their superposition form the packet, the velocity of the packet is $\dfrac{\Delta\nu}{\Delta\sigma}$. When account is taken of de Broglie's fundamental relations (5) and (7), we obtain

$$(16) \qquad \frac{\Delta\nu}{\Delta\sigma} = \frac{\Delta W}{\Delta p} = v.$$

Thus, the packet moves with the velocity of the particle. Furthermore, since the motion of the packet is in the direction of $\vec{\sigma}$, and therefore of $\vec{p}$, we see that the packet will always be moving in the same direction as the particle. In short, the packet and the particle will move together along a trajectory.

Summarizing these results, we are led to the following conclusions:

A particle moving with energy W in a conservative field of force may be represented by a wave-packet of de Broglie waves moving through a medium the dispersive properties of which are determined by the field of force. The optical problem of determining the motion of the packet, and the mechanical problem of obtaining the motion of the particle, are thus equivalent. It remains to be said that, when we wish to consider a swarm of particles moving with the same total energy under the conditions prescribed in the former pages, a single monochromatic wave is taken to represent the wave picture and no wave packets are countenanced.

Although the representation of the particle by means of a wave packet furnishes some information on the particle's position and motion at a given instant, this information is marred by a certain vagueness: first, because the packet occupies a small volume and does not define a point which would represent the particle's exact position; and secondly, because the component waves have different frequencies and wave numbers and hence do not assign a well-determined motion to the particle.

We have here a first inkling of the vagueness and uncertainty which is ubiquitous in the quantum theory and which finds its expression in Heisenberg's Uncertainty Principle.

Hamilton's Wave Mechanics—The de Broglie waves appear at first sight to be something entirely new in science. In a sense this is true; however, if we omit the relativistic refinements introduced by de Broglie, we find that many of his results were anticipated by Hamilton in the earlier part of the last century. Hamilton was not acquainted with the theory of wave packets, developed only much later by Lord Rayleigh, but the general idea of a propagation accompanying mechanical processes was shown by Hamilton to be an immediate consequence of the laws of dynamics.* Indeed, the wave front of a de Broglie wave moves in exactly the same way as a surface of constant Hamiltonian action.

Let us examine this point more fully. We revert to our example of the swarm of non-interacting particles ejected normally with the same total energy W from the points on the same side of a given surface Σ. The field of force in which the particles are moving is assumed to be conservative. The particles describe trajectories starting from the surface Σ at right angles. Let us consider one of the particles. It leaves a point O on the surface Σ at time zero and reaches a current point P on its trajectory at an instant t. The increase in the Maupertuisian action of the particle from O to P is obtained when we integrate the particle's momentum over the trajectory between O and P. If we assume that the action is zero at the point O, the action of the particle at the point P is equal to the foregoing integral. Thus, the particle at each point on its trajectory is associated with some value of the action, and the action increases as the particle describes the trajectory. In this presentation the action appears to be a magnitude attached to the particle, but we may also define it at any point of the trajectory regardless of whether or not the particle is present at that point. With this understanding, the value of the action at any point Q on the trajectory is defined by the action the particle will have (or had) on passing through this point.

Suppose, then, we consider the swarm of particles. We assume that for each particle the Maupertuisian action has the value zero at the point where the corresponding trajectory leaves the surface Σ. Let P be any point on one of the trajectories. At this point the action has some value. We then mark out on all the trajectories the points where the action has this same value. The aggregate of these points defines a continuous

---

* See page 324.

surface, which may be said to represent a surface of equal Maupertuisian action. The equations of dynamics show that the surface is intersected at right angles by all the trajectories. Since the point P, which we selected for the purpose of obtaining the surface, is any point on the trajectory considered, we conclude that the surfaces of equal action form a simply infinite family, and that the trajectories are orthogonal to all these surfaces. One of these surfaces of equal action is the initial surface Σ whence the particles were thrown; it is associated with the value zero for the action. If we proceed along a trajectory in the direction of the motion, we pass through one surface after another, and the value of the action increases progressively. So as to avoid possible confusion, we must mention that if all the particles were to leave the original surface Σ simultaneously, they would not in general pass through any given surface of equal action at the same instant; some would reach it earlier than others.

There is also another kind of action utilized in dynamics; it is called the Hamiltonian action. In a conservative field of force, the increase in the Hamiltonian action of a particle, which moves from a point O to a point P along a trajectory, is obtained when we integrate the instantaneous difference between the kinetic and the potential energy of the particle over the interval of time taken by the particle to pass from O to P. If we assume that the Hamiltonian action of the particle has the value zero at the point O at the instant the particle leaves this point, the value of the Hamiltonian action at P is given by the integral just referred to.

We shall assume that the particle leaves the point O at time zero. Under these conditions, if S represents the Maupertuisian action of the particle at a point P on its trajectory, the Hamiltonian action A of the particle, when the particle reaches this point, can be shown to be

$$(17) \qquad\qquad A = S - Wt.$$

Here W is the total energy with which the particle is moving, and t is the time taken by the particle to pass from O to P. Since the Maupertuisian action S of the particle has already been assumed to vanish at the point of departure O, the formula (17) shows that the Hamiltonian action will also vanish at this point at the initial instant $t = 0$.

Thus far we have argued as though the Hamiltonian action were a magnitude attached to the particle—i.e., a magnitude which would have no meaning except at the point through which the particle happened to be passing. But the concept of the Hamiltonian action is more general, and we may define it at each point of the trajectory and at each instan' of time. Thus let us suppose that at time t the particle is passing through the point P; the formula (17) (in which S is the Maupertuisian action at the point P) defines the value of the Hamiltonian action of the particle at this point. But even after the particle has passed through the point P, or before it has reached this point, the formula (17) still defines a value for the Hamiltonian action at the point P at each instant t. We may therefore view the Hamiltonian action as dissociated from the particle and as existing on its own account at each point of the trajectory and at each instant. Formula (17) then shows that, as time passes, the value of the Hamiltonian action at the point P decreases in value. Also, at a given instant t this action increases in value along the trajectory.

Let us now consider, on the various trajectories, those points at which the Hamiltonian action has at time t the same value A as it has at the point P. The relation (17) shows that, at each one of these points, the Maupertuisian action has the same value $A + Wt$. Consequently, the points at which the Hamiltonian action has the same value at a given instant are points covering a surface of equal Maupertuisian action. A similar situation holds at all instants and for all points on a trajectory. We conclude that the surfaces of equal Hamiltonian action A coincide at any instant with the surfaces of equal Maupertuisian action S. But an important difference distinguishes the two kinds of surfaces. A surface of equal Maupertuisian action, associated with a given value of this action, is a fixed surface which does not change its position in the course of time. On the other hand, a surface of equal Hamiltonian action, associated with a given value of this action, does not remain fixed, for we have seen that the value of the Hamiltonian action at a fixed point varies with time.

We are thus prompted to examine how a surface of equal Hamiltonian action must move in order to remain associated with the same fixed value of the action. This information may be derived from the formula (17). We find that the surface must advance normally to the trajectories with a velocity defined at each point by

(18)
$$\frac{W}{p},$$

where W is the total energy, and p the magnitude of the particle's momentum at the point in question. The velocity (18) is precisely the velocity (8), given by de Broglie for the wave velocity of his wave and hence for the velocity of the de Broglie wave fronts; and since a de Broglie wave front likewise remains perpendicular to the trajectories as it advances, we conclude that de Broglie's wave fronts move along with the surfaces of equal Hamiltonian action.*

De Broglie, whose investigations were undertaken in ignorance of Hamilton's earlier ones, was able to pursue the matter further thanks to his knowledge of the quantum theory. To understand the nature of de Broglie's contributions, we revert to the expression (11) of a de Broglie wave. The argument of the cosine in (11) is $\frac{2\pi}{h}(Wt - S')$; and according to (17), this expression may equivalently be written $-\frac{2\pi\mathcal{A}}{h}$ where $\mathcal{A}$ is the Hamiltonian action. A de Broglie wave may therefore be written

$$(19) \qquad u = C \cos\frac{2\pi}{h}(Wt - S) = C \cos\frac{2\pi}{h}\mathcal{A}.$$

This formula shows that, in a de Broglie wave associated with a doubly infinite family of trajectories, the equiphase surfaces are the surfaces of constant Maupertuisian action, $S = $ constant. Hence the equiphase surfaces at any instant coincide with the surfaces of equal Maupertuisian action, and as a result the de Broglie rays coincide with the trajectories. This peculiarity is already known to us so that we need not discuss it further. Our immediate concern lies in another direction.

In Hamilton's treatment the propagation of the Hamiltonian surfaces does not exhibit any periodicity and has nothing in common with the propagation of a succession of wave crests. But de Broglie, in his formula (19), by taking the Hamiltonian action to be proportional to the argument of a cosine, imposes the wave aspect. It is on this score that de Broglie's treatment goes further than Hamilton's.

A point of interest is that de Broglie's procedure would have been impracticable in Hamilton's days. This statement is readily understood when we note that the argument of a cosine is an angle and hence, like an ordinary number, has no dimensions. The Hamiltonian action, on the other hand, has the dimensions of an action, and, to obtain a number, we must divide it by some other magnitude which also has the dimensions of an action. Now it would have been gratuitous in Hamilton's days to manufacture some magnitude having the dimensions of action for

---

* To be accurate, the velocity (18) of the action wave in Hamilton's theory is not exactly the same as the velocity (8) of a de Broglie wave. The reason is that W, in Hamilton's formula, represents the classical total energy, whereas, in de Broglie's formula (8), it represents the total relativistic energy—the latter differing from the former chiefly because it includes the rest energy m_0c^2. However, when Hamilton's treatment is revised so as to incorporate the theory of relativity, the statement made in the text is found to be correct; namely, the surfaces of equal Hamiltonian action move in exactly the same way as the wave fronts of de Broglie's theory.

the sole purpose of obtaining sinusoidal waves of action, and since no universal unit of action was known to Hamilton, he could not contemplate the possibility of utilizing the Hamiltonian action in the construction of a sinusoidal wave.

The situation changed with the advent of the quantum theory, for Planck's constant h plays the part of a universal unit of action. The Hamiltonian action may now be divided by the unit h so as to yield a dimensionless magnitude, or number, which can serve as the argument of a cosine. The number secured in this way defines the number of units of action h contained in the Hamiltonian action. By this means de Broglie's expression (19) for the wave is arrived at quite naturally.

From the expression (19) of a de Broglie wave, we may easily determine those particular equiphase surfaces over which the wave has the same phase at any given instant. We recall that the equiphase surfaces are the surfaces of equal Maupertuisian action ($S = $ constant). Let $S = S_0$ define one of these surfaces. Then we see from (19) that the equiphase surfaces over which the phase has the same value as it has over S_0, are the surfaces $S = S_0 + nh$, where n is any integer. For instance the surfaces

(20) $$S = 0, h, 2h, \ldots nh, \ldots$$

define equiphase surfaces over which the phase of the wave is exactly the same at any instant. As a result, if one of these surfaces at a given instant defines a wave crest, all the other surfaces (20) will also determine wave crests at the same instant.

Waves and Photons—Since we have associated waves with material particles, it would appear legitimate to associate corpuscles with waves of light. Indeed, many experiments had shown, prior to de Broglie's investigations, that light sometimes behaved like waves and at other times like particles. We shall therefore apply to corpuscles of light the same methods which we have previously applied to material particles.

A first difficulty that confronts any corpuscular theory of light is the fact that in free space (i.e., in empty space when there is no field of force), the corpuscles of light, or "photons," must be moving with the velocity c. But then the theory of relativity requires that their mass, momentum, and energy be infinite. Accordingly, a ray of light falling on our body should crush us. This difficulty was of course unknown to Newton, for it does not arise when classical mechanics is regarded as valid. Inasmuch, however, as relativistic mechanics must certainly be applied for high velocities, some means of removing the difficulty must be found. This difficulty is overcome most simply if we assume that

622   DE BROGLIE'S WAVE MECHANICS

the mass m_0 of a photon at rest is zero. With this assumption, the energy of a photon moving with velocity c is given by the expression $\dfrac{m_0 c^2}{\sqrt{1 - \dfrac{v^2}{c^2}}}$ in which $m_0 = 0$, and $v = c$. This expression then becomes $\dfrac{0}{0}$ and may therefore have any finite value. We shall accept this solution of the difficulty.

According to our assumptions, a photon moving with velocity c in free space may have any energy and also any momentum. If, then, a de Broglie wave is associated with a photon, any frequency and wave length are possible for the wave. In particular, if we have two photons moving in free space, and hence moving with the same velocity c, the two photons may be associated with waves of different frequency.

Let us now consider the wave velocities of the waves. We have seen that, in the case of material particles moving in free space, the wave velocity is $\dfrac{c^2}{v}$, where v is the velocity of the particle. Since the velocity of a photon is $v = c$, the value of $\dfrac{c^2}{v}$ is c, and so we conclude that c is also the wave velocity of the de Broglie wave associated with any photon in free space. In short, all these de Broglie waves (and also the wave packets which may be constructed so as to simulate the photons) move with the same velocity c. This velocity being precisely the velocity of Maxwell's electromagnetic waves, we perceive at least a formal analogy between the de Broglie waves associated with photons and the waves of the electromagnetic theory.

The formal analogy between the de Broglie waves associated with photons and the electromagnetic waves of Maxwell's theory suggests a complete identification of the two kinds of waves. But several difficulties interfere with this view. Firstly, Maxwell's electromagnetic waves are periodic disturbances in the electromagnetic field; and if we identify the de Broglie waves with Maxwell's waves, how are we to account for the stationary electric and magnetic field? Secondly, we must remember that only in the case of photons do the de Broglie waves happen to have the velocity c of Maxwell's waves; in the case of material particles their velocities are considerably greater. Hence the identification of the two kinds of waves in one particular case would not solve the general problem of the significance of the de Broglie waves. The problem of connecting de Broglie waves with those of the electromagnetic theory is as yet unsolved, but a fairly unanimous-interpretation of the de Broglie waves is accepted today by the leading theoretical physicists: These waves are to be regarded as mere mathematical symbols measuring probabilities. We must hasten to add, however, that this view becomes plausible only in the light of further investigation. The point to be kept in mind for the present is that the association of photons and electromagnetic waves is similar to that of material particles and de Broglie waves.

The De Broglie Waves in a Magnetic Field—In our presentation of de Broglie's theory, the word "particle" was used loosely, and we did not specify what kind of particle was intended. To conform with the subsequent development of the theory, we shall suppose that a particle which is associated with a wave is a so-called elementary particle, such as an electron, a proton, or a photon. We refer to these particles as elementary because, so far as we know, they are not composite structures built from simpler constituents. This reservation is not meant to exclude the possibility of waves accompanying composite structures like atoms or grains of sand; it is suggested solely because the complexity of a structure may complicate the representation of the waves. At all events in this chapter our concern will be with elementary particles, chiefly with electrons.

An electron in a mechanical field of force is necessarily acted upon by the field, e.g., by a gravitational field. But an electrostatic field also exerts a force on an electron; and so does a magnetic field provided the electron be in motion. De Broglie was thus led to investigate the propagation of his waves in electrostatic and in magnetostatic fields.

The presence of an electrostatic field does not introduce any novel feature into the situation. As in the case of a conservative mechanical field, space behaves like an isotropic medium having refractive and dispersive properties; the wave associated with an electron is monochromatic, and its rays coincide with the possible trajectories of the electron or with the trajectories of a swarm of non-interacting electrons.

But a novel situation arises when a magnetostatic field is considered. Space is now found to behave like a doubly refractive medium, so that a parallel beam of de Broglie waves on passing through the field will be split into two beams. In optics, we know that the phenomenon of double refraction is connected with the difference in the polarizations of the two refracted waves. But in the case of the de Broglie waves, we can find nothing, for the present at least, to suggest any property similar to polarization. In the subsequent development of the theory, however, the notion of polarization is introduced, and in the particle picture it is connected with the spin of the electron. We shall not elaborate on these points at this stage.

In view of the behavior of de Broglie waves in a magnetic field, and in view of the connection postulated by de Broglie between waves and electrons, we must suppose that a beam of electrons directed through a magnetic field should be split into two beams. This splitting of an electron-beam into two beams is a phenomenon of the same type as the splitting of a beam of silver atoms by a magnetic field in the Stern-Gerlach experiments (see page 601). It is interesting to note that the splitting of the electron-beam, which according to Bohr's theory becomes comprehensible only when we postulate the existence of the electron-spin, is predicted by de Broglie's theory without any reference to this hypothetical spin. This fact affords an indication of the importance which de Broglie's theory may assume in the interpretation of other phenomena.

The fields of force we have considered to this point were permanent and conservative. In them a particle retained the same total energy during its motion, and the shape of its trajectory did not vary with time. The associated de Broglie wave was thus monochromatic, and its rays were fixed lines. But in many cases the field varies with time—an electromagnetic field is of this type. The energy of the particle is now no longer constant, and its trajectories are not fixed. Consequently, the wave is no longer monochromatic, and its rays are constantly changing in shape.

De Broglie's Quantizing Condition—The wave theory suggests an interesting interpretation of Bohr's quantizing conditions. As an example, let us suppose we are dealing with the electron circling inside a

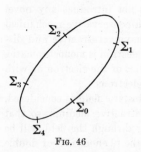

FIG. 46

hydrogen atom. If we disregard the relativistic refinements, the electron is describing an ellipse or a circle. For greater generality we shall assume that the orbit is elliptical. According to de Broglie's theory the electron is accompanied by a wave of definite frequency moving along the orbit. Inasmuch as the individual wave crests advance much faster than the electron, they will soon have moved around the orbit and will have caught up with the electron. If the motion of the electron is to be stable, we may suppose that the wave motion must also be stable. By this we mean that a wave crest which has caught up with the electron must coincide with a wave crest which is just leaving the electron. In any other case the wave crests, on circling time and again around the orbit, would interfere in such a way that the regularity of the wave would be destroyed. The conditions of stability are easily understood from the figure. Thus at any given instant, the wave crests Σ must be so disposed that the wave crest Σ_5, which follows Σ_4, coincides with the wave crest Σ_0.

Now any two wave crests are wave fronts over which the phase of the wave is the same. Hence, if one of the wave crests, e.g., Σ_0, is passing a surface of Maupertuisian action zero at the instant of interest, we see from (20) that the other wave crests will be passing surfaces of action nh at the same instant (here n represents the consecutive positive integers). We conclude that the conditions of stability will be satisfied, whenever the increase in the Maupertuisian action around the orbit has any one of the values nh. This increase in the Maupertuisian action is more simply referred to as the Maupertuisian action along the orbit. Denoting it by S, we see that

$$(21) \qquad\qquad S = nh \qquad\qquad (n \text{ any positive integer})$$

is the condition that must be realized for the de Broglie wave, associated with an orbit, to be stable, and hence for the orbit itself to be stable.

Now, (21) is precisely the quantizing condition originally assumed by Bohr for the purpose of determining the stable orbits, or energy levels, in the hydrogen atom. Hence we have accounted for Bohr's mysterious quantum conditions in terms of a wave phenomenon. As for Bohr's main quantum number n, it is seen to measure the number of wave crests situated along a stable orbit.

De Broglie extended his investigations by showing how the frequencies radiated from the hydrogen atom might be interpreted by means of his waves. He assumed that, when according to Bohr's theory the electron drops from one stable orbit to another emitting radiation in the process, a stable de Broglie wave is associated with each one of the two orbits of interest. These two waves have different frequencies, and their superposition gives rise to a phenomenon of beats, the frequency of which is equal to the difference in the frequencies of the waves. This frequency is precisely that of the radiation which should be emitted in Bohr's theory, and so de Broglie concluded that the radiation waves are closely allied with the beats of the de Broglie waves. We shall not insist on these points, for, as will be seen later, de Broglie's treatment is lacking in rigor. The subject will be resumed in Chapters XXXII to XXXIV, where Schrödinger's investigations will be considered.

In arriving at the stability relation (21), our concern was with the hydrogen atom; but the general method followed in obtaining this relation shows that it expresses the general condition of stability for de Broglie waves following closed paths, and hence for particles describing closed orbits. Since the condition of stability (21) for the waves coincides with

Bohr's quantum condition for the energy levels, we conclude that in all cases de Broglie's theory leads to exactly the same energy levels as Bohr's. This is so, in particular, for the oscillator and for the rotator. In the case of an oscillator of mechanical frequency ν_0, de Broglie's theory, like Bohr's, yields the energy levels

$$(22) \qquad h\nu_0, \quad 2h\nu_0, \quad 3h\nu_0, \quad \ldots \; nh\nu_0, \quad \ldots$$

The two theories also agree in requiring that only periodic motions be quantized. Thus de Broglie's treatment shows that we cannot expect quantizing conditions to appear unless the waves catch up, so to speak, with already existing ones. Hence discrete energy levels cannot occur for a particle describing a trajectory that does not close or at least that does not manifest some periodic feature. As an example, the hyperbolic orbits in the hydrogen atom cannot be quantized according to de Broglie; and this conclusion is in full accord with the tenets of Bohr's theory.

The foregoing illustrations seem to indicate that the theories of Bohr and of de Broglie, though conceptually different, are nevertheless equivalent. If this were so, de Broglie's theory in spite of its interest could not anticipate any phenomenon which was not already predictable on the basis of Bohr's theory. We may hasten to say that this opinion would be unjustified. Indeed, the ability of de Broglie's theory to predict the splitting of a beam of electrons into two beams by a magnetic field suffices to show that the theory may quite well lead to new discoveries.* For the present, however, we are not concerned with new discoveries. What we propose to show is that even if de Broglie's theory had furnished no new discoveries, it would still have the merit of shedding considerable light on the mysterious discontinuities which are characteristic of Bohr's theory of the atom.

For instance, we have seen that, in classical science, all processes were assumed to be continuous, and that the revolutionary aspect of the quantum theory arose from its systematic introduction of discreteness. Now continuity and discreteness seem to be the antithesis of each other and appear incompatible. But de Broglie, by interpreting the discrete stable orbits of Bohr's theory by means of a condition of stability imposed on continuous waves, showed that the two opposites could be reconciled. The fact is that the discrete sequence of stable values resulting from the condition of wave-stability exhibits a striking resemblance with the discrete nodal surfaces that occur in wind instruments. As has long been known, when the air in an organ pipe is set into vibration, station-

---

* To account for this effect in Bohr's theory, we must postulate the electron spin.

ary waves are formed; the air vibrates in some parts of the pipe but remains motionless at the points situated on certain equidistant surfaces. These surfaces, called nodal surfaces, subdivide the length of the pipe into equal intervals, so that integers appear. The appearance of these integers and thereby of discreteness in an otherwise continuous process was in no wise mysterious in classical science, since it could be anticipated on theoretical grounds. De Broglie's investigations indicate that the integral quantum numbers of Bohr's theory are themselves but effects of the continuous waves.

Having stated some of the conceptual advantages of de Broglie's theory, we return to our original question: Does de Broglie's theory account for any phenomena which Bohr's theory was unable to interpret? * The answer is in the affirmative provided, however, de Broglie's theory be refined. The significance of this reservation is clarified by the following considerations. De Broglie's theory, as developed to this point, yields the same quantizing conditions and hence energy levels as Bohr's theory. In particular, it furnishes the energy levels (22) for the oscillator. But, as we have mentioned on other occasions, the levels (22) are certainly wrong. On the basis of the experimental evidence, the correct levels are

$$(23) \quad \frac{h\nu_0}{2}, \quad h\nu_0 + \frac{h\nu_0}{2}, \quad 2h\nu_0 + \frac{h\nu_0}{2}, \quad \ldots \ldots nh\nu_0 + \frac{h\nu_0}{2}, \ldots \ldots$$

Similarly, in the case of the rotator, the energy levels furnished by the theories of de Broglie and of Bohr must be revised. We are thus led to the conclusion that de Broglie's theory, like Bohr's, can be viewed only as an approximation. But it is here that the superiority of the wave theory becomes manifest. Whereas all attempts to refine Bohr's theory were found to require the introduction of additional hypotheses having no obvious connection with those already accepted (e.g., half-quantum numbers), the refinement of de Broglie's theory was obtained not by appealing to new assumptions,† but merely by removing certain restrictions which de Broglie had unwittingly introduced. When these restrictions were removed, a new field of possibilities was opened for the theory to explore.

---

* De Broglie's prediction that a beam of electrons would be split by a magnetic field is not, strictly speaking, a phenomenon which Bohr's theory fails to interpret, for the spinning electron accounts for this effect.

† In later chapters we shall see, however, that de Broglie's original theory has had to be modified and expanded in various ways.

The Refinement of De Broglie's Theory—De Broglie, in developing his theory, assumed for reasons of simplicity that the condition of the space in which the waves were propagated was compatible with the requirements of ray-optics. For instance, he identified the rays of his waves with the mechanical trajectories of the particle. Since the very notion of "ray" is valid only insofar as the conditions of the medium satisfy the requirements of ray-optics for the waves, we conclude that de Broglie's theory in its original form is a wave theory restricted to the domain of validity of ray-optics.

But suppose that in the problem of interest the heterogeneities in the medium are too important to be disregarded over extensions of the order of the wave length of the de Broglie waves. In this case we must reject the methods of ray-optics and rely on the more refined ones of wave-optics. As was explained in Chapter XIX, diffraction phenomena must now be expected. The situation just contemplated is realized in many cases: in particular, whenever the refractive index of space for the de Broglie waves varies greatly from place to place, and hence whenever the field of force differs considerably in magnitude at neighboring points. As Schrödinger was the first to observe, this condition occurs in the hydrogen atom in the vicinity of the proton-nucleus. And so, in dealing with the hydrogen atom, unless we be satisfied with approximate results, we must investigate the behavior of the de Broglie waves by means of the more refined methods of wave-optics.

The methods of ray-optics for the de Broglie waves also cease to be valid when the homogeneity of space is destroyed by the presence of tiny obstacles * or of tiny holes perforated in screens. Diffraction effects and interference patterns will be generated by these heterogeneities of the medium, and in any particular case the precise diffraction effect can be computed from the general theory of wave-optics. Thus, all the information both practical and theoretical collected in wave-optics during the course of the last century becomes immediately available for the study of the diffractions and interferences of de Broglie waves.

We next inquire how the diffractions of the waves can be detected. The waves themselves are mere shadowy existents and we cannot expect their diffractions to be observed directly. But de Broglie's theory, as we know, establishes a connection between the waves and the particles, and since the latter can be observed, the diffractions of the waves may be revealed indirectly through the behavior of the particles.

---

* Obstacles for the de Broglie waves are obstacles for the particles associated with the waves.

The significance of these conclusions is that the behavior of the particles must differ according to whether the conditions of ray-optics or of wave-optics prevail for the waves. If, as de Broglie assumed, the conditions of ray-optics are satisfied, the particles will move in accordance with the laws of classical (or relativistic) mechanics.* Furthermore, the restrictions placed by Bohr on the intra-atomic energy levels are then derived from de Broglie's stability condition for the waves. But suppose the conditions of ray-optics no longer hold for the waves. The behavior of the particles will not be the same, so that the classical (or relativistic) mechanical laws will no longer be valid and will have to be replaced by new mechanical laws. We are thus led to the following conclusion: The passage from the conditions of ray-optics to those of wave-optics for the de Broglie waves will be accompanied by a passage from classical (or relativistic) mechanics to some more refined form of mechanics at present unknown. In short, the relationship between the new mechanics and the classical or even the Einsteinian mechanics will be of the same kind as the one holding between wave-optics and ray-optics. We may therefore refer to the more refined form of mechanics as *wave-mechanics,* and to the classical and the Einsteinian mechanics as *ray-mechanics.*

In all rigor it would be misleading to say that there are two kinds of mechanics in Nature: ray-mechanics and wave-mechanics. The fact is that no clear-cut distinction separates the two mechanical doctrines. If we imagine a situation in which the conditions for the de Broglie waves pass gradually from those of wave-optics to those of ray-optics, wave-mechanics will merge gradually into ray-mechanics. Thus, just as there is only one rigorous doctrine in optics, namely, wave-optics, so also is there only one rigorous mechanics, namely, wave-mechanics. Ray-mechanics, like ray-optics, is a mere degenerate doctrine, never realized rigorously in practice though useful as an approximation.

The conceptual differences which distinguish wave-mechanics from ray-mechanics are fundamental. Indeed, we may question the advisability of retaining the name "mechanics" for the new doctrine, since this name suggests the retention of concepts which must now be discarded. A point of some interest is that the mechanics of the theory of relativity is here classed with classical mechanics, both mechanics being forms of ray-mechanics. Thus relativistic mechanics, once regarded as revolutionary, today appears classical. It is recognized as a great advance over Newtonian mechanics, being a refinement of the latter which cannot be

---

* We may, as we see fit, develop de Broglie's theory in accordance with classical or with relativistic mechanics.

disregarded for bodies moving with high speeds; but in spite of this refinement, relativistic mechanics is not refined enough, for it still ascribes definite trajectories to the particles, and hence will be as incapable of accounting for subatomic phenomena as is Newtonian mechanics.

The advantages of de Broglie's method now become clear. Classical mechanics as applied to subatomic phenomena was known to be invalid, for we saw, especially in the higher atoms, that even when Bohr's quantum restrictions were applied, no adequate model based on classical mechanical laws could be found to account for the results of experiment. On the other hand, there was no means of deciding on the new doctrine that would take the place of the discarded mechanics. De Broglie's theory supplies the answer. We must rely on the methods of wave-optics and apply the wave laws which control the de Broglie waves within the atom; and then, from the behavior of the waves, deduce the corresponding mechanical occurrences and the energy levels. When these refinements were applied by Schrödinger to the hydrogen atom, he obtained the same energy levels as were obtained by Bohr and by de Broglie in his earlier treatment. Thus, the refinements did not yield new results in this case. But with the oscillator and the rotator, the more refined methods gave new values for the energy levels, and these were precisely the values which seemed to be required by the experimental evidence and which Bohr's theory was unable to furnish. A further development of the wave theory by Dirac established the electron spin. Finally, the curious formulae devised by Landé in the anomalous Zeeman effect and the coalescence of the screening-doublet levels in the hydrogen atom were found to be consequences of the wave theory.

But it is not only the theory of the atom that has benefited from wave-mechanics. Thus we have said that when tiny obstacles are placed along the paths of the de Broglie waves, wave-mechanics must be applied in place of the classical ray-mechanics. We shall now be concerned with the phenomena that must be expected under such conditions.

The Diffraction of Electrons—In Chapter XIX we discussed the diffraction of waves of light that pass through a small hole perforated in a screen. We said that, if the diameter of the hole is smaller than the wave length, a divergent beam of waves emerges from the hole, so that a paper placed behind the screen exhibits a uniform illumination. We also mentioned that if two small holes in place of one are considered (as in Young's experiment), a different diffraction effect occurs: the interference of the divergent beams, issuing from the two holes, causes parallel bright and dark bands to appear on the paper.

In our discussion of these effects we were concerned with waves and not with photons. But the same effects may be interpreted when we associate photons with the waves. When this course is followed, the energy formerly ascribed to the waves is transferred to the photons, and so luminous regions must be viewed as regions in which photons are abundant. We must assume therefore that in Young's experiment the bright bands represent regions where the photons collide with the paper, whereas along the dark bands no photons are present. According to this interpretation the density with which the photons are packed is proportional to the intensity of the waves; consequently, where the waves cancel, no photons appear.

Let us return to the de Broglie waves and their associated particles, e.g., electrons. Consider a beam formed of electrons describing parallel courses with the same velocity. We assume that the electrons of the beam exert no interactions. In this case the beam is associated with a plane de Broglie wave, whose frequency and wave length are determined by the energy and momentum of any one of the electrons in accordance with de Broglie's fundamental formulae (5) and (6). Suppose now this beam of electrons falls on a small hole perforated in a screen. If the hole is sufficiently small, the de Broglie waves of the beam, by the mere fact that they are waves, should be diffracted from the hole and give rise to a divergent beam. In view of the association of waves and electrons, the diffraction of the waves should entail the scattering of the electrons through the divergent beam. Equivalently, we may say that the electrons, like the waves, should be diffracted. Thus, by solving the optical problem of diffraction for the waves, we could determine the trajectories of the aggregate of electrons.

The scattering of the electrons would be disclosed if we placed a fluorescent screen behind the hole; the impacts of the electrons against the fluorescent screen would generate scintillations which would indicate the positions of the electrons and hence the deviations they had sustained. According to our earlier explanations the scintillations should be distributed more or less uniformly over a certain area of the screen.

Technical difficulties would make this experiment very arduous to perform, but the results derived from experiments of a similar nature indicate that the diffraction effects we have described would certainly occur. We shall assume therefore that the diffraction effects correspond to reality.

At first sight it might seem possible to account for the diffraction of electrons without postulating mysterious de Broglie waves. For instance, we might suppose that the contour of the hole attracts (or repels) the electrons, deviating them from their straight course. The following experiment, however, shows that this assumption must be discarded: Suppose there are two holes instead of one. According to de Broglie's theory exactly the same interference effect should occur as in the case of Young's optical experiment: parallel bands of maximum and vanishing intensity for the waves should be generated over a plane surface placed in the path of the superposed beams. If, then, by analogy with the association of photons and waves of light, we suppose that the electrons are most densely packed where the waves are most intense, we conclude that the electrons of the swarm will collide with the surface (*e.g.*, a fluorescent screen) along the bands of maximum intensity, and not elsewhere. In short, whereas when there was only one hole the electrons from this hole fell at random all over the fluorescent screen, we find now that by opening the second hole, we prevent these electrons from falling in certain regions.

These results show that the deflections of the electrons on passing through a hole cannot be due to the actions of the contour of the hole; for if such actions were the cause of the deflections, the opening of a second hole would not affect these deflections: the electrons would continue to fall at random over the fluorescent screen, and no discontinuous pattern of bands would arise.

In our discussion of the foregoing experiments, we have assumed a connection between the intensity of the de Broglie waves at a point and the density of the particles around this point. The connection was suggested when we interpreted Young's experiment in terms of photons, and then noted the analogy between optical waves and photons on the one hand, and de Broglie waves and particles on the other. But regardless of its justification, what we wish to stress is that this connection adds a new feature to the wave picture. Thus de Broglie, in his original association of waves with particles (as defined by the fundamental relations (5) and (7)), was only concerned with the frequency and with the wave-number vector of the waves. The intensities of the waves from point to point were not taken into consideration. This omission is now remedied, for henceforth the intensity of the wave at a point P will be supposed to be proportional to the density in the packing of the associated particles around this point. If we are assuming that a single particle is associated with the wave, the intensity at a point P must be regarded as proportional to the probability of our finding the particle in the neighborhood of this point. In the next chapter we shall see that this fundamental postulate constitutes one of Born's basic assumptions.

Suppose, then, we assume that the de Broglie wave is associated with a single electron. When both holes are open, an interference pattern for the waves is formed as before, and if the electron should happen to pass through one of the holes, it will collide with the fluorescent screen at some point situated in a band of maximum intensity. But when one of the holes is closed, the interference pattern does not occur; and the electron, after passing through the open hole, may collide with the screen at any point. If we knew nothing of the waves, this situation would be incomprehensible; for even when both holes are open, the electron can pass only through one of the two holes; and hence it would seem that the phenomenon registered should be the same whether the hole through which the electron does not pass were open or closed. In short, if de Broglie's theory is correct, we must take the wave into consideration and assume that it passes through all the holes that are available; we cannot restrict our attention to the particle and to one hole alone. In this analysis we are assuming that the electron has the attributes which we commonly ascribe to a particle. In the next chapter where the ideas of Heisenberg and Bohr will be discussed, we shall see that these conceptions will need revision. In particular, we shall be led to suppose that the electron and the waves are but different aspects of the same condition; and that the electron is not a strictly localized particle, but a more or less diffuse existent which, in the above experiment, can pass through the two holes simultaneously. In the remaining part of this chapter, however, we shall be concerned only with swarms of particles, and in this case the paradoxical conceptions just mentioned will not be required. Accordingly, we shall continue to regard the waves and the electron as distinct, the latter being viewed as a particle in the ordinary sense of the word.

In the diffraction effects discussed so far, a parallel beam of waves was directed against one or more small holes perforated in a screen. Suppose now we modify the experiment by causing the beam (e.g., of light) to fall on a tiny polished disk. If the diameter of the disk is relatively large, a reflected parallel beam will be produced. But if the diameter is very small, say, half a wave length of the incident light, regular reflection will no longer occur. Instead, a divergent beam will issue from the disk, so that the light will appear to be re-emitted in all directions (on one side of the disk). Let us note that the diffraction effect we are here describing is the same as the one discussed in connection with the hole, except that now a reflection takes the place of a transmission.

Young's interference experiment in optics may be repeated with two tiny disks taking the place of the two tiny holes. The interference effect is enhanced when we utilize a large number of reflecting disks placed on a

634 DE BROGLIE'S WAVE MECHANICS

plane surface. These, however, must be regularly spaced, for if they are situated at random, no interference pattern will occur. Furthermore, the distance between two neighboring disks must be of the same order of magnitude as the wave length of the light. Suppose, then, a parallel beam of monochromatic light is directed against this assemblage of disks. Calculation shows that the light will be diffracted only in certain well-defined directions. If a sheet of paper is made to intercept the diffracted waves, bright and dark bands will appear. The positions of these bands depend on the wave length of the incident light. Consequently, when white light is utilized, the various constituent radiations, which by their superposition yield white light, will undergo different diffractions and a spectrum of colors will appear. The principle of gratings is exhibited in this phenomenon. Similar diffraction effects must occur when the disks are distributed through a volume of space instead of being placed on a plane. But here also the distribution must be regular, and the distance between two neighboring disks must be of the same order as the wave length.

Some thirty years ago a controversy arose on the nature of X-rays The majority of physicists believed that these rays were waves of light of exceedingly high frequency and hence of exceedingly short wave length. But as experimenters had been unable to obtain with X-rays any of the familiar phenomena of reflection, refraction, and diffraction (so easily realized with visible light), the nature of X-rays remained in dispute. Then von Laue accidentally caused X-rays to fall on a crystal, and for the first time diffraction effects were observed. The rays were deflected in a number of privileged directions. Several facts were thereby established. Firstly, X-rays were proved to be waves. Secondly, a crystal was proved to be a regular arrangement of atoms in space, the atoms playing the same rôle as the disks in our previous illustration. Finally, the distances between neighboring atoms were shown to be of the same order of magnitude as the wave lengths of the X-rays. At the same time the reason that crystals did not generate diffraction effects for waves of light belonging to the visible frequencies was seen to be due to the smallness of the distances separating the atoms in a crystal. These distances are far smaller than the wave lengths of visible light, whereas they are of the same order of magnitude as the wave lengths of X-rays.

These preliminary considerations on X-rays and crystals enable us to understand the experiments which afforded the first direct proof of the correctness of de Broglie's theory. De Broglie's fundamental relations show that the waves associated with electrons moving at low speed should have wave lengths of the same order as those of X-rays. Hence the diffraction effects that arise when X-rays fall on crystals should also occur when de Broglie waves are utilized. This implies that a beam of electrons falling on the surface of a crystal ought to be reflected in certain privileged directions and not in others. A crucial test of de Broglie's theory would thus be to submit a crystal to a bombardment of electrons.

The majority of experimenters, however, were unfamiliar with de Broglie's ideas, and his theory was still so speculative that it had not been brought to their attention. It was by accident that Davisson and Germer undertook an experiment which was precisely of the type required to illustrate the diffraction phenomena just discussed. The original aim of these experimenters was to study the scattering (*i.e.*, irregular reflection) of a beam of slow-moving electrons directed against the smooth surface of a metal; a strip of nickel was taken. Now under ordinary conditions, a metal is not in crystalline form, so that the experiment contemplated by Davisson and Germer would not have revealed the diffraction effects anticipated by de Broglie's theory. However, the tube, in which the metal was placed, broke by accident, and to remedy the mishap, the nickel was heated and the experiment resumed. The heating and the subsequent cooling changed the texture of the metal into one of crystalline form, with the result that when the beam of electrons was directed afresh against the metal, Davisson and Germer unintentionally performed an experiment which afforded a crucial test of de Broglie's theory. Privileged directions of rebound were detected for the electrons, and subsequent calculation proved that they were in agreement with de Broglie's theoretical anticipations. The attention of physicists was aroused by these results, and soon after, a number of similar diffraction experiments were devised by G. P. Thomson, Rupp, and others. In every case the previsions of de Broglie's theory were found to be correct, not only in a vague qualitative way, but also to a high degree of accuracy.

In all these experiments, the particles whose diffraction effects have been disclosed are electrons (and also protons). But in theory similar diffraction effects should occur for all particles of matter, and in particular for those particles with which we come into daily contact, *e.g.*, grains of sand or stones. We therefore inquire: Why is it that the wave properties of matter have been detected only in recent years and only in connection with electrons and protons, whereas the wave properties of light have been known for more than a century. The answer to this question is to be found in the microscopic wave lengths of the de Broglie waves (associated with matter) on which we can experiment in practice. In the case of visible light the wave lengths are relatively long, and so we may easily manufacture media exhibiting the measure of heterogeneity necessary to generate diffraction effects. For example, we may perforate sufficiently small holes or slits in screens, or trace sufficiently fine rulings on a piece of glass. But in the case of the de Broglie waves, the wave lengths will usually be far shorter, and our man-made devices will be incapable of producing perceptible diffractions.

Other reasons also militate against an easy detection of the wave properties of matter. We recall that, according to de Broglie's relation (6), the wave length λ of a wave associated with a particle of momentum p is defined by $\frac{h}{p}$. Since Planck's constant h is a very small magnitude, the wave length λ will be exceedingly small unless the momentum p of the particle is extremely minute. In the majority of cases, the wave length of the de Broglie waves will be far shorter than those of X-rays; and even with the help of crystals, we shall be unable to obtain any perceptible diffraction effects. To generate perceptible diffraction effects, we must experiment, therefore, with de Broglie waves having relatively long wave lengths, and hence with particles whose momentum is extremely small. Consequently, we must operate with particles of extremely small mass, moving at low speeds. In view of these requirements, we cannot expect to observe diffraction effects for relatively massive particles, such as grains of sand, and so we must have recourse to electrons. Since electrons have been discovered only recently, and in any case can be observed only indirectly, we understand why the wave properties of matter have been so difficult to detect and so long unsuspected.

A point of interest is the part played by Planck's constant h in these phenomena. The expression $\lambda = \frac{h}{p}$ for the wave length λ of the de Broglie wave shows that, if in our universe the constant h were to become smaller and smaller, the same particle having the same momentum p would be associated with waves of shorter wave length. In such a world the conditions of ray-optics would be more nearly realized, and the wave properties of matter would be even less conspicuous than they actually are. Since the gradual decrease in the value of h corresponds to the gradual disappearance of quantum occurrences, we are led to view the wave properties of matter as being at root a quantum manifestation.

CHAPTER XXX

HEISENBERG'S UNCERTAINTY PRINCIPLE

EVEN before the advent of de Broglie's theory, various optical experiments had shown that radiation betrayed a dual aspect, sometimes wavelike and at other times corpuscular. So long as this dualism was restricted to radiation, it could be assumed to exhibit a new property of the mysterious existent—light. But subsequently, in view of de Broglie's theory and of the corroborative experimental evidence mentioned in the last chapter, physicists were compelled to extend the same dualism to matter. Some general principle of obscure import thus seemed to be involved. Various tentative explanations were advanced by the leading theoretical physicists. The one in favor today is Heisenberg's "Uncertainty Principle." Before discussing it, we shall mention some of the earlier interpretations though they have since been abandoned.

The first interpretation was advanced by de Broglie himself. He did not attribute the dual aspect of matter and of light to the independent existence of waves and of particles in Nature. Instead, he assumed that the waves were the only reality, and that the semblance of particles was due to a peculiarity of the waves. De Broglie's interpretation raises an obvious objection: Particles, so far as we can determine, are well localized and pursue well-defined trajectories, whereas waves and even wave packets are diffuse and hence cannot define tiny localized regions simulating particles. De Broglie overcame this difficulty by supposing that, within the disturbed region occupied by a wave, the intensity of the wave became infinite at a point. This point, having a well-defined position and moving along a definite trajectory, was then identified with the particle. As mathematicians would say, the particle was assumed to be a point-singularity in the wave field. Thanks to this assumption, the particles of physics, though viewed as wave manifestations, could be credited with a definite path and motion. Had de Broglie's interpretation been accepted, we should have had a pure field theory, and the picture of the physical world would have been one in which waves were fundamental and particles mere by-products. This interpretation was subsequently rejected.

A second interpretation, defended by de Broglie at the Solvay Congress of 1927, was similar to the first, in that, as before, definite paths and motions were ascribed to the particles. But in the new interpretation the particles, to the same extent as the waves, were credited with an independent physical existence. De Broglie showed that in a wave manifestation, however complicated, certain expressions could be derived from the wave magnitudes, and that these expressions would define at each point of space an energy and a momentum-vector. As a result, with each wave manifestation, could be associated an aggregate of trajectories susceptible of being described with well-determined motions. A particle present in the region would then follow one of these trajectories with the motion deduced from the wave phenomenon. It would be as though the wave guided the particle along one of the trajectories. For this reason the name "pilot-wave hypothesis" was given to de Broglie's interpretation. The precise trajectory which would actually be followed by a particle could not be determined from the wave, though the relative probabilities of the various trajectories could be calculated. Thus, the probability that the particle would be situated at a given point, at a given instant, was held to be proportional to the intensity of the wave at this space-time point. We may recall that this last assumption was accepted implicitly in our interpretation of the diffraction experiments with electrons; it constitutes one of Born's assumptions, to be mentioned presently.

In spite of its attractiveness, de Broglie's second interpretation was found to entail impossible consequences. For instance, in some situations the permissible trajectories derived from the wave would have to be described by the particle with a velocity exceeding that of light; and this contradicted the theory of relativity. Then again, in the case of a bichromatic wave (to which we shall refer when discussing Born's assumptions), the energy which de Broglie's treatment ascribed to the particle was in conflict with experiment. For these reasons de Broglie abandoned his interpretation. His views were of interest, however, because they revealed the difficulty of ascribing a well-defined path and motion to the particle; and, as a result, they led to the suspicion that a rigorously situated particle having a well-defined motion might be a myth. This conclusion need not conflict with common experience, for only in the shadowy realm of wave mechanics proper does the vagueness to which we have alluded occur.

A third interpretation was at one time defended by Schrödinger. It was suggested to him when he investigated the hydrogen atom by the methods of the new wave mechanics. Schrödinger's interpretation has since been rejected but it is sufficiently interesting to merit mention. We shall discuss it in Chapter XXXIII, and so we need state only its characteristic features here. According to Schrödinger, the electron does not

have the attributes which we commonly ascribe to a particle. Instead
it is a mere wave packet of de Broglie waves. This wave packet is electrically charged, carrying a total charge $-e$. Schrödinger's interpretation differs from both of those previously defended by de Broglie in
that it rejects the idea of a clearly defined particle situated at a definite
point of space and moving with a definite velocity. It exhibits, however,
a certain similarity with de Broglie's first interpretation since it also
requires that we view the waves as fundamental.

An objection which proves fatal to Schrödinger's conception of an
electron is that wave packets eventually spread throughout space. We
cannot therefore identify a wave packet with an electron, which certainly
retains a high degree of permanency over a protracted period of time.
This difficulty was not at first recognized by Schrödinger, because, at
the time, the general theory of the spreading of de Broglie wave packets
had not been worked out mathematically; and Schrödinger, guided as
he was by an investigation he had made in a special case, believed that
wave packets would not disperse.

The dispersing of wave packets, to which we have just referred,
occurs automatically under what might be called the internal forces of
the packet, but a rapid disruption may also be obtained artificially by
the application of external agencies. For instance, if a wave packet falls
on a crystal, the various wave trains which form the packet are sorted
out and are subjected to different deviations. Hence the packet is not
reflected or transmitted as a unit; it ceases to exist. A phenomenon of
this sort occurs in the Davisson-Germer experiments. Obviously, in such
cases the packet cannot behave, even vaguely, like an electron, which
appears to be reflected or transmitted as a unit. Had this more violent
disruption of wave packets impressed itself upon the attention of physicists, the impossibility of Schrödinger's interpretation would have been
recognized even before the mathematical theory of wave packets was
elaborated. But it must be remembered that at the time Schrödinger
was proposing his theory, the Davisson-Germer experiments had not been
performed.

The fourth and latest attempt to elucidate the relation between the
particle and the wave was initiated by Born; it forms the basis of Heisenberg's uncertainty relations.

Born's Assumptions—Born makes two fundamental assumptions.
The first concerns the probability of a particle's presence at a given point
at a given time; the second refers to the probable value of the particle's
momentum and energy.

Let us suppose that de Broglie waves occupy a region of space. The wave disturbance may consist of waves of the same frequency or it may be due to the superposition of waves of different frequencies. In the latter case the wave disturbance may assume the form of a wave packet. Furthermore, the wave motion may exhibit a transfer of the amplitude (as occurs during the advance of a wave packet), or it may be represented by advancing wave crests (as occurs with progressive waves), or again the wave motion may consist in standing waves.

According to Born's first assumption, the following significance must be attached to the wave picture:

> In a region of space in which de Broglie waves are present, the probability that the associated particle will be situated within a given tiny volume $dxdydz$ at time t is proportional to the product of the volume $dxdydz$ and the intensity I of the wave disturbance at any point within this volume at the instant considered.

In the foregoing presentation the volume $dxdydz$ is assumed so small that within it the intensity has approximately the same value at all points.

Now the intensity may have one value or another, depending on the particular tiny volume and the instant of time considered. Hence I will be a function of space and of time, i.e., $I(x, y, z, t)$. Since, according to Born's assumption, the product $I(x, y, z, t)dxdydz$ is proportional to the probability of our finding the particle at the instant t within the volume $dxdydz$, we conclude that the intensity I measures what might be called the density of the probability, or at least a magnitude proportional thereto. We may forego for the present a rigorous definition of the intensity; we shall revert to it when we discuss Schrödinger's theory of radiation. Suffice it to say that the technical meaning of the word is similar to its commonplace significance. For instance, in a stormy sea the greater the height of the waves, the greater the intensity of the disturbance; the intensity would be zero if the sea were perfectly calm. Similarly for a de Broglie wave packet: the intensity is greatest inside the packet and is zero outside.

Born's first assumption may be adapted to the case where a swarm of similar non-interacting particles is associated with a de Broglie wave. Thus, if we are dealing with a swarm of particles, the number of particles in a given tiny volume is obviously proportional to the probability of our finding one special particle in this volume. Consequently, Born's assumption, adapted to the case of a swarm of non-interacting particles, implies that the number of particles in a given tiny volume (and hence the density of the swarm) is proportional to the intensity of the wave disturbance in this volume.

When presented in this form, Born's assumption is seen to be none other than the one we mentioned in the previous chapter in connection with the diffraction of electrons.

Before proceeding, we must stress the fact that Born's assumption, which we have just examined in connection with material particles and de Broglie waves, is also regarded as valid for photons and their associated waves of light. The same remark will apply to Born's second assumption.

We now come to Born's second assumption. It deals with the momentum and energy of a particle associated with de Broglie waves. The basis of the assumption is most easily understood when we examine certain peculiarities of the photo-electric effect. Let us suppose that two trains of plane electromagnetic waves of different frequencies, ν_1 and ν_2, fall on a metallic plate. If the frequencies are sufficiently high, the photo-electric effect will occur; and we shall find that the photo-electrons are ejected, some of them with energy $h\nu_1$, and the others with energy $h\nu_2$. No intermediate values of the energy will be observed. We also find that if the intensity of the first wave train is n times as great as that of the second train, the electrons ejected with energy $h\nu_1$ will be n times as numerous as those ejected with energy $h\nu_2$.

Let us consider the corpuscular interpretation of the foregoing experiment. First we recall what is implied when we say that a particle (photon or electron) and a wave (electromagnetic or de Broglie) are associated. If, at a point P, a wave has frequency ν and wave-number vector $\overrightarrow{\sigma}$, the particle associated with the wave will have, at this point P, an energy W and a momentum-vector $\overrightarrow{p}$ which are related to ν and $\overrightarrow{\sigma}$ in accordance with de Broglie's fundamental relations, viz., $W = h\nu$ and $\overrightarrow{p} = h\overrightarrow{\sigma}$. Having recalled this point we return to the corpuscular interpretation of the experiment with the superposed electromagnetic waves.

We must suppose that the two superposed wave trains are accompanied by a swarm of photons and that the collisions of the photons with the electrons of the metal cause the latter to be ejected with the energies of the photons. Since the ejected electrons are always found to have the energy $h\nu_1$ or $h\nu_2$, we must assume that the photons likewise always have one or the other of these two energies. In other words the photons must be associated with the one or the other of the two wave trains. Furthermore, the relative numbers of photons associated with the two wave trains must be proportional to the intensities of these trains. Consequently, if we single out one photon from the swarm, the probability that

this photon will be associated with the one or the other of the two trains
is proportional to the relative intensities of the trains.*

The conclusions just stated refer to the relations between photons
and electromagnetic waves. But, in view of the general similarity between
photons and electromagnetic waves on the one hand, and material
particles and de Broglie waves on the other, we may extend our former
conclusions to the case of material particles and de Broglie waves. When
this is done, we obtain Born's second assumption in the following form:

> A superposition of different plane, monochromatic wave trains
> (de Broglie waves or electromagnetic ones) is the wave representa-
> tion of a particle which is associated with one or another of these
> wave trains.
> The probability that the particle will be associated with any
> particular wave train of the group is proportional to the relative
> intensity of this train.

Instead of supposing that only one particle is involved we may assume
that we are dealing with a swarm of similar, non-interacting particles.
In this event the relative numbers of particles associated with the respective wave trains are proportional to the intensities of these trains.
Quite generally, as we observed on other occasions, wave mechanics
leaves open the matter of deciding whether a single particle or a swarm
of non-interacting ones is to be considered.

In our presentation of Born's second assumption, we have supposed
that the constituent monochromatic waves were plane and had uniform
intensities. Thanks to this restriction, the intensity of a given wave train
was the same at all points. Usually, however, the intensity of a constituent wave varies from place to place (and also from time to time at
the same place). We must then revise the presentation of Born's second
assumption by taking, in place of the intensity of the constituent wave at
a point, the value of this intensity integrated from point to point throughout space. The revised statement of Born's second assumption is then as
follows:

> If we have a superposition of waves of different frequencies
> and wave lengths, the wave picture indicates that the particle is
> associated with one of these waves. The probability of the particle
> being associated with one particular constituent wave is propor-
> tional to the intensity of this wave *integrated throughout space.*

---

* De Broglie, in his second interpretation (mentioned at the beginning of the
chapter) assumed that, when two different waves were superposed, the particle would
exhibit a partial association with both waves simultaneously. As a result, the particle
might have an energy which would be a compromise between the energies corresponding
to the two waves. However, the experiment we have just discussed soon convinced
de Broglie that his interpretation was untenable.

The possibility of considering in place of a single particle a swarm of non-interacting ones holds as before.

If we compare Born's first and second assumptions, we note that, in the first assumption, we are concerned with the intensity of the resultant wave disturbance at a given point; whereas, in the second assumption, our interest lies in the intensities of the respective constituent waves integrated over the entire domain of existence of the waves. For instance, in the case of a wave packet, the resultant intensity vanishes outside the packet (owing to the interference of the component wave trains outside the packet); and therefore, according to Born's first assumption, the probability of the particle being situated outside the packet is zero. On the other hand, the intensity of any one of the constituent waves does not vanish outside the packet, and, in applying Born's second assumption, we must integrate this intensity over all space and not solely over the volume of the packet. It is the intensity thus integrated which is proportional to the probability of the particle, situated at a current point P, having the momentum and the energy corresponding to the associated wave at this point P.

Let us apply Born's conceptions to a special example. According to the laws of mechanics, the initial position and momentum of a particle of known mass determine the motion of the particle in a given field of force. (The momentum is specified in direction as well as in magnitude, so that it may be represented by a vector.) Instead of stipulating the mass of the particle, we may equivalently state its total energy, and we may then say: The initial space-time position, and the initial momentum and energy of a particle of unspecified mass in a given field of force, determine the particle's motion. Suppose, then, that the initial state of the particle is known only within a certain range of error. For instance, we may know that at a given initial instant the particle is situated somewhere within a given volume, but we do not know at which precise point. We shall also suppose that the momentum and the energy of the particle are known to lie within specified ranges, but that the exact momentum and energy are unknown. We have here a certain amount of information concerning a particle, and we wish now to obtain the associated wave picture. Born's assumptions require that we proceed as follows: We consider all those monochromatic trains of de Broglie waves whose frequencies and wave numbers are comprised between the limits corresponding to those stated for the energy and momentum of the particle; we then imagine that these monochromatic trains are superposed in such a way as to form a wave packet filling the volume of space in which the particle is known to be situated at the initial instant. The wave disturbance thus depicted in space yields the wave picture we are seeking. This wave picture describes our knowledge of the particle's position, momentum, and energy at the initial instant; but it also gives other information, for it enables us to anticipate the future course, momentum, and energy of the particle.

To take the simplest case, let us suppose that there is no field of force, so that we may assume plane de Broglie waves with which to build the packet. We first treat the problem according to classical mechanical ideas. The laws of mechanics tell us that the particle will describe a straight path with uniform speed. If we knew the precise position, momentum, and energy of the particle at the initial instant, we should, according to the laws of mechanics, know the exact path followed by the particle; and hence we should be able to state with certainty the exact position of the particle at any future instant. But, in our present example, we are supposing that the initial conditions are more or less uncertain. Consequently, all that we can predict is a class of possible straight paths contained in a conical region of space, one of these paths (we have no means of telling which) being the actual path followed by the particle. Corresponding to each of the possible paths which the particle may be describing with one or another of the possible motions, there is a well-defined point marking the exact position of the particle at a given instant t after the initial instant. The aggregate of all these possible positions of the particle at time t determines a volume of space, and we are certain that, at time t, the particle will be somewhere within this volume. If, then, we establish the location and the shape of this volume at successive instants of time, we shall find that, as time passes, the volume advances and expands while remaining of course within the conical region of space which comprises all the trajectories. This information is derived solely from commonplace mechanical considerations.

Let us now pass to the changing wave picture which is associated with the evolving particle picture just discussed. The motion and the change in shape of the wave packet from its initial configuration is regulated by a partial differential equation, which is very similar to the equation controlling the flow of heat in a homogeneous medium. The integration of this equation shows that the wave packet will expand while advancing, and that, at any instant t, it will fill the volume which defines the possible positions of the particle, as derived from our former mechanical considerations. Since, according to Born's assumptions, the volume occupied by the wave packet at any instant defines the region of space in which the particle may be situated at the instant of interest, we see that the mechanical picture and the wave picture yield similar information. But the information given by the two pictures is not quite the same, for the wave packet overlaps at each instant the corresponding geometric volume deduced from mechanical considerations. Thus, we must decide which of the two pictures is the correct one. The essence of wave mechanics is to stress the wave picture and to claim that the possible motion of the particle is given correctly by the motion of the wave packet. The wave picture corrects the classical (or relativistic) mechanical anticipations, or, as we may also say, wave mechanics refines classical (or relativistic) mechanics. In the example just discussed, the correction brought about by the wave picture is scarcely perceptible, but, as we saw in the last chapter when discussing the diffraction experiments with electrons, the correction in some cases may be notable.

The situation, as it appears at the present stage of our investigations, is that the motion of the particle is associated with a wave propagation, and that it is the wave picture which predicts correctly the particle's possible motion.

Difficulties soon beset this mode of interpretation. Thus suppose that a beam of ordinary light falls on an imperfectly reflecting plane surface. Part of the light is reflected and part penetrates into the medium and is transmitted. The particle picture which corresponds to this wave picture is that of a swarm of photons advancing initially in the direction of the incident waves, some of the photons subsequently undergoing reflections from the surface and others penetrating through the surface into the medium. Any particular photon of the swarm thus has a certain probability of being reflected and a certain probability of being transmitted. According to Born's assumptions, these two probabilities are proportional to the respective intensities of the reflected and the transmitted beams.

Instead of a swarm of photons, let us consider one photon in isolation. We suppose that the initial position, energy, and momentum of the photon are known within a certain range of approximation. The wave picture in this case is represented by a wave packet advancing in the general direction of the photon's motion. If an imperfectly reflecting surface S is placed in the path of the moving photon, two wave packets will be formed when the original wave packet strikes the surface. One of these wave packets will be reflected and the other transmitted, and the respective intensities of the two packets will be proportional to the probabilities of the photon being reflected or being transmitted. All that has been said of photons and waves of light applies to material particles and de Broglie waves. Hence in our present example we may with equal justification assume that our particle is an electron falling on a surface which it may penetrate or from which it may be reflected, and that the wave packets are formed of de Broglie waves.

Suppose, then, we are dealing with a material particle. We construct the wave packet which represents the initial position, momentum, and energy of the particle within the range of uncertainty assumed. The surface S, on which the particle falls, splits the packet into two others corresponding respectively to the possibility of the particle being reflected or transmitted. We now interpose a fluorescent screen across the paths of the two packets. Both packets will collide with the screen, and the particle, which is contained in one of the two packets, will give rise to a scintillation at the point where it strikes the screen. The scintillation thus determines the position of the particle and hence shows us in which of the two packets the particle is situated. As a result, we know whether the particle has been reflected or transmitted. Suppose we find that the particle has been reflected. In this event the probability that the particle has been transmitted is zero. But then our wave picture becomes unacceptable, for it exhibits the transmitted packet as well as the reflected one, and it thereby assigns a non-vanishing probability to the transmission of the particle. To bring the wave picture up to date, we must strike out the transmitted packet. Thus, a mere accretion to our knowledge of the particle's position has caused the vanishing of the transmitted packet. This circumstance must occur, however far the two packets may be from the reflecting surface and hence from each other when the position of the particle is revealed by the scintillation.

Einstein has pointed out that the instantaneous causal action transmitted from one packet to another in the foregoing illustration violates the requirements of the theory of relativity, and that to set ourselves in agreement with the theory, we must deny all physical reality to the waves and packets of waves. We must therefore assume that the waves and wave packets only represent what may possibly occur, *i.e.*, probabilities. In other words, they are merely symbolic and must be viewed as *probability waves*. With this understanding, the conflict with the theory of relativity disappears, for the sudden vanishing of the transmitted wave packet is no more mysterious than is the vanishing of the probability that tails may turn up when a coin has been tossed and heads has been seen to fall.

We may summarize the situation by saying that the waves do not represent physical existents, and that the evolution of the wave picture merely describes the probabilities of future events which issue from the more or less uncertain knowledge we have of present conditions. If, then, at any instant we perform an experiment, locating the particle and improving thereby our knowledge of the conditions existing at the instant of interest, the wave picture will have to be modified in consequence; and the range of future possibilities exhibited in the new wave picture will be correspondingly restricted. Though these considerations remove all trace of physical reality from the waves, we cannot dispense with these waves, for, as was stated previously, it is the wave picture which furnishes the correct probabilities.

Suppose, then, a train of plane monochromatic de Broglie waves is allowed to fall, as before, on the imperfectly reflecting surface, so that the incident beam is split into a reflected and a transmitted beam. We complicate our former experiment by supposing that a mirror M is placed in the path, say, of the reflected beam; this beam is then reflected by the mirror and made to cross the transmitted beam. In the region of space common to both beams, interference will occur, the waves of the two beams reinforcing one another in some places and canceling in others. The regions of maximum and of minimum intensity form parallel planes.

If we view the wave picture as representing a swarm of particles, we may assert, on the basis of Born's assumptions, that the density of the swarm in the region of interference will be a maximum over the planes of maximum intensity, and a minimum, or zero, over the other planes. The phenomenon anticipated may be rendered visible if we place a fluorescent screen in the region of interference. The screen intersects the planes of maximum and of minimum intensity along parallel lines, so that parallel bands are formed. The particles will collide with the screen at points in the bands of maximum intensity, and the scintillations generated by these encounters will thus determine the positions of the bands. We may also single out, in thought, one particle of the swarm. Then the bands of maximum intensity will define the regions where the particle has the greatest probability of colliding with the screen.

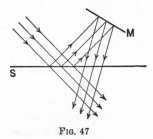

Fig. 47

So long as we assume that the wave picture is associated with a swarm of particles, there is no inordinate difficulty in accounting for the interference effect. We may suppose that some of the particles have followed the transmitted beam, while the others, which have followed the reflected beam, subsequently experience a second reflection at the mirror and are thus deflected into the transmitted beam. The two sets of particles then interfere by reason of their wave properties.

But the situation assumes a revolutionary aspect when we take the wave picture to be associated with a single particle. The interference effect must of course occur as before; and from the physical standpoint this means that if a fluorescent screen is placed in the region of the interference, the particle will collide with the screen and emit a scintillation in one of the bands of maximum intensity. Now the difficulty is that, of the two beams of waves, only that one followed by the particle can be credited with any physical reality; the second beam is a ghost that we retain in the wave picture merely because we do not know which of the two beams the particle will actually follow. Yet, if this be so, how can the two beams of waves interfere? How can ghost waves interfere with real ones? Such is the nature of the paradox.

The Attitude of Heisenberg and Bohr—The paradox may be removed, but only by accepting a further renunciation of our intuitive notions. To confer some measure of symbolic reality on the interference phenomenon, we must necessarily ascribe the same reality to the two beams, at least until such time as the position of the particle is determined. Heisenberg and Bohr suggest, therefore, that we view the particle as situated at one and the same time in the two beams, and hence on either side of the imperfectly reflecting surface. In other words the particle is diffused like a cloud through both beams of waves, and the interference resulting from the superposition of the two beams may therefore be ascribed to the diffused particle, or cloud, interfering with itself. Needless to say, this assumption of a diffused particle conflicts with our understanding of a particle as a discrete existent occupying a well-defined position in space at any given instant.

We must hasten to add that Heisenberg and Bohr do not contend that a particle, say, an electron, is always diffused through an extended region of space occupied by waves. The diffused picture endures only so long as the electron is unobserved. If an experiment is performed for the purpose of locating the electron (*e.g.*, by means of a fluorescent screen), and if the position of the electron is thereby observed, the extended waves immediately condense into a small wave packet, and the electron condenses with the waves, so that it is now diffused only through the small volume of the packet. The more precise our observation of the electron's position, the smaller the wave packet, and hence the more does the corpuscular aspect of the electron manifest itself. We may therefore say that the corpuscular aspect of the electron is brought into existence by our observation of the electron's position.

A further point of interest is that the corpuscular aspect of the electron will not persist for any length of time, for the wave packet which represents the electron will gradually spread; and according to the ideas of Heisenberg and Bohr, the electron continues to fill the packet and hence spreads with the packet and tends to be diffused. If a second position observation is made, the electron will be condensed into some portion of the enlarged packet; and we shall have to cancel the former packet and construct a new one of smaller dimensions. The new packet will spread as it advances, and the same sequence of occurrences will take place each time a new position observation is made. From this discussion we see that the attitude of Heisenberg and Bohr does not conflict with the corpuscular aspect manifested by electrons in physical experiments.

Heisenberg expresses the new ideas by saying:

"In many cases it seems better not to speak of the probable position of the electron, but to say that its size depends upon the experiment being performed." *

In a similar vein, Bohr describes electrons (or particles) as "unsharply defined individuals within finite space-time regions."

The ideas of Heisenberg and Bohr may be extended to the momentum and energy. Before a measurement of momentum has been undertaken, the momentum-vector which we should commonly attribute to a particle must be viewed as pointing in all directions and as having all magnitudes. In other words, the momentum is diffused, much as the position of the particle is diffused and ill-defined before a position measurement has been made. And, just as a position measurement renders the concept of position definite and thereby confers a corpuscular aspect on an electron, so does a measurement of momentum bring the momentum into existence by modifying the wave picture that held prior to the measurement. The change in the wave picture is not, however, of the same kind that occurs for a position observation; for now no wave packet is formed.

At first sight the Heisenberg-Bohr interpretation would appear to renew the conflict with the theory of relativity. For instance, let us suppose that the wave phenomenon extends over a large volume of space. According to the new views a position observation causes an instantaneous condensation of the corresponding diffused particle into a corpuscle. Now, this instantaneous condensation of matter seems to represent a physical change, not merely a change in future probabilities. And, if this be so, we are in open conflict with the theory of relativity, which prohibits in-

---

* Heisenberg. The Physical Principles of the Quantum Theory; Chicago, 1930; p. 34.

stantaneous transfers of matter. However, the conflict is more apparent than real, for we have reasons to suspect that the space-time form of representation, which is valid in the theory of relativity, may lose its validity in wave mechanics. If, despite this fact, for the sake of convenience, we seek to represent quantum occurrences in the space-time frame and picture a particle as being diffused through a region of space, we must not be surprised to find ourselves in conflict with the relativistic laws, which are valid only insofar as a representation in space-time can be made with accuracy. These obscure points will be discussed further when we consider Bohr's "Principle of Complementarity." *

The experiment of the imperfectly reflecting surface is not the only one in which the attitude of Heisenberg and Bohr seems to be in accord with facts. For instance, interference effects with ordinary light may occur even when the intensity of the light is extremely small. If we assume that light is formed of corpuscular photons, we cannot easily understand how such interference effects can arise. But, according to our new views, the situation becomes more comprehensible, for we must assume that a photon is not a corpuscle in the ordinary sense: it may be spread over an extended region and its various parts may interfere among themselves. However strange all these new concepts may appear, no logical inconsistency is attached to them. Our difficulty in comprehending them is due to their unfamiliarity. We must realize, however, that when we penetrate into the world of the physically infinitesimal, we cannot expect to be confronted with familiar notions and concepts.

The views of Heisenberg and Bohr receive support from another direction. In Schrödinger's theory of the hydrogen atom (which we shall discuss in later chapters), an inconsistency arises when we apply Born's assumptions. This inconsistency is dispelled, however, when the views of Heisenberg and Bohr are taken into account. The following explanations clarify the nature of the inconsistency.

Let us suppose that the hydrogen atom is in its state of lowest energy. The Schrödinger wave disturbance corresponding to this state will then be represented by vibrations which are most intense in the immediate neighborhood of the nucleus, but which have a non-vanishing intensity even at extremely distant points. According to our presentation of Born's assumptions, this wave distribution implies that the most probable position of the corpuscular electron at any instant will be in the immediate vicinity of the nucleus, but that there will be a non-vanishing probability of the electron being situated at any distance from the

---

* See page 951.

nucleus. As we shall now see, this statement creates a difficulty if we view the electron as a corpuscle. The fact is that in any given energy state of the atom, the electron has a well-defined energy ; and the principles of mechanics show that, corresponding to this value of the energy, there is a sphere (with centre at the nucleus) within which the electron must remain. When the atom is in its lowest energy state, this sphere is extremely small, so that the electron must always be situated in the immediate vicinity of the nucleus. The probability of the electron's presence outside the tiny sphere being zero, a conflict arises between the requirements of mechanics and those imposed by the wave picture. We hasten to add that the present difficulty is not the inconsistency to which we referred previously.

The difficulty we have alluded to in the previous paragraph is easily removed when Heisenberg's uncertainty principle is taken into account. Though we have not yet explained this principle, its significance in the illustration here discussed is readily understood. The principle informs us that the observation whereby the position of the electron is located, at a given instant, will disturb the momentum and the energy of the electron by unpredictable amounts. Consequently, after the electron's position is observed, the electron no longer has the same energy it had prior to the observation : its energy may now have any value, and, as a result, its position is no longer restricted to a limited region of space. We conclude that the wave picture, insofar as it gives the probabilities of the electron's possible positions *after* the observation, is compatible with the mechanical considerations that follow from our viewing the electron as a corpuscle. In short, thanks to the uncertainty principle, the difficulty vanishes. We are now in a position to understand the nature of the inconsistency mentioned at the beginning of this discussion.

Instead of considering the position of the corpuscular electron *after* the observation has been performed, we concentrate on its position at the instant immediately preceding the observation. We should naturally assume that the two positions were the same. The uncertainty relations corroborate this view, for they show that the observation of the electron's position does not modify this position. Now we have seen that, after the observation, the electron may be found at any distance from the nucleus ; and hence we conclude that this must also be true for the position of the electron *before* the observation. But this conclusion is impossible, for, prior to the observation, the atom was assumed to be in the lowest energy level, so that the electron was necessarily situated within the previously mentioned tiny sphere centered at the nucleus.

The present inconsistency has arisen from our attempt to assign a definite position to the electron before the observation was made: the principles of mechanics require that the electron be situated within the tiny sphere, whereas the wave picture indicates that the electron may be at any point. The source of the inconsistency is thus traceable to our tacit assumption that the electron was a corpuscle before the observation was made. But suppose we accept the attitude of Heisenberg and Bohr. The inconsistency will disappear, for our aforementioned tacit assumption becomes untenable. We must now suppose that the electron comes into existence as a corpuscle only after an observation has been made. Before the observation, the electron is represented by a cloud extending over a certain region of space-time, and hence cannot be credited with any precise position. If, notwithstanding this fact, we erroneously attempt to assign a position to the electron *before* the observation, we shall be attempting to give meaning to a meaningless concept; and so we may expect inconsistencies. Thus, the attitude defended by Heisenberg and Bohr has dispelled the inconsistency.

The general significance of the wave picture now becomes clear. Formerly, we supposed, on the basis of Born's assumptions, that the intensity of the wave at any space-time point was proportional to the probability of the electron being situated at this point. However, when the attitude of Heisenberg and Bohr is taken into consideration, we must revise our former statement and say: The intensity of the wave at a space-time point is proportional to the probability that the electron will be found at this point *immediately after a position observation has been made*; before the observation, the position of the electron is meaningless and hence is not represented by the original wave picture.

The Heisenberg-Bohr interpretation leads to a further conclusion of interest. According to our earlier views, the waves, being mere probability waves, must be regarded as symbolic, whereas the particles represent physical reality. But we must now suppose that particles also should be regarded as symbolic, for they come into existence only when a position observation is performed. In a certain sense this placing of waves and particles on the same footing is more satisfactory, for it would seem strange to view the waves alone as symbolic and not the particles. This argument is particularly convincing in the case of radiation waves and photons. Why indeed should waves of light be regarded as symbolic and photons as real?

The word "symbolic" does not, however, express the situation with sufficient clearness. As Bohr states: the particle aspect and the wave aspect must be viewed as complementary and as exhibiting two different aspects of the same underlying reality. A crude illustration may help us understand what is implied. Under certain conditions we see a circle, under others a triangle, and we conclude that the thing we are seeing has at one and the same time the aspects of a circle and of a triangle. We attempt to combine the two appearances but fail. Eventually, we adapt ourselves to the thought that space has three dimensions and that it contains solid bodies. The two conflicting aspects are now easily reconciled. We have but to postulate the existence of a solid cone, two different orientations of which yield the visual impressions of a circle and of a triangle respectively. In this example the two aspects can of course never appear simultaneously, and we shall see presently that the same is true of the particle aspect and of the wave aspect.

The illustration we have given is a poor one, for the combination of the circular and triangular aspects demands only the introduction of 3-dimensional space, and this notion is assimilated so soon after birth that we are not conscious of its genesis. The theory of relativity affords a more sophisticated example. Prior to Einstein's discoveries, the electric field and the magnetic field were viewed as distinct, in the sense that we could have a field of one type without any trace of the other. Although this situation could occur only when the field was stationary, it was not deemed necessary on this account to view the two fields as mere partial aspects of one and the same underlying entity. The theory of relativity brought about the fusion of the two aspects, no longer by utilizing the background of 3-dimensional space, but by introducing the more refined background of 4-dimensional space-time. The underlying entity, the partial aspects of which are electric and magnetic, was found to be the 4-dimensional electromagnetic tensor situated in space-time. Except for its lesser intuitive appeal, the fusion accomplished by the theory of relativity is of the same type as the one we perform unconsciously when we attribute changing shapes to the various orientations of the same solid body (e.g., a cone) in ordinary space.

But a satisfactory fusion of the wave and of the particle aspects of matter and of radiation has yet to be performed; hundreds of years may elapse before the human mind will be able to comprehend the dualism of waves and particles. One point, however, seems well established: space and time, or even Minkowski's space-time, does not afford the requisite background. To quote Bohr:

"Indeed, we find ourselves here on the very path taken by Einstein of adapting our modes of perception borrowed from the sensations to the gradually deepening knowledge of the laws of Nature. The hindrances met with on this path originate above all in the fact that, so to say, every word in the language refers to our ordinary perception." *

---

* Bohr. Atomic Theory and the Description of Nature; Cambridge, 1934; p. 90.

Although a satisfactory picture of the new developments cannot be attained by means of familiar concepts, yet the mathematical instrument does not fail, for we may predict in any particular case whether it is the wave or the particle aspect that will betray itself.

In view of the impossibility of obtaining a picture in familiar terms, we shall retain side by side the two partial aspects of reality represented by the particle aspect and by the wave aspect. We shall therefore assume that a wave. packet is the wave picture of a particle situated somewhere within the packet, and that the particle has a momentum and energy corresponding to the wave number and frequency of one of the constituent monochromatic de Broglie waves which form the packet. The more complicated ideas of Heisenberg and Bohr may be disregarded for the present. With this understanding we may pass to Heisenberg's uncertainty relations.

Heisenberg's Uncertainty Principle—Born's assumptions show clearly that any wave picture of the mechanical motion of a particle can express only probabilities. We wish now to determine the relationships that connect the various probabilities.

Let us consider the hypothetical case of a particle, whose position, momentum, and energy are accurately known at a given instant t. We wish to obtain the wave picture of this situation. First, let us examine the wave representation of the particle's position. The particle being situated at some point in space, we are called upon to obtain a wave representation of a point. Now we cannot represent a mathematical point by means of a wave picture, for to do so we should have to imagine a wave having a zero intensity everywhere, except at the point in question— and this is impossible. The best we can do is to construct a tiny wave packet having the point as centre. According to Born's first assumption, the particle will be somewhere within that packet; and to this extent its position will be defined. We shall suppose that such a wave packet has been constructed.

Next, we inquire whether our wave picture can describe the exact momentum and energy assumed for the particle. The properties of wave packets show that this is impossible. The fact is that the wave packet can be formed only through the superposition of monochromatic de Broglie waves differing more or less considerably in frequency, in wave length, and in direction of motion. According to Born's second assumption, the particle will have one of the associated sets of values for its energy and its momentum, but the precise set of values is unknown. It is thus impossible to express, by means of a wave picture, the exact position of the particle at a given instant and also its exact momentum and energy. More generally, the peculiarities of wave packets show that, the higher the accuracy with which the position of the particle at a given instant is depicted in the wave picture (*i.e.*, the smaller the wave packet), the more uncertain is the description of the energy and momentum of the particle. The converse also holds: the higher the accuracy with which the momentum and energy are described, the more uncertain the description of the position at a given instant.

Let us obtain the precise quantitative relations which connect these uncertainties. We represent by Δx, Δy, Δz the dimensions of the packet, and by Δt the time taken by the packet to pass from end to end over a fixed point on its line of motion. We set $\Delta \nu$ for the range of the frequencies of the monochromatic waves which by their superposition give rise to the packet, and we set $\Delta \sigma_x$, $\Delta \sigma_y$, $\Delta \sigma_z$ for the ranges of the projections of the wave-number vectors of these constituent waves. In Chapter XIX we saw that, for a wave packet of given space-time dimensions Δx, Δy, Δz, Δt to be formed, the ranges $\Delta \sigma_x$, $\Delta \sigma_y$, $\Delta \sigma_z$, $\Delta \nu$ defining the waves must satisfy the relations

$$\Delta x \cdot \Delta \sigma_x \geqslant 1$$

$$\Delta y \cdot \Delta \sigma_y \geqslant 1$$

(1)

$$\Delta z \cdot \Delta \sigma_z \geqslant 1$$

$$\Delta t \cdot \Delta \nu \geqslant 1.$$

Conversely, if the ranges of the constituent waves are given, the relations (1) determine the smallest possible dimensions of the packet that may be formed.

We multiply both sides of the four inequalities (1) by Planck's constant h. If then we recall that $h\sigma_x$, $h\sigma_y$, $h\sigma_z$ are equal in value to the momentum components p_x, p_y, p_z of the associated particle, and that $h\nu$ equals its energy W, we see that the relations (1) become

$$\Delta x \cdot \Delta p_x \geqslant h$$

$$\Delta y \cdot \Delta p_y \geqslant h$$

(2)

$$\Delta z \cdot \Delta p_z \geqslant h$$

$$\Delta t \cdot \Delta W \geqslant h.$$

These relations are the celebrated Uncertainty Relations of Heisenberg. They show how the space-time dimensions of the wave packet are connected with the ranges in the possible momenta and energies of the particle. We shall elucidate these relations by examining particular illustrations.

According to Born's assumptions a wave picture represents the possible position and motion of a particle. Consequently we must suppose that when a wave picture represents a particle in a volume $\Delta x \Delta y \Delta z$ at some unspecified instant during a given interval of time Δt, the momentum-components and the energy of the particle will be uncertain in accordance with the relations (2). Thus the possible ranges of the momentum and of the energy will be defined by

$$(3) \qquad \Delta p_x \geqslant \frac{h}{\Delta x}; \quad \Delta p_y \geqslant \frac{h}{\Delta y}; \quad \Delta p_z \geqslant \frac{h}{\Delta z}; \quad \Delta W \geqslant \frac{h}{\Delta t}.$$

The smallest possible values, say, of Δp_x and ΔW, are therefore $\dfrac{h}{\Delta x}$ and $\dfrac{h}{\Delta t}$ respectively. On no account can these ranges be any smaller, and usually they will be more extended. For the present we are not attempting to determine whether the vagueness affecting the particle is due to the inadequacy of the wave form of representation or to other causes. However, we shall see later that the uncertainty relations express a fundamental principle of Nature.

In the application of the uncertainty relations two extreme cases are of interest. Thus we may have a wave representation of a particle whose direction of motion, momentum, and energy are accurately known, but whose position is unknown. Here Δp_x, Δp_y, Δp_z and ΔW vanish, whereas Δx, Δy, Δz and Δt are infinite. A mere inspection of the uncertainty relations shows that these conditions are consistent with a wave representation. The wave assumes the form of an infinite train of plane, monochromatic waves of well-determined frequency and wave length, advancing in the direction of the particle's motion. We have already utilized this form of wave representation on many occasions.

The other extreme case arises when the position of the particle is accurately known at a given instant, whereas its direction of motion, momentum, and energy are completely unknown. The wave picture is then furnished by the superposition of an infinite number of concentric, spherical waves of all frequencies and wave lengths. The superposed waves are assumed to generate an infinitesimal wave packet which defines the position of the particle at the instant of interest. In the course of time the wave disturbance spreads in all directions.

Our derivation of the uncertainty relations from the properties of wave packets has shown that the representation of a particle by means of a wave picture inevitably ascribes a certain vagueness to the position of the particle at a given instant, or to its momentum and energy. We therefore return to the question we raised in a former paragraph, and inquire: Is this vagueness due to the inadequacy of the wave form of representation (in which case the uncertainty relations would express properties of waves and not necessarily of particles); or is the vagueness due to some fundamental law affecting particles as well as waves? In other words, should the simultaneous specification of the exact position and of the exact momentum of a particle be denied any physical reality?

At first sight an attitude of reserve towards the uncertainty relations would appear to be warranted by macroscopic observation. Thus in mechanics, simultaneous precise values are ascribed to the momentum coordinates and to the position coordinates of a particle, and hence we must assume that, in mechanics, it is physically significant to speak of the exact position and of the exact momentum of a particle at any instant. Since mechanics, at any rate on the macroscopic level where we are dealing with particles that we can see, has never led to anticipations which were inconsistent with observation, the uncertainty relations, if held valid for particles, would seem to be in conflict with the facts of experience.*

We may readily verify, however, that the acceptance of the uncertainty relations for particles does not contradict common experience. To convince ourselves of this fact, let us suppose that for some unknown reason Planck's constant h were to decrease indefinitely in value. A mere inspection of the uncertainty relations shows that, as a result of this decrease, the uncertainties would become less pronounced and eventually disappear. For instance, in the first uncertainty relation, if we replace h by the vanishingly small magnitude ε, we obtain $\Delta x \Delta p_x = \varepsilon$; and obviously, however small Δx, there is nothing now to prevent Δp_x from being as small as we choose. In other words, the precision in our knowledge of the particle's position would not entail an uncertainty in our knowledge of its momentum. Now this hypothetical decrease in the value of h illustrates, as we have seen time and again, a progressive obliteration of the quantum aspects of phenomena and the gradual approach to the behavior witnessed on the macroscopic level of common experience. Hence, we are perfectly justified in claiming that the uncertainty relations may be valid, and yet not manifest themselves on the macroscopic level (by reason of the small relative importance of h for phenomena viewed macroscopically). This argument does not of course prove that the uncertainty relations for particles *are* valid, but merely that no inconsistency

---

* When we speak of mechanics, we always include the more refined mechanics of the theory of relativity.

is involved between the assumption of their validity and our observation of mechanical processes on the commonplace level of experience.

We now consider some of the reasons which suggest that the uncertainty relations must be valid for particles, and that they are not mere expressions of the uncertainties inherent in waves. In the first place, we have seen that the remarkable predictions obtained from wave mechanics in many situations (*e.g.*, the diffraction of electrons) issue from the assumption that the wave picture furnishes a correct representation of phenomena in which our concern is with particles. It would therefore be unreasonable to accept the predictions derived from the wave picture in the foregoing situations and to reject them in other cases where particles are also involved. Besides, as we shall see later (page 888), the uncertainty relations can be established for particles directly without any reference to waves.

Let us, then, disregard waves entirely and apply the uncertainty relations to the motion of an electron passing through a small hole. We shall see that the diffraction of electrons is an immediate consequence of these relations. Suppose the momentum and direction of motion of an electron are accurately known. According to the uncertainty relations the position of the electron is then completely unknown. A screen punctured with a small hole is now set perpendicularly to the electron's motion. If we assume that the electron passes through the hole, its position after this passage is no longer as uncertain as it was originally. The uncertainty relations require therefore that the momentum of the electron be correspondingly uncertain, so that there is a certain probability of the electron being deflected from its original course. Since such deflections have been confirmed in the experiments on the diffraction of electrons, we may claim that in this case the uncertainty relations are verified for particles.

Let us consider another clue. In the earlier quantum theory, special cases arise where we find that the uncertainty relations are implicitly contained in our assumptions. Thus, in Bohr's theory, the energy levels of the linear harmonic oscillator are obtained when we stipulate that the Maupertuisian action of the particle, taken over a complete oscillation, must be equal to a multiple of Planck's constant h. If Δx represents the total range of the oscillation and Δp_x the range which the momentum will assume during the motion, the total action can be shown to be of the same order of magnitude as $\Delta x \Delta p_x$. Bohr's quantizing condition which determines the stable energy levels may therefore be written

$$\Delta x \cdot \Delta p_x \sim nh,$$

where $\sim$ means "of the same order of magnitude as" and where n is any positive integer. For the lowest energy level we have $n = 1$, so that the quantizing conditions becomes

$$\Delta x . \Delta p_x \sim h.$$

On the other hand, Δx and Δp_x represent the ranges in position and in momentum of the vibrating particle, and thereby define the uncertainties in the position and in the momentum at a given instant. Our quantizing condition thus has the same significance as one of Heisenberg's uncertainty relations.* Now, we cannot ascribe the existence of stable energy levels to the inadequacy of our forms of representation, for the levels have been detected by direct experiment (*e.g.*, those of Franck and Hertz). We must conclude therefore that the uncertainty relations themselves exhibit something real and fundamental, and that the energy levels are connected with them.

The uncertainties exhibited in the Heisenberg relations concern magnitudes which appear in pairs. These are (x and p_x), (y and p_y), (z and p_z), (t and W). According to the first uncertainty relation, if x is known with accuracy, then p_x is uncertain; but nothing prevents any of the other magnitudes, *e.g.*, y or p_y, from being accurately determined. Long before the discovery of Heisenberg's uncertainty relations, arguments of a totally different nature had prompted mathematicians to associate these same magnitudes in pairs. In classical mechanics, x and p_x, y and p_y, z and p_z, always appear together in Hamilton's equations of dynamics, such couples of magnitudes being referred to as *conjugate magnitudes*. Classical mechanics also led to the conclusion that if time were treated as a fourth dimension of space, then t and $-W$ (*i.e.*, time and minus the energy) could also play the part of conjugate magnitudes. This identification of time as a fourth dimension was, however, gratuitous before the theory of relativity proved it to be necessary, so that it is thanks to the theory of relativity that the conjugate nature of t and $-W$ is seen to be as fundamental as that of the other magnitudes.

A better understanding of the conjugate nature of time and energy is obtained when we recall that in the theory of relativity the fundamental

---

* The uncertainty relations (2) refer to free particles. They may be aggravated when the particle is acted upon by a force, as is the case for an electron in the atom of hydrogen. If the electron is known to be in the nth energy level and remains in this level during the measurement, the uncertainty relations are of type

$$\Delta x \; \Delta p_x \sim nh.$$

And this agrees precisely with Bohr's quantizing condition for the nth energy level.

continuum, Minkowski's space-time, is 4-dimensional; and that a vector (arrow) in space-time has four components instead of the three it has in ordinary space. This is but another way of saying that time is viewed as a fourth dimension. The net result is that the four magnitudes (x, y, z, t), which we may write (q_1, q_2, q_3, q_4), now appear as the four components of a space-time vector. Similarly, the four magnitudes (p_x, p_y, p_z, $-W$), or (p_1, p_2, p_3, p_4), are the four components of the so-called momentum-energy vector. The conjugate nature of the three first q's and of the three first p's, which was previously recognized in classical science, is now extended quite naturally to the last ones, namely, to time and energy. These considerations permit a more condensed presentation of the uncertainty relations, for we may say that they express the impossibility of focusing simultaneously the two space-time vectors (q_1, q_2, q_3, q_4) and (p_1, p_2, p_3, p_4). The better the one is focused, the more blurred does the other become.

The method we have followed in deriving the uncertainty relations from the properties of wave packets is due to Bohr. Quantum mechanics furnishes the same relations but in a less intuitive way. Heisenberg has examined a large number of experiments in which the position or the momentum or the energy of a particle is measured, and in each of these experiments the uncertainty relations have been found to be satisfied quantitatively, and not merely qualitatively. No exception has as yet been recorded. This agreement in itself would seem to justify the validity of the uncertainty relations for particles.

Furthermore, Heisenberg's analysis of various experiments affords additional insight into the origin of the uncertainty relations. In every case we shall find that *the uncertainties are generated by the disturbances which our measurements necessarily entail.* The illustration usually given in popular books, because it is the simplest to explain, is the one in which an attempt is made to determine the position of a small particle, *e.g.*, an electron, by viewing it through a microscope. We shall discuss it here.

The theory of the microscope was well known prior to the discovery of wave mechanics. The feature we shall be concerned with is the "resolving power" of the microscope, *i.e.*, the ability of the microscope to yield distinct images of points that are situated very close to one another. Thus let us suppose that we examine two point-like structures through a microscope. The undulatory nature of light prevents the image of a point from being a point; instead a small disk is obtained.* If, then, we view through the microscope two points that are very close together,

---

* More precisely, the image of a bright point, as seen through the microscope, is a bright disk surrounded by bright concentric rings.

we shall observe them as distinct only when the two disk-images of the points do not overlap. The mathematical theory enables us to calculate the diameter of the disk-image of a point when this point is illuminated by light of given wave length λ, and we may thus compute the smallest distance which may separate two points if their images are to be distinct. The smaller this minimum distance, the greater the ability of the microscope to distinguish fine details, and the greater therefore the so-called resolving power of the microscope.

Fig. 48

Let us, then, suppose that a particle is observed at the point P on the axis of the microscope. We denote by α the angle formed by the two lines joining the point P to the two opposite extremities of a diameter of the lens. According to optical theory, the smallest distance between two points which will permit these points to be viewed as distinct through the microscope is defined by

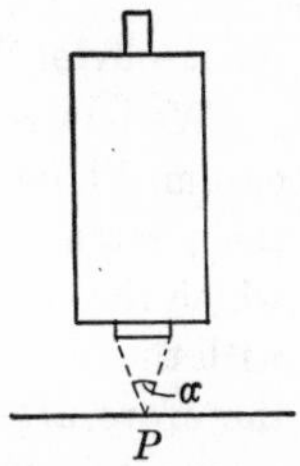

$$(6) \qquad \frac{\lambda}{\sin \alpha},$$

where λ is the wave length of the monochromatic light which illuminates the point. The smaller the wave length λ, the smaller the minimum distance required between the two points; hence finer details will be revealed when the illuminating light is of shorter wave length. An immediate conclusion is that when, on looking through a microscope, we believe a tiny particle to be situated at a point P, we cannot assert that this particle is precisely at P; for even if it were situated slightly to the right or to the left, its image in the microscope would still appear the same. Hence, in attempting to locate the precise position of a particle, we shall always be confronted with an element of uncertainty, the magnitude of which is given by (6).

Suppose, then, that a tiny particle, *e.g.*, an electron of mass m, is known to be moving with specified momentum, $p_x = mv_x$, along a horizontal line Ox passing through the field of vision of the microscope. We may assume that at the instant considered the particle is within the field of vision. For the particle to be visible to us, we must illuminate it; and we must suppose that a photon of the illuminating light strikes the particle and is then deflected through the body of the microscope and reaches our eye. On looking through the instrument, we may, for instance, see the particle at the point P. But, as has just been explained, all we can assert is that the particle is on the Ox axis, somewhere within a small interval comprising the point P, the length of this interval being defined by (6). Calling Δx the magnitude of this length, we have

$$(7) \qquad \Delta x = \frac{\lambda}{\sin \alpha}.$$

Δx is obviously the range of the uncertainty in the particle's position.

We now consider the momentum of the particle. The momentum p_x is assumed known (as rigorously as we choose) prior to the instant at which the photon collides with the particle, and hence prior to the instant at which the observation of the particle's position is made. But the collision with the photon alters the particle's motion, so that after the collision, and therefore after the observation, the momentum of the particle will no longer be the same. The collision of the particle and photon may be likened to a collision between two perfectly elastic billiard balls, the particle sustaining a kick from the photon and the photon being deflected by the particle. We must assume of course that, after the collision, the photon penetrates the microscope, for in the contrary event, nothing would be observed through the instrument. According to Einstein's theory of the photo-electric effect, developed long before the uncertainty relations were known, a photon of wave length λ and frequency ν has energy $h\nu$ and momentum $\dfrac{h}{\lambda}$. It can then be shown, by utilizing the principles of conservation of energy and of momentum for elastic collisions, that, as a result of the kick, the momentum of the electron will no longer be p_x, but will be situated between $p_x - \dfrac{h}{\lambda}\dfrac{\sin \alpha}{2}$ and $p_x + \dfrac{h}{\lambda}\dfrac{\sin \alpha}{2}$. The momentum may thus have any value within a range of magnitude

$$(8) \qquad \Delta p_x = \frac{h}{\lambda}\sin \alpha.$$

The range Δp_x measures the uncertainty in the value of the electron's momentum *after* the observation.

If we multiply the two uncertainties Δx and Δp_x given in (7) and in (8), we obtain

$$(9) \qquad \Delta x \Delta p_x = \frac{\lambda}{\sin \alpha} \cdot \frac{h}{\lambda}\sin \alpha = h;$$

and this is precisely one of Heisenberg's uncertainty relations. The result will always be the same regardless of the microscope used and regardless of the wave length of the illuminating light. If we use light of short wave length, the position of the electron at the instant of the observation will be more accurately known, but then the momentum after the observation will be more uncertain. If we use light of long wave length, only a slight disturbance will occur for the momentum, so that the observation will leave the momentum practically unchanged; but, as against this, the position of the electron will be more uncertain. In any case a simultaneous knowledge of the accurate position and momentum, after the observation, is impossible.

Incidentally, we note that Planck's constant h is introduced in the uncertainty relation because of its appearance in the expression of the uncertainty Δp_x, defined by (8). If, then, the value of h were infinitesimal, as it was assumed to be in classical science, the value of Δp_x would be infinitesimal, and no uncertainty would be attached to the momentum. Consequently, the disturbance in the momentum, and the kick which generates this disturbance, are quantum manifestations.

The experiment with Heisenberg's microscope is one in which a determination of position is attempted, and we see that it leads to one of Heisenberg's uncertainty relations. Other theoretical experiments may be devised with a view of testing the other uncertainty relations, e.g., the one affecting time and energy ($\Delta t \Delta W \geqslant h$). In the experiment of Stern and Gerlach, where an energy measurement is performed, we obtain this latter uncertainty relation.*

In every case that has been considered, the uncertainty relations have been shown to be unvoidable, and each time a means of circumventing them seemed to have been devised, more attentive consideration has shown that some error had crept into our reasoning. We may therefore assume that these relations are of general validity.

If we accept the general validity of the uncertainty relations, we see that the smallest possible value of the uncertainty, say, in the momentum of a particle, will depend solely on the accuracy with which we have measured the particle's position, and will in no wise be affected by the

---

* In the Stern-Gerlach experiments, a parallel beam of atoms moving at high speed is directed normally through an inhomogeneous magnetic field. If the atoms with which we are experimenting have a non-vanishing magnetic moment, any individual atom of the beam will have a certain magnetic energy inside the field. The precise value of this magnetic energy at any instant depends on the position the atom occupies in the beam at the instant considered. The magnetic energy W of any particular atom is thus contained somewhere within an energy range ΔW. If Δt represents the time taken by an atom to cross the field, then Δt measures the vagueness surrounding the precise instant of time at which the atom has a given energy. From an analysis of this experiment Heisenberg shows that, under the most favorable conditions, the following relation must hold:

$$\Delta t \cdot \Delta W \sim h.$$

And this, as we know, is one of the uncertainty relations.

particular method of measurement we may have employed. Let us, then, imagine a particle, the position of which is determined with a high degree of precision. The momentum of the particle is now uncertain. So as to determine this momentum we immediately follow up our position measurement with a momentum measurement, as a result of which the momentum of the particle is known with a high degree of accuracy. But the position of the particle is thereby rendered uncertain. In short, however close the two instants at which the successive measurements are performed, the information we acquire does not yield the simultaneous position and momentum of the particle at the same instant, but only one *or* the other of these two magnitudes. Though the statements we have made are correct, we shall now see that they may prove misleading unless we supplement them with further explanations.

We must be wary of pitfalls when we attempt to understand the deep significance of Heisenberg's uncertainty relations. Thus, suppose we are dealing with a particle in motion. We may determine its position as accurately as we choose, locating it at a point A at time t_A. The uncertainty principle does not interfere with this determination; it merely states that the momentum, or velocity, of the particle is now uncertain. At a later instant t_B, we perform a second accurate measurement of position, and locate the particle at a point B. If the particle is moving in free space (in the absence of a field of force), we may grant that it describes a straight line with constant speed. We may determine this speed from our position measurements; for, since we know that the particle has covered the distance $\overline{AB}$ in a time $t_B - t_A$, we conclude that it must have moved with the constant velocity $\dfrac{\overline{AB}}{t_B - t_A}$. We therefore know the exact position and the exact velocity of the particle at any instant between t_A and t_B. Inasmuch as the velocity of the particle multiplied by its mass is its momentum, we also know the exact momentum of the particle at all instants between t_A and t_B. Hence a simultaneous knowledge of position and of momentum seems to be possible. This result appears to contradict the uncertainty relations, and so we are led to examine the matter more carefully. As we shall now show, the proper interpretation of the uncertainty relations is that a simultaneous knowledge of the position and momentum of a particle may be possible, but that this knowledge can never yield any information which we do not already possess. In other words, it is impossible to obtain a simultaneous knowledge of position and of momentum which will enable us to anticipate the future, or to discover a past that is unknown. These statements will be clarified if we revert to the illustration just discussed.

Thus we determined the position of the particle at time t_A and found that the particle was situated at the point A. We then determined the particle's position at a later instant t_B and located the particle at the point B. These successive determinations enabled us to ascertain the motion of the particle between the instants t_A and t_B, and thereby its exact position and momentum at any instant t_C between t_A and t_B. If this knowledge is to reveal new information, it must enable us to anticipate how the particle will move *after* the instant t_B or *before* the instant t_A; for, as we have said, the motion in the interval of time from t_A to t_B is already known. Obviously, however, the knowledge of the position and momentum of the particle at the instant t_C can give us no such new information, for the position observation at t_B disturbs the subsequent momentum of the particle by an uncertain amount, and the position observation at t_A disturbs the momentum which the particle had prior to this instant. Of course the knowledge of the position and of the momentum of the particle at t_C allows us to determine how the particle will move from t_C to t_B and also how it moved from t_A to t_C; but this is information previously acquired, and, indeed, it was thanks to this prior information that we were able to infer the position and the momentum of the particle at time t_C. In short, the simultaneous knowledge of the position and of the momentum at time t_C enables us to anticipate only what we already know; it does not increase our information.

Heisenberg, commenting on this situation, writes:

"But this knowledge of the past is of a purely speculative nature, since it can never (because of the measurement) be used as an initial condition in any calculation of the future progress of the electron and thus cannot be subjected to experimental verification. It is a matter of personal belief whether such a calculation concerning the past history of the electron can be ascribed any physical reality or not." *

This analysis shows that the uncertainty relations must not be construed to imply that a simultaneous knowledge of the position and of the momentum of a particle is an absurdity. All that the relations convey is that *direct* measurements cannot furnish a simultaneous knowledge of position and momentum. On the other hand the uncertainty relations do not apply to the case where this simultaneous knowledge is obtained indirectly through inference—indeed the wave picture cannot represent such indirect knowledge. The foregoing conclusions do not impair the validity of the uncertainty relations and of the wave form of representation, for we have seen that simultaneous knowledge, obtained indirectly,

---

* Heisenberg. The Physical Principles of the Quantum Theory, p. 20.

is circular and useless. This discussion shows, however, that the uncertainty relations and the wave form of representation have pragmatic connotations and are not concerned with sterile computations.

The uncertainty relations between position and momentum, and between time and energy, pertain more especially to mechanics. But the quantum uncertainties are more general and extend to other cases. Thus, uncertainty relations have been established between the electric, and the magnetic, intensities of the electromagnetic field at the same point of space. We need scarcely add that in these uncertainty relations of electromagnetism, Planck's constant h plays the same important part that it did in the former uncertainty relations. As before, if h is imagined to decrease indefinitely, the uncertainty relations between electric and magnetic magnitudes will vanish. Finally, the source of the uncertainties must be sought as usual in the quantum disturbances generated by all measurements.

The Final Interpretation of the Wave Picture—Heisenberg's analysis of measurements affords a more detailed interpretation of Born's assumptions. Thus suppose we are dealing with a wave picture represented by the superposition of a number of plane monochromatic de Broglie waves. We recall that the resultant intensity I of the wave motion, at any point P and at any instant, is proportional to the probability that, as a result of a position measurement, we shall find the particle to be situated in a tiny volume about this point P at the instant considered. Also, the intensity of any one of the constituent de Broglie waves, integrated over the whole of space, is proportional to the probability that, as a result of a measurement of momentum or of energy, the particle will be found to have the momentum or the energy which correspond to the wave number and frequency of the wave in question.

Born's assumptions seem to imply that a wave picture affords a simultaneous knowledge of the probable position and of the probable momentum of a particle at the instant of interest. Although this conclusion is not incorrect, it might prove misleading unless properly understood. Suppose, for example, we propose to perform a position measurement and follow it up immediately with a momentum measurement. We might believe that correct probabilities would be assigned by the wave picture to the results of our two measurements. But this is not so. It is true that, before any measurement is performed, the picture furnishes correct probabilities for the particle's position; but if now we perform a position measurement, the disturbance which is generated thereby will render the picture useless insofar as the momentum of the particle is concerned. Conversely, before any measurement is performed, the picture furnishes the correct probabilities for the momentum of the particle just as it does for its position; but if now we perform a momentum measurement, the picture no longer yields correct probabilities for the position. In short, we can take advantage of a wave picture to obtain information on the position of a particle at a given instant, *or else* on its momentum and energy; but both kinds of information cannot be derived simultaneously. Hence in every case we must make a choice and decide which of the two types of information is desired. The importance of these considerations will be understood in Chapter XXXIV.

Measurement and Observation—In the foregoing pages we have discussed experiments in which the position or the momentum of a particle is observed, and we have said that such observations generate disturbances. But we wish now to examine more carefully the distinction between a measurement and an observation.

To observe, say, the position of a particle, we must perform a position measurement on it and then observe the result of this measurement. Thus, in all cases an observation is accompanied by a measurement. The converse is not true, however, for we may quite well perform a measurement and yet fail to observe the result. Now in considering the disturbance generated by an observation, we must make clear that the disturbance is caused by the physical measurement, and not by the cognitive act whereby the result of the measurement is comprehended by the percipient. For example, in the experiment with Heisenberg's microscope, a particle is bombarded by a photon which subsequently penetrates into the microscope and thereby reaches our eye. The position of the particle is thus observed. The observation is rendered possible by the collision of the photon with the particle, and hence it is this collision which constitutes the measurement. According to the quantum theory, the collision exerts exactly the same disturbance on the particle's momentum, whether we happen to be peering through the microscope or looking in some other direction. The quantum theory and common sense are thus in agreement on this point. But although the measurement, and not the cognitive act, creates the disturbance, both the cognitive act and the measurement influence the wave picture, albeit in a different way. Furthermore, as will be understood presently, the theoretical possibility of the cognitive act being realized must always be taken into consideration when we wish to gauge the nature of the disturbance generated by a measurement.

In view of the abstruse nature of the matter at issue, we shall proceed in an orderly way by considering the various influences separately. We begin with the measurements. Measurements are classed according to their intent. Thus it may be the position, the momentum, the instant of time, or the energy that we propose to measure. Corresponding to these different intents, we have position measurements, momentum measurements, and so on. In this classification we are not concerned with ascertaining whether the measurements are actually observed or remain unnoticed; what is essential, however, is that the measurements be of a type that we could have observed had we attempted to do so. According to the uncertainty relations, a measurement which determines one species of magnitude (position, momentum, energy, time) disturbs by an unpredictable amount the magnitude which is conjugate to the one measured. In short it is the disturbance generated by the measurement, not by the act of observing, which is controlled by the uncertainty relations.

Next let us examine how these measurements affect a wave picture. To take a special case we shall suppose that the wave picture represents a wave packet. The particle picture then represents a particle situated somewhere within the packet, and having the momentum and the energy corresponding to one of the de Broglie waves which enter into the formation of the packet. We call this wave packet the packet A. We now perform an energy measurement. The physical disturbance caused by the measurement renders the original wave picture useless, and a new one must be constructed. Here a distinction arises according to whether the result of the measurement is, or is not, observed. We first assume that no observation is made. Heisenberg has shown how the new wave picture must be constructed; we call this picture the picture B. Further information will be given in Chapter XXXIV (see page 745).

Next let us suppose that the measurement is observed and that a highly accurate determination of the particle's energy and momentum is obtained. The wave picture corresponding to this situation differs not only from the original picture A but also from the former one, B. We shall call this third picture the picture C. (In the present case the picture C will consist of a train of approximately plane and very nearly monochromatic de Broglie waves.) In short our observation of the measurement has provoked a further change in the wave picture.

From this general discussion we see that progressive changes in the original wave picture A may arise from two separate causes. Firstly, there is the change from the picture A to the picture B brought about by the physical disturbance due to the measurement. Secondly, there is the change from B to C due to our becoming aware of the result furnished by the measurement. In practice, a measurement is usually performed with a view of furnishing information; its result is therefore usually observed.

The change in the wave picture is then from A to C directly, and the intermediate picture B is not considered.

A point that may seem strange is that, according to whether we observe the result of a measurement or look the other way, the wave picture assumes one form or another. But we must remember that a wave picture merely symbolizes the scheme of probabilities which issues from our more or less vague knowledge of actual conditions. Consequently, to say that a wave picture is modified by an observation simply means that the probabilities we can assign to future events are affected by the increased knowledge of present conditions which we have acquired as a result of this observation.

Since in the classical theories we likewise assumed that an increased knowledge of present conditions affected the probabilities of future events, we see that it is not on this score that wave mechanics involves any radical departure from classical ideas. The only new feature introduced by wave mechanics (aside from its wave representation of probabilities) is the disturbance which a measurement provokes on the conjugate magnitudes, and hence on the existing probabilities. In classical science, disturbances caused by measurements were regarded as casual, because physicists believed that they could be decreased indefinitely by the exercise of sufficient care. For these reasons an unobserved measurement did not affect the existing probabilities according to classical science, whereas it does affect them according to the quantum theory.

The following example clarifies the connection between measurements, observations, and probabilities in the classical and quantum theories, respectively. Suppose that a man whose marksmanship is well known shoots at a target. From our knowledge of his marksmanship we may assign a certain probability to his scoring a hit. Now let us assume that a measuring device determines the position of the bullet at some instant during its flight, but that we fail to observe the result of the measurement. According to classical science the measurement need not disturb the bullet's motion and hence need not modify the probability of the target being hit, whereas according to the quantum theory the inevitable disturbance due to the measurement will change the value of the probability. Next let us suppose that the measurement is observed and that the bullet is found to be at a point in line with the target. According to classical science, the bullet is now certain to hit the target and the value of the probability changes from its original value to the value 1. According to the quantum theory the value of the probability is likewise modified by our observation, but we cannot assign the precise value 1 to the new probability, because the measurement has disturbed the bullet's motion in an uncertain way, with the result that the bullet may now be deflected from the target. A more detailed discussion of the various kinds of probability will be given on pages 741-750.

Waves and Corpuscles—The uncertainty relations enable us to predict, in any particular situation, whether it is the corpuscular or the wave aspect which will manifest itself. Hitherto, when we have discussed the uncertainty relations, we have interpreted them on the basis of Born's assumptions and have not taken the attitude of Heisenberg and Bohr into account. Thus we said that if the position of a particle was well known, its momentum was uncertain, and vice versa. But it will now be advisable to consider the uncertainty relations in connection with the views of Heisenberg and Bohr. When we proceed in this way, the uncertainty relations have the following significance:

A particle whose momentum and energy are accurately determined by measurement and observation becomes diffused through an extended region of space; the corpuscular aspect is entirely lost. On the other hand, when the position of the particle is accurately determined at a given instant, the corpuscular aspect of the particle is brought into existence, and then the particle's momentum and energy are not merely uncertain but cease to have any physical significance.

Let us, then, consider an atom while it is emitting radiation. According to the classical principle of conservation of momentum, the total momentum of the isolated system, atom + radiation, will remain constant in magnitude and in direction during the process of emission. But if we take into consideration the ideas of Heisenberg and Bohr, we must recognize that there can be no meaning to this classical principle unless conditions be such that the momentum has a clear significance.

For the momentum to be well defined, we must perform a momentum measurement. In particular, we may measure the precise change in momentum experienced by the atom during the emission of the radiation. If this be done, the classical principle of conservation of momentum is valid, and the momentum of the emitted radiation will be equal and opposite to the change in the momentum of the atom. The radiation will then proceed in the direction opposite to that of the recoil of the atom, and we shall have needle radiation, *i.e.*, radiation emitted in a definite direction. This is the situation which we interpret by saying that a photon is hurled out by the atom. The corpuscular aspect of radiation is then realized. We also note that, in accordance with the uncertainty relations, the exact position of the atom is correspondingly unknown, or even meaningless.

Suppose, on the other hand, we perform a measurement which determines the position of the atom with a considerable degree of accuracy. The uncertainty relations show that the momentum and the change in the momentum of the atom must become correspondingly vague. Consequently, the classical principle of conservation of momentum for the system, atom + radiation, becomes vague to the same extent. In particular, the direction in which the radiation is emitted is clouded in vagueness, and as a result the radiation is emitted in all directions: we then have a spherical wave emitted from the atom. The corpuscular aspect of radiation is lost, and its wave aspect now imposes itself. From this illustration we may infer that both aspects of the radiation cannot be realized simultaneously, and we perceive that the key to the enigma of the dual nature of radiation is to be sought in the uncertainty relations. If ever an experiment could be devised to reveal the two aspects of radiation (or of matter) simultaneously, the uncertainty relations would be at fault.

The Heisenberg Uncertainty Relations, and Errors of Observation—The Heisenberg uncertainty relations are quantum manifestations and must not be confused with the errors of observation that are unavoidable in all human measurements; the two types of uncertainty are utterly distinct and have nothing in common. This is sufficiently obvious when we recall that in a world in which Planck's constant h happened to be infinitesimal, the Heisenberg uncertainly relations would vanish, whereas, so far as we can tell, human errors of observation would still prevail. In many popular writings, however, the uncertainties of the quantum theory have been confused with those that are inseparable from all human measurements. As a result of this confusion, it has been assumed that the Heisenberg-Bohr philosophy is a revival of the ultra-phenomenological outlook of Mach and Ostwald. Yet, only the most superficial similarity holds between the two philosophies. A characteristic view of Mach's was that the kinetic theory of gases should be rejected because molecules had never been observed. However, Mach gave no theoretical argument to support the conclusion that molecules never *would be* observed; he merely based his attitude on the fact that they *had not* been observed. A similar line of thought led him to reject absolute motion. Here, again, Mach never advanced any argument that would have led physicists to anticipate, for instance, the negative result of Michelson's experiment.

Of a very different calibre is the philosophy defended by the quantum theorists. Thus the possibility of a simultaneous accurate experimental determination of position and of momentum is ruled out, not because no experimenter has ever been skilful enough to make a measurement of this kind, but because there appears to be a theoretical reason which precludes any such measurement from being performed successfully, even by a superman. That the clumsiness of the experimenter plays no part in Heisenberg's views is sufficiently demonstrated when we recall that an infinitely accurate measurement of position alone or of momentum alone (which would be ruled out as impossible if practical difficulties were our criterion) is not viewed as impossible in quantum mechanics. Furthermore, according to the uncertainty relations, an ideally perfect measurement, say, of position, does not disturb the position of the particle, for, if we measured the position a second time immediately after the first observation, the same position would still be found—the momentum alone is abruptly changed by the position measurement. In short, there is not the slightest similarity between the uncertainty relations of Heisenberg and the uncertainties that accompany all human observations. Quantum science is thus in agreement with classical thought when it assumes that, in a theoretical discussion at any rate, all inaccuracies due to the clumsiness of the experimenter may be disregarded and that the precision of our measurements may be extended to the limit. The originality of the new views is that this limit, which in practice is usually unattainable, is finite and not infinitesimal; the influence of the finite value of h asserts itself once more.

A numerical example borrowed from de Broglie will make this point clear. Suppose that a tiny ball weighing one milligram is moving along the x-axis. We wish to determine, at an initial instant, the position x of the ball's centre, and also the ball's momentum p_x in the line of its motion. Suppose we succeed in determining the position of the ball's centre within an error Δx of one thousandth part of a millimeter. This is presumably about as accurate a measurement as would be possible in practice. Heisenberg's uncertainty principle states that our measurement of position has disturbed the momentum p_x of the ball, and that the smallest possible value for the uncertainty in the ball's present momentum is $\Delta p_x = \dfrac{h}{\Delta x}$. If we divide this quantity by the mass of the ball, we obtain the uncertainty in the ball's velocity. The numerical value of the uncertainty is found to be $\dfrac{6.55}{10^{20}}$ centimeters per second. Thus, our measurement of the ball's position has generated in its velocity an uncertainty defined by the previous amount. But this theoretical uncertainty in the velocity is far too minute to be detected by any practical measurements, so that practical errors of observation far outweigh, in this example at least, the theoretical quantum uncertainties.

Let us, then, determine under what conditions the uncertainties will become noticeable. In the Heisenberg relation $\Delta x \Delta p_x \geqslant h$, if we adopt the classical expression mv_x for the momentum, we obtain

$$\Delta x \cdot \Delta mv_x \geqslant h, \text{ or } \Delta x \cdot \Delta v_x \geqslant \frac{h}{m}.$$

We conclude that the smaller the mass m of the particle, the more important becomes the uncertainty between position and velocity. Electrons (or still better, photons), owing to their small mass, are among the most favorable kinds of particles on which to detect the influence of the uncertainties. This circumstance is to be connected with the fact that it is for electrons and photons that the wave properties are most accentuated. On the other hand, when the mass m becomes larger, the quantum uncertainty decreases and is soon concealed by the practical uncertainties attendant on all human measurements. For this reason the quantum uncertainties must necessarily fail to manifest themselves on the commonplace level, e.g., in the case of billard balls.

From the foregoing discussion we see that the Heisenberg uncertainties are in many cases far too insignificant to be detected by laboratory experiments, even when performed by a Michelson; and we also see that these uncertainties have nothing in common with human clumsiness. But, whether they be detectible or not, they are assumed in the final analysis to exist. These considerations stress the importance of differentiating between the practically impossible and the theoretically impossible. In a certain sense, the distinction is reminiscent of the discussions we mentioned on the subject of mathematical infinity. There also, according to the intuitionists, a theoretical impossibility prevented us from writing out *all* numbers, as contrasted with a mere practical impossibility of writing out a quadrillion times a quadrillion numbers. In mathematics, as in the quantum theory, it is the theoretical impossibility that is important.

CHAPTER XXXI

VIBRATIONS

WHEN the extremity of a rope receives a sudden jerk, a wave crest proceeds along the rope. But suppose both extremities of the rope, or string, are fixed. If we pull the string and then release it, a vibratory motion is set up; the various points vibrate in a direction perpendicular to the string's length, and no wave crest is transmitted. Thus, we are dealing with a standing wave. This wave is generated by wave crests propagated along the string and reflected at one of the fixed ends: the reflected wave crests interfere with those that are advancing, and the superposition of the two wave trains generates the standing wave. In what follows we shall be concerned with the standing wave itself, regardless of the process by which it comes into existence.

The standing waves, or vibrations, that can be set up in the string are many and varied. It may happen that the string vibrates harmonically and that the only points which remain fixed are the two end points A and B. When this situation occurs, the string is said to be vibrating

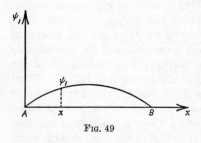

Fig. 49

in its *fundamental mode*. The ear detects a pure sound, the pitch of which is determined by the frequency ν_0 of the vibration. When the string experiences its maximum displacement, its general shape will be of the type illustrated in the figure. The maximum displacement of the string differs from point to point, and if we take the line AB as the Ox axis and call ψ_1 the maximum displacement at a current point x, we see that ψ_1 will be a function of x. We may therefore represent the displacement by

$$(1) \qquad\qquad \psi_1(x).$$

Next we consider the motions of the various points of the string. All these points vibrate up and down harmonically, experiencing their maximum deviations simultaneously, and passing through the straight line AB at the same instant. Let us call u the displacement of a point x

674

of the string at an instant t; then u is a function of x and t. If we assume that, at the instant $t = 0$, the string lies along the line of maximum distortion above the horizontal AB, the motions of its various points may be represented by

$$(2) \qquad u = \psi_1(x) \cos 2\pi\nu_0 t.$$

After each period of time $\dfrac{1}{\nu_0}$ the string resumes its position of maximum displacement; and it assumes the straight horizontal position at the successive instants

$$t = \frac{1}{4\nu_0}, \; 3\frac{1}{4\nu_0}, \; 5\frac{1}{4\nu_0}, \; \cdots$$

Other simple vibrations that may be set up occur when, besides the fixed points A and B, additional intermediary points remain motionless. Some of the situations that may occur when the string suffers its greatest distortion are illustrated in Figure 50. These various elementary

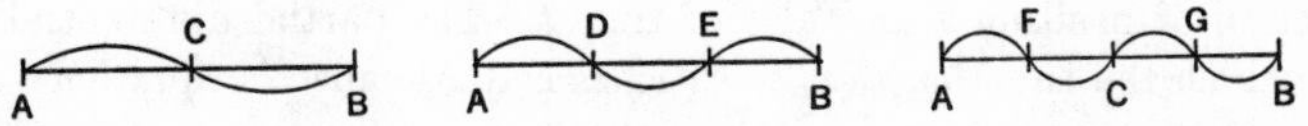

Fig. 50

vibrations are called the *normal modes* of vibration, or more simply the *modes* of vibration. The points, such as C, D, E, F, G, which remain fixed in the various figures, are called the *nodal points*. The extreme points A and B are, of course, always nodal points. In any one of these modes of vibration, the nodal points are equidistant and they divide the string into equal parts. The frequencies of the vibrations differ from one mode to another, but they are always whole multiples of the fundamental frequency ν_0. For the modes illustrated in the figure, the frequencies are $2\nu_0$, $3\nu_0$, and $4\nu_0$. All these modes yield pure sounds. For instance, if the fundamental vibration gives rise to the musical note C, the first mode will give the upper octave C, the second mode will give the following G, and the third mode will give the second octave C. These pure notes are called the harmonics of the fundamental note.

We may follow for each one of the modes of vibration the method outlined for the fundamental mode. Thus the mode of frequency $3\nu_0$ (the second one in Figure 50) will be expressed by

$$(3) \qquad u_3 = \psi_3(x) \cos 2\pi(3\nu_0)t,$$

where $\psi_3(x)$ defines the maximum amplitude from point to point for the mode in question. We see from the figure that $\psi_3(x)$ vanishes not only at the end points A and B, but also at the nodal points D and E. In the general case the string will not be vibrating in a normal mode. But mathematical analysis shows that whatever the vibration may be, it is equivalent to a superposition of the normal modes, each mode being credited with an appropriate amplitude and with a suitable difference in phase. In these more complicated cases, the ear detects no longer a pure sound, but a superposition of the various harmonics. This procedure of viewing a complicated motion as due to the superposition of simple harmonic ones illustrates the physical significance of Fourier series.

Let us now examine the problem from the standpoint of the mathematician. Suppose a disturbance is propagated along a line Ox with velocity V; the velocity may be constant or variable. The state of the disturbance at any point x will vary with the time t, and at any instant t it will depend on the point x. If, then, we assume that the disturbance is measured by some magnitude u, we see that this magnitude u is a function of position x and also of time t. The partial differential equation which the function $u(x, t)$ satisfies is d'Alembert's equation

$$(4) \qquad \frac{\partial^2 u}{\partial x^2} - \frac{1}{V^2}\frac{\partial^2 u}{\partial t^2} = 0.$$

Suppose the one-dimensional medium along which a disturbance is propagated is a string with fixed extremities. D'Alembert's equation (4) will regulate the motion, the propagation velocity V which appears

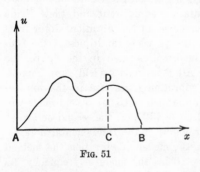

FIG. 51

in the equation being determined by the properties of the string itself. If we assume that the string is homogeneous, V is a constant. We now proceed to determine how the string will be affected by a disturbance. In this attempt we shall pay no heed, of course, to the knowledge we have already gained from watching strings vibrate. In particular, we shall not assume from the outset that standing waves will be set up. Our purpose is to rediscover all these experimental results by purely mathematical means. The method of procedure is as follows:

We first assume that the function $u(x, t)$, which measures at time t the displacement CD of any point x of the string above or below the straight line AB (Figure 51), is of the general form

(5) $$u(x, t) = \psi(x)f(t).$$

Here $\psi(x)$ and $f(t)$ are two unknown functions of x and of t respectively. The unknown function u, and hence the motion of the string, will be determined if the two unknown functions, ψ and f, can be found. We may also note that the equation of the curve defined by the string's position at any specified instant t_0 is given by

(6) $$u = \psi(x)f(t_0).$$

Next let us discover the two unknown functions, ψ and f. To do this, we observe that the function $u(x, t)$ must be a solution of (4); and that, consequently, if in (4) we substitute (5), the equation (4) must be satisfied. Making the substitution, we are led to the two following equations, of which ψ and f must be solutions:

(7) $$\frac{d^2\psi}{dx^2} + \lambda\psi = 0,$$

(8) $$\frac{d^2f}{dt^2} + \lambda V^2 f = 0.$$

In these equations, λ is any constant of arbitrary value.

First we examine the equation (7), which is the relation that our unknown function $\psi(x)$ must satisfy. As occurs for all differential equations, different functions, infinite in number, are solutions; and only if additional restrictions are imposed can one definite solution be singled out. In the present case the very nature of our problem introduces restrictions. Thus the end points A and B of the string are to remain fixed during the motion. This requires that when x corresponds to the point A or B, the magnitude u (which at any instant measures the distance of a point x of the string from the straight line AB) must retain the value zero throughout time. For simplicity, we may suppose that the string is of unit length and that the points A and B correspond respectively to the values $x = 0$ and $x = 1$. Under these conditions, when $x = 0$ or $x = 1$, $u(x, t)$ must vanish for all values of t. The formula (5) then shows that the same must be true of $\psi(x)$. We must therefore impose the restrictions

(9) $$\psi(0) = 0 \text{ and } \psi(1) = 0$$

on the solution $\psi(x)$ of the equation (7). These conditions are called the boundary conditions for the solution of (7).

Further restrictions on the solution function $\psi(x)$ are imposed by the physical nature of the problem. Thus the string is continuous (*i.e.*, it has no gaps); and since the equation of the curve defined by the string at any instant t_0 is given by (6), we conclude that the function $\psi(x)$ must be continuous between $x = 0$ and $x = 1$. We must also assume that the function $\psi(x)$ is one-valued, *i.e.*, has a well-determined value for each value of x. This last requirement, however, happens to be satisfied automatically by all the solution functions of the present problem. Finally, we shall suppose that the first and second derivatives of the solution function $\psi(x)$ are continuous. This means that there will be no angular points in the string during its motion, and that the radius of curvature will not change abruptly from place to place.

We now revert to the equation (7) and seek to discover the solutions $\psi(x)$ which satisfy the boundary conditions (9) and also the conditions just imposed. There will always be the solution $\psi(x) \equiv 0$, namely, the vanishing solution. Physically, this solution corresponds to the string remaining at rest; its existence teaches us nothing new, for we know perfectly well that the string may remain at rest. Let us, then, consider the non-vanishing solutions. Mathematical analysis yields the following important conclusions:

For the equation (7) to admit non-vanishing solutions $\psi(x)$, satisfying all our requirements, the arbitrary constant (or parameter) λ must have any one of the values

(10) $$\pi^2, \quad 2^2\pi^2, \quad 3^2\pi^2, \ldots n^2\pi^2, \ldots,$$

where n is any positive integer.

In mathematics we often meet with a situation of this kind. Quite generally, if a differential equation contains an undetermined parameter, and if the equation admits solutions (on which certain restrictions are imposed) only when particular values are credited to the parameter, these values are called the *eigenvalues,* and the corresponding solutions are called the *eigenfunctions.** In the problem of the string, which we are here considering, the numerical values (10) that λ must receive are the eigenvalues of the equation (7); we refer to them in the order written as the first, second, third . . . eigenvalue. The nth eigenvalue is $\lambda = n^2\pi^2$. Replacing λ by this value in the equation

---

* The names *proper* values and *proper* functions are often used; also the names *characteristic* values and functions or *fundamental* values and functions. We have preferred to use the German word *eigen*, so as to draw attention to the importance of these values and functions.

(7) and solving the equation, we obtain the nth eigenfunction $\psi_n(x)$. It is found to be

$$(11) \qquad \psi_n(x) = C \sin n\pi x,$$

where C is any constant.

We now consider the equation (8), the office of which is to determine the other function, $f(t)$. In the equation (8), the same parameter λ appears as in (7). Hence to obtain the solution function $f(t)$ which is associated with the nth eigenfunction (11), we must replace λ in (8) by the same eigenvalue, $\lambda = n^2\pi^2$, as before. When this is done and we restrict $f(t)$ to be a continuous function, a solution $f(t)$ can be shown to be $\cos nV\pi t$. Altogether then, utilizing (6), we have for the nth solution $u_n(x, t)$ of (4), and hence for the nth normal motion of the string,

$$(12) \qquad u_n(x, t) = C \sin n\pi x \cdot \cos nV\pi t.$$

This solution defines a standing wave represented by a harmonic vibration of frequency $\nu = n\dfrac{V}{2}$. The first part of (12), namely (11), defines the amplitude of the vibration at any point x of the string. In other words, it gives the maximum displacement which the point x of the string assumes during the vibration. It will be noted that (11) contains the arbitrary constant C, so that the amplitude at a point x may have any value. This circumstance corresponds to the physical fact that, when vibrating in a given mode, the string may be oscillating violently or imperceptibly. The points x at which $u_n(x, t)$ always vanishes, and hence at which $\sin n\pi x$ vanishes, are the nodal points: they remain at rest when the string is vibrating in a given mode. The end points are always nodal points, and the other nodal points (if any there be) are seen to divide the length of the string into equal parts.

The solution (12) determines the nth normal mode of vibration, where n is any positive integer. When we give to n successive integral values, we obtain all the normal modes of vibration. In particular the value $n = 1$ yields the fundamental mode. In the fundamental mode the only nodal points are the two extremities of the string, and the frequency of the vibration is

$$(13) \qquad \nu_0 = \frac{V}{2}.$$

Any number of solutions of the equation (4) may be added together, and the result will still be a solution. Physically, this implies that the string may be vibrating in its various modes at one and the same time; the resultant vibration is then given by the superposition of the elementary vibrations. In short, we have obtained by mathematical means all the results known to us from the experimental study of vibrating strings. We have discussed this vibration problem in some detail, because it is typical of the more complicated problems that arise when, in place of strings, membranes or elastic solids are set into vibration.

Certain features of the more complicated cases may be mentioned. Suppose we are concerned with the vibrations of a drum. The edges of the membrane, being fixed, will not vibrate and will define so-called *nodal lines*. When the drum is vibrating in its fundamental mode, there are no other nodal lines. But when we pass to the higher modes of vibration, additional nodal lines appear. These nodal lines are the two-dimensional analogues of the nodal points which we mentioned in connection with the vibrations of strings. We may easily demonstrate the existence of the nodal lines by sprinkling sand over the surface of the membrane: the particles of sand are thrown off the vibrating portions and cluster along the motionless nodal lines; the positions of the nodal lines are thus revealed immediately. Since the rim of the drum is one of the nodal lines, we may well imagine that the shape of the drum (rectangular, square, round) will affect the disposition of these lines.

Similar problems occur for the elastic vibrations of solid bodies. Here also we have nodal regions but they are now represented by surfaces. The mathematical method of determining the various modes may easily be understood from our analysis of the vibrations of strings. The wave equation in space is the d'Alembert equation

$$(14) \qquad \frac{\partial^2 u}{\partial x^2} + \frac{\partial^2 u}{\partial y^2} + \frac{\partial^2 u}{\partial z^2} - \frac{1}{V^2} \frac{\partial^2 u}{\partial t^2} = 0,$$

in which, for a homogeneous solid, V is a constant characteristic of the body. The physical significance of V is to define the velocity with which waves are propagated in the medium.*

To simplify matters, we shall assume from the outset that standing waves are formed when the body is set into vibration. This assumption implies that the solution functions $u(x, y, z, t)$ of (14), which repre-

---

* More precisely V is the propagation velocity; it is the velocity with which a disturbance is propagated. Waves are propagated with the wave velocity; and only when the conditions of ray-optics are realized, do the propagation velocity and the wave velocity coincide.

sent the state of the vibration at any point x, y, z and at any time t, will be of the form

(15) $$u = \psi(x, y, z)\cos 2\pi\nu t.$$

Here ν is the frequency and $\psi(x, y, z)$ the amplitude of the vibration at the point x, y, z. If now in the equation (14) we replace u by the expression (15), we obtain the equation

(16) $$\frac{\partial^2\psi}{\partial x^2} + \frac{\partial^2\psi}{\partial y^2} + \frac{\partial^2\psi}{\partial z^2} + \frac{4\pi^2\nu^2}{V^2}\psi = 0;$$

and our problem will be solved if we can determine the solutions $\psi(x, y, z)$ of (16).

Just as in the case of strings and membranes, boundary conditions must be considered. For instance, if we agree to hold the surface of the body fixed, we must assume that our solution $\psi(x, y, z)$ vanishes whenever x, y, z are the coordinates of any point on the surface. For physical reasons, we also require that $\psi(x, y, z)$ be one-valued and continuous, with continuous first and second derivatives. Finally we shall agree to disregard the vanishing solution, $\psi \equiv 0$, which corresponds to the case where the solid is not vibrating. As a specific example, we shall suppose that the homogeneous solid is a rectangular parallelepiped of sides a, b, and c.

For the present, the value of ν in the equation (16) is undetermined, and we shall find that only when ν receives specific values forming a discrete set, will our equation have solutions $\psi(x, y, z)$ satisfying the imposed restrictions. For this reason the privileged values which the parameter ν must assume are the eigenvalues of the equation (16). The solution function $\psi(x, y, z)$, which exists when some particular eigenvalue is given to ν in (16), is the corresponding eigenfunction. The physical significance of this statement is easily understood. Thus ν represents the frequency of the vibrations. Hence to say that the equation (16) has solutions only when ν has privileged values is to say that only those modes of vibration which have these privileged frequencies ν can arise in the solid. As for the corresponding eigenfunction $\psi(x, y, z)$ associated with some specific eigenvalue ν, it represents the amplitude of the vibration from point to point when the solid is vibrating in the mode of frequency ν.

We now proceed to obtain the eigenvalues and eigenfunctions of our problem. We take as coordinate axes Ox, Oy, Oz the axes defined by the edges of the solid. We may assume that the unknown solution function $\psi(x, y, z)$, which we wish to determine, can be represented by the product of three functions $X(x)$, $Y(y)$, $Z(z)$, each one of which depends on one variable alone. Thus we get

(18) $$\psi(x, y, z) = X(x)Y(y)Z(z).$$

We then substitute (18) in the equation (16) and obtain the three following equations:

(19) $$\frac{d^2X}{dx^2} + \lambda X = 0$$

(20) $$\frac{d^2Y}{dy^2} + \mu Y = 0$$

(21) $$\frac{d^2Z}{dz^2} + \left(\frac{4\pi^2\nu^2}{V^2} - \lambda - \mu\right)Z = 0,$$

where λ and μ are arbitrary numbers. The frequency ν of the vibrations is unknown for the present and hence is left unspecified.

The restrictions placed on $\psi(x, y, z)$ impose similar ones on $X(x)$, $Y(y)$, and $Z(z)$. In particular, all these functions must vanish over the surface of the solid, and they must be one-valued and continuous. With these restrictions taken into account, we proceed to solve the equations (19), (20), and (21). We find that (19) can admit non-vanishing solutions satisfying our requirements only if λ has one of the values

(22) $$\lambda = \frac{m^2\pi^2}{a^2},$$

where m is any non-vanishing integer. We may suppose without loss in generality that m must be a positive integer. The values (22) are the eigenvalues of (19) ; and when we introduce one of these values for λ in (19), and integrate the equation we obtain the corresponding solution, or eigenfunction, $X(x)$. Integration yields

(23) $$X(x) = \sin\frac{m\pi x}{a}.$$

Similarly, solving (20), we obtain the eigenvalues

(24) $$\mu = \frac{n^2\pi^2}{b^2},$$

where n is any positive integer. The corresponding eigenfunction is

(25) $$Y(y) = \sin\frac{n\pi y}{b}.$$

Finally, passing to the equation (21) and setting in it the eigenvalues (22) and (24) for λ and μ, we find that the expression in the bracket of (21) must have as values $\dfrac{p^2\pi^2}{c^2}$, where p is any positive integer. The equation (21) thus becomes

$$(26) \qquad \frac{d^2Z}{dz^2} + \frac{p^2\pi^2}{c^2}Z = 0.$$

The eigenfunctions of (26) are

$$(27) \qquad Z(z) = \sin\frac{p\pi z}{c}.$$

We may now derive the eigenvalues ν and eigenfunctions ψ of the original equation (16). We have said that, for the equation (21) to have solutions satisfying our requirements, the quantity in the bracket of (21) must have the value $\dfrac{p^2\pi^2}{c^2}$ where p is any positive integer. Hence we must have

$$(28) \qquad \frac{4\pi^2\nu^2}{V^2} - \frac{m^2\pi^2}{a^2} - \frac{n^2\pi^2}{b^2} = \frac{p^2\pi^2}{c^2},$$

whence we derive for the eigenvalues ν of (16)

$$(29) \qquad \nu = \frac{V}{2}\sqrt{\frac{m^2}{a^2} + \frac{n^2}{b^2} + \frac{p^2}{c^2}},$$

in which the numbers m, n, and p may have all positive integral values.

The eigenfunctions of (16), being defined by (18), are given by the product of the three expressions (23), (25), and (27). Thus the eigenfunctions are

$$(30) \qquad \psi(x, y, z) = A \sin\frac{m\pi x}{a} \cdot \sin\frac{n\pi y}{b} \cdot \sin\frac{p\pi z}{c},$$

where A is an arbitrary constant.

The presence of the arbitrary constant A in the eigenfunctions, or amplitudes, (30) indicates that the amplitudes of the vibrations may be large or small, and hence that the vibrations may be violent or scarcely perceptible.

Having obtained the expression of $\psi(x, y, z)$, we may now derive from (15) the expression of the vibratory disturbances. We get

$$(31) \quad u(x, y, z, t) = A \sin\frac{m\pi x}{a} \cdot \sin\frac{n\pi y}{b} \cdot \sin\frac{p\pi z}{c} \cdot \cos 2\pi\nu t,$$

where the frequency ν is the corresponding eigenvalue (29).

If in (29) and (30), we take in succession all possible combinations of positive integers for m, n, and p, we obtain the frequencies and the corresponding amplitude distributions in the various modes. Thus, each triplet of integers assigned to m, n, and p defines a particular mode of vibration. Obviously, there is an infinite number of different modes. The fundamental mode is obtained when we set $m = n = p = 1$. Its frequency of vibration ν_0 (the fundamental frequency) is

$$(32) \qquad \nu_0 = \frac{V}{2}\sqrt{\frac{1}{a^2} + \frac{1}{b^2} + \frac{1}{c^2}}.$$

The nodal surfaces that occur when the solid is vibrating in a given mode (m, n, p) are determined by the points x, y, z at which the vibration function (31) persistently vanishes. Hence the nodal surfaces are also given by the points at which the amplitude function, or eigenfunction, (30) vanishes. In the fundamental mode, there are no nodal surfaces other than the faces of the solid. In the higher modes, intermediary nodal surfaces appear; they are planes parallel to the sides of the solid, and they divide the solid into a number of juxtaposed equal rectangular parallelepipeds.

When we compare the vibrations of strings with those of membranes and solids, several differences appear. With strings, the frequencies of vibration of the various modes are always multiples of the fundamental frequency, so that the higher modes of vibration are the harmonics of the fundamental mode. This situation no longer occurs for membranes and solids. The statement may be verified in the case of solids by contrasting the general expression (29) of the frequency of a mode with the expression (32) of the fundamental frequency.

A further difference between strings on the one hand, and membranes and solids on the other, concerns the existence of distinct modes having the same frequency. With strings, no two normal modes have the same frequency. This also is usually true for vibrating membranes and solids. For example, if the sides of a vibrating parallelepiped have random lengths and hence are incommensurable, the normal modes will all differ in frequency. The foregoing statement no longer applies, however, if the sides are commensurable, and as a special case if they are equal, so that the parallelepiped becomes a cube. In such cases different normal modes may have the same frequency. To clarify these points, let us consider a cube with sides of length a. The frequencies (29) of the various normal modes are then

$$(33) \qquad \nu = \frac{V}{2a}\sqrt{3(m^2 + n^2 + p^2)}.$$

Each triplet of values ascribed to the three integers m, n, and p defines a different mode. But we see from (33) that different modes (m, n, p) and (m', n', p') will have the same frequency whenever

$$(34) \qquad m^2 + n^2 + p^2 = m'^2 + n'^2 + p'^2.$$

The fundamental mode, $m = n = p = 1$, is thus the only one that is single.

The considerations just developed are of considerable importance in wave mechanics and also in pure mathematics. Thus we recall that the frequencies ν of the normal modes are the eigenvalues of the equation (16), and that the amplitude functions (30), which determine the amplitudes of the vibrations in the various modes, are the corresponding eigenfunctions. Consequently, when each mode has a different frequency (as happens in the cases of a string and of a parallelepiped with incommensurable sides), each eigenvalue is connected with a single eigenfunction. But when two or more different modes have the same frequency ν (as in the case of a vibrating cube), the eigenvalue ν will be associated with more than one eigenfunction. In wave mechanics we describe this situation by saying that the problem is degenerate.*

The normal modes of vibration we have discussed are not the only motions a vibrating solid may betray. Just as in the case of the vibrating string, different normal modes may be superposed. In a vibrating string different modes always have different frequencies, and hence, when a superposition of modes occurs, the vibration is no longer simple harmonic and a pure musical note is not emitted. But in a solid, where degeneracy may be present, different modes having the same frequency may be superposed. The vibrations of the points of the solid are then still harmonic and a pure musical note is heard. The superposition of different modes associated with the same frequency entails a considerable change in the nature of the nodal surfaces. Thus in the case of the parallelopiped the nodal surfaces which are formed when the solid is vibrating in a normal mode are always plane; but when there is degeneracy and the superposed modes have the same frequency, the nodal surfaces may become curved.

The most important facts to be retained from this summary treatment of vibrations are that, when we impose on the vibrations certain

---

* In Bohr's theory, degeneracy was said to occur whenever different energy levels of the atom happened to coalesce. In the wave-mechanical treatment of the atom, this coalescence of energy levels occurs whenever different modes of vibration happen to have the same frequency; and so the word degeneracy is retained in this situation (see page 702).

conditions required by physical considerations and when boundary conditions are also specified, the possibility of vibrations occurring is determined by the existence of the eigenvalues. These eigenvalues usually form a discrete set of magnitudes, and we thus see how discreteness is automatically introduced into problems which in the final analysis deal with continuous waves.

The physical problems we have here investigated by mathematical means are problems that have attracted the attention of mathematicians ever since the eighteenth century. The problem of the rectangular solid discussed in the text is extremely simple, but according to the shape of the vibrating solid and the boundary conditions imposed, the methods of solution vary considerably. For instance, if we are dealing with a sphere in place of a parallelepiped, the solution of the problem requires the introduction of the functions discovered by Legendre, Laplace, and Bessel. When the shape of the vibrating solid is arbitrary, more general methods of approach involving the theory of integral equations must be utilized.

CHAPTER XXXII

SCHRÖDINGER'S WAVE MECHANICS OF THE ATOM

WE have seen that de Broglie sought to account for Bohr's energy levels in the hydrogen atom by assuming that the planetary electron was associated with waves. The waves advanced normally along the orbit of the electron, and the wave motion was stable only if no overlapping of wave crests occurred around the orbit (see page 624). The condition of stability was thus expressed by the following requirements: A wave crest which, having circumscribed the orbit, is passing through any point P on the orbit must coincide with the wave crest that is leaving this point. This implies that there must always be an integral number of wave lengths set end to end around the orbit. De Broglie's method may be extended to all situations in which an electron undergoes periodic motions. When consideration is given to the wave lengths of the waves associated with the electron, de Broglie's stability condition for the wave furnishes the following stability condition for the mechanical motion of the particle: The Maupertuisian action of the particle, or electron, over a cycle of the motion must be an integral multiple of h. Since this requirement is precisely the one expressed in Bohr's quantizing condition, we infer that the quantizing conditions of Bohr and the stability conditions of de Broglie are equivalent. As a result, the energy levels will be the same in either theory.

We know, however, from the experimental evidence that Bohr's quantizing condition does not always furnish the correct energy levels. For instance, in the linear harmonic oscillator, we must modify the quantizing condition by introducing half-integral values for the quantum numbers, and in the rotator, still more complicated changes seem to be necessary. And so we conclude that de Broglie's theory in its original form, by the mere fact that it leads to Bohr's quantizing conditions, cannot always account for the correct energy levels. At this juncture, Schrödinger remarked that de Broglie had oversimplified the problem of quantization when he assumed that the conditions of ray-optics held for the waves within the atomic volumes. And Schrödinger expressed the hope that, if the problem of the wave motions were treated rigorously and de Broglie's oversimplification remedied, the correct values for the

687

energy levels would be obtained. We shall see presently that Schrödinger's hope has been gratified.

The reason why the conditions of ray-optics cannot be valid within the atoms is easily understood. In de Broglie's treatment the rays of the waves lie along the Bohr orbits. The curvature of these rays may be ascribed to the varying refractive index of space around the nucleus, generated by the nuclear field of force. Particularly for the inner orbits, the curvature of the rays must be very great, a fact which implies that the refractive index must differ considerably in magnitude from one point to another within the atom. Numerical calculation then shows that this heterogeneity in the distribution of the refractive index is far too great for ray-optics to be possible. It is true that as we pass to the outer orbits, the conditions of ray-optics increase in approximation and that de Broglie's simplified theory tends to become correct; but even so, if we wish to treat the problem rigorously, we must operate according to the methods of undulatory optics. A first point to be noted is that Schrödinger's more refined treatment will deprive the Bohr orbits of any precise significance. Thus we recall that de Broglie identified the rays of the waves with possible Bohr orbits. But in undulatory optics, rays are no longer clearly defined, and so a corresponding degree of vagueness must necessarily be communicated to the orbits of the electron.

Owing to the greater complication of undulatory optics, de Broglie's original method of establishing the stability conditions for the waves ceases to be available. Let us, then, investigate how Schrödinger proceeded. We consider the case of the hydrogen atom, and we shall suppose that the nucleus of the atom is fixed, so that the energy W of the electron will also be the energy of the atom. De Broglie's method of associating a wave with the circling electron is retained. But Schrödinger, instead of confining his attention to the wave in the vicinity of the orbit, considers its distribution throughout space. In this respect, we note that there is no barrier preventing the wave from extending in all directions, and as a result it fills all space. Certain restrictions, however, are placed on the wave for physical reasons; we shall examine them later.

The problem which Schrödinger set himself was: What kinds of de Broglie waves, which satisfy the foregoing restrictions, can exist permanently in the field of force surrounding the nucleus. Schrödinger confined his attention, at first, to standing waves, or vibrations. The wave distributions he was seeking constituted therefore the modes of vibration of the de Broglie waves in the nuclear field of force. These modes represent the stable states of the atom. To obtain the modes, we must start as usual from d'Alembert's wave equation, which controls wave motions generally.

Let $u(x, y, z, t)$ be a function of x, y, z, t, which defines, at each point x, y, z of space and at each instant of time t, the state of vibration of the wave disturbance. The function u must satisfy the relation expressed by d'Alembert's wave equation

$$(1) \qquad \frac{\partial^2 u}{\partial x^2} + \frac{\partial^2 u}{\partial y^2} + \frac{\partial^2 u}{\partial z^2} - \frac{1}{V^2}\frac{\partial^2 u}{\partial t^2} = 0,$$

where V is the velocity of propagation of the waves at the point x, y, z. To shorten the writing we shall set Δf for the sum of the three second-order partial derivatives of any function f with respect to the Cartesian coordinates x, y, z. With this convention, the wave equation (1) becomes

$$(2) \qquad \Delta u - \frac{1}{V^2}\frac{\partial^2 u}{\partial t^2} = 0.$$

If now we wish to determine the standing sinusoidal waves that can be formed, we must assume that u represents a standing wave and so may be defined by

$$(3) \qquad u = \psi(x, y, z)\cos 2\pi\nu t,$$

where ν is the frequency of the waves in question. As for $\psi(x, y, z)$, it defines the maximum swing, or the amplitude, of the vibration at the point x, y, z. As yet the function $\psi(x, y, z)$ is unknown, but by substituting (3) in (2), we obtain the equation that ψ must satisfy, namely,

$$(4) \qquad \Delta\psi + \frac{4\pi^2\nu^2}{V^2}\psi = 0.$$

The velocity V in the equation (4) represents the propagation velocity of the de Broglie waves of frequency ν, but in order to derive Schrödinger's wave equation we shall have to assume that V represents the wave velocity. The approximation involved in this assumption will prevent our treatment from being rigorous, but it will be sufficiently rigorous for the object we have in view.

Now the waves which we are considering are associated with the planetary electron that is moving in the field of force generated by the atom's nucleus; and we have seen that a de Broglie wave in a field of force behaves as though it were propagated in a medium of variable refractive index μ. The refractive index is determined, at each point x, y, z of space, by the field of force and by the energy W of the electron with which the wave is associated. The expression of the refractive index was given in formula (13), Chapter XXIX. We recall this formula; it is

$$(5) \qquad \mu = \frac{\sqrt{[W - E_{pot}(x, y, z)]^2 - m_0^2 c^4}}{W},$$

where m_0 is the rest-mass of the electron and $E_{pot}(x, y, z)$ is the potential energy which the electron would have, as a result of the nuclear attraction, if it were situated at the point x, y, z. This potential energy, which varies in value from place to place, serves to define the field.

On the other hand we know that the wave velocity V is defined by $\frac{c}{\mu}$; and we also know that the frequency ν of the wave is $\frac{W}{h}$. Collecting all these results we may express the ratio $\frac{\nu^2}{V^2}$ of the wave magnitudes, which appears in equation (4), in terms of the mechanical categories. We find

$$(6) \qquad \frac{\nu^2}{V^2} = \frac{W^2 \mu^2}{h^2 c^2} = \frac{[W - E_{pot}(x, y, z)]^2 - m_0^2 c^4}{h^2 c^2}.$$

If now we replace $\frac{\nu^2}{V^2}$ by this expression in the equation (4) we obtain Schrödinger's wave equation in the relativistic form. As a first step, however, we may disregard the relativistic refinements and be content with the classical approximation. In this case the expression (6) of $\frac{\nu^2}{V^2}$ is simplified into

$$(7) \qquad \frac{\nu^2}{V^2} = \frac{2m_0[E - E_{pot}(x, y, z)]}{h^2},$$

where E is the sum of the electron's potential energy and of its classical kinetic energy.* Thus E differs from W mainly in that it does not include the rest energy $m_0 c^2$.

If we substitute the value (7) for $\frac{\nu^2}{V^2}$ in (4) we get Schrödinger's celebrated wave equation corresponding to the classical approximation. Schrödinger's equation is

$$(8) \qquad \Delta \psi + \frac{8\pi^2 m_0}{h^2}[E - E_{pot}(x, y, z)]\psi = 0.$$

Schrödinger's equation (8) controls the relative magnitude of the ampli-

---

* According to the classical approximation we have $W = m_0 c^2 + E$. If in (6) we replace W by the foregoing expression and neglect $(E - E_{pot})^2$ owing to its relatively small importance, the result (7) is obtained.

tude function $\psi(x, y, z)$ of the waves, at each point x, y, z, when the total energy of the electron or atom has the value E.

Let us observe that the value of $E_{pot}(x, y, z)$ in this equation is determined at each point by the nuclear field of force. On the other hand, the value of the total energy E is undetermined, for we are supposed to have no knowledge of the stable energy levels of the atom.

We have rendered the derivation of Schrödinger's wave equation less abstract by discussing it in connection with a particular system, i.e., the hydrogen atom. The same equation applies, however, to all systems in which the waves are associated with a single electron moving in 3-dimensional space, but of course the expression of the potential energy, E_{pot}, differs from one system to another. If the electron is moving along a straight line (as in the linear oscillator), the wave equation will be simplified in consequence.

We must now examine the restrictive conditions which Schrödinger imposes on the permissible waves. Physical considerations indicate that the amplitude of the vibrations cannot exhibit sudden discontinuities from place to place, and that it can never be infinite. Since a solution ψ of Schrödinger's equation represents the amplitude from place to place, we conclude that a solution function must be continuous (as also its first and second derivatives) and that it must never become infinite at any point in the region occupied by the waves, even on the boundary of this region. Furthermore a solution function must have a well-determined value at each point, i.e., it must be one-valued.* This last condition is obviously necessary, because otherwise the wave would be indeterminate.

In the particular case of the hydrogen atom, the region occupied by the waves consists of the whole of space, the boundary of this region being the surface of an infinite sphere. We shall find that in the hydrogen atom, the waves that satisfy the restrictive conditions not only remain finite at spatial infinity on the bounding surface but actually vanish there.

Formally, at least, the problem of determining the modes of vibration of the deBroglie waves is similar to those we mentioned in the last chapter in connection with the vibrations of elastic solids. For example, the problem of the hydrogen atom is analogous to that of determining the modes of vibration of an infinitely extended elastic sphere, the surface points of which do not vibrate. The hydrogen problem is more complicated, however, because space behaves as a dispersive medium of variable refractive index for the de Broglie waves, so that the wave velocity depends on the frequency of the waves and varies from place to place.

* This condition is equivalent to the one imposed by de Broglie.

The method of obtaining the de Broglie modes of vibration is now clear. We must integrate the Schrödinger wave equation which corresponds to the problem of interest, and then retain those solution functions ψ which satisfy the restrictive conditions. For the purpose of discussion we shall suppose that the waves occur in 3-dimensional space, so that the wave equation has the form (8).

Schrödinger showed that solution functions $\psi(x, y, z)$ satisfying all these requirements are in general non-existent, but that when the parameter E in the equation (8) happens to have certain privileged values, solutions exist, and stable waves can be formed. These privileged values which E must assume constitute the *eigenvalues* of the problem, and from a physical standpoint they represent the energy levels of the system. The corresponding solution functions $\psi(x, y, z)$ are the *eigenfunctions*.

The eigenvalues usually form a discrete set

$$(9) \qquad E_1, E_2, E_3, \ldots E_n, \ldots,$$

but they may also form a continuous range of values.

When the eigenvalues are obtained, we may derive the corresponding eigenfunctions $\psi(x, y, z)$. Thus in Schrödinger's equation (8), we replace the parameter E by one of the eigenvalues, e.g., E_n. The equation is now known to have a solution which satisfies the restrictive conditions. This solution function is then the eigenfunction $\psi_n(x, y, z)$ corresponding to the eigenvalue E_n. The eigenfunction determines the amplitude of the de Broglie wave from place to place; and the eigenvalue, or energy, E_n determines the frequency of the wave. Consequently, the eigenvalue and the eigenfunction, considered jointly, furnish all particulars on the corresponding mode of vibration.

The frequency ν_n of the wave is defined by $\dfrac{W_n}{h}$, where W_n, the energy of the particle, has as value $E_n + m_0 c^2$. Consequently, the frequency ν_n of the de Broglie wave is given by

$$(10) \qquad \nu_n = \frac{E_n + m_0 c^2}{h}. \text{ *}$$

---

* Sometimes the rest-energy of the particle is dropped, and in this case the frequency of the de Broglie wave associated with a particle of energy E_n (without the rest energy) is assumed to be $\dfrac{E_n}{h}$, instead of $\dfrac{E_n + m_0 c^2}{h}$. This change in the frequency credited to the wave does not, however, change its wave length λ; for $\lambda = \dfrac{h}{p}$, where p, the momentum of the particle, has the same value in either case.

Finally, then, if $\psi_n(x, y, z)$ is the eigenfunction corresponding to the energy value E_n of the atom, the normal mode of vibration of the de Broglie waves in the atom is given by (3) and by (10), and thus by

$$(11) \qquad u_n(x, y, z, t) = \psi_n(x, y, z) \cos 2\pi \left(\frac{E_n + m_0 c^2}{h} t \right).$$

In the foregoing discussion, we assumed that there was only one eigenfunction corresponding to each eigenvalue. But in some problems, notably in the hydrogen atom, several eigenfunctions are associated with the same eigenvalue. This association implies that different normal modes of vibration have the same frequency. The system is said to be degenerate. A situation of this kind was mentioned in the last chapter in connection with the vibrations of an elastic cube.

The Linear Harmonic Oscillator—The considerations we have developed for the hydrogen atom apply to any atomic system in which a single electron is involved. One of the simplest cases is afforded by the linear harmonic oscillator. According to classical mechanics, if a particle, abandoned at rest, is attracted towards a fixed centre with a force proportional to the distance (elastic force), the particle will vibrate with a simple harmonic motion along the straight line passing through the centre of attraction. The potential energy of the particle, when situated at a distance x from this centre, is

$$(12) \qquad E_{pot} = 2\pi^2 \nu_0^2 m_0 x^2,$$

where m_0 is the mass of the particle and ν_0 is the mechanical frequency of the vibration.

Let us see what information Schrödinger's equation (8) furnishes on the energy levels of the oscillator. In the oscillator we must assume that the waves are situated solely along the line of motion of the particle. We take this line as the Ox axis. Under these conditions the amplitude function ψ is a function of x alone, and the expression $\Delta\psi$ in Schrödinger's equation becomes $\dfrac{d^2\psi}{dx^2}$. If we take this change into account in the wave equation (8) and insert the value (12) for the potential energy E_{pot}, we obtain

$$(13) \qquad \frac{d^2\psi}{dx^2} + \frac{8\pi^2 m_0}{h^2}(E - 2\pi^2 \nu_0^2 m_0 x^2)\psi = 0.$$

This is Schrödinger's equation for the linear oscillator. In it appears the magnitude E, which represents the sum of the kinetic and potential energies of the vibrating particle and hence the energy of the oscillator. The value of E is left undetermined for the present.

We must now obtain the solutions $\psi(x)$ of the equation (13). Any one of these solutions will give the amplitude of the corresponding stable wave from place to place along the line of oscillation. We must remember, however, that restrictions are placed on the solution $\psi(x)$; namely, the function $\psi(x)$ must be one-valued, continuous (together with its first and second derivatives), and it must be finite everywhere even on the boundary. In the present case the boundary is represented by the two infinitely distant points to the right and left on the line of oscillation Ox.

The equation (13) may easily be transformed into an equivalent one studied by Hermite and often called Hermite's equation. Its peculiarities are therefore well known. The equation has in general no solutions of the type required, but solutions exist when the undetermined constant E happens to have any one of the infinite aggregate of values

$$(14) \qquad E = \frac{h\nu_0}{2} + nh\nu_0, \quad \text{or} \quad \left(n+\frac{1}{2}\right)h\nu_0,$$

where n is any positive integer or zero. In particular, there will be a single solution for each one of the foregoing permissible values of E. These privileged values of E are what we have called the *eigenvalues, i.e.,* the privileged values for which the problem is possible. As for the corresponding solutions $\psi(x)$, they are the *eigenfunctions.*

Passing from the mathematical to the physical standpoint, we recall that E is the total energy of the particle and hence of the oscillator. We therefore conclude that stable waves are possible only if the energy of the oscillator has any one of the values (14). In other words, the eigenvalues (14) are the energy levels of the linear oscillator.

This result is remarkable, for Planck's original theory, and also de Broglie's method of approach, gave the values

$$(15) \qquad E = nh\nu_0$$

for the energy levels; the part $\frac{h\nu_0}{2}$ was missing. But, to account for experimental measurements, notably in connection with the specific heats of solids at low temperatures, physicists found it necessary to accept the form (14). Accordingly, the modified form (14) involving half-quantum numbers $\left(n+\frac{1}{2}\right)$ was suspected of being correct, though no adequate theoretical basis was advanced in support of the modification. It is true that Planck subsequently attempted to account for the correct form (14) by assuming that the oscillator could absorb and store energy continuously while radiating it by jerks as before. But Planck's assumption was in conflict with what was known of the process of absorption, for it indicated that all energy states were stable and that the so-called stable energy states were merely those for which radiation might occur. Schrödinger's treatment has the advantage of giving the correct energy levels without any additional assumptions. The correct form (14) of the energy levels for the linear harmonic oscillator shows that the energy can never fall below the minimum value $\frac{h\nu_0}{2}$; in particular, it can never vanish entirely. At the absolute zero, where the energy should have its smallest value, we must conclude that there will still be some energy; this energy is called the "zero point energy $\frac{h\nu_0}{2}$." In the particle picture, the above conclusion implies that even at the absolute zero all motion does not cease.

When our aim is merely to obtain the energy levels, the discovery of the eigenvalues solves our problem. But when we wish to investigate the nature of the radiations that the oscillator will emit, we must calculate the stable de Broglie waves, and hence determine the amplitude functions $\psi(x)$ associated with the various eigenvalues, or energy levels. These amplitude functions are the solutions of Schrödinger's equation (13) for the oscillator, and are therefore the eigenfunctions. They are complicated functions known as Hermite functions; we shall not write them out explicitly. If we call $\psi_n(x)$ the eigenfunction that corresponds to the nth eigenvalue $E_n = \left(n + \frac{1}{2}\right)h\nu_0$, the corresponding stationary de Broglie wave is expressed by

$$(16) \qquad u_n(x, t) = C \, \psi_n(x) \cos \frac{2\pi}{h}(E_n + m_0 c^2) t,$$

where C is an arbitrary constant.

The nodal points are determined by the values of x which make $\psi_n(x)$ vanish. It can be shown that all the eigenfunctions vanish at infinity to the right and left. In addition, the nth eigenfunction $\psi_n(x)$, which corresponds to the nth energy level, also vanishes at n different points along the line of oscillation. There are therefore n nodal points at finite distance when the oscillator is in the nth energy level. The quantum number n of Bohr's theory thus receives a simple interpretation. Incidentally, we may note that, on account of the presence of the arbitrary constant C in (16), the exact amplitude of the wave is undetermined. All that is definite is the ratio of the amplitude from one point to another.

Schrödinger's method for deriving the radiation from the stable waves will be examined in Chapter XXXIV.

The Rotator—The name rotator is given to a solid body which can rotate about an axis fixed in space. For example, a rotating molecule may be viewed as a rotator. Bohr's quantizing condition shows that, for a given rotator, only special rates of rotation, forming a discrete set of values, are possible, or at least stable. Each rotational motion is associated with a corresponding energy, and hence Bohr's theory indicates the presence of a discrete set of energy levels. If we call A the solid's moment of inertia about the axis of rotation, Bohr's energy levels are

$$(17) \qquad E = \frac{h^2 n^2}{8\pi^2 A},$$

where n is any positive integer or zero. But the observation of band spectra, which are caused by the rotations of the molecules in a gas, showed that Bohr's energy levels for the rotator could not be quite correct, and that a more complicated sequence of energy levels would have to be assumed. Bohr's theory thus failed to furnish correct results. Now, if we treat the problem of the rotator by means of Schrödinger's equation, we find that the privileged values which E must assume (the eigenvalues) are

$$(18) \qquad E = \frac{h^2}{8\pi^2 A} n(n+1).$$

These values determine the energy levels of the rotator and, in contradistinction to the levels (17), they are in agreement with the measurements on band spectra.

The Hydrogen Atom—The wave equation of the hydrogen atom is derived from (8) by ascribing to E_{pot} the expression of the electron's potential energy in the nuclear field of the atom. Now in the derivation of (8), we assumed it permissible to disregard the relativistic refinements; furthermore, no attempt was made to introduce a possible wave counterpart of the electron spin. For these reasons the wave theory of the hydrogen atom, which we are about to consider, is the wave analogue of Bohr's original theory in which the relativistic refinements were disregarded and in which none of the complications discussed in Chapter XXVII were considered. This point being kept in mind, we pass to an examination of the wave equation for hydrogen.

The potential energy of the planetary electron in the hydrogen atom has as value at a point x, y, z

$$(19) \qquad E_{pot}(x, y, z) = - \frac{e^2}{\sqrt{x^2 + y^2 + z^2}} = - \frac{e^2}{r},$$

where $-e$ represents the charge of the electron and $+e$ that of the proton. The two charges are the same in absolute value but differ in sign. The variable r in (19) represents the distance of the electron from the proton. Schrödinger's equation (8) for the hydrogen atom is thus

$$(20) \qquad \Delta\psi + \frac{8\pi^2\mu}{h}\left(E + \frac{e^2}{r} \right)\psi = 0,$$

where μ is the mass of the electron. The magnitude E in (20) represents the total energy of the electron, and therefore of the atom. Its value for the time being is left undetermined.

To solve the hydrogen problem, we must obtain the normal modes of vibration of the waves, for these modes will define the stable states of the atom. As usual, we are required to determine the amplitude functions that are solutions of Schrödinger's equation (20) and which, in addition satisfy the conditions of being one-valued, continuous, and finite everywhere including the boundary points at spatial infinity. The equation (20) is solved by the same general method that was explained in connection with the equation (16) in the previous chapter. The method consists in splitting the equation (20) into three simpler ones, which we attempt to solve separately by obtaining solutions that satisfy the restrictive conditions. In this way three eigenvalues are introduced. The product of three solutions (each one of which is a solution of one of the three equations) then defines a solution of the wave equation (20), and this solution automatically satisfies the restrictive conditions. When we proceed as has just been outlined and obtain from the equation (20) the three subsidiary equations, we find that the first of these is exceedingly simple; that the second is the well-known equation of the associated Legendre functions; and that the third is more complicated, but may be thrown into the form of an equation studied by Laguerre. The eigenvalues connected with these three subsidiary equations are integers, which we denote by m, l, and n; we shall find that they are Bohr's quantum numbers for the hydrogen atom.

Mathematical analysis shows that solutions ψ of the wave equation (20), which satisfy the restrictive conditions, exist only when the parameter E is credited with certain particular values, the eigenvalues. In the present case these eigenvalues are:

1. All real positive values of E, whether integral or not;
2. The discrete set of negative values

$$(21) \qquad E = -\frac{2\pi^2 \mu e^4}{h^2 n^2}.$$

In this formula, n is any positive integer.

The eigenvalues in our earlier examples formed a discrete aggregate of numbers; the present situation differs in that we now have a continuous range as well as a discrete aggregate. This more complicated situation often occurs when the regions of space at which the boundary conditions are imposed are infinitely distant. The physical interpretation of these results is that the various stable modes of vibration of the de Broglie waves can occur only when the total energy E of the atom has a positive value, or when it has any one of the discrete negative values defined by (21). Let us consider the latter values. They define exactly the same energy levels as result from Bohr's theory of the hydrogen atom. Hence, we conclude that Bohr's energy levels are merely the eigenvalues of Schrödinger's equation (20).

The continuous range of positive values for E also receives a simple interpretation. We recall that, according to mechanics, positive values of the total energy E indicate that the planetary electron is describing a hyperbolic orbit. Inasmuch as only periodic motions were subjected to quantization in Bohr's theory, the hyperbolic orbits were not quantized. This implied that all hyperbolic orbits were possible, and that none of them was singled out as more stable than the others. Schrödinger's treatment furnishes the same result; for since whenever E is positive, Schrödinger's equation always admits solutions satisfying our requirements, we must conclude that any positive value is permissible for E.

We now consider the solutions ψ, or eigenfunctions, of Schrödinger's equation. But first we must recall how Schrödinger's equation was solved. We split the equation into three subsidiary equations, the solutions of which had to satisfy certain restrictions. An eigenfunction ψ of Schrödinger's equation was then given by the product of any triplet of solutions pertaining to the three subsidiary equations, respectively. Now each one of the subsidiary equations contains a parameter which must receive appropriate values (eigenvalues) if the equation is to have a solution satisfying the imposed restrictions. The parameters of the three equations may be designated, respectively, by l, m, and by some other letter which we subsequently connect with the n of (21).

The first subsidiary equation will have acceptable solutions only if the parameter l has one of the n values

(22) $\qquad l = 0, 1, 2 \ldots (n-1).$

The second subsidiary equation will have appropriate solutions only if m has one of the $2l + 1$ values

(23) $\quad m = -l, -(l-1), \ldots -1, 0, 1, \ldots (l-1), l.$

As for n, we have seen that it may have any positive integral value.

If we call a_l, b_m, c_n the solutions of the three subsidiary equations, we see that an eigenfunction ψ_n of Schrödinger's equation will be of form

(24) $\qquad \psi_n = a_l b_m c_n.$

Inasmuch as the solutions a_l and b_m differ according to the values ascribed to l and m, there will be different eigenfunctions ψ_n connected with the same eigenvalue E_n. So as to distinguish these latter eigenfunctions ψ_n, we shall designate them by $\psi_{n,l,m}$, and then ascribe to l and m all their permissible values (22) and (23). In this way all the different eigenfunctions $\psi_{n,l,m}$ will be represented. We may easily verify from (22) and (23) that there are n^2 different combinations for the values of l and m; hence there are n^2 different eigenfunctions $\psi_{n,l,m}$ connected with the same eigenvalue E_n. All these eigenfunctions vanish at infinity.

Each eigenfunction $\psi_{n,l,m}$ defines a possible distribution of the amplitude of the standing waves from place to place and therefore determines one of the normal modes of the de Broglie waves. The standing wave connected with an eigenfunction $\psi_{n,l,m}$ is defined by

(25) $\qquad \psi_{n,l,m} \cos 2\pi \dfrac{E_n + m_0 c^2}{h} t.$

It will be noted that the modes connected with the n^2 eigenfunctions which are associated with the same eigenvalue E_n all have the same frequency. However, the manner in which the vibrating regions form patterns differs from one of these modes to another.

We now consider the nodal surfaces associated with these modes of vibration, i.e., those surfaces over which no vibrations occur and therefore over which the eigenfunctions of interest vanish. There are three different kinds of nodal surfaces for the normal modes of vibration. These are: concentric spheres with centre at the nucleus O; cones of revolution about the Oz axis, having the nucleus O as vertex; and, finally, planes passing through the Oz axis. In particular, if we consider the mode of vibration $\psi_{n,l,m}$, where n, l, and m are given definite numerical values, we obtain the following results:

1. There are $n - l$ nodal spheres, one of which is always the spherical surface at infinity. The existence of a nodal surface at infinity expresses the fact that the vibration always dies down at infinity.

2. There are $l - |m|$ nodal cones of revolution having as vertex the point O where the nucleus is situated, and the common axis of these cones lies along a direction we may call the Oz axis. Here $|m|$ signifies the absolute value of the integer m. If $l \neq 0$, one of these cones always collapses into the Oz axis itself. But if $l = 0$, we also have $m = 0$, so that all the nodal cones vanish.

3. Finally, there are $|m|$ nodal planes variously situated, but all of which pass through the Oz axis. If $m = 0$, no nodal planes will arise.[*] Thanks to these nodal surfaces, we may obtain a graphical description of the different modes of vibration represented by the eigenfunctions, or amplitude distributions, of type $\psi_{n,l,m}$.

We must emphasize, however, that when the atom is in the energy level E_n, the vibration of the de Broglie waves need not be represented by any individual one of the independent eigenfunctions $\psi_{n,l,m}$. The vibration may very well result from a superposition of these various modes occurring simultaneously. The reason is that, owing to the linear form of Schrödinger's wave equation, any linear combination of its different solutions (corresponding to the same eigenvalue E_n) is itself a solution and is therefore an eigenfunction defining a possible state of vibration. The superposition of modes previously referred to is a mere physical interpretation of this fact. Suppose, then, that we consider one of these more complicated types of vibration associated with the energy level E_n. This state of vibration, being due to a superposition of some or of all of the simple modes $\psi_{n,l,m}$, will no longer have the nodal surfaces we have just listed. By selecting the combination of modes in a suitable way, we may obtain nodal surfaces that are paraboloids of revolution, for instance. We may also obtain moving waves and wave packets. We mention these points to show that the precise distribution of the de

---

[*] Our mathematical calculations show that the line we have taken for the Oz axis is the axis of the nodal cones and is the line through which all the nodal planes pass. But it must be noted that the direction Oz is selected arbitrarily, so that the precise orientations of the nodal cones and planes is undetermined. This is, of course, to be expected, for since the hydrogen atom is radially symmetrical, there is no reason why the axis of the nodal cones should lie in one direction rather than in another. In any case, however, the nodal planes will always pass through the axis of the cones; and the relative situations of the cones and planes are thus determined. No ambiguity arises for the nodal spheres, because their centres are always at the centre of the atom and their radii are well determined.

Broglie vibrations or waves, when the atom is in the energy state E_n, is not well defined. On the other hand, the frequency of the vibrations is always the same; it is defined by (10) and (21).

Let us contrast Schrödinger's results with those obtained by Bohr. We have already observed that Schrödinger's present theory does not take into account the relativistic refinements or the electron spin. Consequently, if we wish to compare Schrödinger's theory with Bohr's, the comparison must be made with Bohr's original theory in which the above complications were likewise ignored.

Schrödinger's stable energy levels (21) coincide with Bohr's. The number n of the wave theory thus plays the same rôle as Bohr's quantum number n, though its physical significance, which we shall investigate presently, is very different.

In addition to the number n, Schrödinger's eigenfunctions $\psi_{n,l,m}$ involve other numbers, l and m. Since these numbers can assume only the values (22) and (23), which are also the values to which Bohr's quantum numbers l and m are restricted, we conclude that the l and m of Schrödinger are the wave expressions of Bohr's quantum numbers l and m.

Bohr's original treatment of the undisturbed atom necessitated only the main quantum number n. But we explained on page 528 that three quantum numbers could be introduced if so desired, and that in that case the orbits were quantized more completely. These three quantum numbers were n, k, m, but at a later stage Bohr replaced k by $l = k - 1$; and since the wave treatment involves l and not k, we shall consider Bohr's theory from the standpoint of n, l, and m. When these three numbers are utilized in the quantizing of Bohr's undisturbed atom, each energy level E_n is seen to be associated with n^2 different orbits (n, l, m). Consequently, when the atom is in the energy state E_n, the electron may be describing any one of these associated orbits.

If, now, we pass to Schrödinger's atom, we find that each energy level E_n is associated with n^2 different eigenfunctions, or modes of vibration, $\psi_{n,l,m}$; and when the atom is in an energy level E_n, it will be vibrating in one of the corresponding modes (or in a superposition of these modes). If we neglect for the present the complication issuing from a superposition of modes, we see that the results of Schrödinger and of Bohr coincide, provided we view each mode of vibration $\psi_{n,l,m}$, in which the atom may be vibrating, as the wave equivalent of each orbit (n, l, m) which the electron may be describing.

Finally let us compare the two atoms from the standpoint of degeneracy. We saw that Bohr's undisturbed atom was degenerate because each energy level E_n was associated with several (*i.e.*, n^2) different orbits (n, l, m). This same degeneracy is exhibited by Schrödinger's atom; it is expressed by the association of each eigenvalue E_n with n^2 different eigenfunctions, or modes of vibration, $\psi_{n,l,m}$.

The phenomenon of degeneracy is very important in Bohr's theory, because whenever a certain number of energy levels coalesce, the action of a perturbing force may affect the various levels differently and thereby bring about their separation. The degeneracy is then said to be removed in part, or in whole, by the perturbation. As an example, in Bohr's theory of the hydrogen atom in which the electron spin is ignored, all sub-sublevels (n, l, m) connected with the same value of n coalesce. But if we introduce the relativistic refinements (the effect of which is equivalent to that of a small perturbing force), those of the sub-sublevels (n, l, m) in which the value of l differs, are associated with different energies and hence appear as distinct. The degeneracy is thus decreased, though it is not completely removed, because the sub-sublevels in which m has different values coalesce as before. This latter degeneracy may be removed in turn by the application of a magnetic field, as in the Zeeman effect. We have recalled these points because precisely the same situation occurs in the wave theory. Thus a perturbing force will remove the degeneracy, in whole or in part, by causing the various modes $\psi_{n, l, m}$ which have the same frequency (and hence energy) to assume different frequencies and thereby to become separate.

The foregoing explanations may be expressed in a different, though equivalent, form. We may say that each sub-sublevel (n, l, m), or each mode of vibration $\psi_{n,l,m}$, is associated with one definite energy value $E_{n,l,m}$. When the degeneracy is complete, all those energy values $E_{n,l,m}$, in which n retains the same value (while l and m receive all permissible values), have the same magnitude. But when the degeneracy is removed in part, or in whole, under the action of perturbing influences, some or all of the various energy values $E_{n,l,m}$ in which n is fixed become distinct. The frequencies of the various modes $\psi_{n,l,m}$ connected with the fixed value n then undergo corresponding changes.

The Interpretation of the Quantum Numbers, *n*, *l*, *m* in the Wave Theory—In Bohr's treatment, the quantum numbers n, l, m are associated with mechanical magnitudes. In the wave theory, they are connected with the wave distributions, and it is of interest to ascertain what characteristics of the modes of vibration these quantum numbers represent.

We first consider the number m. On reverting to our discussion of the nodal surfaces, we see that, in any given mode of vibration determined by the eigenfunction $\psi_{n,l,m}$, the absolute value of m, i.e., $\mid m \mid$, defines the number of nodal planes. In Bohr's theory of the hydrogen atom (in which the spinning electron is disregarded) the quantum number m is associated with the inclination of the plane of the orbit with respect to the Oz direction. We now pass to the quantum number l, which in Bohr's theory is associated with the angular momentum of the electron on its orbit. To obtain its wave interpretation, we recall that the number of nodal cones associated with a mode of vibration is $l - \mid m \mid$. If, then, we consider a mode in which $m = 0$ (no nodal planes), the number of nodal cones will be l. The quantum number l may then be interpreted as defining the number of nodal cones in this mode of vibration. Finally, let us obtain a wave interpretation of Bohr's main quantum number n. The number of nodal spheres including the sphere at infinitely is $n - l$. Hence in a mode of vibration for which $l = 0$ (i.e., the most eccentric orbit), we see that there will be n nodal spheres. The quantum number n thus receives a wave interpretation. In short, we have obtained wave interpretations for Bohr's quantum numbers. We shall defer for the present the wave interpretation of the planetary electron moving on a well-defined orbit; it will be considered in the following chapter.

An interesting feature of the wave treatment is that it imposes the quantum number l in place of Bohr's earlier number k. We recall that in the original treatment of Bohr's atom, the quantum number k represented the angular momentum (in units $\frac{h}{2\pi}$) of the electron on its orbit. According to mechanics the permissible quantized values of k were

(26) $$k = 0, 1, 2, \ldots \ldots n.$$

The value $k = 0$ implied that the electron had no angular momentum, and hence was oscillating through the nucleus. This possibility was excluded for physical reasons, and as a result the smallest possible value of k was assumed to be 1. We pointed out at the time that the exclusion of the value $k = 0$ was unsatisfactory since it was not based on any theoretical necessity. Besides it led to difficulties in other cases.

Subsequently, in order to interpret the multiplet structure, Bohr replaced the number k by the number $l = k - 1$, and in this way l and not k was seen to measure the angular momentum of the electron. The theoretically possible values of l were then

(27) $$l = 0, 1, 2, \ldots \ldots (n - 1).$$

Since $l = 0$ was the equivalent of $k = 1$, the possibility of the electron oscillating through the nucleus was excluded automatically. Thus the introduction of l removed our original difficulty. On the other hand, the identification of l with the electron's angular momentum was utterly incomprehensible from the standpoint of Bohr's mechanical model, for it implied that an electron describing a circle had no angular momentum.

The wave theory is not confronted with these difficulties, because now the number l is introduced quite naturally as an eigenvalue of one of the subsidiary equations, and its physical significance is expressed clearly by means of the nodal surfaces. To this extent the wave treatment has the advantage over Bohr's.

The Relativistic Hydrogen Atom—If we except the replacing of k by l, we see that Schrödinger's wave theory of the undisturbed hydrogen atom yields exactly the same results as Bohr's earlier theory: the energy levels and the degree of degeneracy are the same in either case. We know, however, that the correct energy levels of hydrogen are not the simple ones which were obtained by Bohr originally; and so the failure of the wave theory to refine Bohr's earlier results is somewhat of a surprise.

To understand the reason for this failure, we recall that the refinements finally introduced into Bohr's treatment of the hydrogen atom issued from the application of relativity mechanics and from the hypothesis of the spinning electron. It was by this means that we discovered the sub-sub-sublevels (n, l, j, m), or (n, l, m_l, m_s), where m_l is our former m of the previous pages and m_s refers to the orientation of the axis of spin of the electron. Now Schrödinger's wave equation (20) is derived from the general wave equation (8), and this latter equation is obtained by disregarding the relativistic refinements and taking the approximate semi-classical expression (7) for the ratio $\dfrac{v^2}{V^2}$ *. For this reason, the Schrödinger wave equation (20) with which we are here dealing can be valid only to the same extent as classical mechanics is valid.

A relativistic form of Schrödinger's wave equation may, however, be obtained by utilizing, in place of (7), the relativistic value (6) of the ratio $\dfrac{v^2}{V^2}$. When this is done, the wave theory predicts more or less the same fine structure of lines that was discovered by Sommerfeld when he introduced the relativistic refinements into Bohr's atom but was unaware of the electron's spin. We know, however, that Sommerfeld's theory

---

* See page 690.

of the fine structure of lines had subsequently to be revised when the spinning electron was accepted. These considerations show that, even when Schrödinger's wave equation is refined so as to be in agreement with the theory of relativity, we cannot expect it to furnish the correct energy levels of the hydrogen atom until the wave analogue of the spinning electron is incorporated into the equation. Pauli was the first to attempt this incorporation. The theory has since been revolutionized by Dirac, who has shown that the electron spin is a manifestation which is a direct consequence of the relativistic wave equation when the latter is modified in a suitable way.* Thanks to this discovery, Dirac was able to obtain the refined theory of the hydrogen atom by means of wave mechanics. The doublet nature of the hydrogen spectrum was disclosed, and the separations between the sub-sublevels were calculated. In particular, it was shown that the screening levels must coalesce, as mentioned on page 583. For the present we shall defer consideration of the more refined treatment, and confine ourselves to the results obtained by Schrödinger on the basis of his classical wave equation without the spinning electron.

The Normal Zeeman Effect for Hydrogen—The study of Schrödinger's wave equation shows that the n^2 independent eigenfunctions $\psi_{n,l,m}$ (in which n has a fixed value while l and m are given all possible values) are associated with the same eigenvalue E_n. As has been pointed out previously, we may express the situation in an equivalent way by associating one eigenvalue $E_{n,l,m}$ with each eigenfunction $\psi_{n,l,m}$, and by saying that, for a fixed value of n and all possible values of l and of m, the n^2 eigenvalues $E_{n,l,m}$ coalesce. It is this coalescence which constitutes the degeneracy of the problem. Let us now suppose that a magnetic field is applied to the atom. Schrödinger's wave equation is modified in consequence, and the eigenvalues $E_{n,l,m}$ (in which m receives different values, while n is fixed and l receives all possible values) become distinct. In more physical language, this implies that the magnetic field splits the energy levels E_n into sublevels $E_{n,m}$. The separation between the consecutive sublevels depends on the intensity of the applied magnetic field and has the same value as was found in Bohr's earlier theory. Consequently, the wave treatment yields the normal Zeeman effect for the hydrogen atom just as did Bohr's earlier treatment. We know, however, that the effect actually exhibited by hydrogen is the anomalous Zeeman effect. But we must not be surprised at the failure of the wave theory to predict the correct effect for hydrogen; for we have seen that this effect is intimately connected with the relativistic refinements and with the spin

---

* See Chapter XXXIX.

of the electron, and in our present wave treatment these complications have not been taken into account.

The Hydrogen Atom when the Motion of the Nucleus Is Taken into Account—In Bohr's semi-mechanical theory, we may assume as a first approximation that in the hydrogen atom the proton-nucleus, owing to its large mass, remains at rest when the electron circles around it. But in all rigor, the nucleus, being attracted by the electron, also describes an orbit; and the motion of the nucleus should be taken into account. When this is done in Bohr's theory, we obtain the same energy levels (21) as before, except that now the mass μ of the electron must be replaced by

$$(28) \qquad \frac{\mu M}{\mu + M}, \text{ or } \frac{\mu}{1 + \dfrac{\mu}{M}},$$

where M is the mass of the nucleus. Only a very small change results from this refinement, since the value of $\dfrac{\mu}{M}$ in (28) is very small.

The foregoing refinement may be taken into consideration in the wave theory. It must be noted, however, that, if we no longer view the nucleus as a fixed point developing a field of force in which the electron is moving, we must associate a de Broglie wave with it, just as we did for the electron. The correct method of procedure when we are dealing with two or more particles that interact is to consider, not a number of de Broglie waves in the same 3-dimensional space, but instead a single wave in a configuration space. In particular, our hydrogen atom must now be associated with a single de Broglie wave in a configuration space of $2 \times 3 = 6$ dimensions. We are thus led to generalize Schrödinger's wave equation, so that it will apply to waves propagated in a configuration space. This complication in itself proves the symbolic nature of the waves. In the particular case of the hydrogen atom, the solution of the wave problem is quite simple, and we obtain the same results as were obtained by Bohr on the basis of his mechanical model.

The energy levels in the Stark effect may also be determined by Schrödinger's method.

Alternative Methods of Obtaining the Wave Equation—We mentioned previously that the method employed in obtaining the wave equation (8) was not rigorous. Accordingly, Schrödinger devised other methods. The first involves a principle of variation and is similar in its significance to the principles of Least Action in mechanics.

Let us suppose that some magnitude ψ is distributed throughout infinite space, so that at each point x, y, z of space, ψ has a certain value. To make the presentation more concrete we may assume that the square of ψ, i.e., ψ^2, defines at the point x, y, z the density of some gas at this point. The distribution is not entirely arbitrary, for we shall restrict it in such a way that, if ψ^2 is integrated throughout infinite space, the same numerical value 1 will be obtained, regardless of the distribution considered. In our simile this means that the total amount of gas must be the same, regardless of the exact distribution.

We now consider an appropriate mathematical expression F, the value of which depends on the distribution ascribed to ψ, and therefore to the gas. We then seek to determine the distribution which satisfies our previous requirements and which, at the same time, makes F a minimum or rather an extremum. This second condition imposes of course still further restrictions on the function ψ. We find that when the expression F is chosen suitably, the function ψ we are seeking turns out to be a solution of a certain partial differential equation which is none other than Schrödinger's wave equation (8). We conclude that Schrödinger's wave equation expresses the restriction that the function ψ must satisfy if an appropriate mathematical expression F is to be economized. The analogy with the mechanical principles of Least Action is obvious. We have mentioned this deduction of the wave equation, because it shows that the economizing principles of mechanics, which in Maupertuis' opinion revealed some divine guidance, have their analogues in many other departments of theoretical physics. No theological interpretation can be placed on these economizing principles, for, by varying our choice of the entity F to be economized, we may obtain any result we choose.

Finally, we must mention still another method of obtaining the wave equation, also due to Schrödinger. Suppose we have a particle of mass m_0 moving in space under the action of a conservative force. The total energy E of the particle is given by the sum of its kinetic and potential energies. The potential energy $E_{pot}(x, y, z)$ depends on the particle's position x, y, z in the field of force. The kinetic energy may be expressed in terms of the particle's momenta p_x, p_y, p_z. In particular, we find for the kinetic energy the expression

$$(29) \qquad \frac{1}{2m_0}(p_x^2 + p_y^2 + p_z^2).$$

The total energy E is thus

$$(30) \qquad E = \frac{1}{2m_0}(p_x^2 + p_y^2 + p_z^2) + E_{pot}(x, y, z).$$

The function E of x, y, z, p_x, p_y, p_z thus obtained is the Hamiltonian function, which we discussed in the chapter on analytical mechanics. It is usually represented by $H(x, y, z, p_x, p_y, p_z)$ for short. Our preceding equation may then be written

(31) $$H(x, y, z, p_x, p_y, p_z) - E = 0,$$

and as such it expresses the conservation of energy.

Schrödinger now observes that, if in this equation we replace the momenta p_x, p_y, and p_z by the operators

(32) $$\frac{h}{2\pi i}\frac{\partial}{\partial x}, \quad \frac{h}{2\pi i}\frac{\partial}{\partial y}, \quad \frac{h}{2\pi i}\frac{\partial}{\partial z},$$

and if we assume that the operator thus formed operates on a function ψ, the wave equation (8) is obtained. Thus, when the calculations are effected, the equation

(33) $$\left[H\left(x, y, z, \frac{h}{2\pi i}\frac{\partial}{\partial x}, \frac{h}{2\pi i}\frac{\partial}{\partial y}, \frac{h}{2\pi i}\frac{\partial}{\partial z}\right) - E\right]\psi = 0$$

is found to be the wave equation (8). The method just described for obtaining the wave equation is extremely symbolic and appears to have no physical interpretation. But we shall understand, when we discuss the matrix method, why the symbolic substitution (32) gives the correct results.

We mentioned on a previous page that the wave equation may be refined so as to take into account the theory of relativity. To obtain this more refined equation, we have merely to replace the classical Hamiltonian by the relativistic one in the equation of energy (31), and then proceed by the same method as before. In Chapter XXXIX we shall find, however, that the relativistic wave equation thus secured must be modified further.

The Generalized Wave Equation—We revert to Schrödinger's wave equation (8) for a particle of mass m_0 moving in three dimensional space x, y, z. This equation is

(34) $$\Delta\psi + \frac{8\pi^2 m_0}{h^2}(E - E_{pot}(x, y, z))\,\psi = 0.$$

If we assume that the field of force acting on the particle does not vary with time, the expression of the potential energy in (34) will not depend on time and will thus be of form $E_{pot}(x, y, z)$. The equation (34) now contains no reference to time, so that its solutions $\psi(x, y, z)$ may also be independent of time. We know that solutions which are finite, continuous, and one-valued exist only when an eigenvalue is credited to the parameter E. These eigenvalues may form a discrete set

(35) $$E_1, E_2, \ldots E_n, \ldots .,$$

or a continuous set, or both. If the eigenvalues form a continuous set, there will obviously be a solution satisfying the restrictions whenever E, in (34), is credited with any value within the continuous range of the set.

Let us suppose that the eigenvalues form a discrete set. The solution function corresponding to the nth eigenvalue E_n is then, say, $\psi_n(x, y, z)$. This solution may, but need not, be unique. In all rigor $\psi_n(x, y, z)$ merely defines the amplitude from place to place of the de Broglie vibrations; the wave, or the vibration, is represented by (25). If we write the wave in the complex form, the nth vibration u_n is

(36) $$u_n = \psi_n e^{\pm 2\pi i \frac{E_n + m_0 c^2}{h} t} .$$

We shall drop the rest energy $m_0 c^2$ in (36) shall use the minus sign. **Thus**

(37) $$u_n = \psi_n e^{- 2\pi i \frac{E_n}{h} t} .$$

The function (37), though it depends on time, is also a solution of the equation (34) (provided E be replaced by E_n in this equation). In short, the presence of the time factor does not affect results. Hence we may write the wave equation

(38) $$\Delta u + \frac{8\pi^2 m_0}{h^2}(E - E_{pot}) u = 0,$$

where a solution $u(x, y, z, t)$ is of form (37) when E is E_n in (38).

Suppose now we consider a superposition of different possible waves corresponding to different energy values. Such a superposition may be expressed by

(39)
$$u = c_1 u_1 + c_2 u_2 + \ldots + c_n u_n,$$

or
$$c_1 \psi_1 e^{\frac{-2\pi i}{h} E_1 t} + c_2 \psi_2 e^{\frac{-2\pi i}{h} E_2 t} + \ldots + c_n \psi_n e^{\frac{-2\pi i}{h} E_n t},$$

where the c_k's are constants. A function of this sort is no longer a solution of (38), for the first term is a solution only when $E = E_1$; the second term is a solution only when $E = E_2$; and so on. But we can obtain an equation of which the function (39) is a solution. We may derive it by eliminating E between (38) and $u = \psi e^{-\frac{2\pi i}{h} Et}$. When this elimination is performed, there results the equation

$$(40) \qquad \Delta u - \frac{8\pi^2 m_0}{h^2} E_{pot} u + \frac{4\pi i m_0}{h} \frac{\partial u}{\partial t} = 0.$$

We shall call this equation the generalized wave equation. When we are dealing with a superposition of waves, as occurs in a wave packet, it is to this equation (40) that we must revert.

Schrödinger shows how his formal method of substitution gives the more general equation (40) directly. He starts, as before, from the equation of energy.

$$(41) \qquad H(x, y, z, p_x, p_y, p_z) - E = 0,$$

and, as before, he replaces p_x, p_y, and p_z by the operators (32). But in addition, he replaces E by

$$(42) \qquad -\frac{h}{2\pi i} \frac{\partial}{\partial t}.$$

Altogether then the substitutions to be effected in the energy equation are

$$(43)\ p_x = \frac{h}{2\pi i} \frac{\partial}{\partial x}, \quad p_y = \frac{h}{2\pi i} \frac{\partial}{\partial y}, \quad p_z = \frac{h}{2\pi i} \frac{\partial}{\partial z}, \quad E = -\frac{h}{2\pi i} \frac{\partial}{\partial t}.$$

Their significance will be understood in Chapter XXXVIII. The operator obtained is assumed to operate on a function $u(x, y, z, t)$. The resulting equation is

$$(44) \left\{ H\left(x, y, z, \frac{h}{2\pi i} \frac{\partial}{\partial x}, \frac{h}{2\pi i} \frac{\partial}{\partial y}, \frac{h}{2\pi i} \frac{\partial}{\partial z}\right) + \frac{h}{2\pi i} \frac{\partial}{\partial t} \right\} u = 0.$$

And this is none other than the generalized equation (40).

The equation (40), like (34), corresponds to the case where the relativistic refinements may be disregarded. The relativistic generalized equation (in the case of a particle moving in free space) is

$$(45) \qquad \Delta u - \frac{4\pi^2}{h^2} m_0^2 c^2 u - \frac{1}{c^2} \frac{\partial^2 u}{\partial t^2} = 0.$$

CHAPTER XXXIII

SCHRÖDINGER'S THEORY OF RADIATION

In Bohr's theory, the frequencies of the radiations that may be emitted from an atom are deduced from the energy levels by an application of Bohr's frequency condition. The particle picture which illustrates the process of radiation is represented by an electron dropping from a higher to a lower orbit. In Schrödinger's wave theory of the hydrogen atom, the energy levels turn out to be the privileged values which the energy of the atom must have if stable de Broglie vibrations (satisfying certain restrictive conditions) are to be possible. These de Broglie vibrations have nothing in common with the electromagnetic waves which may be radiated from the atom; this is seen from the fact that their frequencies are much greater than those of the electromagnetic waves. The de Broglie vibrations, as such, do not therefore furnish the radiation waves; and Schrödinger's theory, as developed to this point, cannot anticipate the radiated frequencies, or even suggest under what conditions radiation will occur. Schrödinger might have pursued the same course as Bohr, and utilized Bohr's frequency condition to deduce the electromagnetic frequencies from the energy levels. But the originality of Schrödinger's treatment is that it furnishes the electromagnetic frequencies directly from the wave picture without having recourse to Bohr's frequency condition. The theory devised by Schrödinger for this purpose is called Schrödinger's theory of radiation.

Schrödinger's radiation theory was advanced before Born's assumptions were made, and the significance which Schrödinger ascribed to the de Broglie waves differs entirely from the one accepted today. The modern interpretation of the waves is that based on Born's assumptions and on the views of Heisenberg and Bohr. It is the interpretation which was explained in Chapter XXX, and it leads to the assimilation of de Broglie waves with probability waves. But although Schrödinger's original interpretation of the waves has been abandoned, his theory of radiation has the advantage of furnishing a simple explanation of the phenomenon of radiation from an atom. Furthermore, without having to rely on the correspondence principle, it accounts for the selection rules, for the forbidden drops of Bohr's theory, and also for the polarizations and intensities of the radiations. These remarks justify our presenting Schrödinger's discarded theory. In the next chapter we shall see how the results obtained by Schrödinger have been revised in accordance with the views of Born, Heisenberg, and Bohr.

De Broglie, in his treatment of the hydrogen atom, assumed that the conditions of ray-optics were valid. This implied that well-defined rays were associated with the waves in the atom and that some of these rays could define the Bohr orbits. Side by side with the waves, the concept of a corpuscular electron accompanying the waves was retained from the particle theory. Subsequently, Schrödinger showed that in the interior of the hydrogen atom the conditions of ray-optics could not be realized, so that the rays, and with them the concept of well-defined Bohr orbits, became vague. Furthermore, Schrödinger abandoned the idea of a corpuscular electron and sought to interpret the electron by means of the waves. To understand how this interpretation was made possible, we must recall certain features of the de Broglie vibrations which occur when the atom is in a stable state.

We saw in the previous chapter that, corresponding to each eigenvalue E_n of Schrödinger's wave equation, there were n^2 independent solutions, or eigenfunctions, $\psi_{n,l,m}$. The n^2 different eigenfunctions were obtained by keeping fixed the value of n in $\psi_{n,l,m}$ and by giving all permissible values to l and m. The physical significance of these mathematical results was that each stable energy level E_n, in which the hydrogen atom might be situated, was associated with n^2 independent modes of vibration of the standing de Broglie waves. All these n^2 standing waves vibrated with the same frequency

(1) $$\frac{E_n + m_0 c^2}{h}, *$$

but the modes differed from one another because of the different distributions of the amplitudes of the vibrations. In any given mode, the amplitude of the vibration at a point P was measured by the numerical value of the corresponding eigenfunction $\psi_{n,l,m}$ at that point. Hence the eigenfunction determined the amplitude of the vibration from point to point throughout space.

Nevertheless, as the following discussion will show, a considerable measure of arbitrariness is connected with the amplitude. Thus, if $\psi_{n,l,m}$ is an eigenfunction of Schrödinger's equation corresponding to the energy value E_n, then $C\psi_{n,l,m}$ likewise is such an eigenfunction, where C is any constant. More generally, if ψ, ψ', ψ'', . . . represent different

---

* See note page 692.

eigenfunctions associated with the same energy level E_n, and if C, C', C'',
. . . are arbitrary constants, then

(2) $$C\psi + C'\psi' + C''\psi'' + \ldots$$

is also an eigenfunction connected with the same energy level E_n. These
considerations show that even when the energy level of the atom is specified, the amplitude of the wave phenomenon from place to place is not
well determined; and we may still select the distribution of the amplitude
in many different ways. The arbitrariness with which we are here faced
need not disturb us for the present; its significance will be understood
later. In what follows we shall call ψ_n any one of the eigenfunctions, or
amplitude functions, which we may have selected for the definition of
the wave phenomenon when the atom is in the energy level E_n.

A point which might cause confusion should, however, be mentioned.
The fundamental eigenfunctions $\psi_{n,l,m}$ are the amplitude functions of
standing de Broglie waves; but when we superpose various modes belonging to the same energy level, the resultant wave disturbance may betray
moving wave crests. In such cases, even though the wave is no longer
a standing wave, the amplitude from place to place does not vary with
time.* In short, when the atom is in a definite energy level E_n, the distribution of the amplitude function remains fixed. This remark will
be helpful presently.

Next, we consider the intensity of the de Broglie wave at any point
x, y, z. The connection between the intensity and the amplitude of a
vibration may easily be understood from the following simple example.
Suppose a tuning fork is vibrating. The ear detects a pure musical note
which gradually dies down in intensity without sustaining any change
in pitch. Now the frequency with which the fork is vibrating does not
vary, but the amplitude of the vibration decreases progressively, owing
to frictional resistances and to the communication of energy to the surrounding air. We conclude therefore that the frequency of the vibration
determines the pitch of the musical note, and that the amplitude governs
its intensity. When we are dealing with vibrations of a familiar kind,
the intensity is proportional to the square of the amplitude. We shall
assume that the same is true for the de Broglie vibrations.

Let us, then, suppose that the atom is in the stable energy state E_n,
and let ψ_n represent the corresponding eigenfunction which we have

---

* This permanence of the amplitude distribution occurs because the superposed
standing waves have the same frequency. If the superposed waves had different
frequencies, a transference of the amplitude would result.

selected. The amplitude of the de Broglie wave phenomenon at a point P is measured by the value of ψ_n at this point. The intensity of the de Broglie wave at P is therefore proportional to the square of the value of ψ_n at the same point. Inasmuch as the eigenfunction ψ_n may be a complex magnitude, we must write, in place of the square ψ_n^2, the expression

$$(3) \qquad \psi_n \psi_n^*,$$

where ψ_n^* is the conjugate complex of ψ_n. Thus, (3) is proportional to the intensity of the wave.

The Charge Cloud—Schrödinger now makes the assumption that the whole region of space occupied by the de Broglie wave is filled with a continuously distributed charge of negative electrification, the density of which at any point is proportional to the intensity of the wave at this point. Since we are supposing that the atom is in a given energy state E_n, the amplitude of the wave at each point does not vary with time. Hence the intensity of the wave disturbance is also permanent, and so then is the distribution and the density of the electrified cloud. The density of the cloud, being proportional to the intensity of the wave disturbance must necessarily vanish over the nodal surfaces where the intensity of the wave is zero. In particular, the cloud vanishes at spatial infinity.

According to Schrödinger, this electrified cloud, or *charge cloud*, will take the place of the corpuscular electron of Bohr's theory. The total charge of the cloud in the hydrogen atom must therefore be $-e$, the charge commonly attributed to a corpuscular electron. Schrödinger next supposes that the density ρ of the charge cloud at any point P (when the atom is in a stable energy state E_n) is equal to $-e$ times the intensity $\psi_n \psi_n^*$ of the de Broglie vibration at this point. Thus we have

$$(4) \qquad \rho = -e\psi_n \psi_n^*.$$

The total electric charge of the cloud is obtained when we integrate the electric density ρ over the whole region of space where the cloud is present (*i.e.*, over infinite space). Since the total charge of the cloud must have the value $-e$, we must assume that we have

$$(5) \quad -e\iiint \psi_n \psi_n^* dx dy dz = -e, \text{ and hence } \iiint \psi_n \psi_n^* dx dy dz = 1.$$

Our eigenfunction ψ_n, being any one of the eigenfunctions connected with the value E_n of the energy, does not necessarily satisfy the relation (5). But we may always multiply ψ_n by an appropriate constant C, so that by taking $C\psi_n$ as eigenfunction in place of ψ_n, the relation (5) will be satisfied. In what follows we shall assume that an eigenfunction is always so modified as to fulfill this condition. In mathematical language the eigenfunction is said to be *normalized to* 1.†

In the foregoing exposition we have been concerned with the amplitude functions ψ, but we may, if we prefer, utilize the wave functions u. In order to simplify matters, we shall assume, as before, that the de Broglie wave is a standing wave. Now, quite generally, a harmonic motion of frequency ν and of amplitude A can be represented by the complex magnitude

$$(7) \qquad A e^{2\pi i\nu t}.$$

Here A is a real quantity. When we are dealing with a region of space undergoing vibrations, the amplitude A may be a function of position, and hence of x, y, z. The de Broglie vibrations which arise when the atom is in the energy state E_n have the frequency ν defined by (1); and, if we utilize (7), we see that they are expressed by

$$(8) \qquad u_n = \psi_n e^{\frac{2\pi i}{h}(E_n + m_0 c^2)t},$$

where ψ_n is the amplitude function (assumed normalized). The vibrations are of course real, whereas (8) is complex; but, in writing (8), we are supposed to consider only the real part of this expression. We see immediately from (8) that

$$(9) \qquad u_n u_n^* = \psi_n \psi_n^*,$$

so that $u_n u_n^*$ can be taken to define the intensity of the de Broglie vibrations. Also, from (4) we see that the density ρ of the charge cloud at any point P is

$$(10) \qquad \rho = -e\psi_n \psi_n^* = -e u_n u_n^*.$$

Obviously $u_n u_n^*$, like $\psi_n \psi_n^*$, when integrated over all space, yields the value 1. The vibration function u_n, defined by (8) (in which ψ_n is normalized), is itself said to be normalized.

---

† If ψ_n is a solution of Schrödinger's equation, so also is $C\psi_n$, where C is any constant, real or complex. Suppose, then, that when the intensity $\psi_n \psi_n^*$ is integrated over all space, the value obtained is K, i.e.,

$$(6) \qquad \iiint \psi_n \psi_n^* dx dy dz = K.$$

If we take as new eigenfunction $\dfrac{\psi_n}{\sqrt{K}}$ in place of ψ_n, the value of our integral with the new eigenfunction will be 1. It is this new eigenfunction which will be taken in Schrödinger's theory. The factor $\dfrac{1}{\sqrt{K}}$, thanks to which the new eigenfunction has been obtained, is called the *normalizing factor*.

Our formulæ show that, although a vibratory condition exists throughout space owing to the de Broglie vibrations, no motion is apparent in the charge cloud. The fact is that the density of the cloud, being given by (10), does not involve time; and so the cloud, though differing in electric density from place to place, remains permanent. Now, according to classical electromagnetism, if the charge cloud were to be in a state of vibration, electromagnetic radiation would occur, whereas no radiation could be expected if the charge cloud were to remain fixed. From this we see that when the atom is in a given energy state E_n, so that the charge cloud remains permanent, no electromagnetic waves can be emitted.

Radiation—Schrödinger's theory, so far, seems unable to account for the phenomenon of radiation. To overcome this defect, Schrödinger assumed that the atom might vibrate at one and the same time in two or more different modes which differed in frequency. In other words a superposition of modes might occur. Elementary theory shows that in this case the charge cloud would be set into vibration, with the result that radiation would be emitted. But before examining Schrödinger's assumption further, we must mention an obvious objection to it.

In the ordinary theory of elastic vibrations, an elastic solid may very well be vibrating with different frequencies simultaneously. But in the wave theory of the hydrogen atom, where each mode of vibration is associated with a definite energy level, Schrödinger's superposition of modes implies that the atom must be situated simultaneously in different energy states. This conclusion is in conflict with Bohr's theory and with experimental results, according to which an atom is in one energy state at a time. Such objections, however, did not disturb Schrödinger, presumably because he felt that the new wave theory might open vistas undreamed of in the older particle theory. Indeed, as subsequent events have proved, the superposition of states postulated by Schrödinger is valid, though the modern interpretation placed on this phenomenon is that of Born, Heisenberg and Bohr; and it differs entirely from Schrödinger's.

Let us accept the hypothesis of a superposition of states and see how it will account for the Bohr frequencies radiated from the atom. We consider two modes of vibration associated with the distinct energy levels E_n and $E_{n'}$, viz.,

$$(11) \qquad \psi_n e^{\frac{2\pi i}{h}(E_n + m_0 c^2)t} \quad \text{and} \quad \psi_{n'} e^{\frac{2\pi i}{h}(E_{n'} + m_0 c^2)t}$$

In these expressions the amplitude functions ψ_n and $\psi_{n'}$ are, as usual, normalized eigenfunctions. If one of the two modes of vibration occurred alone, the vibratory motion would be given by one of the two functions (11). But, when both modes occur simultaneously, we must superpose the two vibratory motions. The resultant motion, however, is not obtained by merely adding the two expressions (11); instead, for reasons which will appear presently, we must first multiply these expressions by appropriate constants C_n and $C_{n'}$. These constants, which may be real or complex, are in a large degree arbitrary; but they are not entirely arbitrary, for they must satisfy certain restrictions (see (15)). The analytical expression of the resultant vibration is thus

$$(12) \qquad u = C_n \psi_n e^{\frac{2\pi i}{h}(E_n + m_0 c^2)t} + C_{n'}\psi_{n'}e^{\frac{2\pi i}{h}(E_{n'} + m_0 c^2)t}.$$

The superposition (12) indicates that the amplitudes of the two constituent waves are now $C_n\psi_n$ and $C_{n'}\psi_{n'}$, in place of ψ_n and $\psi_{n'}$. Since the intensities of the constituent waves at a point P are given by the squared amplitudes at this point, we obtain for the expression of the intensities

$$(13) \qquad C_n C_n^* \psi_n \psi_n^* \text{ and } C_{n'} C_{n'}^* \psi_{n'} \psi_{n'}^*.$$

As for the intensity of the resultant wave at a point P, it is determined by the value of uu^* at this point. The resultant intensity uu^*, in contradistinction to the intensities (13) of the constituent waves, varies with time. The reason for this variation may readily be understood when we note that, at the point P, the two superposed de Broglie vibrations have different frequencies, so that the vibrations will sometimes be in phase and thus reinforce each other, whereas at other instants they will be opposite in phase and will tend to cancel. The intensity of the resultant de Broglie wave at a point P will thus rise and fall periodically. It can be shown that the frequency of this vibration is defined by the difference in the frequencies of the two constituent vibrations.

A phenomenon similar to the one just described occurs in acoustics; it is known under the name of *beats*. If two piano chords, vibrating simultaneously, emit very nearly, but not exactly, the same musical note, we notice a periodic rise and fall in the intensity of the sound. The consecutive maxima of the intensity are referred to as "beats." The frequency of these beats is equal to the difference in the frequencies of vibration of the two chords. Obviously, the rise and fall in the intensity of the resultant sound is a phenomenon of the same nature as the rise and fall in the intensity of our resultant de Broglie wave.

Suppose, then, that we are dealing with the superposition (12). Schrödinger applies his general relation (10), which connects the density ρ of the charge cloud at a point P with the intensity of the wave at this point. In the present case this intensity is uu^*, where u is defined by (12). Thus

$$(14) \qquad \rho = -euu^*.$$

As in the simpler case, the charge cloud is supposed to represent the electron, and hence the total electric charge of the cloud must be equal to the charge $-e$ commonly attributed to the electron. We conclude that when the intensity is integrated over the whole of space, the value of the integral must be 1; *i.e.*, the resultant function u of (12) must be normalized. The normalization is brought about by crediting appropriate values to the two heretofore arbitrary constants C_n and $C_{n'}$.†

We shall leave in abeyance the significance of the constants C_n and $C_{n'}$. What we wish now to examine is the behavior of the charge cloud. When we were dealing with a single mode of vibration, we noted that the density ρ of the cloud at any point P was given by (10) and that it did not vary with time. The charge cloud thus remained motionless. But when we are dealing with superposed modes, the density ρ at a given point is defined by the expression (14), in which uu^* varies with time. Consequently, the density ρ also varies with time and the charge cloud no longer remains permanent. As was explained in connection with the phenomenon of beats in acoustics, the intensity vibrates with a frequency defined by the difference in the frequencies of the two constituent de Broglie waves. Since the frequencies of the two constituent monochromatic de Broglie vibrations are

$$(17) \qquad \frac{E_n + m_0 c^2}{h} \quad \text{and} \quad \frac{E_{n'} + m_0 c^2}{h},$$

their difference is

$$(18) \qquad \frac{E_n + m_0 c^2 - E_{n'} - m_0 c^2}{h}, \quad i.e., \quad \frac{E_n - E_{n'}}{h}.$$

---

† As will be explained in the next chapter, the normalizing condition imposed on u requires that c_n and $c_{n'}$ should satisfy the relation

$$(15) \qquad c_n c_n^* + c_{n'} c_{n'}^* = 1.$$

The condition (15) shows that we cannot assume $c_n = c_{n'} = 1$, so that constants differing from 1 must be introduced in the expression (12). We may, however, assume that c_n and $c_{n'}$ are equal by setting

$$(16) \qquad c_n = c_{n'} = \frac{1}{\sqrt{2}}.$$

At any given point, the intensity uu^* and the electric density ρ will therefore vibrate with the frequency (18).

Now, the frequency (18) is precisely the one furnished by Bohr's frequency condition for the electromagnetic wave when the atom drops from the energy level E_n to the level $E_{n'}$. We conclude that when de Broglie waves are vibrating in two different modes simultaneously, corresponding to the two energy levels E_n and $E_{n'}$, the density ρ of the charge cloud, at any point, will vibrate with the Bohr frequency (18).

Schrödinger's interpretation of radiation may now be understood. He views the charge cloud and its vibrations or wave motions from place to place as physically real, at least as a working hypothesis. He then assumes that the classical electromagnetic laws apply to the charge cloud. It follows that the vibrating electric cloud should emit electromagnetic waves having exactly the same frequency as the vibrations of the cloud.

Summarizing these results, we see that, when the de Broglie waves are vibrating in two different modes connected with the energy values E_n and $E_{n'}$, the atom will emit electromagnetic radiations having the Bohr frequency (18). This frequency is the one which, according to Bohr's corpuscular theory, is emitted when the electron drops from the energy level E_n to the level $E_{n'}$. Thus, in Schrödinger's theory, the coexistence of two modes of vibration pertaining to two different energy levels is the wave picture of a drop of the electron from one energy level to another. On the other hand, when the de Broglie waves are vibrating in a single mode, the charge cloud remains motionless and no radiation arises. We have here the wave analogue of the non-radiating atom of the corpuscular theory, in which the electron is describing some stable orbit and is not dropping to a lower level.

The possible frequencies of the radiated light are thus accounted for in Schrödinger's theory, and so also is the fact that the atom does not radiate when it remains in a given energy level. The new theory is superior to Bohr's, because Bohr's frequency condition is no longer imposed as a separate hypothesis, but instead appears to be a necessary consequence of the original postulates. Schrödinger's theory has also other advantages; these concern the selection rules and the calculation of the intensities and of the polarizations of the radiated light.

We know that the various spectral lines are not equally intense and that some of the lines, theoretically possible according to Bohr's theory, have zero intensities and so do not arise at all. These are the lines, or radiations, that would be emitted during the forbidden energy drops if such drops were possible. Furthermore, the various radiations emitted by an atom may have characteristic polarizations, and here also Bohr's theory can yield no information. In Bohr's theory these defects were remedied by an appeal to the correspondence principle, according to which the intensities and polarizations of the radiations had to be calculated classically. The results thus obtained were then utilized to supplement the deficiencies of the quantum treatment. We pointed out at the time that this procedure did not yield rigorous results in the matter of the intensities, and that, besides, Bohr's appeal to the classical correspondence principle was unsatisfactory, for it showed that his quantum theory was not self-sufficient.

One of the triumphs of Schrödinger's wave theory is that it permits the calculation of the intensities and polarizations of the emitted radiations without introducing new postulates. To understand how this information can be obtained, we must revert to the vibrating charge cloud.

Classical electromagnetism shows that, at a sufficiently distant point, the waves emitted by a vibrating electrified cloud are the same in frequency, intensity, and polarization as the waves which would be emitted by a hypothetical charged particle fixed to the mobile centre of charge of the cloud, the particle having a charge equal to the total electric charge of the cloud. In the case of Schrödinger's charge cloud, which we are here considering, the total charge is equal to the charge $-e$ of an electron, and the centre of charge of the cloud is the analogue of the centre of mass of an extended massive body. The net result is that the radiation to be expected is the one which would be emitted by an electron attached to the vibrating centre of charge of the Schrödinger charge cloud. The rule we have just given for obtaining the radiation due to the vibrating charge cloud is valid only when the wave length of the light radiated is considerably greater than are the linear dimensions of the region of space in which the density of the electric charge is appreciable. In all the examples we shall discuss, this situation will be realized.

The Selection Rules—We have said that the vibrating charge cloud radiates as would a corpuscular electron attached to the centre of charge of the cloud. It follows that whenever for some reason or other the centre of charge remains motionless, no radiation can occur. The centre of charge of course remains at rest in the trivial case where the atom is in a given energy state E_n, for we know that in this case the entire cloud is motionless. We have here the wave expression of the fact that an atom which remains in a stable energy state does not radiate. But even when modes of vibration corresponding to two different energy levels E_n and $E_{n'}$ are superposed, so that the cloud is vibrating with the frequency (18), the centre of charge may still remain motionless, and hence no radiation occur.* The situation here contemplated is the wave analogue of the particle picture in which the electron cannot drop from the higher orbit E_n to the lower one $E_{n'}$ and emit radiation. We are thus dealing with a forbidden drop. Now Schrödinger's theory enables us to determine under what conditions the centre of charge will or will not vibrate. Consequently, we see that it will allow us to anticipate the forbidden drops without having recourse to the correspondence principle.

Let us apply these considerations to the hydrogen atom. We recall that in Bohr's theory, when the relativistic refinements and the spinning electron were disregarded, the levels n of the atom were subdivided into n^2 coalescing sub-sublevels (n, l, m). Exactly the same results are obtained in Schrödinger's theory, as was explained in Chapter XXXII. In Bohr's theory we also saw that during an energy drop the quantum number n might decrease from any one of its values to any other, and that the quantum number l always changed by ± 1. Finally, as was made clear when we studied the Zeeman effect, the quantum number m either changed by ± 1 or else remained fixed. Inasmuch as these selection rules, derived from the correspondence principle, appear to be in agreement with the facts observed, we must demand of Schrödinger's theory that it yield the same results.

Let us, then, assume two modes of vibration which correspond to the two Bohr sub-sublevels (n, l, m) and (n', l', m'), i.e., the modes defined by $\psi_{n,l,m}$ and by $\psi_{n',l',m'}$. We find that when these two modes occur simultaneously, the centre of charge remains motionless, unless m and m' differ by ± 1, and unless l and l' differ by ± 1 or else do not differ at all. We thus obtain exactly the same results as were obtained from the correspondence principle, so that Schrödinger's theory furnishes the correct selection rules, or what comes to the same, it tells us what drops are forbidden.

Similar successes are registered with the oscillator and the rotator. In the case of the linear oscillator, the charge cloud is no longer spread through infinite space; it is distributed along the line of oscillation and vanishes at infinity in either direction. In spite of this difference, the general procedure is the same, and, to account for radiation, we must assume that the de Broglie waves are vibrating simultaneously in two different modes of energies E_n and $E_{n'}$. We then find that the centre of charge of the charge cloud does not move unless the quantum numbers

---

* It would be as though the radiations from the different parts of the vibrating cloud cancelled by interference.

n and n' differ by ± 1. This implies that, in the oscillator, only drops between consecutive energy levels are possible. We thus obtain the same results that were derived from an application of the correspondence principle. Since the wave treatment also gives the correct energy levels, we see that the required selection rules and the correct energy levels for the oscillator are both obtained in Schrödinger's theory. The application of the charge-cloud method to the rotator also gives correct results, for we find that, as is the case with the oscillator, only transitions between contiguous energy levels are possible.

The Polarizations—Next we examine what is to be expected when the centre of charge moves. It will always be found vibrating with the frequency of vibration of the charge cloud, so that light will be radiated with the required Bohr frequency. But the vibration may be of different kinds: rectilinear or circular. In the former case the light radiated will be plane polarized, and in the second case circularly polarized.

If we apply these considerations to the linear oscillator and to the rotator, we find that, for the oscillator, the radiation will be plane polarized, and that, for the rotator, the polarization will be circular. These are precisely the results obtained in Bohr's theory when the correspondence principle is applied. We may also consider the polarizations for the hydrogen atom. The charge-cloud theory shows that when the quantum number m does not change during a drop, the centre of charge vibrates along the Oz axis, so that the radiation is plane polarized. On the other hand, when m changes by ± 1, the light is circularly polarized in one sense or the other, with the plane of the vibration perpendicular to the Oz axis. But we must remember that when we examine the radiations emitted by hydrogen gas, we are dealing with an emission from a large number of atoms; and since the Oz direction is arbitrary, the radiations will not manifest definite polarizations. We may, however, make the polarization conspicuous by imposing a common direction for the axes of the various atoms. This may be done by applying a magnetic field, as in the Zeeman effect. In this event, the polarizations given by Schrödinger's theory are in agreement with those derived from the correspondence principle and are verified by experiment in the normal Zeeman effect.*

The Intensities—Finally, we must examine what information Schrödinger's theory furnishes for the intensities of the electromagnetic

---

* Throughout this discussion, we are treating the hydrogen atom without introducing the spinning electron (or indeed without any of the relativistic refinements). For this reason, the polarizations that hold for the normal Zeeman effect are the ones to be expected.

radiations emitted. The intensity of the radiation is defined by the amount of energy that is emitted from the radiating system per unit time. Now we have seen that the radiation from the atom is the same as would be generated by an electron of charge $-e$ vibrating with the centre of charge of the cloud. Hence we have but to apply the classical formula which gives the intensity of the radiation emitted by a vibrating electron. It can be shown that if A is the amplitude of the vibration of the centre of charge, and ν the Bohr frequency with which the centre of charge is vibrating, the intensity radiated is

$$(19) \qquad e^2 \frac{(2\pi\nu)^4}{3c^3} A^2.$$

Schrödinger's theory thus allows us to calculate the intensities of the spectral lines without utilizing the correspondence principle.

The Electron in Schrödinger's Theory—In Schrödinger's theory of radiation, the particle picture of well-defined Bohr orbits described by a corpuscular electron is lost sight of entirely. Nevertheless, it is possible to obtain wave representations which suggest this picture. We first consider the wave representation of the Bohr orbits.

When the hydrogen atom is in a stable state E_n, the electron in the particle picture is describing some orbit (n, l, m). In the corresponding wave representation, a de Broglie wave is vibrating throughout infinite space with the amplitude $\psi_{n,l,m}$ from place to place, the numerical values to be assigned to the quantum numbers n, l, and m in the eigenfunction $\psi_{n,l,m}$ being the same as in the definition of the orbit. Thus, to each Bohr orbit, corresponds a definite mode of vibration of the standing wave. In the wave theory we may also consider a wave resulting from the superposition of two or more of the n^2 amplitude functions $\psi_{n,l,m}$ associated with the same energy level E_n. The amplitude function thus obtained is still an eigenfunction of Schrödinger's wave equation and is connected with the eigenvalue E_n; hence it determines a possible wave phenomenon when the atom is in the stable state E_n. In Bohr's particle theory, this condition of the atom would be represented by associating the electron simultaneously with different orbits of the same energy E_n. An association of this sort would of course be incomprehensible. We must remember, however, that the undisturbed atom is totally degenerate, so that considerable latitude is afforded in the choice of the exact arrangement of the orbits; and if we modify the orbits in a suitable way (by utilizing the principle of adiabatic invariance) it may be possible in certain cases to obtain a simple particle representation of the foregoing superposed modes.

Our present aim is to obtain the wave analogues of the various particle pictures of Bohr's theory. We shall therefore set aside the wave pictures of the more complicated type just mentioned and confine our attention to the normal modes $\psi_{n,l,m}$, which are the analogues of the Bohr orbits (n, l, m). In particular, let us examine the wave pictures of the circular orbits. According to Bohr's theory, one, and only one, circular orbit is connected with every main energy level (at least if we do not establish a distinction between the various possible orientations of the plane of the orbit). The circular orbits are those for which $l = n - 1$, while m may have any one of its permissible values. The modes of vibration corresponding to the circular orbits are determined by the same allotments of quantum numbers; and so we may single out the modes which correspond to the circular orbits.

Bohr's lowest orbit $(n = 1; \ l = 0; \ m = 0)$ is a circle, and the corresponding mode of vibration $\psi_{1,0,0}$ is characterized by the absence of nodal surfaces other than the sphere at infinity (which is always a nodal surface). The amplitude of the standing de Broglie wave does not vanish at finite distance; it is greatest at the centre, where the proton nucleus is situated, and it decreases when we move away from the nucleus. Since the intensity of the vibration, and therefore the density of the charge cloud, is proportional to the square of the amplitude, the electric density of the charge cloud in the lowest stationary state will also be greatest at the nucleus and will decrease progressively in all directions, finally vanishing at infinity. Now it can be shown that the lowest circular orbit of Bohr's theory lies in a region where the intensity of the vibrations, and hence the density of the charge cloud, has a mean value contained between the maximum value at the nucleus and the zero value at infinity. This region may then serve to indicate the position of the orbit. On the other hand, even a vague connection between the wave picture and the particle picture ceases to become apparent when we seek to link the wave representation to that of a corpuscular electron circling on the orbit. The wave picture in this case affords no means of representing the particle.

We now pass to the higher circular orbits, say the nth $(n > 1)$. The mode of vibration is here characterized by the vanishing of the amplitude at the nucleus and at infinity; the amplitude assumes a maximum value over the surface of a certain sphere whose centre is the nucleus. Consequently, the density of the charge cloud is greatest over this sphere. Calculation shows that the radius of the sphere is approximately the same as that of the nth circular Bohr orbit. In other words, Bohr's circular orbit lies very nearly along an equator, or great circle, of the sphere. Except for the fact that Bohr's orbit is assumed to be a definite circle, whereas the region of greatest density for the charge cloud is a spherical surface, we seem here to have a wave interpretation of a Bohr orbit. Let us note that the conflict between a sphere and a circular orbit is not as important as it might appear, for the orbit may have any orientation and may therefore be any one of the great circles of the sphere.

We next inquire whether we can obtain a wave picture of the planetary electron moving along an orbit. The only way to secure a wave picture which simulates a corpuscle is to construct a wave packet of tiny dimensions. This requires that we superpose a large number of de Broglie waves differing only slightly in their wave lengths and frequencies. In free space, we can always construct a wave packet because no restriction is placed on the de Broglie waves which we may utilize. But in the hydrogen atom, the only waves available are those which are associated with the various energy levels, *i.e.*, the waves which constitute the various modes of vibration; and the differences in the frequencies of the consecutive modes may be too great to permit the construction of packets. This is precisely the situation which occurs in the lower orbits, and for this reason we cannot construct a wave packet simulating a corpuscular electron on a lower orbit; the nearest approach to a corpuscular electron that the wave picture can furnish in this case is a more or less extended wave disturbance moving around the nucleus.

However, as we pass to the higher orbits we find that the frequencies of the waves associated with consecutive energy levels differ by progressively smaller amounts, so that wave packets can be constructed. In particular, if we superpose in a suitable way the vibrations associated with a high energy state E_n and with other energy states slightly higher and slightly lower than E_n, we can obtain a wave packet which will follow very approximately the nth Bohr orbit with the motion prescribed to the planetary electron in Bohr's theory. Furthermore, since the resultant intensity of the wave motion is non-vanishing only within the packet, the charge cloud will be concentrated inside the packet, and so the packet will carry the charge $-e$ commonly attributed to a corpuscular electron. Thus a satisfying wave picture of the circling corpuscular electron would appear to be obtainable for the higher orbits.

It must be observed, however, that the construction of a wave packet involves a superposition of modes belonging to different energy levels, so that the atom must be assumed to be in different energy states simultaneously. This unsatisfying conception of a superposition of states is not new, for we encountered it when we examined Schrödinger's interpretation of radiation. Nevertheless, it seems strange that the possibility of constructing a wave picture of a corpuscular electron should be denied us when the atom is in one definite energy state. Besides, we have seen that, in any case, the wave representation of the corpuscular electron becomes vaguer and vaguer as we pass from the higher to the lower orbits.

At the time Schrödinger was developing his theory, he did not view this conflict between the wave picture and the particle picture as an argument against the wave theory. Instead, he claimed that it merely proved the particle picture to be untenable. According to Schrödinger the waves were the only physical reality, and the electric charge carried by the waves was a wave manifestation; the electron, whether it betrayed the corpuscular aspect or not, was essentially a wave phenomenon. In the higher energy levels and to a still greater extent outside the atom, the wave theory showed that the charge cloud might condense into a packet and simulate a corpuscular electron. But in the lower levels near the nucleus, the corpuscular aspect was impossible. In short, there was no such thing as a corpuscular electron, but only a more or less diffuse charge cloud accompanying the de Broglie waves.

Schrödinger's ideas are not necessarily incompatible with the experimental results which seem to reveal the electron as a corpuscle, for in physical experiments we are always dealing with electrons situated well outside an atom; and in such cases Schrödinger's theory shows that the charge cloud will be condensed into a tiny volume and hence will have the appearance of a corpuscle. As for Schrödinger's conception of a diffused electron in the immediate neighborhood of the nucleus, it is not in contradiction with experiment since the electron in the innermost regions of the atom is inaccessible in practice.

Schrödinger's views appear plausible when we recall that, according to wave mechanics, the classical conceptions tend to become valid only when the conditions of ray-optics are fulfilled. Now in the atom, the conditions of ray-optics are far from being realized in the lower orbits, but they tend to be realized when we pass to the outer orbits. We may therefore reasonably suspect that if revolutionary revisions are required in our classical concepts, these revisions will be needed more particularly in the lower orbits. Bearing this in mind, we need not be surprised to find that the corpuscular aspect of electrons may lose its validity in the lower orbits.

When we examine the problem from the standpoint of the atom's radiation, Schrödinger's views also appear consistent. Thus, in our discussion of Bohr's atom, we saw that the frequency of the radiation emitted, though always given by Bohr's frequency condition, tended nevertheless to approximate to the mechanical frequency of the electron on its orbit when the higher orbits were considered. The higher the orbit, the better was the approximation. This observation constituted one of the foundations of the correspondence principle. Now in the wave picture, the possibility of representing the electron by means of a charged wave packet is enhanced as we pass to the higher orbits, and in such cases the charge cloud becomes a circling packet of electric charge. The centre of charge of the cloud coincides with the centre of the wave packet, and hence, according to Schrödinger's theory, the radiation emitted should have the frequency of the packet describing the orbit. Since the packet moves as does the corpuscular electron in Bohr's theory, the frequency radiated is the mechanical frequency of the electron, so that we obtain results which are in agreement with Bohr's particle picture. On the other hand, when in Bohr's theory the electron is describing a lower orbit and the frequency radiated differs considerably from that of the electron's motion, the wave theory leads to no inconsistency, because it is now impossible to construct a wave packet, and the centre of charge of the cloud no longer moves as Bohr supposed the electron to move. Accordingly, the wave theory clarifies the reason why in this latter case the mechanical frequency of Bohr's electron is no longer the same as the frequency of the radiation emitted.

The ubiquitous influence of Planck's constant h is obvious in Schrödinger's theory. If h were to have a considerably smaller value, the eigenvalues of Schrödinger's wave equation, and hence the energy levels, would differ by smaller amounts. We could then construct a wave packet simulating a corpuscular electron even in the lowest orbit: the electron would always have the appearance of a corpuscle, and the frequency radiated would always be the classical frequency of the electron's motion. Thus the classical concepts would be correct.

The Objections to Schrödinger's Theory—Schrödinger's theory of radiation, based on the conception of the continuous charge cloud, is a great advance over Bohr's treatment of the atom. Not only does it furnish Bohr's frequency condition, instead of compelling us to introduce this condition as a separate postulate, but in addition it yields the polarizations and the intensities of the emitted radiations. Inasmuch as radiations of zero intensity are those that cannot occur, Schrödinger's theory automatically determines the forbidden energy-transitions. All this information is obtained directly from the postulates of the theory, and we need not appeal to the correspondence principle.

But there are numerous objections to Schrödinger's charge cloud. One of these was mentioned in Chapter XXX: it concerns the spreading of wave packets. A wave packet, and hence the charge it contains, eventually spreads all over space and does not betray that degree of permanence which is essential if we are to mistake a charged wave packet for a corpuscular electron. Schrödinger did not at first realize this tendency of wave packets to spread, because in the particular example which he submitted to mathematical analysis (the oscillator), the packet happens to remain coherent. Heisenberg proved later that the example of the oscillator was an exception, and that the reason its wave packets remained coherent was that the successive energy levels of the oscillator happened to differ by equal amounts. Heisenberg then showed that the majority of wave packets, especially those in the atom, must spread very rapidly. This discovery showed that Schrödinger's wave-packet representation of a corpuscular electron in a higher orbit was untenable, even over a relatively short period of time.

Other objections to Schrödinger's theory may be mentioned. For instance, in his wave equation there enters a perfectly definite expression for the potential energy of the electron at a point x, y, z. So long as we assume that an electron is an electrified particle occupying at each instant a definite position in space, no objection can be raised against our specifying a well-defined potential energy. But when the electron is replaced by a diffuse charge cloud, we cannot readily understand how a precise value of the electron's potential energy can be defined at each point of space. Besides, if the charge of the electron is diffused throughout space, why not also its mass, its energy, and the other magnitudes connected with it?

A further difficulty concerns the degree of physical reality that we should attribute to the de Broglie waves and hence to the charge cloud. Schrödinger, in his radiation theory for the hydrogen atom, treats the charge cloud as though it were physically real. Nevertheless, when he deals with a system of N degrees of freedom, he assumes that the waves and the charge cloud must be represented in a configuration space of N dimensions. So long as N does not exceed 3, the representation can be made in ordinary space, but for all greater values of N, a hyperspace must be considered. Strictly speaking, this situation occurs even for the hydrogen atom, because the hydrogen atom has three degrees of freedom only when we agree to disregard the motion of the nucleus; in reality, there are six degrees of freedom, and the waves and the cloud should be represented in a 6-dimensional space.* In any case, when we pass to the higher atoms with their several electrons, the introduction of the hyperspace cannot be avoided.

Now a hyperspace is obviously a mathematical fiction; and waves that can be represented only in a fictitious space must themselves be unreal.

---

* See page 706.

Under these conditions, we must suppose that even when the problem we are considering has three degrees of freedom (so that the waves and the charge cloud may be represented in ordinary 3-dimensional space), we should still view the waves and the cloud as fundamentally symbolic. That the de Broglie waves are symbolic is not a new discovery, since it has already been impressed on us on other occasions. But the symbolic nature of the charge cloud is more disturbing, for Schrödinger, in his interpretation of the process of radiation from the hydrogen atom, treats the cloud as though it were physically real. However, the possibility of an unreal charge cloud generating the effects to be expected from a real one cannot be said to furnish an argument against Schrödinger's theory. Rather does it afford one more illustration of the conflicting notions to which the wave theory leads.

Schrödinger's theory is in conflict with experiment on another score. According to his theory, we must suppose that an atom which is emitting radiation is at one and the same time in two different energy states; a monochromatic radiation is then emitted. But if we grant the possibility of this superposition of states, there is no reason why the superposed states should be limited to two. Indeed, when we discussed the formation of a wave packet simulating an electron in the atom, we had to assume that the superposition involved several different states. Now if more than two states are superposed, Schrödinger's theory would require that a corresponding number of different radiations be emitted simultaneously by the atom. We are here in conflict with Bohr's theory, according to which only one radiation can be emitted at a time; and on this point Bohr's theory appears to be corroborated by experiment.

Difficulties of this sort led to the conclusion that something was radically wrong, not only in Schrödinger's interpretation of his theory of radiation, but also in the theory itself. In a certain sense this conclusion is not surprising, for the idea of a charge cloud emitting radiation according to the laws of classical electromagnetism does not seem convincing in view of the revolutionary changes which the quantum theory has introduced in the classical ideas. Guided by past experience, we know that the classical conceptions tend to yield correct results only under limiting conditions, and it is therefore reasonable to suppose that Schrödinger's semi-classical theory is in the nature of an approximation. This view has since been vindicated by the more refined theores of radiation developed by Dirac and others. Nevertheless, in spite of its defects Schrödinger's theory of radiation is of considerable value as a first approximation, and the idea of the charge cloud is often retained in a formal capacity as a working hypothesis.

CHAPTER XXXIV

BORN'S INTERPRETATION OF SCHRÖDINGER'S THEORY

The Atom in a Stable State—Let us first suppose that the hydrogen atom is in the energy level E_n and that no radiation occurs. A standing de Broglie wave is then vibrating throughout space with a frequency

$$\frac{E_n + m_0 c^2}{h},$$

and the amplitude of the vibration from point to point is defined by one of the nth eigenfunctions ψ_n (normalized). The vibration itself may be represented by the wave magnitude

$$(1) \qquad u_n = \psi_n e^{\frac{2\pi i}{h}(E_n + m_0 c^2)t}.$$

According to Schrödinger, the density ρ of the charge cloud at any point P is defined by $-e$ times the intensity $u_n u_n^*$, or $\psi_n \psi_n^*$, of the wave at this point. Thus,

$$(2) \qquad \rho = -eu_n u_n^* = -e\psi_n \psi_n^* \qquad \text{(at } P\text{).}$$

If $dxdydz$ represents a tiny volume enclosing the point P, then $\rho dxdydz$ is the electric charge contained within this volume. In terms of the wave magnitudes, we thus obtain

$$(3) \qquad \rho dxdydz = -eu_n u_n^* dxdydz = -e\psi_n \psi_n^* dxdydz.$$

In these formulae (2) and (3), the wave functions ψ_n and u_n are assumed normalized to 1; and we recall that Schrödinger introduced this normalization in order that the total electric charge of the entire cloud be $-e$.

Next, we examine how Schrödinger's interpretation must be revised when Born's assumptions are accepted. According to Born, the probability that the electron will be found in a tiny volume $dxdydz$ surrounding a point P is proportional to the product of the volume and the intensity $u_n u_n^*$, or $\psi_n \psi_n^*$, of the wave at the point P. Thus

$$(4) \qquad u_n u_n^* dxdydz, \quad \text{or} \quad \psi_n \psi_n^* dxdydz, \qquad \text{(at } P\text{)}$$

is proportional to this probability.

Let us determine under what conditions the expression (4) will define the exact probability of the electron being located in a given volume $dxdydz$, instead of merely defining a quantity proportional thereto. To obtain this information, we note that the electron is necessarily situated somewhere in space, and hence that the probability of our finding it in some volume $dxdydz$ is a certainty. This implies that when the probability is integrated throughout space, the integral must have the value 1. Consequently, for the expression (4) to measure the probability itself, the amplitude function ψ_n must be chosen in such a way that we have

(5) $$\iiint \psi_n \psi_n^* \, dxdydz = 1.$$

We recognize in this restrictive condition (5) the normalization condition imposed by Schrödinger for other reasons. We are thus led to the following conclusion : When Born's ideas are accepted and the normalized eigenfunctions are utilized, the probability that the electron will be found in a given volume $dxdydz$ about a point P is equal to the product of the volume and the intensity of the wave at the point P. Thus (4) expresses the probability provided u_n and ψ_n be normalized.

Schrödinger's expression (2) for the electric density ρ of the charge cloud at a point P may likewise be interpreted on the basis of Born's ideas. The interpretation differs according to whether or not we accept the views of Heisenberg and Bohr. We first assume the correctness of these views. We must then suppose that, before a position observation is made, the electron is diffused throughout the region occupied by the waves. This conception of the electron appears similar to Schrödinger's, and indeed the two interpretations have much in common if we suppose that, in Heisenberg's picture, the diffused electron is distributed in the same way as in Schrödinger's charge cloud. In this event, Schrödinger's expression (2) of the electric density ρ of the charge cloud also gives the density of charge of Heisenberg's diffused electron. However, the views of Schrödinger and of Heisenberg are far from being the same in all respects. Thus in Schrödinger's conception, the charge cloud is a physical reality which endures. But according to Heisenberg and Bohr, the diffuse picture of the electron is a temporary manifestation, which arises from the impossibility of picturing events with precision in the space-time frame, prior to the position observation. After the observation, the electron becomes a corpuscle, and the charge cloud ceases to exist.

Next let us suppose that the Heisenberg-Bohr conception of a diffused electron is rejected and that we view an electron as corpuscular under all conditions. An interpretation of Schrödinger's density ρ consistent with Born's ideas may still be given, provided we ascribe a statistical significance to Schrödinger's theory. Thus we assume that the charge cloud and the de Broglie wave describe a condition which characterizes the hydrogen atom, not at a single instant, but throughout a protracted period of time. In other words, the charge cloud represents the average picture which would result from the superposition of a number of snap-shot pictures taken at consecutive instants. Suppose, then, an experiment be repeated time and again with a view of determining whether the electron is in the particular volume $dxdydz$. We assume that before each successive observation is made, the atom is brought back to the same energy level E_n. The observations will show that on certain occasions the electron is situated in the volume of interest, but that in the majority of cases this volume is unoccupied. In the first case, the volume will contain an electric charge $-e$ (i.e., the charge of an electron), whereas in the second case it will have no charge. Taking the average charge within the tiny volume after a large number of experiments have been performed, we shall obtain the value (3), so that ρ defined by (2) is the average, or probable, density of charge at the point P.

When Schrödinger's charge cloud is interpreted in this statistical way, its existence at an instant is deprived of physical reality. It becomes a mere probability cloud which defines the probability that, as a result of continual rapid circlings, the electron will be found at one or at another point of space. Nevertheless, if we consider even a short period of time, we may suppose that the electron has passed through the various points of space, so that over this period of time the charge cloud may be credited with some measure of physical reality.

Another statistical interpretation, also consistent with Born's ideas, may be given to Schrödinger's charge cloud. We suppose that the cloud describes conditions at an instant, not for a single hydrogen atom, but for a large aggregate of non-interacting atoms in the same energy state. If we follow this interpretation, we must reason as though each atom were situated in a space of its own, with the result that, by a definite tiny volume $dxdydz$, we now mean the aggregate of volumes similarly situated with respect to the various atoms. Let us select such an aggregate of volumes $dxdydz$, and let us suppose that we perform position observations simultaneously with a view of locating each electron in each atom. For some of the atoms the corresponding volumes $dxdydz$ will be occupied by the respective electrons at the instant considered, and hence will contain the common charge $-e$. For the majority of atoms, the corresponding volumes will be unoccupied. The average density of charge in all these volumes will then define Schrödinger's density ρ.

We now pass to Born's second assumption, and we shall suppose that we are dealing with a single atom. Born's second assumption requires that, in a wave picture formed by the superposition of different waves, the probability that the electron will be found to have the energy and the momentum corresponding to one or another of the waves should be proportional to the respective intensities of the individual waves integrated throughout space. Since in our present wave picture there is only one wave, Born's assumption requires that if an energy observation is made, the electron will always be found to have the energy corresponding to this wave. Consequently, if the intensity $u_n u_n^*$, or $\psi_n \psi_n^*$, is integrated throughout space, the value of the integral should be 1. Now, this restriction is none other than the normalization condition (5), which has already been imposed. Hence, when the amplitude function ψ_n is normalized, the wave picture of the hydrogen atom, in an energy state E_n, furnishes the exact probabilities of the electron's position, and it also determines the electron's energy.

Let us verify that the wave picture in the hydrogen atom is in agreement with the uncertainty relations. The frequency of the vibrations being well determined, the energy of the associated electron is well known. But then, according to the uncertainty relations, time becomes uncertain, and hence the position of the electron at a given instant must be completely indeterminate. The wave picture confirms this expectation since the wave disturbance fills all space.

The Electron—Having discussed the significance of the waves and of the charge cloud when Born's assumptions are accepted, we now pass to the electron itself. Suppose that the atom is in the lowest energy state and that we wish to observe the position of the electron at any instant. The intensity of the de Broglie wave is greatest in the immediate neighborhood of the nucleus, and it has a non-vanishing value even at distant points; it vanishes only at infinity. Interpreting these results on the basis of Born's assumptions, we conclude that the electron will most probably be found near the nucleus, but that there is a non-vanishing probability of our finding it at any point of space. As we mentioned in Chapter XXX, this conclusion entails a paradox, for according to the principles of mechanics, the electron, having the energy which corresponds to the lowest state of the atom, must necessarily be situated within a certain tiny sphere surrounding the nucleus. The explanation of the paradox was obtained when we took into account the disturbance generated in the electron's momentum and energy by the position observation. We shall, however, examine the matter further.

To simplify the discussion, we shall accept the particle picture as a working hypothesis. According to this picture, the electron is describing the lowest orbit, and hence at any specified instant it will be situated somewhere on this orbit. If, then, our position observation is to yield more precise information, this observation must be refined enough to determine on what particular segment of the orbit the electron is situated at the instant considered. We may attempt to locate the electron by illuminating it, as was done in the experiment of Heisenberg's microscope. A photon is shot against the electron; and if, subsequently to its collision with the electron, the photon enters our eye, the position of the electron will be observed. Optical theory shows that, for this illumination to yield the position of the electron with an accuracy sufficient to locate it on a definite segment of its orbit, the wave length of the radiation associated with the incident photon must be considerably less than the diameter of the orbit. Since Bohr's theory furnishes the diameter of the orbit, we can evaluate the order of magnitude of the wave length which the illuminating radiation must have.

The foregoing conclusions may be presented in another form. Thus, in order that the incident light should reveal the position of the electron on a small segment of the lowest orbit which it is describing, the process whereby this position is disclosed must occur with such rapidity that the electron will not have time to circumscribe the orbit. Now the process referred to involves a period of time measured by the time taken by the incident light to execute a complete vibration; and this time is the inverse of the frequency of the incident light. On the other hand, Bohr's theory informs us on the time required by the electron to describe the lowest orbit. We may thus compute immediately the frequency which the incident light must have if it is to disclose the position of the electron on its orbit. Either one of the two methods suggested shows that, for the hydrogen atom in the lowest energy state, the incident light that must be utilized is of the X-ray variety.

In our discussion we have assumed the correctness of Bohr's particle picture as a working hypothesis. We know of course that the particle picture illustrated by Bohr's atom is incompatible with the wave picture. For instance, the particle picture indicates that the position observation should always locate the electron somewhere on Bohr's lowest orbit, whereas the wave picture shows that the electron may be found at any point. However, though Bohr's atomic model is too crude, the minimum frequency it has enabled us to assign to the incident light in the previous discussion is correct.

Let us then direct X-rays against the unexcited hydrogen atom so as to locate the electron. The light will ionize the atom, and the electron will be ejected with an energy which exceeds by a considerable amount the energy it had in its lowest orbit. Our position observation may quite well locate the electron outside the tiny sphere mentioned previously, and no paradox will be involved thereby since the electron now has a greater energy than it formerly had.

Suppose now that we perform the same experiment when the electron is moving along a highly excited orbit, say, the orbit E_{1000}. Owing to the larger dimensions of the present orbit, it will be possible to illuminate the electron with light of lower frequency. Calculation shows that the electron will no longer be removed from the atom entirely; instead it will merely be displaced to some neighboring orbit comprised, say, between E_{950} and E_{1050}. The position of the electron will be more or less accurately known as a result of our observation, this position being represented by a small region limited by the two extreme orbits E_{950} and E_{1050}. Finally, the energy of the electron will be uncertain within the energy range of these two orbits.

Next, let us construct the wave picture which describes our knowledge of the electron's position and energy. To obtain this picture, we must superpose monochromatic de Broglie waves having frequencies corresponding to the range in the possible energy values of the electron, *i.e.*, the quantized energy values between E_{950} and E_{1050}. The superposition must be performed in such a way as to yield a wave packet occupying the tiny volume within which the electron has been located. Now this wave packet, is none other than the one which Schrödinger believed to constitute the electron in the orbit E_{1000}. Our present understanding of the packet is entirely different, however. The packet is now seen to be a mere probability packet, which represents our more or less exact knowledge of the electron's position, momentum, and energy at the instant following the observation. The rapid spreading of the packet implies that our knowledge of the electron's future behavior rapidly becomes vaguer and vaguer at successive instants of time. Another result of importance which is revealed by this analysis is that, in a superposition of states, the atom must not be viewed as being in different energy states simultaneously. Instead, a superposition of two or more energy states is now seen to indicate that the atom is in one *or* in the other of these states.

Schrödinger's Superposition of States—According to Born's interpretation, which is today generally accepted, a superposition of states for an atom indicates that the atom is in one *or* in another of these states, and to this extent the superposition is a symptom of uncertainty. Several points which were mentioned incidentally in the preceding chapter will now be examined with greater attention.

We first recall some of Schrödinger's results. When the atom is vibrating, say in the kth mode, the vibration of the de Broglie wave at any point P is represented by

$$(6) \qquad u_k = \psi_k e^{\frac{2\pi i}{h}(E_k + m_0 c^2)t} \qquad (\text{at } P),$$

in which ψ_k is a normalized eigenfunction. When we are dealing with a superposition of states (*e.g.*, the states E_1, E_2, E_3, . . .), the resultant wave disturbance, at a point P and at the time t, is expressed by

$$(7) \qquad u = \sum_k c_k u_k = \sum_k c_k \psi_k e^{\frac{2\pi i}{h}(E_k + m_0 c^2)t},$$

computed at this point P. The constants c_k may be real or complex, and, except for a certain restriction to be explained presently, they are arbitrary. Their office is to determine the relative intensities of the various component vibrations. For instance, the intensity of the kth mode at a point P is

$$(8) \qquad c_k c_k^* u_k u_k^*, \quad \text{or} \quad c_k c_k^* \psi_k \psi_k^* \qquad (\text{at } P).$$

It will be noted that these intensities do not vary with time. On the other hand, the intensity of the resultant disturbance at a point P, namely, the value of uu^* at the point P, does vary with time. Calculation gives

$$(9) \qquad \begin{cases} uu^* = \sum_k c_k c_k^* u_k u_k^* + \sum'_{lj} c_l\, c_j^* u_l\, u_j^* \\[2mm] = \sum_k c_k c_k^* \psi_k \psi_k^* + \sum'_{lj} \left[c_l c_j^* \psi_l \psi_j^* e^{\frac{2\pi i}{h}(E_l - E_j)t} \right], \end{cases}$$

where Σ' implies that, in the summation, the values $l = j$ are to be excluded.[†] The second term in (9) expresses the variation with time of the intensity uu^* at the point P.

[†] If we represent the modulus of a complex quantity x by $|\, x\, |$, we may write (9) in the following equivalent way:

$$(9') \qquad \sum_k |\, c_k\, |^2 \,|\, \psi_k\, |^2 + \sum'_{lj} |\, c_l\, |\cdot|\, c_j\, |\cdot|\, \psi_l\, |\cdot|\, \psi_j\, |\cos \frac{2\pi}{h} \left[(E_l - E_j)t - \delta_{lj} \right].$$

Schrödinger assumes that the density ρ of the charge cloud at a point P is defined at any instant by the value at P of $-euu^*$ at this instant. Thus,

$$(10) \qquad \rho = -euu^* \qquad \text{(at } P\text{)}.$$

Obviously, the density ρ now varies with time at a given point, so that the charge cloud is no longer permanent. Schrödinger further assumes that the charge cloud represents the electron. Consequently, the density ρ when integrated throughout space must have the value $-e$. This condition requires that the wave function u should satisfy the relation

$$(11) \qquad \iiint uu^* dxdydz = 1.$$

In other words, the wave function u must be normalized to 1.

Let us examine what restrictions this normalization will entail. If we integrate the expression (9) throughout space and take into consideration the normalization and the orthogonality † of the functions ψ_k, we obtain

$$(12) \qquad \iiint uu^* dxdydz = \sum_k c_k c_k^*, \text{ or } \sum_k |c_k|^2.$$

Consequently, the normalization of u requires that the constants c_k, which are otherwise arbitrary, should satisfy the condition

$$(13) \qquad \sum_k c_k c_k^* = 1.$$

The constants c_k are then said to be normalized to 1

We must now reinterpret these results in terms of Born's assumptions. According to Born's first assumption (see page 731), when u is normalized to 1,

$$(14) \qquad uu^* dxdydz \qquad \text{(at } P \text{ and at time } t\text{)}$$

---

† The eigenfunctions of Schrödinger's wave equation exhibit a property which is expressed by the statement that the functions are orthogonal. By this is meant that, if we consider the product of any eigenfunction and of the conjugate complex of another eigenfunction pertaining to a different eigenvalue, or energy level, and if we integrate this product throughout space, then the integral will vanish. Thus

$$(15) \qquad \iiint \psi_k \psi_l^* \, dxdydz = 0, \text{ whenever } k \neq l.$$

In the particular case of Schrödinger's wave equation for the hydrogen atom, the eigenfunctions are real, so that the distinction between an eigenfunction and its conjugate complex no longer arises. The orthogonal property (12) now becomes

$$(16) \qquad \iiint \psi_k \psi_l \, dxdydz = 0, \text{ whenever } k \neq l.$$

is the probability that, as a result of a position observation at the instant t, the electron will be found in the volume $dxdydz$ surrounding the point P. Equivalently, we might say that the intensity uu^* gives the density of the probability.

Born's second assumption is expressed by the statement that the electron will be found to be associated with one or with another of the superposed modes. The probability that the electron will be associated with the kth mode, and hence have the energy $E_k + m_0c^2$, is proportional to the value obtained when we integrate the intensity of the kth mode throughout space. Now the intensity of the kth mode is

$$(17) \qquad c_k c_k^* \psi_k \psi_k^*.$$

If we integrate this intensity throughout space, we are left with

$$(18) \qquad c_k c_k^*$$

(owing to the normalization of ψ_k). We conclude that $c_k c_k^*$ is proportional to the probability that the electron will have the energy $E_k + m_0c^2$. In point of fact, (18) is not only proportional to this probability, but is equal to it. To see this, we note that according to Born's assumptions the electron necessarily has the energy corresponding to one of the component modes. Hence, if we add the probabilities of all the various modes, we must obtain the value 1. The result is that (18) will be equal, and not merely proportional, to the probability provided the constants c_k be such that

$$(19) \qquad \sum_k c_k c_k^* = 1,$$

i.e., provided they be normalized to 1. Since this condition was already imposed in (15) to secure the normalization of u, we obtain the following conclusion:

The expression (18) represents the probability that, in a superposition of states, the electron will be associated with the kth mode and hence will have the energy E_k.† Equivalently, we may say that (18) measures the probability that, in a superposition of states, the atom will be found to be in the energy state E_k.

The Statistical Interpretation of a Superposition of States —As we observed on an earlier page, Born's assumptions enable us to place a statistical interpretation on Schrödinger's theory and to suppose that

---

† We have omitted the rest energy m_0c^2, and shall continue to do so in the future.

the wave picture describes average conditions holding for a large aggregate of non-interacting similar atoms. When this statistical view is adopted, a superposition of different standing waves, pertaining to different energy levels E_1, E_2, E_3, $\ldots E_k \ldots$, implies that some of the atoms of the aggregate are in the energy state E_1, others in the state E_2, still others in the state E_3, and so on. According to Born's second assumption, if an atom is selected at random from the aggregate, the probability that it will be in the state E_k is equal to the value obtained when we integrate the intensity of the corresponding wave ψ_k throughout space; and this value, as we have seen, is $c_k c_k^*$. If, then, we assume that all stable energy states have the same a-priori probability, we must suppose that the probability just considered is proportional to the number of atoms in the state E_k. Collecting these results, we conclude that, when the statistical interpretation is accepted for a superposition of states (7), the magnitude $c_k c_k^*$ is proportional † to the number of atoms in the state E_k.

One of the difficulties which confronted Schrödinger's interpretation of the process of radiation is removed by the present statistical interpretation. Thus we noted in the previous chapter that, according to Schrödinger, a superposition of more than two different states should occasion the simultaneous emission of different frequencies from a single atom. We pointed out that this simultaneous emission conflicted with the accepted idea that an atom could emit only one radiation at a time. Thus, a contradiction was involved. But if we apply Born's statistical interpretation and assume that a superposition of states refers to an aggregate of atoms in different energy states, the contradiction no longer appears. We have but to suppose that the simultaneous emission of different radiations is due to the various atoms undergoing different drops and emitting thereby different monochromatic radiations at more or less the same instant.

There are, however, other objections to Schrödinger's theory of radiation, which the present statistical interpretation is unable to answer. Thus according to Schrödinger, radiation cannot occur unless we are

---

† In our former treatment the constants c_k were normalized to 1. But, if we wish $c_k c_k^*$ to be equal to the number of atoms in the state E_k, instead of merely proportional thereto, we must normalize the constants c_k to N (where N is the total number of atoms in the aggregate). The new normalization is expressed by the restriction

$$\sum_k c_k c_k^* = N.$$

dealing with a superposition of states. On the basis of our statistical interpretation, this implies that an aggregate of excited atoms cannot radiate when all the atoms are in the same state. To take a particular example, we may suppose that the wave picture indicates a superposition of the two states E_n and E_k, so that some of the atoms of the aggregate are in the state E_n and others in the state E_k. If we assume that the level E_n is the higher, the atoms in this level will drop to the lower state E_k, emitting radiation of frequency $\dfrac{E_n - E_k}{h}$. (Drops can occur only to the precise lower state E_k, because other lower states are not represented in the wave picture.) The intensity of the radiation emitted is, according to Schrödinger, proportional to the squared amplitude of the vibrating centre of the charge cloud, and can be shown to be proportional to $c_n c_n^* c_k c_k^*$. In Born's statistical interpretation, the intensity of the radiation is thus proportional to the product of the numbers of atoms in the two states, E_n and E_k, between which the drops occur. We are thus led to the conclusion that when all the atoms are in the same state E_n, no radiation can arise.

All these expectations are in conflict with facts, for experiment shows that the intensity of the radiation emitted is proportional solely to the number of atoms in the higher state, and hence to $c_n c_n^*$. Obviously, something is fundamentally wrong in Schrödinger's theory of radiation, even when it receives the statistical interpretation. Dirac has since remedied Schrödinger's treatment. In Dirac's analysis the intensities of the radiations emitted by an aggregate of atoms are proportional to the numbers of atoms in the higher states whence the drops occur. The numbers of atoms in the lower states do not affect the radiation, so that in an aggregate of atoms all of which are excited to the same higher state, the atoms will fall to lower states and will radiate. Thus, the difficulty mentioned is removed entirely. Dirac's theory, which we shall not elaborate here, is a highly abstract mathematical scheme; it cannot be interpreted in terms of familiar notions.

The waves and the superpositions of waves considered to this point were discussed in connection with the hydrogen atom. But the general conclusions we have derived, and the normalizations we have imposed, will be retained in all cases where we are dealing with a superposition of de Broglie waves, whether the superposition yields diffuse regions or concentrated wave packets.

The Various Probabilities—In the last few paragraphs we stressed the statistical interpretation of a wave picture, and so we connected the waves with a swarm of atoms, electrons, or particles. In the following pages we shall revert to our original method of connecting a wave picture with a single particle. Both methods are acceptable, and, except for minor changes in terminology, the conclusions derived from the application of one method remain valid when the second method is applied.

Several subtle points must be mentioned in connection with a wave picture. Some of these were explained in Chapter XXX. We remarked that a wave picture furnishes the probability that the particle will be found at a given instant in a given volume $dxdydz$, and that the picture also yields the probability that the particle will be found to have the momentum and energy corresponding to this, or to that, component wave. But we stressed that the two kinds of information were never valid simultaneously. Thus, if a position observation is performed at a specified instant, we shall find that the position of the particle is in agreement with the requirements of the wave picture; but, in this event, the information the picture gives on the momentum and energy of the particle becomes incorrect (and vice versa). We traced this situation to the disturbance which measurements generate. A further point, that was mentioned referred to the difference between a measurement which is observed and one which is not observed. We noted that the disturbance suffered by the wave picture was not the same in the two cases. We shall now examine these points in greater detail.

Let us revert to the wave phenomenon depicted in (7). It represents various stationary de Broglie vibrations having different frequencies, and also various intensities from place to place. According to the rules given by Born,

$$(20) \qquad\qquad uu^*dxdydz \qquad\qquad (\text{at } P \text{ and at } t)$$

is the probability that the particle, or electron, will be found at time t in the volume $dxdydz$ surrounding the point P. The explicit expression of uu^* for the wave superposition (7) is given by the formula (9). Accordingly, the probability (20) is

$$(21) \qquad \left[\sum_k c_k c_k^* \psi_k \psi_k^* + \sum'_{lj} c_l c_j^* \psi_l \psi_j^* e^{\frac{2\pi i}{h}(E_l - E_j)t} \right] dxdydz.$$

The manner in which this probability at a given point P varies with time is expressed by the last term in (21).

The probability (21) is correct, however, only when no measurements, whether observed or not, are performed on the electron prior to the position observation. All such measurements would necessarily disturb the original wave picture (7) and would compel us to reject it in favor of a new one constructed in an appropriate way. The original picture (7) could then no longer be utilized for the purpose of computing probabilities.

As a particular illustration of a disturbance imposed on the wave picture (7), let us suppose that we perform an energy observation and find that the electron has the energy E_k. Our observation disturbs the original picture and we must replace it by a new one. The new picture is easily constructed when we take the uncertainty relations into account and recall that $c_k c_k^*$ is the probability that the electron will be found to have the energy E_k. According to the uncertainty relations, an energy measurement does not disturb the electron's energy. Hence we may be sure that if a second energy measurement is repeated immediately, the same energy E_k will still be found. The new wave picture must therefore express the fact that the value E_k for the energy of the electron is a certainty. This requires that the value of $c_k c_k^*$ in the wave picture be 1, *i.e.*, the expression of a certainty. If we assume that the constants c are real, this condition implies $c_k = 1$. Also, since the probability that the energy of the electron has a value E_l differing from E_k is zero, we must set $c_l c_l^* = 0$, and hence $c_l = 0$. In short, the new wave picture will be one in which $c_k = 1$ and all the other constants vanish.

We also note that, according to the uncertainty relations, an energy observation disturbs solely the magnitude which is conjugate to energy (*i.e.*, time); in particular, it does not disturb position. As a result, the energy observation will not modify the various amplitude functions $\psi(x, y, z)$ which appear in the wave expression (7), and which are independent of time. The value of the time t in (7) will, however, become uncertain; but this last feature does not affect the problem we have in view. Collecting these results, we conclude that the new wave picture will be obtained from (7) by setting $c_k = 1$ and by cancelling all the other constants. The new wave picture is thus

$$(22) \qquad \bar{u} = \psi_k(x, y, z) e^{\frac{2\pi i}{h}(E_k + m_0 c^2)t} .$$

The symbol $\bar{u}$ is taken for the wave function in place of u, so as to stress that the wave pictures (22) and (7) are not the same.

From the new wave picture, we obtain the probability that the electron will be found in a volume $dxdydz$ about a point P. According to Born's assumptions, this probability is

$$(23) \qquad \bar{u}\bar{u}^* dxdydz \quad \text{or} \quad \psi_k \psi_k^* dxdydz \qquad \text{(at } P\text{)}.$$

In contradistinction to the probability (21), the present one does not depend on time.

The physical significance of the two probabilities (21) and (23) is entirely different. Thus (21) expresses the probability that the electron will be found in the neighborhood of a point P at time t, when only probable values of the electron's energy (as illustrated in the wave picture) are known. On the other hand, (23) determines the probability that the electron will be found in the neighborhood of the point P at any instant *when we know that its energy is* E_k.

A third kind of probability which also deals with the electron's position may be mentioned. To understand how it arises, we must recall a few elementary notions of the calculus of probabilities. Suppose that one among a number of different events A_1, A_2, A_3, . . . A_n is certain to occur; and suppose that a second event B may, but need not, arise after one of the events A has taken place. The event B is thus not certain to occur, though there is a certain probability that it will occur. We represent this probability by b.

Let a_1, a_2, . . . a_n be the probabilities of the events A_1, A_2, . . . A_n and let b_1, b_2, . . . b_n represent the probabilities of B taking place after A_1 or A_2 or . . . or A_n has occurred. In other words, *if we know* that the event A_k has occurred, the probability that B will arise is b_k. But suppose we do not know whether or not A_k has occurred. Since this event has a probability a_k of taking place, the calculus of probabilities entitles us to assert that $a_k b_k$ measures the probability that A_k will occur and be followed by B. Since one of the events A necessarily precedes B in any case, we conclude that the probability of the event B taking place, regardless of which of the events A has preceded it, is given by the sum of the previous probabilities of type $a_k b_k$. This probability is of course none other than the probability b of the event B taking place. We thus obtain the equality

$$(24) \qquad\qquad b = \sum_{k=1}^{k=n} a_k b_k.$$

Let us apply these considerations to the wave picture (7). We have seen that the expression (21) measures the probability that the electron will be found in the neighborhood of a point P. We now propose to obtain the same probability in a different way by following the method just outlined. Thus we have already shown that if the electron is found to have the energy E_k, the probability that it will be located in a given volume $dxdydz$ is (23). Since the probability that the electron will be found to have the energy E_k is known to be $c_k c_k^*$, the probability that it will be found to have the energy E_k and also be situated in the volume $dxdydz$ about the point P is given by the product of $c_k c_k^*$ and (23). The probability is thus

$$(25) \qquad\qquad c_k c_k^* \psi_k \psi_k^* dxdydz \qquad\qquad \text{(at } P).$$

Hence, according to the general probability considerations mentioned in the preceding paragraph, the probability that the electron will be located in the volume $dxdydz$, regardless of the value that may have been found for its energy, is

$$(26) \qquad\qquad \sum_k c_k c_k^* \psi_k \psi_k^* dxdydz \qquad\qquad \text{(at } P).$$

On reverting to the example of the events A and B, we note that the probability b there considered is the analogue of the probability (21), and the probability $\sum_k a_k b_k$ the analogue of (26). The equality (24) therefore entails the equality of the two probabilities (21) and (26). But a mere inspection of these two expressions shows that they are different. In particular, (26) lacks the last term of (21). Thus, we appear to be faced with a contradiction. The paradox is easily explained, however, for we shall find that the situations contemplated in the calculations of the two probabilities (21) and (26), though equivalent from the standpoint of classical science, are different in the quantum theory. This prompts us to examine more carefully the significance of the probability (26).

Let us suppose that we start from the wave picture (7) and that an energy measurement is made. The result of the measurement is observed, and we find that the energy of the electron is E_k, for instance. The wave picture must then be modified, and the picture (22) is obtained. But suppose that when the energy measurement is made, we happen to be looking in some other direction. As we have no idea of what value the energy may have, we no longer obtain the picture (22). On the other hand, the measurement necessarily disturbs the picture (7). According to the uncertainty relations, its effect is to render time uncertain. Since no other modification is entailed we conclude that, after the energy measurement has been performed, we shall still be left with the original wave picture (7), but that in this picture the value of the time variable t is now uncertain.

Let us apply Born's first assumption to this wave picture (7) in which time is uncertain. According to Born's assumption, the probability of the electron being found in a volume $dxdydz$ is $uu^*dxdydz$, where uu^* is the intensity of the wave in the above picture. Since the new picture is the same as (7), except for the uncertainty in time, we might expect the probability just written to be the same as (21). But this is not so, for as a result of the uncertainty in time the new probability is determined not by (21), but by (21) averaged over a long period of time. The average value of the last term of (21) is zero, and so the average value of (21) is simply the first term—and this first term is precisely the probability (26). We conclude that the expression (26) measures the probability that the electron will be found in a volume $dxdydz$ after an energy measurement *which has not been observed*, has been performed on the electron.

We are now in a position to compare the meanings of the probabilities (21), (26), and (23), each one of which is claimed to define the probability that, as a result of a position observation, the electron will be located in the tiny volume $dxdydz$ surrounding the point P.

I. The probability (21) applies to the case where no energy measurement is performed.

II. The probability (26) applies to the case where an energy measurement is made, but is not observed.

III. The probability (23) corresponds to the situation where an energy measurement is not only made, but is also observed.

The only difference between the situations I and II is that, in the latter, an energy measurement (unobserved) is made. We conclude that the physical disturbance caused by this measurement is responsible for the difference in the two wave pictures, and hence in the two probabilities. In the situations II and III, exactly the same energy measurement is made, so that the difference in the two wave pictures does not arise from the measurement; it arises solely from the fact that, in III, the energy measurement is observed. This observation, and therefore the cognitive act, entails the difference in the two corresponding wave pictures, and hence in the two probabilities. The foregoing illustrations bring out clearly the different meanings that must be ascribed to the various changes in our wave picture, *i.e.*, to those changes generated by the disturbing effects of a measurement as such, and to those entailed by an observation of the measurement, namely, by a cognitive act.

We have seen that the change in the probability, when we pass from the situation I to the situation II, is due entirely to the physical disturbance generated by the energy measurement. On the other hand, this change in the probability is expressed in the wave picture by a change which we obtain by applying the uncertainty relations (as between energy and time) to the original picture (7). Thus, the wave form of representation, supplemented by the uncertainty relations, affords a self-contained scheme for calculating the probabilities. We need no added correction to allow for the disturbances due to the measurements; these disturbances are automatically taken into account when we apply the uncertainty relations to the wave picture.

The intrusion of probability considerations in the interpretation of physical phenomena is not peculiar to the quantum theory, for probabilities were also utilized in classical physics. In the quantum theory, however, these probabilities are regarded as essential, whereas in classical science they were viewed as makeshifts introduced to overcome our ignorance of the fundamental processes. But we now propose to establish a comparison between the probabilities of the two theories from a totally different point of view.

We recall that in classical physics a measurement, as such, could have no disturbing effect on the course of events. Physicists recognized, of course, that a measurement performed in a clumsy way would occasion a disturbance; but a disturbance of this kind was regarded as casual, and the accepted belief was that by exercising sufficient care, we could reduce it to a vanishing point. On the other hand, it was admitted that if a measurement were observed, the added knowledge derived from the observation would affect the probabilities we could assign to future events. The classical view may be summarized by the statement that a measurement which is unobserved can exert no effect on the probabilities, but that a measurement which is observed will affect them.

These considerations show that, from the classical standpoint, the situations I and II are equivalent and should be associated with the same value of the probability. Since, in the quantum theory, these two situations and their attendant probabilities are different, the classical and the quantum theories are in disagreement. The situation III need not detain us, for both theories recognize that the observation of a measurement injects a new element and must modify the probability. Thus, the main difference between the classical and the quantum theories resides in the distinction which the quantum theory establishes between the situations I and II.

We are thus led to inquire which of the two quantum probabilities (21) and (26) is the one most in harmony with classical anticipations. As we shall see, the classical probability is furnished by (26), whereas (21) is revolutionary and incomprehensible from the classical standpoint. The reason for this statement becomes clear when we examine the wave pictures more attentively. But first we must recall the acoustical problem of "beats," mentioned in the last chapter.

Suppose we place side by side two diapasons vibrating with frequencies ν_1 and ν_2 respectively. Each diapason emits sound waves, so that the two wave trains issuing from the diapasons are superposed. At any given point P, the waves of the two trains may happen to be in phase at the instant considered, and in this event the resultant amplitude due to the superposition of the two waves will be a maximum. But since the frequencies ν_1 and ν_2 of the two sets of waves are not the same, the concordance in the phases cannot last, and after a certain time the phases at the point P will be opposite. The resultant amplitude is then a minimum. A simple calculation shows that the resultant vibration is harmonic, that its frequency is the mean of the frequencies ν_1 and ν_2, and that its amplitude oscillates with frequency $\cdot\,|\,\nu_1 - \nu_2\,|$ between its maximum and its minimum values. In acoustics the intensity of a musical note is proportional to the square of the amplitude of the vibration. Hence at the point P, the ear will detect a periodic rise and fall in the intensity of the sound. The phenomenon is particularly noticeable when the frequencies ν_1 and ν_2 differ only little. At any instant the regions of space where a reinforcement of the sound occurs define concentric spheres having as centre the point where the diapasons are situated. As time passes, these spheres expand with the group, or the amplitude, velocity.

The phenomenon of beats which we have just described illustrates a wave manifestation, and we might therefore be tempted to expect its occurrence for optical waves. Thus, if two monochromatic point-sources emitting light of different color were to replace the two diapasons, the superposition of the two wave trains might be expected to give rise to luminous and obscure spherical surfaces which would alternate and would expand from the point occupied by the sources. A sheet of paper placed in the region of interference would then be covered at any instant with brilliant and with dark bands. The bands would, however, be moving across the paper, so that the interference pattern produced would be a mobile one, and hence would differ from the permanent pattern mentioned in connection with Young's optical experiment. The bands would move with tremendous speed, but by giving a suitable inclination to the sheet of paper, we could reduce this speed and the mobile bands might be observed.

If the two wave trains emitted from the two point-sources were regular (*i.e.*, without abrupt changes in the phases), the interference phenomenon just described could be realized. But, as we mentioned when discussing Young's experiment (Chapter XIX), wave trains emitted from a point-source invariably present discontinuities owing to the phases of the waves undergoing sudden jumps. Because of these abrupt jumps, the interference bands will not advance with uniform motion across the paper, but will be subjected to shifts thousands of times every second, so that only a uniform illumination will be observed. The intensity of this uniform illumination will have an average value, which will be a compromise between the brilliancy of the bright bands and the obscurity of the dark ones. Thus, in practice, the interference effect is destroyed by the abrupt changes in the phases,[†] and the wave attributes of light cease to be conspicuous. As we shall now see, the phenomena we have just discussed also occur for superpositions of de Broglie waves of different frequencies.

Let us revert to the original wave picture (7) in which de Broglie vibrations of different frequencies are superposed. The vibrations are assumed regular, *i.e.*, they exhibit no sudden changes in their phases. This assumption is plausible, since the discontinuities in the phases, noted for optical waves, are due to the discontinuity in the emission of radiation from matter; whereas de Broglie waves are merely symbolic, and we do not have to consider by what process they come into existence. We shall also assume that no energy measurement is performed, so that no disturbance arises in the wave picture.

The intensities of the individual monochromatic stationary waves at a point P are defined by the values of the various expressions $\psi_k \psi_k^*$ at this point. Obviously these intensities do not vary with time. But the superposition of the different monochromatic waves will generate the interference phenomenon of beats, so that the resultant intensity uu^* of the superposed waves will vary with time at any fixed point. This variation of uu^* is expressed analytically in the formula (21). It is true that the present interference phenomenon is more complicated than the one we discussed in the acoustical case, because we are now dealing with a large number of superposed de Broglie waves instead of only two; the nature of the phenomenon is essentially the same, however.

---

† In discussing Young's experiment in optics, we saw that the abrupt jumps in the phases of the superposed waves did not necessarily preclude the formation of an interference pattern. All that was required for an interference pattern to be produced was that the abrupt jumps in the phases of the two superposed vibrations at a common point P should occur simultaneously and be the same. In Young's experiment we were dealing with superposed waves having the same frequency. But when we pass to the phenomenon discussed in the text, where the frequencies of the superposed waves are not the same, the situation is different, and an interference pattern can arise only when the two superposed wave trains are regular.

When we recall that the resultant intensity of the superposed de Broglie waves, at a point P, represents the density of the probability that the electron will be found at this point, we understand the justification for referring to this probability as an *interference of probabilities.*

Next let us suppose that an energy measurement is performed but is not observed. This measurement renders time uncertain, and hence the phases of the vibrations also become uncertain; the measurement does not affect the intensities of the individual component waves, for these are independent of time. On the other hand, the uncertainty in the phases prevents the constituent vibrations from having well-defined phases at every instant. It would be as though the regularity of the vibrations were destroyed by abrupt jumps in the phases, these jumps occurring at random. The phenomenon of beats no longer occurs, and the resultant intensity uu^* does not vary with time; instead, it assumes a permanent average value at each point, and its mathematical expression is given no longer by (21), but by (26). Thus, the interference effect illustrated in the beats is destroyed by the energy measurement, and the probability (26) does not express an interference of probabilities.

Now, the revolutionary disclosures of wave mechanics concern the wave aspects of matter. Since these wave aspects manifest themselves in an interference of probabilities, whereas they cease to be conspicuous when the interference effect is destroyed by the disturbances provoked by a measurement, we must conclude that the probability (26), not (21), is the one which is in closest agreement with classical ideas.

The foregoing considerations show that the energy measurement, by destroying the wave aspect of the probability, has transformed the probability into one with which classical thought was acquainted. The circumstance, however, must be regarded as accidental, for the quantum disturbances responsible for this transformation were unsuspected in classical science.

In the above examples we have examined the disturbing effect of an energy measurement on an interference pattern. All measurements, however, generate disturbances of one kind or another, so that in any case when a measurement (even unobserved) is performed, the wave picture is modified in some important respect. As an example, let us suppose that a position measurement is made at some specified instant, but that the result of the measurement is not observed. Insofar as we are performing a space-time measurement, the magnitudes conjugate to space and to time will be disturbed in some unpredictable way. The average values of the disturbances are governed by the uncertainty relations, and they increase with the precision of our measurement. The wave picture will thus be modified, the waves being affected in their wave lengths and frequencies. The original interference effect will be destroyed. Suppose, on the other hand, that the position measurement is observed and that we locate the particle in a volume $dxdydz$. The wave picture undergoes a more drastic change, for we must now replace it by a wave packet occupying the volume $dxdydz$ and formed by the superposition of appropriate waves.

CHAPTER XXXV

THE HELIUM ATOM

ONE of the most important features of the helium spectrum, which Bohr's theory was unable to explain, was accounted for by Heisenberg when he applied his matrix method to the theory of radiation. Heisenberg's findings may be transcribed into the language of wave mechanics and it is in this form that we shall examine them here. Their interest lies not so much in the solution they afford of a difficulty encountered in the helium atom, as in the light they shed on the significance of Pauli's Exclusion Principle.

Let us recall briefly the theory of the helium atom as we left it in Bohr's theory. In the helium atom two electrons are circling around an alpha particle which forms the nucleus. The electrons are assumed to be spinning on their axes. The axes of spin may set themselves parallel or antiparallel, so that the spins may be in the same or in opposite directions. In the former case, the resultant spin vector $\vec{s}$ has the value $s = 0$, and the energy levels are single. In the latter case, $s = 1$, and the levels are triple. In the basic singlet level, both electrons are circling in the lowest orbit, the four quantum numbers attached to the electrons being respectively

$$
\left\{
\begin{array}{l}
n_1 = 1; \; l_1 = 0; \; j_1 = \dfrac{1}{2}; \; m_1 = \dfrac{1}{2} \\[2mm]
\text{and} \\[2mm]
n_2 = 1; \; l_2 = 0; \; j_2 = \dfrac{1}{2}; \; m_2 = -\dfrac{1}{2}.
\end{array}
\right.
$$

But in the basic triplet level, where the spins must be in the same direction, Pauli's exclusion principle prevents both electrons from being situated in the lowest level, for if they were, the four quantum numbers connected with the two electrons would be the same. We must assume therefore that in the basic triplet level one of the two electrons, which in Bohr's theory plays the part of the optical electron, is situated in the next higher level, *i.e.*, $n = 2$. If we refer to the optical electron as the first electron and indicate the quantum numbers accordingly, the four quantum numbers associated with each of the two electrons (now spinning in the same direction) are

$$
\left\{
\begin{array}{l}
n_1 = 2;\ l_1 = 0;\ j_1 = \dfrac{1}{2};\ m_2 = \dfrac{1}{2} \\[2ex]
\text{and} \\[2ex]
n_2 = 1;\ l_2 = 0;\ j_2 = \dfrac{1}{2};\ m_1 = \dfrac{1}{2}.
\end{array}
\right.
$$

When the optical electron drops from one singlet level to another, spectral lines of the singlet type are produced. And when the drops occur between triplet levels, we obtain triplets for the spectral lines or triplets with satellites. The spectrum of singlets is referred to as the *para-spectrum* and the spectrum of triplets is called the *ortho-spectrum*. These appellations were coined at a time when the two different kinds of spectra were attributed to two different species of the helium atom, *i.e.*, para-helium and ortho-helium. Later it was recognized, thanks chiefly to the introduction of the spinning electron, that the same helium atom was responsible for both kinds of spectra, but the original prefixes *para* and *ortho* were still retained for the two different kinds of energy levels and spectral series.

A peculiarity of the helium spectrum is that the lines which would be caused by drops from a singlet level to a triplet level, or from a triplet level to a singlet one, do not occur. Such drops appear to be forbidden. This implies that the quantum number s cannot change during an energy transition; and in the particle picture it means that, during a drop, the spins of the two electrons must retain their relative directions (parallel or antiparallel). Now a satisfactory theory of the helium atom should provide a means for anticipating the impossibility of the drops connected with a change in the value of s. Bohr's theory, we recall, relies on the correspondence principle to determine forbidden drops. Unfortunately, this principle yields no information on the possible changes of the quantum number s; it deals only with the quantum numbers l and j, and so is of no use in our present problem. We cannot circumvent the difficulty by claiming as a general rule that changes in the value of s are excluded in principle, for in many atoms such changes are known to occur. One of the outstanding successes of the new quantum theory has been to show why, in the case of the helium atom, the quantum number s cannot change during a drop.

Before examining the wave-mechanical treatment of the helium atom, certain features of configuration spaces must be recalled. In classical mechanics, when we are dealing with a system having N degrees of freedom, we can represent the motion of this complicated system by means of the motion of a single point-mass (say, of unit mass) in a configuration space of N dimensions. The configuration space may be Euclidean or non-Euclidean, *i.e.*, curved. The complexity of the mechanical system is thus reduced through the device of the configuration space, but, as against this, the number of dimensions of the space in which the motion occurs is increased.

In mechanics, this procedure is recognized as artificial, though it is adopted in theoretical discussion when it simplifies a demonstration. But in wave mechanics, the introduction of a configuration space is a necessity, for when this course is not followed, incorrect results are obtained. Of course, when we are dealing with a single particle moving in space, and when we assume that the particle is incapable of rotating, so that it has only three degrees of freedom, the configuration space coincides with ordinary space; in such cases the de Broglie waves are represented in ordinary space. This situation is illustrated in the problem of the hydrogen atom when we view the nucleus as fixed. There is then but one particle (the electron) to be considered, and the de Broglie modes of vibration are therefore pictured in ordinary space. On the other hand, when the nucleus is no longer regarded as fixed, we have two particles to consider; our system now has six degrees of freedom, and the de Broglie waves must be represented in a 6-dimensional configuration space. A similar situation is encountered in the problem of the helium atom, for, owing to the presence of two electrons, the system has more than three degrees of freedom, even when we view the nucleus as fixed.

We shall investigate the helium problem by a method called the "method of perturbations." The word *perturbation* is borrowed from celestial mechanics. In this science the mechanical problems we are called upon to solve are extremely complicated because of the large number of planets which exert mutual gravitational attractions. But in many cases we may neglect as a first approximation some of the mutual attractions, and when this is done, we obtain a problem which can be solved more readily. This simplified problem, when solved, serves as a basis for the higher approximations.

For example, let us consider the motions of the earth and of the planet Jupiter around the sun; we assume that the sun is fixed, and we disregard the presence of the other planets. If it were not for the mutual attraction between the earth and Jupiter, the problem of determining the motions of the two planets would be an easy one : in this simplified problem both planets would describe ellipses around the sun. But because of the mutual attraction, the motions are more complicated and the problem is difficult to solve. We note, however, that the mutual attraction is small in comparison with the action of the sun on either planet. Consequently, the actual motions of the two planets will not differ greatly from their elliptical motions in the simplified problem. To obtain the actual motions, we therefore start from the simplified problem and then regard the mutual attraction between the earth and Jupiter as introducing a small perturbation. The actual motions are thus viewed as resulting from a perturbation imposed on the simple motions. For this reason the simplified problem is called the unperturbed problem, whereas the problem of ascertaining the actual motions is named the perturbed problem. Our aim is to determine the effect of the perturbation on the unperturbed motions. The precise effect of the perturbation is not obtained immediately, however, but only as a result of a series of successive approximations. In other words, we pass from one stage of approximation to the next, refining our results at each step. This method of solution constitutes the method of perturbations.

In Bohr's theory of the atom, where the problems to be solved were often of considerable difficulty, the method of perturbations was frequently applied. As may well be imagined, the corresponding problems, when treated by wave mechanics, are also extremely difficult; and it therefore appeared advisable to construct a wave theory of perturbations. A wave theory of perturbations was first developed by Schrödinger and at a later date by Dirac and others. Schrödinger's theory of perturbations has been of great assistance in the problem of the helium atom. In dealing with the helium atom, we shall assume that the alpha particle (nucleus) is fixed; furthermore, we shall treat the problem by means of the classical wave equation, all relativistic refinements being disregarded. The spins of the two electrons will, however, be taken into account. The omissions we are here making do not affect results perceptibly, so that we are not simplifying the situation unduly.

The solution of the helium problem is divided into three steps:

I. We suppose that the two electrons, which in the particle picture circle around the nucleus, exert no mutual repulsions due to their electric charges. We also disregard the existence of any spin. This model of the helium atom is of course incomplete, but it serves as a first approximation. We shall refer to this incomplete atom as the "unperturbed helium atom."

II. We next supplement our model by taking into consideration the mutual electrostatic repulsion between the two electrons; but we still disregard the spins. This mutual repulsion is assumed to act as a perturbation on the first model, and so we call our supplemented model "the perturbed helium atom."

III. Finally, we assume that the electrons are spinning. The spins generate magnetic fields, and the two electrons are subjected to inter-actions occasioned by these fields. In this picture the model of the helium atom is completed, and we may refer to it as "the completed helium atom."

The First Step—We call the two circling electrons a and b. Sup-pose one of the two electrons is removed, so that, say, the electron a is the only one present. The situation is now the same as in the once-ionized helium atom. The problem differs from that of the hydrogen atom only by the fact that the electric charge of the helium nucleus is twice that of the proton-nucleus of hydrogen.

If, then, we treat the problem by the methods of wave mechanics and disregard the relativistic refinements and also the electron spin, we obtain practically the same results as we did for the hydrogen atom in Chapter XXXII. The only difference will be that the eigenvalues of Schrödinger's present wave equation will have four times their former values, and this implies that the energy levels will have four times the numerical values of the energy levels of hydrogen. We shall also find that, corresponding to each energy level, or eigenvalue E_n, there are n^2 independent eigen-functions $\psi_{n,l,m}$ (represented by giving all permissible values to l and to m, while the value of n is kept fixed). These eigenfunctions, or amplitude functions, determine the distributions of the amplitude of the de Broglie waves in the modes of vibration associated with the various stationary states. No special peculiarity enables us to differentiate one electron from the other; and, if instead of the electron a, we had selected the electron b as planetary electron in the ionized helium atom, exactly the same energy levels and eigenfunctions would have resulted. Let us represent by E_n^a and E_n^b the nth energy level of the ionized helium atom when it is the electron a or the electron b which is assumed to be circling round the nucleus. Obviously we have

$$(1) \qquad\qquad E_n^a = E_n^b.$$

Similarly for the eigenfunctions, we may write

$$(2) \qquad\qquad \psi_{n,l,m}^a = \psi_{n,l,m}^b.$$

We must now consider the wave-mechanical treatment when the two *non-interacting* electrons a and b are present simultaneously in the atom. According to the usual methods, we replace our system of two electrons by a single point-mass moving in 6-dimensional space; the coordinates of a point in this configuration space are, for instance, the Cartesian co-ordinates $x_a, y_a, z_a, x_b, y_b, z_b$ of the two electrons in ordinary 3-dimensional space. The Schrödinger wave equation which illustrates this situation is an equation controlling an amplitude function in 6-dimensional space. Its eigenvalues are merely the sums of the eigenvalues we have just considered. Our new eigenvalues are thus

$$(3) \qquad\qquad E^a_n + E^b_{n'}, \qquad \begin{cases} n = 1, 2, 3 \ldots \\ n' = 1, 2, 3, \ldots \end{cases},$$

where n and n' may assume independently all positive integral values. The corresponding eigenfunctions are the products of our former eigenfunctions, and are therefore of type

$$(4) \qquad\qquad \psi^a_{n.l.m} \, \psi^b_{n',l',m'},$$

where n and n' have the same values as in (3), but where l, m, l', and m' may have all permissible values (as explained in connection with the hydrogen atom).

These results may easily be interpreted in the particle picture in ordinary 3-dimensional space. Thus the eigenfunction (4) corresponds to the situation where the electron a is in the orbit (n, l, m), while the electron b is in the orbit (n', l', m'). The energy (3) of the unperturbed atom is then the sum of the energies of the two electrons. No additional energy comes into consideration, because we are assuming that the electrons do not interact, and so have no mutual potential energy. The various eigenfunctions (4) define the modes of vibration of the de Broglie waves in 6-dimensional space for our unperturbed helium atom; hence we may also say that the particular eigenfunction (4) determines the mode of vibration when the electron a is in the sub-sublevel (n, l, m) while the electron b is in the sub-sublevel (n', l', m').

Let us determine what kind of spectrum would be emitted by a helium atom constituted as we are now supposing. For simplicity, we shall assume that one of the two electrons, e.g., the electron a, remains in the lowest orbit $n = 1$. The second electron b will be raised to higher orbits. Our energy levels are now

$$(5) \qquad\qquad \begin{matrix} E^a_1 + E^b_{n'} \\ \cdot \quad \cdot \\ \cdot \quad \cdot \\ \cdot \quad \cdot \\ \cdot \quad \cdot \\ \cdot \quad \cdot \\ E^a_1 + E^b_2 \\ E^a_1 + E^b_1. \end{matrix}$$

The lowest of these levels is the one at the base of the column. Let us, then, consider the various drops in the energy when the atom drops from the higher levels to the lowest one. According to our scheme, these energy drops will be accompanied by the drops of the electron b from the various orbits n' to the lowest orbit 1 while the electron a continues to circle in this lowest orbit. Let us consider the frequencies that will be radiated. Since Bohr's frequency condition is known to give the correct frequencies, we may apply it in the present case; and we thus find that the frequencies radiated will be

$$(6) \qquad \frac{E_{n'}^b + E_1^a - E_1^b - E_1^a}{h} = \frac{E_{n'}^b - E_1^b}{h} \qquad (n' = 2, 3, 4 \ldots).$$

Since the energy levels $E_{n'}^b$ and E_1^b have four times the values of the hydrogen levels, the frequencies (6) will be four times greater than are those of the corresponding hydrogen series (*i.e.*, the Lyman series). Exactly the same frequencies would be radiated if we assumed that the electron a dropped while the electron b remained on the lowest orbit.

Our present scheme indicates that the spectral lines of helium will all be single, whereas the observation of the helium spectrum shows that many of the lines are triple. This discrepancy should not surprise us, however, for in our present treatment the interaction between the two electrons of the atom has been disregarded.

We must now draw attention to the degeneracy of the problem. We explained in Chapter XXXII that an atomic system is said to be non-degenerate if to each eigenvalue (energy level) corresponds only one eigenfunction (mode of vibration); and that the system is degenerate if more than one eigenfunction is association with each eigenvalue. Equivalently we may say: Whether the system be degenerate or not, only one mode of vibration is associated with each energy level; but in the event of degeneracy two or more of the energy levels coalesce, so that different modes of vibration (having the same frequency) appear to be associated with the same level.

The consequences of degeneracy are particularly noticeable when the atomic system is subjected to some perturbing influence; for this influence may modify in different degrees the various energy levels, with the result that the coalesced levels may become separate. The degeneracy is then said to be removed partly or entirely, and the number of distinct energy levels is thus increased. On the other hand, in the absence of degeneracy, a perturbation, though modifying the energy values, cannot increase the number of distinct energy levels. These considerations play an important part in the theory of the helium atom, because we shall find that our present unperturbed atom is degenerate.

Let us examine the nature of the degeneracy in the unperturbed helium atom (*i.e.*, when the mutual repulsions of the two electrons are disregarded). The expressions (3) and (4) show that each energy level $E_n^a + E_{n'}^b$ will be associated with a number of eigenfunctions $\psi_{n,l,m}^a \psi_{n',l',m'}^b$ (in which n and n' have the same values as in the energy level). The theory of the hydrogen atom shows that there are n^2 independent functions $\psi_{n,l,m}^a$ and n'^2 independent functions $\psi_{n',l',m'}^b$. Accordingly, there are $n^2 n'^2$ independent eigenfunctions (4) associated with each energy level (3). Equivalently we may associate an energy level with each one of these $n^2 n'^2$ eigenfunctions and say that the corresponding $n^2 n'^2$ energy levels coalesce. The unperturbed helium atom is thus obviously degenerate.

When we proceed to the second step in the solution of the helium problem, taking the mutual repulsion of the two electrons into consideration, the perturbation generated by the mutual repulsion removes the degeneracy in part; and some of the coalesced energy levels are separated. Each one of the energy levels $E_n^a + E_{n'}^b$ will then be split into $n^2 n'^2$ sublevels defined by the different values that may be given to the quantum numbers l, m, l', and m'. Consequently, a much larger variety of spectral lines may be expected. Calculation shows, however, that the separations between the sublevels issuing from the same coalesced level will be so small that no perceptible complication of the spectrum can ensue. Under these conditions the form of degeneracy we have been discussing might as well be inexistent; and in the sequel we shall disregard it and treat the levels as though they were single. This simplified treatment implies that we are not distinguishing those eigenfunctions $\psi_{n,l,m}^a \psi_{n',l',m'}^b$ for which the numerical values of n and of n' remain fixed. Accordingly, we shall represent the eigenfunctions $\psi_{n,l,m}^a \psi_{n',l',m'}^b$ by $\psi_n^a \psi_{n'}^b$, the quantum numbers l, m, l', and m' being omitted. In short, with each energy level (3), namely,

$$(7) \qquad\qquad E_n^a + E_{n'}^b,$$

is associated one eigenfunction, or mode of vibration,

$$(8) \qquad\qquad \psi_n^a \psi_{n'}^b.$$

But there is another type of degeneracy over which we cannot pass so lightly. It arises for the following reasons:

Consider the two energy levels

$$(9) \qquad\qquad E_n^a + E_{n'}^b \text{ and } E_n^b + E_{n'}^a.$$

They are associated respectively with the eigenfunctions

$$(10) \qquad\qquad \psi_n^a \psi_{n'}^b \text{ and } \psi_n^b \psi_{n'}^a.$$

As will be explained presently, the two eigenfunctions (10) are different. Consequently, the energy levels (9), with which these eigenfunctions are associated, must be viewed as different. On the other hand, the equalities (1) show that the numerical values of the two energy levels (9) are exactly the same, so that the two levels coalesce. The coalescence of the two energy levels (9), associated with the two different eigenfunctions (10), implies a state of degeneracy. For reasons which will be understood later, Heisenberg refers to this form of degeneracy as *resonance degeneracy*, or *exchange degeneracy*. We shall find that it plays a fundamental rôle in the theory of the helium atom.

We must emphasize that we are entitled to view the levels (9) as *different* only because the two corresponding eigenfunctions (10) are themselves different. If these two eigenfunctions were identical, the two levels (9) would constitute a single level; there would be no coalescing of different levels and hence no resonance degeneracy. Let us, then, make clear for what reason the two eigenfunctions (10) are different.

At first sight the eigenfunctions (10) appear to be exactly the same. Thus the first eigenfunction (10) expresses the amplitude distribution of the mode of vibration when the electron a is in the orbit n and the electron b in the orbit n', whereas in the second eigenfunction (10) the electrons are interchanged; and inasmuch as we have no means of distinguishing one electron from the other, we might expect the two modes of vibration, and therefore the two eigenfunctions, to be identical. Now in point of fact, the two eigenfunctions would indeed be identical if they were to define amplitude distributions in the same ordinary 3-dimensional space. But we have said that the eigenfunctions and waves must be pictured in the 6-dimensional configuration space, and, as explained in the note, this feature causes the two eigenfunctions to be different.*

These considerations show that different results are obtained according to whether we represent the waves of the two electrons in ordinary space or in the 6-dimensional configuration space. Only in the latter case does

---

* If we call x, y, z, x', y', z' the six coordinates of a point in 6-dimensional space, the value of one of the eigenfunctions at this point is

$$\psi_n(x, y, z)\psi_{n'}(x', y', z')\,;$$

whereas the value of the second eigenfunction at the same point is

$$\psi_n(x', y', z')\psi_{n'}(x, y, z).$$

These two values differ for the same reason that $5^3 \cdot 4^2$ differs from $5^2 \cdot 4^3$. Thus, at any point of 6-dimensional space the two functions differ in value and so are different functions. The amplitude distributions of the two modes of vibration in 6-dimensional space therefore change when the electrons are interchanged. The energy, and hence the frequency of the de Broglie vibrations is, however, the same in either case.

resonance degeneracy occur. Since the assumption of resonance degeneracy is essential to the interpretation of the helium spectrum, we recognize that, formally at least, the configuration space must be taken seriously. We pointed out in Chapter XXXIII that the fictitious nature of a configuration space of more than three dimensions deprives of all physical reality the de Broglie waves and the charge cloud represented in this space. The helium atom affords an illustration in which this symbolic nature of the de Broglie waves imposes itself.

We note that not all the energy levels coalesce in pairs. Thus, if $n' = n$, the two eigenfunctions (10) are the same, even in the configuration space, and their respective energy levels must be viewed as defining only one level.

We may summarize our findings by the following statements:

The energy levels

$$(11) \qquad\qquad E_n^a + E_n^b \qquad\qquad (n = 1, 2, 3 \ \ldots)$$

are each of them single.

The energy levels

$$(12) \qquad\qquad E_n^a + E_{n'}^b = E_n^b + E_{n'}^a \qquad\qquad (n' \neq n)$$

are double.

Corresponding to an energy level (11), there is but one eigenfunction, viz.,

$$(13) \qquad\qquad \psi_n^a \, \psi_n^b \qquad\qquad (n = 1, 2, 3 \ \ldots) ;$$

and corresponding to an energy level (12), there are two different eigenfunctions, viz.,

$$(14) \qquad\qquad \psi_n^a \, \psi_{n'}^b \text{ and } \psi_n^b \, \psi_{n'}^a \qquad\qquad (n' \neq n).$$

We explained in connection with Schrödinger's theory of radiation that the eigenfunctions must be normalized to 1. In the present case, the eigenfunction $\psi_n^a \, \psi_{n'}^b$, for example, will be normalized if the product $\psi_n^a \, \psi_{n'}^b . \psi_n^{*a} \psi_{n'}^{*b}$ integrated throughout the 6-dimensional configuration space yields the value 1. This normalization can always be secured; and we shall suppose that the eigenfunctions (13) and (14) are so normalized.

We have also seen that, in the case of degeneracy, where two or more different eigenfunctions have the same energy value, we may select as eigenfunction any linear combination of these eigenfunctions. If we take a linear combination of the two different eigenfunctions (14), associated with the level (12), we obtain

$$C_1 \psi_n^a \psi_{n'}^b + C_2 \psi_n^b \psi_{n'}^a.$$

(where C_1 and C_2 are arbitrary constants). According to our former statement, this linear combination (15) is also an eigenfunction connected with the same energy level. For this eigenfunction to be normalized to 1, the constants C_1 and C_2 must be restricted by the relation

$$(16) \qquad C_1 C_1^* + C_2 C_2^* = 1.$$

The relation (16) may be satisfied in many ways; two permissible ways are illustrated by

$$(17) \qquad C_1 = \frac{1}{\sqrt{2}}, \quad C_2 = \frac{1}{\sqrt{2}}$$

and by

$$(18) \qquad C_1 = \frac{1}{\sqrt{2}}, \quad C_2 = -\frac{1}{\sqrt{2}}.$$

When these choices are made for the constants C, we obtain from (15) the two normalized eigenfunctions

$$(19) \qquad \frac{1}{\sqrt{2}}(\psi_n^a \psi_{n'}^b + \psi_n^b \psi_{n'}^a)$$

and

$$(20) \qquad \frac{1}{\sqrt{2}}(\psi_n^a \psi_{n'}^b - \psi_n^b \psi_{n'}^a).$$

The eigenfunctions (19) and (20) are mutually independent, though they are not independent of the eigenfunctions (14). The new eigenfunctions are associated with exactly the same coalesced levels (12) as the simpler eigenfunctions (14); and we may, if we choose, select them in place of the eigenfunctions (14). Let us note, however, that if we are dealing with the energy level (11), which is single, we have only one eigenfunction at our disposal, namely, (13). As a result, the more complicated eigenfunctions of type (19) and (20) do not occur, and we retain the simple eigenfunction (13) as before. In our present unperturbed problem, we need not consider the more involved kinds of eigenfunctions, but we mention them here, because they will impose themselves when we take into account the mutual repulsion of the two electrons.

The simpler eigenfunctions $\psi_n^a \psi_{n'}^b$ may be given an immediate interpretation in the particle picture represented in ordinary 3-dimensional space. Let us see how this can be done. The wave determined by the mode of vibration $\psi_n^a \psi_{n'}^b$ extends throughout the 6-dimensional configuration space. It has two regions of maximum intensity. When the representation is made in ordinary 3-dimensional space, the two regions of maximum intensity are found to lie along two Bohr orbits of quantum numbers n and n' respectively. The waves exist, however, throughout 3-dimensional space and vanish only at infinity. Since the intensity of a wave at a point measures the probability that a position observation will reveal the presence of an electron at this point, we conclude that the mode of vibration $\psi_n^a \psi_{n'}^b$, has the following significance:

If a position observation is made, the electrons a and b may be found anywhere in space; however, the electron a will most probably be found somewhere on the orbit n, and the electron b somewhere on the orbit n'.

If we were basing our judgment on Bohr's theory, where the electrons of an atom describe definite orbits, the vagueness of the wave picture could not be comprehended; and the nearest description of the present wave picture would be to say that the electrons a and b were describing the orbits n and n' respectively. For the sake of convenience, we shall accept this inaccurate description of the mode of vibration $\psi_n^a \psi_{n'}^b$. With this understanding, the other mode $\psi_n^b \psi_{n'}^a$ represents a mere interchange in the orbits of the two electrons, and a mode such as $\psi_n^a \psi_n^b$ indicates that the two electrons are in the same orbit n.

The particle interpretation of the more involved eigenfunctions (19) and (20) is not so simple. In these latter functions the electron a is associated at one and the same time with the orbit n and with the orbit n'; and likewise for the electron b. This would seem to imply that each electron is at one and the same time on two different orbits—a situation which is incomprehensible when electrons are viewed as particles. However, we may also suppose that each one of the two electrons occupies a single orbit n or n'; but that we have no means of deciding whether it is the electron a or the electron b which is, say, on the orbit n.

The Second Step—We now consider the perturbed problem, in which the two electrons exert a mutual repulsion in accordance with the known laws of electrostatics. Schrödinger's wave equation which corresponds to this situation is more difficult to solve, and so we shall apply the method of perturbations. The eigenvalues and eigenfunctions will then be obtained by successive approximations. We shall consider only the first approximation, for it suffices to clarify the theoretical points of interest. For simplicity, let us suppose that, in the unperturbed problem, the only energy levels considered are those in which $n' = 1$ while n has any permissible value. If the eigenfunctions (14) are taken, the restriction we are here making implies that one of the two electrons is always in the lowest orbit. The energy levels of the unperturbed problem are thus of type $E_n^a + E_1^b = E_n^b + E_1^a$. Changing the order of 1 and n for convenience, we may represent the various levels by the table (21). The level $E_1^a + E_1^b$ at the base of the table is the level of lowest energy.

(21)

$$E_1^a + E_n^b = E_1^b + E_n^a$$
$$\vdots \qquad \vdots$$
$$E_1^a + E_3^b = E_1^b + E_3^a$$
$$E_1^a + E_2^b = E_1^b + E_2^a$$
$$E_1^a + E_1^b$$

The equality signs indicate the levels which coalesce.

We now consider the effect of the perturbation. The perturbation causes each one of the energy levels (21) to undergo a small change, and the changes affecting any two of the coalescing levels are different. The effect of the perturbation is thus to split the former coalesced levels. For instance, the coalesced level $E_1^a + E_n^b = E_1^b + E_n^a$ is split into two slightly different levels, which we shall denote respectively by

$$E_{1n}' \text{ and } E_{1n}''.$$

The perturbed levels which take the place of the unperturbed ones (21) may then be represented by the following table:

(22)

$$E_{1n}' \qquad E_{1n}''$$
$$\vdots \qquad \vdots$$
$$E_{13}' \qquad E_{13}''$$
$$E_{12}' \qquad E_{12}''$$
$$E_{11}'$$

We shall understand presently why the basic level E_{11}' is written with one dash, like the levels in the left-hand column.

We have now to consider the eigenfunctions which correspond to the perturbed levels (22). When we follow the method of perturbations, the eigenfunctions (and also the energy levels) are obtained by a series of successive approximations. To a first approximation we find that the eigenfunctions of the perturbed problem are very nearly the same as the complicated eigenfunctions (19) and (20) of the unperturbed problem. They only differ from (19) and (20) by small additive functions which we may agree to neglect; this omission will not affect the major results of the problem. The normalized eigenfunctions corresponding to the energy levels (22) of the perturbed problem may thus be expressed by the table

(23)

$$\frac{1}{\sqrt{2}}(\psi_1^a\psi_n^b + \psi_1^b\psi_n^a) \qquad \frac{1}{\sqrt{2}}(\psi_1^a\psi_n^b - \psi_1^b\psi_n^a)$$

$$\vdots \qquad\qquad\qquad \vdots$$

$$\frac{1}{\sqrt{2}}(\psi_1^a\psi_3^b + \psi_1^b\psi_3^a) \qquad \frac{1}{\sqrt{2}}(\psi_1^a\psi_3^b - \psi_1^b\psi_3^a)$$

$$\frac{1}{\sqrt{2}}(\psi_1^a\psi_2^b + \psi_1^b\psi_2^a) \qquad \frac{1}{\sqrt{2}}(\psi_1^a\psi_2^b - \psi_1^b\psi_2^a)$$

$$\psi_1^a\psi_1^b$$

Symmetric and Antisymmetric Eigenfunctions—The eigenfunctions of the table (23) are of two kinds. All those placed in the left-hand column, and also the one at the bottom of the table, are called *symmetric*. Those in the right-hand column are called *antisymmetric*. The reason for this distinction results from the following considerations: If, in any one of the eigenfunctions, we interchange the positions of the two letters a and b, the eigenfunctions which we have called symmetric do not change. In this sense they are symmetric in the indices a and b, which refer to the two electrons. If we perform the same operation on an antisymmetric eigenfunction, the function is reproduced, but its sign is reversed. The name antisymmetric is thus justified.

By analogy we may extend the same qualifications of symmetry and antisymmetry to the energy levels (22) which correspond to the eigenfunctions (23). The basic energy level and those in the left-side column, being associated with symmetric eigenfunctions, may be called symmetric energy levels, and those in the right-hand column may be referred to as antisymmetric. In the table (22), the symmetric energy levels are represented by one dash, and the antisymmetric ones by two. Inasmuch as these eigenfunctions determine modes of vibration of the de Broglie waves (in the 6-dimensional configuration space), we may also speak of symmetric, and of antisymmetric, modes of vibration.

In the particle picture, the physical significance of the eigenfunction $\psi_1^a\psi_1^b$ of the table (23) is that both electrons a and b are in the lowest orbit $n = 1$. But, as pointed out previously, the other eigenfunctions of (23), whether symmetric or antisymmetric, are more difficult to interpret. The best course is to assume that, for either one of the two eigenfunctions

$$\frac{1}{\sqrt{2}}(\psi_1^a\psi_n^b \pm \psi_1^b\psi_n^a),$$ each electron occupies one of the two distinct

orbits 1 and n; but that we cannot specify which electron is on the orbit 1.

At all events, let us accept the latter interpretation and consider on this basis the symmetric and the antisymmetric eigenfunctions associated with the orbits 1 and 2 in the table (23). According to our interpretation, one of the electrons is on the orbit $n = 1$ and the other on the orbit $n = 2$. The two modes are different and are associated with the different energy values E_{12}' and E_{12}''. Now we cannot easily understand why the energy values should be different, since in either case the two electrons are on the same two orbits. Some light is shed on the matter when we recall that, if it were not for the mutual repulsion of the two electrons, the energy levels would be the same. It is therefore the energy of interaction of the two electrons which must be assumed different in the two cases. A simple explanation of this situation in the particle picture is to suppose that the relative positions of the electrons on the two orbits are not the same when the atom is in the symmetric or in the antisymmetric state. However, scant information can be derived from the particle picture, and so we shall not pursue the subject any further.

Radiation and the Selection Rule—When we discussed Schrödinger's theory of radiation for the hydrogen atom, we mentioned that the electron was assumed to be spread over space, forming a charge cloud of total charge $-e$. In any definite mode of vibration of the de Broglie waves, the electric density of the charge cloud at a point P was then defined by $-e$ times the intensity of the de Broglie vibrations at the point of interest. The phenomenon of radiation was accounted for by supposing that two different modes of vibration were active simultaneously. The superposition of the two modes produced an interference effect, so that the electric density of the charge cloud at a point vibrated with a frequency equal to the difference in the frequencies of the two modes. The electromagnetic waves which this vibrating charge cloud radiated were those which would be radiated by an electron permanently attached to the centre of charge of the cloud. If this electric centre vibrated at all, it would always be with the frequency just mentioned, and the electromagnetic waves would have this frequency. If in spite of the vibration of the cloud, the centre of charge failed to vibrate, we were certain that no radiation would be emitted by the total charge cloud and hence by the atom. When we applied these considerations to specific cases, the selection rules were obtained.

We pursue a similar course in the present helium problem. But the situation is more complicated here, for we now have two electrons instead of one. Besides, the de Broglie waves occur in the 6-dimensional configuration space, and so it is in this space that the charge cloud must be represented. However, we may simplify our understanding of the situation by appealing to a separate 3-dimensional space for each of the two electrons, and by considering two charge clouds, one in each of the two spaces. Each electron is thus represented by a distinct charge cloud of total charge $-e$ in a distinct 3-dimensional space.* If now we assume that the two spaces are superposed, we obtain a single change cloud in ordinary 3-dimensional space, and we may then proceed as we did in the simpler one-electron problem of the hydrogen atom. These general indications suffice to show how the emission or non-emission of radiation may be predicted in accordance with Schrödinger's postulates.

Calculation shows that, when one of the modes is symmetric and the other antisymmetric, no resultant emission is to be expected from the two clouds. We conclude that a drop in the energy of the atom from a symmetric to an antisymmetric level (or vice versa) is forbidden. This result, as we shall see, furnishes the selection rule which accounts for the absence of intercombination lines in the helium spectrum (*i.e.*, lines that would be generated by transitions between para, and ortho, levels). The methods of the correspondence principle, utilized by Bohr, were unable to yield the foregoing rule, so that we must view its discovery as illustrating a signal success for the new quantum theory. The selection rule just mentioned does not prohibit drops between symmetric levels

---

* We call $x_a, y_a, z_a, x_b, y_b, z_b$ the coordinates of the 6-dimensional configuration space, and $\psi(x_a, y_a, z_a, x_b, y_b, z_b)$ the eigenfunction (normalized). The electric density ρ_a of the charge cloud at a point x_a, y_a, z_a, due to the electron a, is obtained by integrating $-e\psi\psi^*$ over the range of the coordinates b, the coordinates x_a, y_a, z_a being kept fixed. Similarly for the density ρ_b due to the electron b.

or between antisymmetric ones. But we must not interpret this statement as implying that such drops are always possible, for, in addition to
our former selection rule, others (*i.e.*, those connected with the quantum
numbers l and m) must be taken into account.

The Exchange Phenomenon—We now propose to investigate how
the particle picture will express the simultaneous activity of a symmetric
and of an antisymmetric mode. We mentioned, when discussing the
particle representation of the individual modes, that if an atom is in a
given mode, for instance in the symmetric mode of energy E'_{12}, we may
assume that the electrons are situated in the orbits 1 and 2 respectively,
but we cannot determine which of the two electrons is in the orbit 1 or
in the orbit 2. Similar conclusions hold when we consider the corresponding antisymmetric mode of energy E''_{12}. The difference between the two
modes may be interpreted in the particle picture when we suppose that
the relative positions of the electrons in the two orbits are not the
same.

We now assume that the two modes occur simultaneously. As before,
we may grant that either orbit contains one electron. But, now, a new
phenomenon arises: the two electrons no longer remain attached to their
respective orbits; instead, they exchange positions incessantly with a
frequency equal to the difference in the frequencies of the two superposed
modes.* The frequency of the exchange is thus

$$\frac{E'_{1n} - E''_{1n}}{h}.$$

Furthermore, as occurs in all quantum transitions, the exchanges are
sudden. The atom emits no radiation during these exchanges; for while
energy is released by the drop of one electron from the orbit 2 to the
orbit 1, the same amount of energy is absorbed by the second electron
when it is raised from the orbit 1 to the orbit 2. An internal emission
and an internal absorption thus take place simultaneously and counteract
each other. The exchange phenomenon we have just described furnishes
the particle representation of the superposition of a symmetric mode and
of the corresponding antisymmetric one. It explains why no radiation
is generated.

Let us note that the exchange phenomenon would not occur if the
atom were vibrating in a single mode, whether symmetric or antisym-

---

* A more detailed analysis of this exchange phenomenon is given in the Appendix
at the end of the chapter.

metric. The superposition of the two kinds of modes is thus essential. Consequently, we may trace the possibility of the exchange phenomenon to the existence of the symmetric and antisymmetric modes. Now these modes were obtained when we combined the two independent eigenfunctions of type (10); and we mentioned that the independence of these two eigenfunctions was responsible for the degeneracy of the original unperturbed model of the helium atom. Hence to this original degeneracy must be ascribed the exchange phenomenon just studied. The reason Heisenberg qualified this degeneracy by the name *exchange degeneracy* is thus obvious. The name *resonance degeneracy* which is sometimes preferred, results from the analogy between the exchange phenomenon and the phenomenon known as resonance in acoustics. If a diapason is set into vibration and if an identical diapason which is not vibrating is placed near the first, the second diapason gradually enters into vibration and necessarily emits the same musical note that is emitted by the first diapason. This phenomenon occurs only when the two diapasons are susceptible of emitting the same musical note, and for this reason the name resonance is given to it.

The Spectrum of the Perturbed Model—The type of spectrum to be expected from our theory of the helium atom (as developed up to this point) may be understood when we combine the selection rule with the table of the energy levels (22). In this table the lowest of the energy levels is the symmetric level E'_{11}; it corresponds to the case in which both electrons are in the lowest orbit $n = 1$. The lowest of the antisymmetric levels is E''_{12}.

Suppose, then, the atom is excited to some higher symmetric level. It may drop spontaneously to any one of the lower symmetric levels and in particular to the lowest one E'_{11}, but it cannot drop to a lower antisymmetric level. If the atom is excited to a higher antisymmetric level, it can drop only to lower antisymmetric levels. Since E''_{12} is the lowest of the antisymmetric levels, it constitutes the basic level of the antisymmetric series. An immediate consequence is that if the atom is in the lowest antisymmetric level, it can drop no further, even though this level is in reality an excited one. (Incidentally, we have here an illustration of a metastable level.) These considerations show that our present perturbed model is consistent with the emission of two distinct series of spectral lines, corresponding to drops between symmetric and between antisymmetric levels; combination lines caused by drops from one type of level to the other will never occur. All the lines will be singlets.

Let us compare this spectrum of the perturbed model with the spectrum of helium that is revealed by experiment. The peculiarities of the helium spectrum show that there must be two series of levels, called the para levels and the ortho levels, the former being single and the latter triple. The lowest level of all is a para level. Transitions between para and ortho levels do not occur. If, then, we identify the symmetric and the antisymmetric levels of the perturbed model with the para and the ortho levels of helium respectively, we see that the perturbed model accounts for some of the peculiarities of the helium spectrum. In particular it accounts for the impossibility of transitions between para and ortho levels; and in this way one of the most mysterious features of the helium spectrum is explained. On the other hand, our model yields only singlet levels, and hence does not account for the triplet nature of the ortho levels.

We cannot be surprised at the failure of the wave mechanical treatment to anticipate triplets for the ortho levels. The fact is that the triplet levels are intimately connected with the electron spin (as we mentioned in dealing with Bohr's theory), and in our present wave-mechanical treatment, we have disregarded the spins. From this standpoint the two separate series of single lines, which wave mechanics predicts at this stage for the helium spectrum, represent the spectrum that would be correct if the electrons had no spin.

The Third Step—The wave theory does not suggest at this point any reason for supposing that the electrons, in the particle picture, are in a state of spin. If, then, we seek a wave analogue for the spins and incorporate it into the present theory, our procedure will be just as arbitrary as it was in Bohr's theory. Dirac's more recent discoveries in the wave theory show, however, that when relativistic considerations are introduced, a condition arises which can be expressed in the particle picture by the statement that the electrons are spinning. The hypothesis of the spinning electron thus becomes a necessary consequence of the wave theory and ceases to appear as a postulate *ad hoc*. We shall defer a study of Dirac's relativistic considerations for the present and shall merely incorporate bodily the hypothesis of the spinning electron into the wave-mechanical treatment so far outlined., This is the procedure that was followed by the first investigators in the field.

We divide the third step into two parts, which we shall call the step *A* and the step *B*. In th step *A* we assume that the two electrons in our former model are spinning, and that they are subjected to the action of the internal magnetic field of the atom. On the other hand, we disregard the mutual magnetic actions which the electrons exert as a result of their spins. Calculation shows that each energy level E'_{1n} and E''_{1n} of the former model must be replaced by four levels, but that in each quadruplet of levels two of the levels coalesce, so that only three distinct levels appear. The problem is therefore degenerate.

In the step B, we take into consideration the magnetic interaction due to the two spins and we view it as a perturbation. The former degeneracy is thereby removed and the coalesced levels separate. The net result is that each level E'_{1n} and E''_{1n} of the original atom is replaced by four *distinct* levels.

Before proceeding, we recall that, according to the hypothesis of Uhlenbeck and Goudsmidt, an electron is assumed, by reason of its spin, to have a half-unit of angular momentum $\left(i.e., \dfrac{1}{2} \dfrac{h}{2\pi}\right)$ and a magnetic moment of one magneton. The rules of space-quantization require that a spinning electron placed in a magnetic field should set its axis of spin either parallel or antiparallel to the field. The energy of the electron in the field is of course different in the two cases.

We now revert to the step A just mentioned. The two electrons in the perturbed model of the helium atom are supposed to be spinning in a magnetic field, but we disregard the interaction between the two electrons brought about by their spins. Under the action of the atom's internal magnetic field, the axes of the two electrons will then set themselves either parallel or antiparallel to the direction of the internal field. Here three situations may arise:

(α) The north poles of both electron-magnets may point in the direction of the field.

(β) The south poles of both electron-magnets may point in the direction of the field.

(γ) The north poles of the two electron-magnets may point in opposite directions along the field.

We note that, whereas the situations (α) and (β) can be realized in one way only, the situation (γ) can occur in two different ways. Thus in the situation (γ) it may be the north pole of the electron a, or else of the electron b, which points in the direction of the field.

We now consider the energy states of our former perturbed atom when the electron spins are taken into account (the mutual magnetic action of the two electrons being disregarded). Let us suppose that, if it were not for the electron spins, the atom would be in the symmetric energy level E'_{1n}. The action of the magnetic field on the spinning electrons modifies the value of the energy, and the precise change to be expected depends on which one of the three situations (α), (β), or (γ) is realized. If we denote by ε_a, ε_β, ε_γ, the magnetic energies corresponding to the three situations respectively, the total energy of the atom has one of the three values

$$(24) \qquad E'_{1n} + \varepsilon_a, \quad E'_{1n} + \varepsilon_\beta, \quad E'_{1n} + \varepsilon_\gamma.$$

Similarly, if the energy of the atom (without spins) is antisymmetric, (24) must be replaced by

$$(24') \qquad E''_{1n} + \varepsilon_a, \quad E''_{1n} + \varepsilon_\beta, \quad E''_{1n} + \varepsilon_\gamma.$$

Inasmuch as the situation (γ) may occur in two different ways, the last energy level of (24) and also that of (24') must be double; whereas the two first levels of (24) and of (24') are single. The presence of a coalesced level indicates that the problem is degenerate. The degeneracy is of the same kind that we encountered earlier, namely, a form of resonance degeneracy.

The general results just outlined may be obtained in a more rigorous way when we apply the methods of wave mechanics. We first turn our attention to the simple problem of two electrons spinning in a magnetic field. The electrons are considered independently of the atom in which they may be situated, and their mutual magnetic actions are disregarded. To investigate the problem, we view the spinning electrons as rotators, and we avail ourselves of the restriction postulated by Uhlenbeck and Goudsmidt for the angular momentum of spin. Schrödinger's wave equation for the two electrons is found to have the three eigenvalues

$$(25) \qquad \varepsilon_a, \ \varepsilon_\beta, \ \varepsilon_\gamma.$$

These are the three energy levels mentioned previously. The corresponding eigenfunctions are four in number; we denote them by

$$(26) \qquad \phi_a, \ \phi_\beta, \ \phi_{\gamma_1}, \ \phi_{\gamma_2}.$$

The last two of these eigenfunctions are associated with the same energy level ε_γ. Since two independent eigenfunctions are associated with this energy level, or eigenvalue, the problem is degenerate.

Next, let us suppose that our two spinning electrons are those of the helium atom. They are then spinning in the atom's internal magnetic field. As before, we disregard their magnetic interaction due to the spins. The physical system here considered may be viewed as resulting from a simultaneous consideration of the perturbed helium atom and of the system of two electrons spinning in a magnetic field. The eigenvalues of the wave equation are given by the sums of the energy values (22) and (25). We thus obtain the energy values (24) and (24'). The eigenfunctions connected with these eigenvalues are defined by the products of the corresponding eigenfunctions. Consequently, the eigenfunctions of (24) are

$$(27) \qquad \frac{1}{\sqrt{2}}(\psi_1^a \psi_n^b + \psi_1^b \psi_n^a) \times \phi_a \text{ or } \phi_\beta \text{ or } \phi_{\gamma_1} \text{ or } \phi_{\gamma_2},$$

and those of (24') are

$$(28) \qquad \frac{1}{\sqrt{2}}(\psi_1^a \psi_n^b - \psi_1^b \psi_n^a) \times \phi_a \text{ or } \phi_\beta \text{ or } \phi_{\gamma_1} \text{ or } \phi_{\gamma_2}.*$$

Our present system has more than six degrees of freedom, because now we have to take into consideration the degrees of freedom connected with the spins. The configuration space in which the eigenfunctions, or modes of vibration, (27) and (28) are represented has therefore more than six dimensions. But the general results we have obtained through the wave-mechanical treatment are seen to agree with those we derived from the particle picture. In particular, the same degeneracy noted previously occurs here also; it issues from the fact that the two last eigenfunctions of (27) and the two last of (28) are associated respectively with the last eigenvalue of (24) and of (24').

The problem we have just considered may be treated as the new unperturbed problem. In the step B, which we now propose to examine, we take into account the magnetic interaction of the two electrons. It is this interaction which plays the part of a perturbation. Now the perturbation, though scarcely affecting the energy values of type E' and E'', modifies in a marked degree and in different ways the various energy values ε. As a result, the two coalesced levels $E'_{1n} + \varepsilon_\gamma$ become separate, assuming new values $E'_{1n} + \varepsilon'_\gamma$ and $E'_{1n} + \varepsilon''_\gamma$. Similarly, the two coalesced levels $E''_{1n} + \varepsilon_\gamma$ are split into the two different levels $E''_{1n} + \varepsilon'_\gamma$ and $E''_{1n} + \varepsilon''_\gamma$. The degeneracy is thus removed by the perturbation. If we call ε'_a and ε'_β the energy values into which ε_a and ε_β are changed respectively, the three energy levels (24) change, under the influence of the perturbation, into the four distinct levels

$$(29) \qquad E'_{1n} + \varepsilon'_a, \quad E'_{1n} + \varepsilon'_\beta, \quad E'_{1n} + \varepsilon'_\gamma, \quad E'_{1n} + \varepsilon''_\gamma,$$

whereas the three energy levels (24') become

$$(30) \qquad E''_{1n} + \varepsilon'_a, \quad E''_{1n} + \varepsilon'_\beta, \quad E''_{1n} + \varepsilon'_\gamma, \quad E''_{1n} + \varepsilon''_\gamma.$$

---

* If $n = 1$, (27) is replaced by $\psi_1^a \psi_1^b \times \phi_a$ or ϕ_β or ϕ_{γ_1} or ϕ_{γ_2}; whereas (28) vanishes.

The eigenfunctions ψ connected with E'_{1n} and E''_{1n} are not modified by the perturbation; but those of type ϕ are affected, and to distinguish the perturbed eigenfunctions ϕ from the former ones (26), we denote them by

$$(31) \qquad \phi'_\alpha, \quad \phi'_\beta, \quad \phi'_\gamma, \quad \phi''_\gamma.$$

In this notation the functions ϕ' are symmetric and the function ϕ'' is antisymmetric.* The total eigenfunctions corresponding to the energy levels (29) and (30) are then, by analogy with (27) and (28),

$$(31) \qquad \frac{1}{\sqrt{2}}(\psi_1^a\psi_n^b + \psi_1^b\psi_n^a) \times \phi'_\alpha \text{ or } \phi'_\beta \text{ or } \phi'_\gamma \text{ or } \phi''_\gamma$$

and

$$(32) \qquad \frac{1}{\sqrt{2}}(\psi_1^a\psi_n^b - \psi_1^b\psi_n^a) \times \phi'_\alpha \text{ or } \phi'_\beta \text{ or } \phi'_\gamma \text{ or } \phi''_\gamma.$$

We have here the eigenfunctions for our completed helium atom, in which the electrons are spinning and are exerting a mutual magnetic action. Now the product of a symmetric and an antisymmetric function is an antisymmetric function; and the product of two symmetric or of two antisymmetric functions is a symmetric one. Consequently, the first three eigenfunctions of (31) are symmetric and the last is antisymmetric. Also, the first three eigenfunctions of (32) are antisymmetric and the last is symmetric. As before, we refer to the corresponding energy levels as symmetric or antisymmetric according to whether they are associated with symmetric or with antisymmetric eigenfunctions.

We are thus led to the following conclusions: The effect of the electron spins will be to cause each symmetric energy level E'_{1n} of (22) ($n = 1, 2, 3, 4 \ldots$) to be replaced by three symmetric and one antisymmetric energy level; also, each antisymmetric level E''_{1n} of (22) ($n = 2, 3, 4 \ldots$) will be replaced by three antisymmetric and one symmetric energy level. If we represent the symmetric eigenvalues by ● and the antisymmetric

---

* To a first approximation, the new eigenfunctions ϕ'_γ and ϕ''_γ are equal to the sum and to the difference of the original ones ϕ_{γ_1} and ϕ_{γ_2}. Thus

$$\phi'_\gamma = \phi_{\gamma_1} + \phi_{\gamma_2}$$

and

$$\phi''_\gamma = \phi_{\gamma_1} - \phi_{\gamma_2}.$$

ones by —, we may picture the distribution of energy levels by the following table:

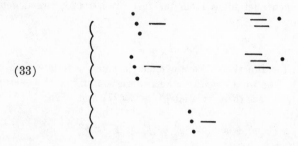

(33)

The table (33) depicts the splitting of the simpler levels which were illustrated in the table (22), these simpler levels corresponding to the case where the electron spins were disregarded.

Whereas the table (22) did not furnish a sufficient number of energy levels to account for all the lines of the helium spectrum, the table (33) yields too many levels. Hence we conclude that some of the levels of (33) must be impossible. To ascertain which of these levels do not occur in practice, we compare the spectrum which must be expected according to the table of levels (33) with the helium spectrum which is actually observed in the laboratory.

Let us consider the spectrum which we should expect from the table (33). Owing to the spins which we are now attributing to the electrons, our completed model of the helium atom has an increased number of degrees of freedom. Hence Schrödinger's charge cloud, which we discussed before the electron spins were introduced, must be represented in a configuration space having an increased number of dimensions. This change in the configuration space does not, however, affect the results obtained in the simpler treatment. As before, we find that, if a symmetric and an antisymmetric mode of vibration occur simultaneously, the charge cloud as a whole emits no resultant radiation. Drops between levels of opposite symmetry are thus excluded. If, then, (33) gave the true picture of the energy levels for the helium atom, we should expect two different types of spectra according to whether drops occurred solely between symmetric levels or solely between antisymmetric ones. The drops between antisymmetric levels — would yield a spectrum in which the basic level was single; whereas the drops between symmetric levels • would yield a spectrum in which the basic level was triple. Now the empirical observation of the helium spectrum proves that the basic level is single, and we must therefore exclude as impossible all the symmetric levels ● of the table (33). Our table of energy levels thus becomes

(34)

para   ortho

We appear to have here an allotment of energy levels which is in agreement with the peculiarities of the helium spectrum, the singlet levels being the para levels, and the triplet levels being the ortho levels.

We have yet to show that transitions cannot occur between the singlet and the triplet levels, for we know that this restriction is a characteristic feature of the helium spectrum. Inasmuch as in (34) all the levels concerned are antisymmetric, our former selection rule would not appear to prohibit drops from a para to an ortho level and vice versa. But more attentive consideration shows that such drops are necessarily excluded. This point is understood when we examine the eigenfunctions which are associated with the various levels.

Thus the antisymmetric eigenfunctions of a para level are of type

$$(35) \qquad \frac{1}{\sqrt{2}}(\psi_1^a \psi_n^b + \psi_1^b \psi_n^a) \times \phi_\gamma''.$$

The first factor in (35) is a function which is symmetric in the coordinates of the two electrons in the original 6-dimensional configuration space; and the second factor ϕ_γ'' is antisymmetric in the coordinates which define the orientations of the electronic axes. On the other hand, the antisymmetric eigenfunctions connected with an ortho level are of type

$$(36) \qquad \frac{1}{\sqrt{2}}(\psi_1^a \psi_n^b - \psi_1^b \psi_n^a) \times \phi_\alpha' \quad \text{or} \quad \phi_\beta' \quad \text{or} \quad \phi_\gamma'.$$

Here the conditions of symmetry and of antisymmetry are reversed.

If, then, we consider the situation where a mode (35) and a mode (36) occur simultaneously, we see that the first parts of the functions (35) and (36) are of opposite symmetry, and so also are the second parts. This opposition in the symmetry excludes the possibility of radiation. The theory we have developed is thus able to anticipate the characteristics of the helium spectrum.

Pauli's Exclusion Principle—An unsatisfactory feature of the present theory is that it affords no theoretical justification for omitting the symmetric energy levels. These levels were rejected only because their retention would have entailed a more complicated spectrum for helium than is observed in practice. But we shall now see that the rejection of the symmetric levels is a consequence of Pauli's exclusion principle and is thereby connected with a large body of facts; it no longer stands as an isolated phenomenon.

We recall that, according to Pauli's exclusion principle, no two electrons in the same atom can be associated with the same quadruplet of numerical values for the four quantum numbers n_i, l_i, j_i, and m_i; or, if we prefer, for the quantum numbers n_i, l_i, m_{l_i} and m_{s_i}.* The latter choice of quantum numbers will be taken in the present discussion. Here n_i is the main quantum number of an electron, l_i its angular momentum due to its circling, m_{l_i} the projection of this angular momentum along the direction of an applied magnetic field, and m_{s_i} the projection of the spin vector $\left(\text{all these angular momenta being expressed in units } \dfrac{h}{2\pi}\right)$.

Let us revert to the model of the helium atom considered in the first step (page 756). In this model, the electron spins were disregarded and no interaction between the two electrons was taken into consideration. The two electrons were then associated with the quantum numbers n, l, m and n', l', m', these numbers being in effect the quantum numbers n_i, l_i, m_{l_i} of Pauli's principle (with $i = 1$ and 2). Pauli's number m_{s_i} did not appear because we were neglecting the spins. In our earlier treatment we agreed to omit the quantum numbers of type l and m because they were superfluous for our purpose; but since we now propose to take Pauli's principle into consideration, we must reinstate these quantum numbers. So as to simplify the symbolism, we shall omit the indices i and represent the quantum numbers of the two electrons by n, l, m_l, m_s and n', l', $m'_{l'}$, $m'_{s'}$.

We now pass to the perturbed model of the helium atom, in which the spins are still disregarded but the mutual electrostatic repulsion of the two electrons is taken into account. No change occurs in the significance of the three quantum numbers. The normal modes of vibration connected with this model are determined by the symmetric and the antisymmetric eigenfunctions of type (19) and (20). The eigenfunctions (19) and (20) (in which the normalizing factor $\dfrac{1}{\sqrt{2}}$ is omitted for simplicity), may be written

$$(37) \qquad \psi^a_{n,l,m_l}\psi^b_{n',l',m'_{l'}} + \psi^b_{n,l,m_l}\psi^a_{n',l',m'_{l'}}$$

and

$$(38) \qquad \psi^a_{n,l,m_l}\psi^b_{n',l',m'_{l'}} - \psi^b_{n,l,m_l}\psi^a_{n',l',m'_{l'}};$$

* See note on page 580.

the former are symmetric, the latter antisymmetric. These functions indicate that the electrons a and b are associated with the orbits (n, l, m_l) and $(n', l', m'_{l'})$.

When the electron spins are taken into account, the eigenfunctions are (31) and (32) and hence will now be written

$$(39) \quad (\psi^a_{n,l,m_l}\psi^b_{n',l',m'_{l'}} + \psi^b_{n,l,m_l}\psi^a_{n',l',m'_{l'}}) \times \phi'_\alpha \quad \text{or} \quad \phi'_\beta \quad \text{or} \quad \phi'_\gamma \quad \text{or} \quad \phi''_\gamma,$$

and

$$(40) \quad (\psi^a_{n,l,m_l}\psi^b_{n',l',m'_{l'}} - \psi^b_{n,l,m_l}\psi^a_{n',l',m'_{l'}}) \times \phi'_\alpha \quad \text{or} \quad \phi'_\beta \quad \text{or} \quad \phi'_\gamma \quad \text{or} \quad \phi''_\gamma.$$

Let us examine the behavior of the eigenfunctions (39) and (40) when we assume that the two electrons have the same quadruplet of quantum numbers, *i.e.* when

$$(41) \quad n' = n, \, l' = l, \, m'_{l'} = m_l, \, m'_{s'} = m_s.$$

We first observe that, in view of the last quality (41) (spins in the same sense), only those eigenfunctions (39) and (40) which contain ϕ'_α or ϕ'_β need be considered. Accordingly, we restrict our attention to the two first eigenfunctions of (39) and of (40). Let us, then, impose the first three equalities (41) on these eigenfunctions. We see that the functions (40) vanish whereas the functions (39) retain non-vanishing values. Now the eigenfunctions define modes of vibration, and the latter constitute the wave representations of the various allotments of quantum numbers to the two electrons. Consequently, the vanishing of the first two eigenfunctions (40), when the equalities (41) are imposed, implies that these eigenfunctions are inconsistent with these equalities and hence with the allotment of the same quadruplet of quantum members to the two electrons. On the other hand, since the first two functions (39) do not vanish, they are consistent with this allotment.

These preliminary considerations show that if we wish to obtain the wave representation of Pauli's principle, we must reject the first two eigenfunctions of (39) and retain the first two of (40). We next consider the four remaining eigenfunctions (39) and (40). Since they contain ϕ'_γ and ϕ''_γ, they correspond to the case in which the electrons are spinning in opposite senses. Pauli's principle does not interfere with the presence of the two electrons on the same orbit when they are spinning in opposite senses. Hence these eigenfunctions do not appear to be excluded by Pauli's principle. We must remember, however, that our introduction of the electron spins has been highly artificial, for we have merely postulated the spins as an additional hypothesis, and then sought to incorporate their wave representation into the wave picture.*

In view of the crudeness of our present procedure, we cannot be certain that the remaining four eigenfunctions, mentioned above, should be retained; and so we must be guided by other clues. In this connection, we note that the two eigenfunctions which are certainly incompatible with Pauli's principle (*i.e.* the first two of (39)) are symmetric; and we may assume that their rejection entails that of the symmetric eigenfunctions generally. If this assumption is made, the only possible eigenfunctions are the antisymmetric ones, represented by the last of (39) and by the three first of (40). The wave expression of Pauli's principle consists therefore in the statement that all symmetric eigenfunctions, or modes of vibration, must be excluded. Since the exclusion of the symmetric eigenfunctions necessarily entails that of the corresponding symmetric eigenvalues, or energy levels, we see that in the completed helium atom we must retain only the antisymmetric energy levels. We thus obtain the same scheme as was required by the spectroscopic evidence.

In our analysis we have merely accepted Pauli's principle and examined the wave requirements which it entails. This procedure does not of course demonstrate the necessity of the principle, for to establish this necessity, we should have to show that theoretical reasons oppose the existence of symmetric modes in an atom. The interest of the analysis lies more particularly in the fact that it has afforded a wave interpretation of Pauli's principle.

When we discuss the New Statistics in Chapter XL, we shall find that Pauli's empirical principle, originally restricted to the arrangements of the electrons in Bohr's atom, has a far wider significance than was at one time suspected. The principle, when generalized, is valid for any aggregate composed of similar particles which carry an odd multiple of the fundamental electric charge e (Fermi-Dirac statistics). This is precisely the situation that holds in atoms containing more than one electron. On the other hand, a diametrically opposite principle, represented in the wave picture by the retention of the symmetric functions and the rejection of the antisymmetric ones (the Bose-Einstein statistics), may also be realized in Nature. It will be applicable whenever the electric charges of the similar particles are even multiples of the fundamental charge e or, as a particular case, when they have no charge at all (*e.g.*, photons or the neutral molecules of a gas).

---

* The proper treatment of the spins is afforded by Dirac's theory of the electron (Chapter XXXIX).

The Exchange Phenomenon

CONSIDER two identical linear harmonic oscillators vibrating with the same frequency ν_0. We assume that the vibrating particles are electrons. If these two oscillators are so far apart that they exert no mutual actions, the vibrations will proceed independently with the common frequency ν_0. We now place the oscillators side by side. Owing to the proximity of the two electrons, their mutual repulsive action ceases to be negligible, and each oscillator exerts a perturbing influence on the motion of the other. The two oscillators are said to be *coupled*, the so-called *coupling force* being illustrated by the mutual repulsion of the two electrons. We must now view the two oscillators jointly as forming a single system. If we disregard the energy loss due to radiation, the system is conservative and hence its total energy remains constant during the motion.

Two particular types of motion corresponding to two normal modes may occur. In one of these, both electrons vibrate together with the same amplitude and frequency, and the electrons are always moving simultaneously in the same direction. We call this mode the symmetric mode. In the other normal mode, the electrons are also vibrating with the same amplitude and frequency, but this time in opposite directions. We call this mode the antisymmetric mode. Calculation shows that the frequencies ν' and ν'' of the symmetric and antisymmetric mode, respectively, are not the same, and that both frequencies differ from the original frequency ν_0 of the unperturbed motion. The unperturbed frequency ν_0 has a value half-way between the values ν' and of ν''. It can also be shown that the difference $\nu' - \nu''$ in the frequencies of vibration will increase in absolute value if the coupling force between the two oscillators is increased.

When the two oscillations are started from arbitrary initial conditions, neither of the two normal modes occurs. Nevertheless, in any case the motion of the system may be viewed as due to an appropriate superposition of the two normal modes. From this fact we may anticipate the nature of the motion in the general case. As an example, let us suppose that at the initial instant the electron a is at rest at the centre of oscillation, while b is oscillating. We shall then find that the oscillation of the electron b will gradually decrease, while the electron a enters into oscillation with its amplitude of swing increasing progressively. Eventually the electron b will come to rest at its centre of oscillation, and a will then be vibrating with maximum amplitude. The total energy of the system, originally concentrated in the oscillator b, has thus been transferred to the oscillator a. After this new state of affairs is realized, the reverse process will occur: the electron a, now oscillating violently, will gradually come to rest and the electron b will resume its original oscillatory motion of maximum amplitude. In short, the energy of the system will be transferred back and forth periodically from one electron to the other. The frequency of this periodic transference of energy is equal to the difference $\nu' - \nu''$ (in absolute value) in the frequencies of the symmetric and of the antisymmetric modes. The stronger the coupling between the two electrons, the greater the absolute value of $\nu' - \nu''$, and hence the more frequently will the energy be transferred back and forth.

Instead of supposing that one of the electrons is at rest at the initial instant, we may suppose that both electrons are vibrating, but with different amplitudes. The exchange phenomenon will occur as before, though in the general case neither of the two electrons will come to rest at the centre of oscillation, so that only a part of the total energy will be transferred back and forth from one electron to the other. If we call E_1 and E_2 the maximum and the minimum energy of one oscillator, the energy which will be transferred periodically is $E_1 - E_2$.

The exchange phenomenon we have considered for two coupled oscillators arises whenever two identical vibrating systems are coupled. Let us understand how it applies to the model of the perturbed helium atom (*i.e.*, to the model in which the electron spins are disregarded but the mutual repulsion of the two electrons is taken into account). To examine the matter, we must first revert to the unperturbed atom, in which the mutual repulsion of the electrons is not taken into account. According to Bohr's particle theory, the electrons are describing definite orbits. For instance, we may suppose that one electron is describing the orbit 1 and the other the orbit 2. In the wave representation, we must differentiate between two cases. Thus it may be the electron a or the electron b which is describing the first orbit. To these two different situations correspond two independent eigenfunctions, or modes of vibration, for the de Broglie waves; they are represented by $\psi_1^a \psi_2^b$ and $\psi_1^b \psi_2^a$ respectively. The two modes have the same frequency, which we may designate by ν_0.

Each one of these two modes depicts a condition which concerns both electrons simultaneously, but in a purely formal way each mode may be represented by a single oscillator. The two modes may then be represented by two identical oscillators having the same frequency of vibration ν_0. When the two modes are superposed, we must assume that the two oscillators are vibrating simultaneously. In the model of the helium atom which we are here considering, the mutual repulsion of the two electrons is disregarded, so that the two oscillators are uncoupled. But if we pass to the perturbed model, in which the mutual perturbation is taken into account, we must assume that the oscillators are coupled; and the system of the two oscillators will then be in a symmetric mode, or in an antisymmetric mode, or in a superposition of the two. From the standpoint of the de Broglie waves, the frequency will be $\nu' = \dfrac{E'_{12}}{h}$ in the symmetric mode, and $\nu'' = \dfrac{E''_{12}}{h}$ in the antisymmetric one. When these two modes are superposed, the exchange phenomenon takes place, and energy is transferred back and forth from one oscillator to the other with the frequency $\nu' - \nu''$.

Let us construct the particle picture of this condition. The two oscillators portray the original modes $\psi_1^a \psi_2^b$ and $\psi_1^b \psi_2^a$. Hence, when the energy is transferred back and forth from one oscillator to the other with frequency $\nu' - \nu''$, the respective situations represented by the two individual modes must follow each other with this frequency. Since the first mode implies that the electron a is in the first orbit, whereas the second mode implies that the electron b is in this orbit, we conclude that the two electrons will exchange orbits back and forth with the frequency $\nu' - \nu''$. In the quantum theory, only sudden transitions can occur, and so the exchanges in the orbits will take place brusquely after recurrent intervals of time.

Our analysis also enables us to understand why the coupling force between the electrons splits the two coalesced levels of the unperturbed problem. Primarily, the coupling splits the original frequency ν_0 of the de Broglie waves into the two frequencies ν' and ν''; and since in wave mechanics the frequency of the wave determines the energy, the coupling automatically splits the original energy level $h\nu_0$ into the two levels $h\nu'$ and $h\nu''$ (i.e., E'_{12} and E''_{12}).

CHAPTER XXXVI

MATRICES

Matrices—The name matrix is given in mathematics to an aggregate of magnitudes arranged in a certain order. We shall consider only the so-called square matrices. Thus

$$(1) \qquad \left\{ \begin{array}{ccc} a_{11} & a_{12} & a_{13} \\ a_{21} & a_{22} & a_{23} \\ a_{31} & a_{32} & a_{33} \end{array} \right\}$$

is a square matrix with three rows and columns. Though the terms a_{ik} in the matrix stand for numbers (real or complex), the matrix itself is not a number in the ordinary sense; it is an array of numbers.

From the mathematical standpoint, matrices can be given a definite significance only when we have agreed on their rules of combination. Inasmuch as they are not ordinary numbers, we must not suppose that their associative, distributive, and commutative properties are necessarily those of ordinary numbers. In fact at the present stage, we may postulate any rules of combination we choose (provided they be consistent).

The introduction of matrices into mathematics did not result, however, from a mere desire to construct new mathematical beings exhibiting strange properties. Matrices were introduced by Cayley for the definite purpose of expressing in condensed form the well-known properties of linear transformations. For this reason the associative, distributive, and commutative * properties of matrices were imposed from the start.

Linear Transformations—The linear transformations which lead to the consideration of square matrices are of a type called homogeneous. These alone will be examined in this chapter. Furthermore, for our present purpose, it will be sufficient to consider the transformations of the plane. We shall assume that x_1 and x_2 are the Cartesian coordinates of any point P in the plane; and we shall suppose that, as a result of the transformation, a point P is transformed into a point P' of coordinates x'_1 and x'_2.

---

* We shall see that the commutative property does not hold in the case of multiplication.

The general linear homogeneous (or "linear" for short) transformation of the plane may be written

$$(2) \qquad \begin{cases} x_1' = a_{11}x_1 + a_{12}x_2 \\ x_2' = a_{21}x_1 + a_{22}x_2, \end{cases}$$

where the coefficients a_{ik} are constants. When the values of these constants are not specified, the transformation (2) furnishes no definite information. We shall suppose therefore that the constants a_{ik} represent known real values.

When the values of the constants a_{ik} are specified, the transformation (2) supplies us directly with the coordinates x_1', x_2' of the transformed point P' in terms of the coordinates x_1, x_2 of P. Thus to each point P corresponds a definite point P'. Usually, the reverse is also true, so that a one-to-one correspondence is set up between the points P and P'. In this case, however, the coefficients a_{ik} cannot be entirely arbitrary; they must satisfy the condition that the determinant * of the transformation be non-vanishing. When this condition is satisfied, the linear transformation is called *non-singular*. In the sequel, only non-singular transformations will be considered.

What is known as the *matrix* of the transformation (2) is the assemblage of coefficients a_{ik} written in the same order as they appear in the transformation. We agree to represent this matrix by the letter *a*. The matrix *a* of the transformation (2) is thus

$$(3) \qquad a = \begin{Bmatrix} a_{11} & a_{12} \\ a_{21} & a_{22} \end{Bmatrix}.$$

The constants a_{ik} in the expression (3) are called the *terms*, or the *elements*, of the matrix. Each different transformation is thus associated with a corresponding matrix; and since an interchange of the coefficients a_{ik} of the transformation (2) would alter the transformation, we conclude that an interchange in the positions of the elements of the matrix (3) would modify the matrix.

It may happen that the constants a_{ik} in the transformation (2) have such values that when P is any point in the plane, the transformed point

---

* The determinant is $\begin{vmatrix} a_{11} & a_{12} \\ a_{21} & a_{22} \end{vmatrix}$, which is a symbolic way of writing $a_{11}a_{22} - a_{12}a_{21}$. In contradistinction to a matrix, a determinant has a well-defined numerical value.

P' is always obtained by submitting the point P to a definite rotation about the origin O. The linear transformation in this case is equivalent to a rotation about the origin; it is called an *orthogonal transformation*. Thus

$$(4) \qquad \begin{cases} x_1' = x_1 \cos \theta + x_2 \sin \theta \\ x_2' = - x_1 \sin \theta + x_2 \cos \theta, \end{cases}$$

in which θ is a fixed angle, is an orthogonal transformation. Its effect is to rotate any point P by an angle θ about the origin in a clockwise direction. The matrix of an orthogonal transformation is called an *orthogonal matrix*. Thus the matrix of (4), namely,

$$(5) \qquad \begin{Bmatrix} \cos \theta & \sin \theta \\ -\sin \theta & \cos \theta \end{Bmatrix}.$$

is an orthogonal matrix.

For the present we shall be concerned with the general linear transformation. Suppose, then, we consider a second transformation

$$(6) \qquad \begin{cases} x_1' = b_{11} x_1 + b_{12} x_2 \\ x_2' = b_{21} x_1 + b_{22} x_2. \end{cases}$$

Its matrix b is

$$(7) \qquad b = \begin{Bmatrix} b_{11} & b_{12} \\ b_{21} & b_{22} \end{Bmatrix}.$$

If all the corresponding elements a_{ik} and b_{ik} of the two matrices (3) and (7) are equal two by two, the matrices are necessarily identical. This prompts us to inquire whether two matrices can be equal without being identical. Two determinants, for instance, may have the same numerical value and hence be equal, and yet the elements in the determinants may be different so that the determinants are not identical. Thus the two determinants

$$\begin{vmatrix} 2 & 5 \\ 2 & 6 \end{vmatrix} \quad \text{and} \quad \begin{vmatrix} 1 & 0 \\ 0 & 2 \end{vmatrix}$$

are equal, for their common value is 2; yet they are not identical. But with matrices, this situation cannot arise, for equality involves ordinary magnitudes, whereas matrices are not numbers in the ordinary sense; they express a definite ordering of ordinary magnitudes. The only significance that can be attached to the statement that two matrices are equal is that the matrices are identical; *i.e.*, that their corresponding elements are equal two by two. When, therefore, we write down a matrix equation such as

$$a = b,$$

we mean that the following equalities are satisfied:

$$a_{11} = b_{11}, \quad a_{12} = b_{12}, \quad a_{21} = b_{21}, \quad a_{22} = b_{22}.$$

The features we have just stressed are clarified when we revert to linear transformations. For instance, the two transformations (2) and (6) are equivalent if the correspondence they establish between points P and P' is the same. A situation of this sort will be realized when, and only when, the coefficients which occupy similar positions in the expressions of the two transformations are equal two by two. But then the transformations are necessarily identical since there is no means of distinguishing one from the other. We conclude that two transformations are either identical or else different. The same conclusions must therefore be extended to the matrices which define the transformations.

The Addition of Matrices—The additive properties of matrices may be derived from the properties of linear transformations. Let us suppose that (2) and (6) are different transformations. The first transformation (matrix a) transforms a point P into a point P', whereas the second transformation (matrix b) transforms the same point P into a point P''. To make the situation clearer, we shall write the transformation (6) as

$$(8) \quad \begin{cases} x_1'' = b_{11}x_1 + b_{12}x_2 \\ x_2'' = b_{21}x_1 + b_{22}x_2. \end{cases}$$

We now consider the point P''' of coordinates

$$(9) \quad \begin{aligned} x_1''' &= x_1' + x_1'' \\ x_2''' &= x_2' + x_2'', \end{aligned}$$

and we wish to determine the transformation which will transform the point P into this new point P'''. The transformation in question is obviously obtained by adding the transformations (2) and (8). It is thus

$$(10) \quad \begin{cases} x_1''' = (a_{11} + b_{11})x_1 + (a_{12} + b_{12})x_2 \\ x_2''' = (a_{21} + b_{21})x_1 + (a_{22} + b_{22})x_2. \end{cases}$$

Its matrix is

$$(11) \quad \begin{Bmatrix} a_{11} + b_{11} & a_{12} + b_{12} \\ a_{21} + b_{21} & a_{22} + b_{22} \end{Bmatrix}.$$

Inasmuch as the two transformations have been added, we will agree to say that their respective matrices have been added. Consequently, the matrix (11) is the matrix

$$(12) \qquad \boldsymbol{a + b}.$$

We are now in possession of the rule of addition for matrices. It states that the corresponding elements of the two matrices must be added.

A similar rule holds for the subtraction of matrices. The difference $a - b$ of two matrices a and b is expressed by a matrix the elements of which are the differences of the corresponding elements of the two matrices.

In particular, if the two matrices a and b are equal and hence identical, their difference is expressed by a matrix in which every element is zero. This matrix is called the zero matrix.* Obviously, if the zero matrix is added to any matrix a, the result is still the matrix a.

We see from (11) that the equality

$$a + b = b + a$$

holds just as it does for ordinary numbers. Thus matrices exhibit the commutative property for addition. They also exhibit the associative property for addition, e.g.,

$$a + (b + c) = (a + b) + c.$$

Our next step will be to examine the multiplication of matrices. We shall find that the associative and the distributive properties hold, but that the commutative property is invalid.

---

* The matrix all of whose elements vanish is the only matrix that can be connected with the value zero. This fact illustrates one of the differences between matrices and determinants, for a determinant may quite well vanish even though all its elements do not vanish, e.g., the determinant $\begin{vmatrix} 1 . 2 \\ 2 . 4 \end{vmatrix}$.

The Multiplication of Matrices—Let us revert to the first linear transformation (2) of matrix a. It transforms any point $P(x_1, x_2)$ into a corresponding point $P'(x_1', x_2')$. We now consider a second linear transformation which we apply to the former transformed point P'. This point will then be transformed to some other point $P''(x_1'', x_2'')$. We represent the second transformation by

$$(14) \qquad \begin{cases} x_1'' = b_{11}x_1' + b_{12}x_2' \\ x_2'' = b_{21}x_1' + b_{22}x_2', \end{cases}$$

its matrix is (7), i.e.,

$$(15) \qquad b = \begin{Bmatrix} b_{11} & b_{12} \\ b_{21} & b_{22} \end{Bmatrix}.$$

We shall agree to express this sequence of transformations by saying that the two transformations have been multiplied the one by the other.

We now propose to determine a third transformation which will transform our initial point P directly into the final point P'', without causing it to pass through the intermediary point P'. Suppose the required transformation is

$$(16) \qquad \begin{cases} x_1'' = c_{11}x_1 + c_{12}x_2 \\ x_2'' = c_{21}x_1 + c_{22}x_2, \end{cases}$$

where the quantities c_{ik} are appropriate constants. The matrix of (16), which we will call c, is represented by

$$(17) \qquad c = \begin{Bmatrix} c_{11} & c_{12} \\ c_{21} & c_{22} \end{Bmatrix}.$$

We may easily determine from (2) and (14) the values of the constants c_{ik} in terms of the constants a_{ik} and b_{ik}. We find

$$(18) \qquad \begin{cases} c_{11} = b_{11}a_{11} + b_{12}a_{21}, \quad c_{12} = b_{11}a_{12} + b_{12}a_{22} \\ c_{21} = b_{21}a_{11} + b_{22}a_{21}, \quad c_{22} = b_{21}a_{12} + b_{22}a_{22}. \end{cases}$$

Since we have agreed to say that the two transformations (2) and (14) have been multiplied, we shall extend the same convention to their matrices. Consequently, we shall view the matrix c as resulting from the multiplication of the matrix a by the matrix b.

If matrices were ordinary numbers, we should obtain exactly the same result whether we were to multiply *a* by *b*, or *b* by *a*; and it would be unnecessary to devise a notational scheme defining the order in which the multiplication was effected. But when we are dealing with entities which, like matrices, are not ordinary numbers, we cannot presume without further inquiry that the order of multiplication has no importance. In other words, we cannot take for granted that the commutative property holds for multiplication. Indeed, as we shall see presently, this property does not hold in the case of matrices. Accordingly, we must agree on some notational scheme for the multiplication of matrices. We shall agree that when a matrix *a* is multiplied by a matrix *b*, the product will be written *ba*. On this basis the matrix *c* mentioned above will be denoted by

(19) $c = ba.$

From the standpoint of linear transformations, the equality (19) implies that the transformation of matrix *a* has been applied first, that it has been followed by the transformation of matrix *b*, and that the result of the two successive transformations is the same as the one we should have obtained had we applied the single transformation of matrix *c*.

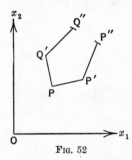

Fig. 52

Suppose now we reverse the order of the transformations. We first apply the transformation of matrix *b* and follow it with the transformation of matrix *a*. We operate as before on the point *P*. The transformation *b* displaces *P* to some point *Q'*; and then, if we operate with the transformation *a* on the point *Q'*, we obtain a point *Q''*. We propose to demonstrate that, in general, the point *Q''* differs from our former point *P''*. Our demonstration will show thereby that the effect of applying the two transformations *a* and *b*, in one order of succession or in the other, does not usually yield the same result.

To prove this statement, we proceed as before by seeking the transformation which will transform the point *P* to the point *Q''* directly. Let

(20) $$\begin{cases} x_1'' = d_{11}x_1 + d_{12}x_2 \\ x_2'' = d_{21}x_1 + d_{22}x_2 \end{cases}$$

be this transformation, of matrix d. The numerical values of the coefficients d_{ik} in (20) are given analogously to (18) by the relations

$$(21) \quad \begin{cases} d_{11} = a_{11}b_{11} + a_{12}b_{21}, & d_{12} = a_{11}b_{12} + a_{12}b_{22} \\ d_{21} = a_{21}b_{11} + a_{22}b_{21}, & d_{22} = a_{21}b_{12} + a_{22}b_{22}. \end{cases}$$

We observe that the expressions (21), determining the coefficients d_{ik}, differ from the expressions (18), defining the coefficients c_{ik}. Hence only under exceptional conditions will the numerical values of the corresponding coefficients c_{ik} and d_{ik} be the same. This shows that only under special conditions will a reversal of the order, in which the transformations a and b are applied, yield the same resultant transformation. The two resultant transformations (16) and (20) are thus usually different. The difference will betray itself in the graphical representation of Figure 52 by the failure of the final points P'' and Q'' to coincide. From the standpoint of matrices, the difference in the two resultant transformations will entail a difference in the corresponding matrices c and d.

The expression of the matrix c in terms of the matrices a and b was given by (19). And since the matrix d is obtained by reversing the order of the transformations a and b, we must set

$$(22) \qquad d = ab.$$

The matrices c and d being usually different, we conclude that in general the matrices ba and ab are not the same. Thus, in general

$$(23) \qquad ba \neq ab, \text{ or } ba - ab \neq 0.$$

This feature is expressed by the statement that the multiplication of matrices is not usually commutative.

In special cases, however, matrices may commute. Thus, if the numerical values of the constants a_{ik} and b_{ik} in (17) and (21) happen to confer the same numerical values on the constants c_{ik} and d_{ik} two by two, the transformations (16) and (20) will be the same; and the corresponding matrices c and d will be identical. We should then have $ba = ab$, and the matrices a and b would commute. In the graphical representation of Figure 52, this situation would be expressed by the coincidence of the transformed points P'' and Q'', regardless of our choice of the original point P. Other special cases of commutation will be mentioned later.

Let us examine a simple example in which, according to circumstances, two matrices may or may not commute. Suppose we have two plane mirrors placed perpendicularly to the page. They are represented by

the straight lines OA and OB in Figure 53. Consider any point P on
the paper. Its image in the mirror A is the symmetrically situated
point P', and in the mirror B its image is the point Q'. If we displace
the point P over the sheet of paper, its image P' will move, and a
one-to-one correspondence is set up between the points P and P'. The
pairs of points are connected by a linear transformation which we shall
call the transformation of matrix a. Similar arguments apply to the

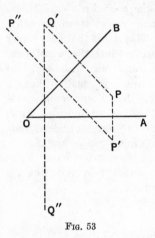

point P and to its image Q' in the mirror B.
These two points likewise are connected by
a linear transformation, but it is not the
same transformation as the previous one.
We shall refer to the new transformation
as the transformation of matrix b. When
we pass from P to its image P' in the mirror A, we are transforming the point P by
the transformation a; and when we pass
from P to its image Q' in the mirror B, we
are transforming P by the transformation b.

Suppose, then, we consider the reflection
P' of any given point P in the mirror A,
and then take the reflection of P' in the
mirror B. These two reflections are the
physical expressions of the two transforma-

FIG. 53

tions a and b. Thus a is applied to P, yielding P', and then b is applied
to P', yielding P''. The direct passage from P to P'' is thus represented
by a transformation of matrix c, where

$$c = ba.$$

Let us reverse the order of the reflections. We first reflect P in the mirror B, obtaining thereby Q', and then reflect Q' in the mirror A, obtaining the point Q''. The direct passage from P to Q'' may be accomplished
by applying the transformation of matrix d, where

$$d = ab.$$

The figure shows immediately that the points P'' and Q'' will not
coincide in general. This implies that the resultant transformations of
matrices c and d, respectively, are not the same, so that we have

$$ba \neq ab.$$

The two matrices do not commute.

But we also see from the figure that if the two mirrors happened to
be placed at right angles, the two points P'' and Q'' would always coincide

regardless of the position of the point P. In this event the two transformations c and d would always have the same effect and would therefore be identical. We should have $ba = ab$, and hence the two matrices would commute.

This example shows in an elementary way why matrices do not necessarily commute, and it warns us that the commutative property, to which we have become accustomed through our daily experience with ordinary numbers, is by no means a general property which must be expected to hold in all situations. Vector analysis furnishes another example in which multiplication is non-commutative. Thus, if $\vec{x}$ and $\vec{y}$ represent two vectors, their "vector products" $\vec{x} \times \vec{y}$ and $\vec{y} \times \vec{x}$ are not the same, for we have

$$\vec{x} \times \vec{y} = -(\vec{y} \times \vec{x}),$$

so that the order in which the multiplication is effected modifies results.

Diagonal Matrices—A variety of matrix of considerable importance in the mathematics of Heisenberg's method is the so-called *diagonal matrix*. In a matrix of this sort, only the diagonal elements extending from the upper left-hand corner are non-vanishing. Thus, *e.g.*,

$$(23) \qquad \left\{ \begin{array}{cc} a_{11} & 0 \\ 0 & a_{22} \end{array} \right\}$$

is a diagonal matrix. Obviously it is the matrix of the transformation

$$(24) \qquad \left\{ \begin{array}{l} x'_1 = a_{11} x_1 \\ x'_2 = a_{22} x_2. \end{array} \right.$$

In the particular case where the elements on the diagonal have the same numerical value, a diagonal matrix is called a *scalar matrix*. If the elements on the diagonal have the value unity, the matrix obtained is called the *unit matrix*. The unit matrix is represented by **1**. It is defined by

$$(25) \qquad 1 = \left\{ \begin{array}{cc} 1 & 0 \\ 0 & 1 \end{array} \right\},$$

and it corresponds to the identical transformation

$$(26) \qquad \left\{ \begin{array}{l} x'_1 = x_1 \\ x'_2 = x_2. \end{array} \right.$$

In such a transformation each point P is left unchanged; or, if we prefer, each point is transformed into itself.

The following commutation rules may be mentioned. Two diagonal matrices always commute, and their product is itself a diagonal matrix. A diagonal matrix (when it is not a scalar matrix) does not commute with a non-diagonal matrix; the product of the two matrices is non-diagonal. Finally, a scalar matrix commutes with any matrix a; the product of the two matrices is the matrix we obtain when we multiply each element of the matrix a by the scalar number corresponding to the scalar matrix.

Inverse Matrices—Suppose we have a transformation, such as (2), of matrix a. The transformation establishes a one-to-one correspondence between the pairs of values x_1 and x_2, defining a point P, and the pairs of values x'_1 and x'_2, defining a point P'. As a result, if P is given, P' is determined, and conversely. Let us first suppose that the point P and hence its coordinates x_1 and x_2 are given. The transformation (2) furnishes immediately the coordinates x'_1 and x'_2 of the point P'. But if the opposite procedure is followed, *i.e.*, if the coordinates x'_1 and x'_2 are given, a certain amount of calculation is required before we can derive the corresponding values x_1 and x_2 from the transformation (2). It is therefore convenient to perform these latter calculations once and for all so as to obtain the expression of x_1 and x_2 in terms of x'_1 and x'_2. The relations we are led to are of the form

$$(27) \qquad \begin{cases} x_1 = a'_{11} x'_1 + a'_{12} x'_2 \\ x_2 = a'_{21} x'_1 + a'_{22} x'_2 \end{cases}$$

in which the a'_{ik}'s are appropriate constants depending on the values of the a_{ik}'s in the original transformation.

The transformation (27) is called the "inverse" of the transformation (2).* Usually the coefficients a'_{ik} differ in value from the coefficients a_{ik} of the direct transformation. Hence the matrix of the inverse transformation (27), which we may represent by a', is not the same as the matrix a of (2). The matrix a' is called the "inverse" of the matrix a.

Suppose, then, that, to a point P of coordinates x_1, x_2, we apply the transformation (2). We thus obtain a point P' of coordinates x'_1, x'_2. To

---

* We are assuming that the transformation (2) is non-singular and hence that it establishes a one-to-one correspondence between the points P and P'. If this condition were not realized, the inverse transformation could not be obtained in the form (27).

this point P' we apply the inverse transformation (27). We shall obviously be led back to the point P whence we started. Consequently, the transformation (2), when followed by its inverse (27), changes nothing at all and is equivalent to the identical transformation. Inasmuch as the two transformations we have just effected are equivalent to a single transformation of matrix

$$(28) \qquad\qquad a'a,$$

we conclude that (28) is equivalent to the unit matrix $\mathbf{1}$. Exactly the same results would have ensued had we first operated with the inverse transformation a' and then followed it with the direct one. Altogether, then, we see that

$$(29) \qquad\qquad a'a = aa' = \mathbf{1}.$$

The inverse matrix of a matrix a is usually represented by a^{-1}. We thus obtain for all matrices the general relation

$$(30) \qquad\qquad a^{-1}a = aa^{-1} = \mathbf{1}.$$

Obviously a matrix and its inverse commute.

Orthogonal Transformations—A particular kind of linear transformation which plays a prominent part in pure mathematics and in theoretical physics is the *orthogonal transformation*. It corresponds to a rigid rotation about the origin. An example of an orthogonal transformation in the plane has already been mentioned. We shall now consider the three-dimensional case.

Suppose that all points P in space are subjected to the same rotation about the origin O. Each point P will be displaced to a corresponding point P', and the origin O alone will remain fixed. If we call x_1, x_2, x_3 and x'_1, x'_2, x'_3 the coordinates of the two points P and P', these triads of coordinates will be related in a definite way in any given rotation. The relationship expresses the orthogonal transformation. Instead of supposing that the points are rotated about the origin O, we may equivalently imagine that the points of space remain fixed, but that we rotate the coordinate axes rigidly around the origin to some new position. Calling Ox_1, Ox_2, Ox_3 the original axes, we shall represent the rotated axes by Ox'_1, Ox'_2, Ox'_3. If this rotation of axes is assumed, a point P of coordinates x_1, x_2, x_3 in the original system of axes will occupy a new position relative to the new axes and will thus have new coordinates x'_1, x'_2 x'_3. As before, the two triads of coordinates are connected by an orthogonal transformation. In the applications we have in view, we shall find it convenient to interpret an orthogonal transformation as a rotation of axes.

To understand some of the analytical peculiarities of orthogonal transformations and of their matrices, we first consider the general linear transformation in three variables. We may write it

(31)
$$\begin{cases} x_1 = s_{11}x_1' + s_{12}x_2' + s_{13}x_3' \\ x_2 = s_{21}x_1' + s_{22}x_2' + s_{23}x_3' \\ x_3 = s_{31}x_1' + s_{32}x_2' + s_{33}x_3', \end{cases}$$

where the coefficients s_{ik} are constants. We have placed the primed quantities on the right instead of on the left. This change, however, is of no importance, for our transformations merely connect triads of coordinates, or points; and it matters not which point we agree to call P or P'.

The matrix of the transformation (31) is

(32)
$$S = \begin{Bmatrix} s_{11} & s_{12} & s_{13} \\ s_{21} & s_{22} & s_{23} \\ s_{31} & s_{32} & s_{33} \end{Bmatrix}.$$

The inverse transformation is obtained from (31) by expressing x_1', x_2', x_3' in terms of x_1, x_2, x_3. The matrix of the inverse transformation is written S^{-1}. In the case of the general linear transformation, the inverse transformation is rather complicated, but in the particular case where the transformation (31) is an orthogonal one, its inverse has a very simple aspect. It is defined by

(33).
$$\begin{cases} x_1' = s_{11}x_1 + s_{21}x_2 + s_{31}x_3 \\ x_2' = s_{12}x_1 + s_{22}x_2 + s_{32}x_3 \\ x_3' = s_{13}x_1 + s_{23}x_2 + s_{33}x_3. \end{cases}$$

Its matrix is thus

(34)
$$S^{-1} = \begin{Bmatrix} s_{11} & s_{21} & s_{31} \\ s_{12} & s_{22} & s_{32} \\ s_{13} & s_{23} & s_{33} \end{Bmatrix}.$$

We observe that this matrix S^{-1} is merely the matrix S with rows and columns interchanged. It is called the "conjugate matrix" of S and is written $\tilde{S}$. Hence if S represents the matrix of an orthogonal transformation, we have

(35) $$\tilde{S} = S^{-1}.$$

Multiplying both sides before, or after, by S, we get

(36) $$S\tilde{S} = \tilde{S}S = 1 \quad (\text{since } SS^{-1} = S^{-1}S = 1).$$

The relation (35), or (36), is characteristic of orthogonal transformations and of their matrices (called *orthogonal matrices*). The numerical values of the coefficients in the matrices S and S^{-1}, or $\tilde{S}$, will of course depend on the particular rotation performed; but, in all cases, provided we be dealing with a rotation and hence with orthogonal matrices, the general relations (35) and (36) will be satisfied.

Let us then assume that (31) is an orthogonal transformation. The geometrical significance of the constants s_{ik} is easily obtained. Those on the first line of (32), *i.e.*, s_{11}, s_{12}, s_{13}, measure the cosines of the angles that the original axis Ox_1 makes with the three new axes Ox'_1, Ox'_2, Ox'_3; and those in the first column, namely, s_{11}, s_{21}, s_{31}, are the cosines of the angles the new axis Ox'_1 makes with the three former axes Ox_1, Ox_2, Ox_3.* Similarly for the other coefficients.

---

* The relations (36) show that the determinant of an orthogonal transformation must have the value ± 1, a fact which signifies that volumes are not altered by the transformation. The geometrical reason for this becomes obvious when we note that orthogonal transformations represent rigid rotations. The possibility of the determinant having the value -1 implies that an orthogonal transformation may also

Fig. 54

represent a rotation followed by a reflection. For instance, in the two-dimensional case, *i.e.*, in the plane, an orthogonal transformation may rotate the coordinate axes as in the first figure. But it may also change the coordinate axes to the positions indicated in the second figure. Clearly, no mere rotation can secure the second result. This second result may, however, be obtained by executing a rotation and following it by a reflection of the axis Ox'_2 with respect to the axis Ox'_1.

Matrices with Complex Terms—The matrices we have so far considered were assumed to contain only real terms. But situations arise where our transformations involve complex coefficients, and in this event the terms in the matrix are necessarily complex.

Two complex numbers, such as

$$a + ib \quad \text{and} \quad a - ib$$

(where a and b are real), are said to be conjugate complex. The two numbers differ merely in the sign preceding the imaginary magnitude i. By analogy, if a is a matrix containing complex terms, we shall agree that when the sign of i is changed in all the terms, the new matrix obtained is to be called the conjugate complex of a. Conjugate complex magnitudes are denoted by an asterisk, so that a^* will represent the conjugate complex matrix of a.

Some of the complex matrices exhibit a striking similarity with the real orthogonal matrices. In common with the latter, these special complex matrices are associated with rotations, but this time in an imaginary space. We shall refer to such complex matrices as complex orthogonal matrices; and the corresponding rotations in the imaginary space will be called complex orthogonal transformations.

Suppose, then, that (31) is a complex transformation of the orthogonal type. It can be shown that if S represents its matrix, the relations (35) and (36) must be replaced by the similar ones

(37) $$\tilde{S}^* = S^{-1},$$

and

(38) $$S\tilde{S}^* = \tilde{S}^*S = 1.$$

These are the relations that are characteristic of complex orthogonal matrices. Let us observe that should our orthogonal transformation S happen to be real, the asterisk would lose all significance in the matrices of (37) and (38); and the relations (37) and (38) would then pass over into the former ones, (35) and (36), that hold for real orthogonal matrices.

All these properties of matrices, whether real or complex, may be extended to matrices having any number of rows and columns. Thus, if we have a linear transformation connecting a set of n variables $x_1, x_2, \ldots x_n$ with a set of n variables $x'_1, x'_2, \ldots x'_n$, the geometrical representation of the transformation will be a transformation of points $P(x_1, x_2, \ldots x_n)$ in a space of n dimensions. The matrix of this transformation will contain n rows and n columns.

An illustration of linear transformations in a hyperspace is afforded by the special theory of relativity. The Lorentz transformations are linear transformations in 4-dimensional space-time. Indeed, as was shown by Minkowski, they are orthogonal transformations and hence correspond to rotations of the axes of coordinates about the origin. (This is made particularly clear when we render space-time Euclidean by viewing the time dimension as imaginary.) The matrix of a Lorentz transformation is thus an orthogonal matrix.

In the quantum theory, we shall have to consider orthogonal transformations in a space having an infinite number of imaginary dimensions; the corresponding matrices will then contain an infinite number of rows and columns (infinite matrices). In passing from the finite to the infinite, we must always proceed with caution and verify that extrapolations are justified; however, all that we have said so far in connection with matrices holds also for infinite matrices.

Symmetric Matrices and Quadratic Forms—When the elements of a matrix that are symmetrically situated with respect to the diagonal are equal, the matrix is said to be symmetric. For instance, the matrix

$$(39) \qquad \left\{ \begin{array}{ccc} a_{11} & a_{12} & a_{13} \\ a_{21} & a_{22} & a_{23} \\ a_{31} & a_{32} & a_{33} \end{array} \right\}$$

is symmetric if

$$(40) \qquad a_{12} = a_{21}, \quad a_{23} = a_{32}, \quad a_{31} = a_{13}.$$

A symmetric matrix may, if we choose, be associated with a linear transformation. But there are other ways of utilizing matrices. Symmetric matrices in particular can be associated with so-called quadratic forms. A quadratic form in three variables x_1, x_2, x_3 is

$$(41) \qquad \left\{ \begin{array}{l} a_{11}x_1^2 + a_{22}x_2^2 + a_{33}x_3^2 + a_{12}x_1x_2 + a_{21}x_2x_1 \\ + a_{23}x_2x_3 + a_{32}x_3x_2 + a_{31}x_3x_1 + a_{13}x_1x_3. \end{array} \right.$$

The coefficients a_{ik} are constants, which we shall here assume real. Without any loss in generality, we may suppose that the symmetric relations (40) hold between these coefficients.*

---

* Suppose that in (41) the coefficients a_{ik} are not symmetric. Without affecting the value of the quadratic form, we may replace $a_{12}x_1x_2 + a_{21}x_2x_1$ by

$$\left(\frac{a_{12} + a_{21}}{2} \right) x_1x_2 + \left(\frac{a_{12} + a_{21}}{2} \right) x_2x_1.$$

We may proceed in a similar manner with the other terms and by this means obtain the same quadratic form but with symmetric coefficients.

Suppose then that the form (41) be written in three lines as indicated below:

(42)
$$\begin{cases} a_{11}x_1^2 + a_{12}x_1x_2 + a_{13}x_1x_3 \\ + a_{21}x_2x_1 + a_{22}x_2^2 + a_{23}x_2x_3 \\ + a_{31}x_3x_1 + a_{32}x_3x_2 + a_{33}x_3^2. \end{cases}$$

A matrix, associated with the form, now suggests itself immediately; it is the symmetric matrix (39). We conclude that an appropriate symmetric matrix may be connected with each quadratic form. If the matrix is given, the quadratic form is determined, and vice versa.

To proceed, it is advantageous to give a geometric interpretation of a quadratic form. Let us view the variables x_1, x_2, x_3 as defining the Cartesian coordinates x, y, z of a point in space; and let us consider the equation we obtain by equating the quadratic form (41) to a constant, e.g., to 1. We shall write the quadratic form briefly $\Sigma a_{ik}x_ix_k$. The equation we wish to consider is then

(43)
$$\Sigma a_{ik}x_ix_k = 1.$$

It can be shown that those points P whose coordinates x_1, x_2, x_3 satisfy the equation (43) lie on a quadric surface having the origin as centre.

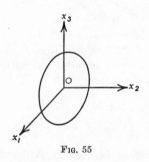

FIG. 55

The precise nature of the quadric surface depends on the values of the coefficients a_{ik}. If the a_{ik} are real (as we here assume), and if the determinant of the matrix associated with the quadratic form is non-vanishing, the quadric surface is an ellipsoid (real or imaginary) or a hyperboloid, situated in ordinary 3-dimensional space.* The numerical values of the coefficients in the quadratic form enable us to determine which of the two situations is realized. We shall suppose that the equation (43) defines a real ellipsoid. This ellipsoid, having the point O as centre, will then be represented by some such drawing as is indicated in the figure. The *principal axes* of the ellipsoid are the three mutually perpendicular axes of symmetry that pass through the centre

---

* This determinant is called the *discriminant* of the quadratic form. If it vanishes, the quadric surface (43) degenerates into a cylinder of the elliptic or hyperbolic type, or else into two parallel planes symmetrically situated with respect to the origin.

O. These axes are uniquely determined when they differ in length, as we shall here assume.

The Transformation of a Quadratic Form into a Sum of Squares by Means of an Orthogonal Transformation—Suppose we wish to determine the lengths of the principal axes of the ellipsoid defined by (43). These lengths depend on the particular ellipsoid considered and hence on the coefficients a_{ik} in (43). But a mere inspection of this equation does not enable us to determine the lengths of the principal axes, and so we are compelled to resort to indirect methods. We proceed as follows: we establish the equation of the ellipsoid in a system of axes which coincide with the ellipsoid's principal axes; the lengths of the principal axes are then obtained immediately.

Let us investigate this procedure. The coordinate axes $Ox_1x_2x_3$ are assumed to be rotated about the origin O so as to coincide with the principal axes of the ellipsoid. If we call $Ox_1'x_2'x_3'$ the rotated axes, a current point P on the surface of the ellipsoid will have coordinates x_1', x_2', x_3' in the new axes. The equation of the ellipsoid in the new system of axes will be given by some relation linking the coordinates x_1', x_2', x_3'. For the present this equation is unknown, but we do know that it will be of the same general type as (43), and hence that it will be expressed by equating to 1 some appropriate quadratic form in x_1', x_2', x_3'. Furthermore, since the new axes coincide with the principal axes, we know that the ellipsoid's surface will be symmetrically situated with respect to them. This symmetry requires that the quadratic form, which enters into the equation of the ellipsoid, should degenerate into a sum of squares, and hence that the equation of the ellipsoid in the new axes should be of form

$$(44) \qquad b_1 x_1'^2 + b_2 x_2'^2 + b_3 x_3'^2 = 1.$$

Here b_1, b_2, b_3 are positive constants,* but at this stage their numerical values are unknown. For reasons which will appear presently, they are called the *eigenvalues* of the problem.

Let us assume that the exact equation (44) of the ellipsoid has been obtained, so that the eigenvalues b_1, b_2, and b_3 are known. The lengths of the principal axes, which it was our original aim to discover, are then measured by

$$(45) \qquad \frac{2}{\sqrt{b_1}}, \ \frac{2}{\sqrt{b_2}}, \ \frac{2}{\sqrt{b_3}}.$$

Our problem is thus to determine the eigenvalues b_1, b_2, and b_3.

---

* The constants b_i are certainly positive because our quadric surface is assumed to be a real ellipsoid.

The problem we have been discussing is equivalent to that of deriving the equation (44) of an ellipsoid, referred to its principal axes, from the equation (43) of the same ellipsoid, referred to an arbitrary system of Cartesian axes. A more analytical interpretation of this problem will now be given.

The change in the form of the ellipsoid's equation when we pass from (43) to (44) is brought about by rotating the axes of coordinates in a suitable way. Now, a rotation of axes is an orthogonal transformation, and hence the equation (44) is derived from (43) when we submit the coordinates x_1, x_2, x_3 of (43) to a suitable orthogonal transformation. The transformation exerts no effect on the number 1 on the right hand side of (43); it merely transforms the quadratic form $\Sigma a_{ik}x_ix_k$ of (43) into a sum of squares, such as is exhibited in (44). From this standpoint the problem is equivalent to the following one:

To transform by means of an orthogonal transformation a given quadratic form $\Sigma a_{ik}x_ix_k$ into a sum of squares

$$(46) \qquad b_1x_1'^2 + b_2x_2'^2 + b_3x_3'^2.$$

If we were apprised of the required orthogonal transformation, or rotation of axes, we could apply it to the quadratic form, and the sum of squares would be obtained immediately. Conversely, if we knew the values of the constants b_i in the sum of squares (i.e., the eigenvalues), we could obtain the orthogonal transformation without difficulty. But for the present, we are ignorant of the eigenvalues and also of the orthogonal transformation.

To understand how the problem can be solved, let us suppose that the quadratic form $\Sigma a_{ik}x_ix_k$ is submitted to any arbitrary orthogonal transformation. For instance, if (31) represents an orthogonal transformation, the variables x_1, x_2, x_3 in the quadratic form will have to be replaced by their values (31) in terms of x_1', x_2', x_3' and of the coefficients s_{ik} of the transformation. When this transformation is performed and the various terms are regrouped, we find that a new quadratic form $\Sigma a_{ik}'x_i'x_k'$ is obtained, the new coefficients a_{ik}' being appropriate combinations of the original ones and of the coefficients s_{ik} of the orthogonal transformation.*

---

* The coefficients a_{ik} in the original quadratic form are the components of a "tensor" of the type called "twice covariant." When an orthogonal transformation (or any linear transformation) is applied to the quadratic form, we obtain a new quadratic form $\Sigma a_{ik}'x_i'x_k'$. The tensor components a_{ik}' of the new form may be derived, by means of well-known rules, from the original tensor components a_{ik} and from the coefficients s_{ik} of the transformation that has been applied.

The transformation is most easily expressed in terms of matrices. We recall that the original quadratic form has a known matrix a, and that the orthogonal transformation (31) to which we submit it has a matrix which we may call S. Now, quite generally it can be shown that, if the transformation (31) were the general linear transformation instead of an orthogonal one, the transformed algebraic form would still be quadratic and its matrix would be

$$(47) \qquad\qquad \tilde{S}aS.$$

But since the transformation (31) is assumed orthogonal, we know, by (35), that $\tilde{S} = S^{-1}$; and hence the matrix of the new quadratic form will be

$$(48) \qquad\qquad S^{-1}aS.$$

We wish the transformed quadratic form to be a sum of squares $b_1 x_1'^2 + b_2 x_2'^2 + b_3 x_3'^2$, whose coefficients b_i are, however, undetermined. The matrix b of this sum of squares is the diagonal matrix

$$(49) \qquad\qquad b = \left\{ \begin{array}{ccc} b_1 & 0 & 0 \\ 0 & b_2 & 0 \\ 0 & 0 & b_3 \end{array} \right\}.$$

Consequently, our problem is to select the orthogonal transformation of matrix S in such a way that the following matrix equation holds

$$(50) \qquad\qquad S^{-1}aS = b, \quad \text{or} \quad aS = Sb.$$

To this equation must be added the matrix equation

$$(51) \qquad\qquad \tilde{S}S = 1,$$

which expresses that S is an orthogonal matrix.

In short, our geometrical problem leads to the matrix equations (50) and (51), and it is solved if we can solve these equations. We may, however, disregard the geometrical aspect of the problem and concentrate on the matrix aspect. From this standpoint our problem may be expressed as follows:

To transform by means of an orthogonal transformation a given symmetric matrix a in such a way that it becomes a diagonal matrix.

This is the problem which is expressed by the matrix equations (50) and (51). Now the equality of two matrices, *i.e.*, a matrix equation, implies the equalities of the corresponding elements in the two matrices. Hence an equation of type (50) entails a number of equations connecting the elements a_{ik} of the original matrix a, the coefficients s_{ik} of the unknown orthogonal transformation S, and the unknown elements (or eigenvalues) b_i of the diagonal matrix b. We propose to show that the equation (50) suffices to determine the unknown diagonal matrix b, and that the orthogonal matrix S may then be derived from (50) and (51).

To prove this assertion, we revert to the geometrical significance of the problem. The orthogonal transformation rotates the coordinate axes, bringing them into coincidence with the principal axes of the ellipsoid. We have seen that when this rotation is performed so that the equation of the ellipsoid becomes (44), the coefficients b_1, b_2, and b_3 in the sum of squares are connected by (45) with the lengths of the principal axes. These coefficients b_i are thus determined by the shape and size of the ellipsoid. We conclude that, whatever rotation may be utilized to bring the original coordinate axes $Ox_1x_2x_3$ into coincidence with the principal axes of the ellipsoid, the coefficients b_1, b_2, and b_3 in the sum of squares will always have the same numerical values. Furthermore, these values, being determined by the dimensions of the ellipsoid, will depend on the coefficients a_{ik} of the quadratic form of matrix a.

Suppose, then, we were to impose three arbitrary values for the coefficients b_1, b_2, and b_3 in the sum of squares, and we then sought to obtain the orthogonal transformation which would transform the quadratic form into the sum of squares selected. According to what has just been said, no orthogonal transformation could secure the required transformation, and our problem would be impossible. Thus our problem is possible only when the coefficients b_1, b_2, and b_3 have appropriate values.

These results can be translated into the language of matrices. We may say : The matrix equation (50), in which a is a given symmetric matrix, S some orthogonal one, and b some diagonal one, is possible when, and only when, the diagonal elements b_1, b_2, and b_3 of the diagonal matrix b have appropriate values, *i.e.*, the eigenvalues. These eigenvalues are determined in some way (as yet unknown) by the matrix a and hence by the elements a_{ik} of this matrix.

According to the above explanations, we shall know the eigenvalues b_1, b_2, b_3 if we can establish the conditions which render the matrix equation (50) possible. Now, it can be shown that these conditions are realized when b_1, b_2, and b_3 are the three roots of a certain algebraic equation.* The equation is easily constructed, its coefficients being well-determined combinations of the elements a_{ik} of the matrix a; the three roots of this equation are then the three eigenvalues we are seeking. Of course, when these eigenvalues are obtained, the particular sum of squares into which the quadratic form will be transformed is known.

In some cases we may rest content with the discovery of the eigenvalues, and hence of the diagonal matrix b into which the symmetric matrix a can be transformed by an orthogonal transformation. But in other cases we may wish to proceed further and determine the orthogonal matrix S, which brings about this transformation. From the geometrical standpoint, this means that we may wish to determine the rotation of axes which brings the original axes $Ox_1x_2x_3$ into coincidence with the principal axes of the ellipsoid. A knowledge of this rotation of axes will automatically apprise us of the angles that the original coordinate axes $Ox_1x_2x_3$ made with the principal axes of the ellipsoid defined by (43).

The determination of the rotation (or of the orthogonal matrix S) is easily secured, for after we have found the eigenvalues b_i, the diagonal matrix b is known, and hence the only unknown magnitude in the matrix equation (50) is the orthogonal matrix S. When account is taken of (51), which expresses the orthogonality of the matrix S, the matrix equation (50) yields the required matrix S.

In point of fact there will always be some measure of indeterminacy in the orthogonal transformation. It arises from the existence of different rotations, each of which brings about the coincidence of the coordinate axes with the principal axes of the ellipsoid. For instance, we may rotate the coordinate axes in such a way that the new axis Ox'_1, coincides with

---

* This equation is expressed by equating to zero the determinant

$$(52) \qquad \begin{vmatrix} a_{11} - b_i & a_{12} & a_{13} \\ a_{21} & a_{22} - b_i & a_{23} \\ a_{31} & a_{32} & a_{33} - b_i \end{vmatrix}$$

in which b_i is the unknown magnitude to be determined and the terms a_{ik} are the symmetric coefficients of the original quadratic form. It will be noted that the determinant (52) differs from the determinant of the matrix a solely in that the unknown magnitude b_i is subtracted from each diagonal element of the latter determinant. When the determinant (52) is expanded, we obtain an algebraic equation of the third degree in the unknown b_i. A theorem due to Sylvester states that, since the coefficients a_{ik} are real and symmetric, the three roots of our equation are always real. These three roots are the three eigenvalues b_1, b_2, b_3, which play the part of coefficients in the sum of squares.

Algebraic equations of the kind just discussed are of frequent occurrence in pure mathematics, in celestial mechanics, and in physics.

one or another of the three principal axes of the ellipsoid, so that the three axes Ox'_1, Ox'_2 and Ox'_3 may coincide in different ways with the principal axes. Obviously the precise rotation will differ according to the way in which we choose to operate. This indeterminacy is, however, of minor importance, for the sole effect of the different rotations is to interchange the columns of the orthogonal matrix, and hence also to interchange the eigenvalues b_1, b_2, and b_3 in the sum of squares, or in the diagonal matrix. But the eigenvalues, as such, remain the same.

We mentioned on a previous page, that the elements in the kth column of the orthogonal matrix S define the cosines of the angles formed by the kth axis Ox'_k of the rotated axes with each of the three original coordinate axes Ox_1, Ox_2, Ox_3. Similarly, the elements in the kth line of the matrix S define the cosines of the angles formed by the kth axis Ox_k with each of the rotated axes Ox'_1, Ox'_2, Ox'_3. Inasmuch as the new coordinate axes coincide with the principal axes of the ellipsoid, we are apprised of the relative position of the ellipsoid with respect to the original coordinate axes. As for the lengths of these principal axes, we know that they are given by the expressions (45) in terms of the eigenvalues b_i. We are thus fully informed of the dimensions and the orientation of the ellipsoid defined by (43).

Thus far we have assumed that the principal axes of the ellipsoid are of unequal length. But when two or all three of the principal axes of the ellipsoid have the same length (*i.e.*, when two or all three of the eigenvalues b_i are equal), the orthogonal matrix becomes indeterminate. The geometric picture enables us to understand the reason for this indeterminateness. For example, if our ellipsoid is a spheroid and we rotate the axis Ox_1, bringing it into a position Ox'_1 which coincides with the axis of symmetry of the spheroid, the axes Ox'_2 and Ox'_3 will always coincide with two other principal axes, however much we pivot our coordinate axes about Ox'_1. The exact rotation imposed on our initial axes is then arbitrary to a large extent, and it is this arbitrariness which betrays itself in the indeterminateness of the matrix S.

In the special example treated, we were dealing with 3-dimensional ellipsoids and rotations in 3-dimensional space. But exactly the same problem can be considered in a space of any number n of dimensions, though the geometrical representation now fails us unless we are able to visualize ellipsoids and rotations in a space of n dimensions. From the standpoint of matrices, the problem consists in transforming a symmetric matrix of n dimensions into a diagonal one by means of an orthogonal transformation. The n diagonal elements of the diagonal matrix are, as before, the eigenvalues, and they are the roots of an algebraic equation of the nth degree.

Hilbert investigated the problem when the matrices contain an infinite number of rows and columns (infinite matrices). The geometrical representation must now be extended to a space of an infinite number of dimensions. Here, as in many cases where we pass from the finite to the infinite, a new situation develops. When the number of dimensions is finite, the diagonal terms of the diagonal matrix (*i.e.*, the eigenvalues) form a discrete set of terms. But when we are dealing with infinite matrices, the eigenvalues, necessarily infinite in number, need not form a denumerable set (like the class of all integers); they may form a continuous set (like the class of all real numbers), or they may form a set, part continuous and part discrete.* The name spectrum was given to any such aggregate of the eigenvalues, and in the case of infinite matrices the eigenvalues were said to exhibit a discrete spectrum or a continuous spectrum or both. The word "spectrum" was borrowed from optics. At the time these mathematical investigations were taking place, the new quantum theory of the atom was unknown; but it so happens that the eigenvalues represent in certain cases the energy levels of the quantum theory, so that a discrete or a continuous spectrum of eigenvalues entails

---

* In the 3-dimensional case we obtained for the sum of squares the expression

$$b_1 x_1'^2 + b_2 x_2'^2 + b_3 x_3'^2 ,$$

where the eigenvalues b_1, b_2, and b_3 were the roots of an algebraic equation of the third degree. When our original quadratic form contains an infinite number of terms, the eigenvalues should therefore be the roots of an algebraic equation of infinite degree. The consideration of equations of this type requires, however, the introduction of new mathematical instruments, such as integral equations. It was as a result of the study of integral equations that Hilbert established the possibility of the eigenvalues forming a denumerable set or a continuous set or both.

If the eigenvalues form a denumerable set, the sum of squares contains an infinite number of terms and is therefore represented by

$$\sum_{n=1}^{n=\infty} b_n x_n'^2 .$$

The matrix of this form is a diagonal matrix containing a denumerable set of non-vanishing diagonal terms.

If there is also a continuous infinity of eigenvalues, we must adjoin to the previous sum of squares the integral

$$(53) \qquad \int b(\lambda) x'^2(\lambda) d\lambda.$$

The diagonal matrix $\boldsymbol{b}$ of the total form is then part continuous and part discrete.

an optical spectrum of discrete lines or a continuous spectrum for the atoms of interest. Thus, by a curious coincidence the introduction of the word "spectrum" in the mathematical theory has shown itself to be a singularly appropriate one.

Hermitean Matrices—Suppose that in the matrix a of (39) the terms are complex, and suppose that terms symmetrically situated with respect to the diagonal are conjugate complex, the diagonal terms alone being always real. In a matrix of this type the relations (40) are replaced by the relations

$$(55) \qquad a_{12}^* = a_{21}, \quad a_{23}^* = a_{32}, \quad a_{13}^* = a_{31},$$

where the asterisk denotes the conjugate complex of a magnitude. These matrices are called *Hermitean* in honor of Hermite who investigated their properties. Their determinants are real.

Quadratic forms with complex coefficients and variables may be associated with Hermitean matrices. Thus, in place of the real quadratic form written in three lines in the arrangement (42), we consider the quadratic form

$$(56) \qquad \left\{ \begin{array}{l} a_{11}x_1^*x_1 + a_{12}x_1^*x_2 + a_{13}x_1^*x_3 \\ + a_{21}x_2^*x_1 + a_{22}x_2^*x_2 + a_{23}x_2^*x_3 \\ + a_{31}x_3^*x_1 + a_{32}x_3^*x_2 + a_{33}x_3^*x_3, \end{array} \right.$$

in which the coefficients a_{ik} satisfy the relations (55). The matrix of this quadratic form is the Hermitean matrix a just discussed. As we have said, the diagonal terms are real, and those situated symmetrically with respect to the diagonal are complex conjugate. Hence the sum of all the terms in (56) is real, and we may proceed as we did with real quadratic forms and equate the form (56) to a real constant, *e.g.*, to 1.

From a geometrical standpoint, the equation obtained may be viewed as defining a quadric surface having the origin as centre and situated in an imaginary space. The problem of determining the lengths of the principal axes of the quadric surface (*e.g.*, ellipsoid) occurs here as it did for real quadratic forms. The method of solution is practically the same; the only difference is that the orthogonal transformation will usually be of the more general complex kind. We shall therefore assume that the orthogonal matrix S is complex.

It can be shown that if we transform the quadratic form by an arbitrary complex linear transformation of matrix S, the transformed form is still a quadratic form, and that its matrix is (analogously to (47))

$$(57) \qquad \tilde{S}^*aS,$$

where a is the matrix of the original quadratic form. Since, however, we are assuming that the transformation is orthogonal, we may utilize the relation (37), *i.e.*,

$$(58) \qquad \tilde{S}^* = S^{-1},$$

which holds for orthogonal complex matrices; and the matrix (57) thus becomes

$$(59) \qquad S^{-1}aS.$$

The matrix equation which expresses that the orthogonal transformation transforms the quadratic forms into a sum of squares is thus

$$(60) \qquad S^{-1}aS = b,$$

where b is some diagonal matrix. The diagonal matrix b corresponds to a sum of squares; its general form is

$$(61) \qquad b_1 x_1^* x_1 + b_2 x_2^* x_2 + b_3 x_3^* x_3.$$

The equation (60) may equivalently be written

$$(62) \qquad aS = Sb, \text{ or } aS - Sb = 0.$$

Hence we obtain exactly the same results as in (50).

As before, the problem is possible only when the diagonal terms of the diagonal matrix b have privileged values. These are the eigenvalues of the problem and they are the roots of an algebraic equation of the third degree. Though we are here dealing with complex magnitudes, these eigenvalues may be shown to be real. From the eigenvalues the lengths of the principal axes may be deduced; these lengths are also real. Finally, the orthogonal transformation may be obtained. There is therefore no essential difference between the present treatment and the one we outlined more fully in connection with real quadratic forms.

Hermitean matrices are of considerable importance in the quantum theory because Heisenberg's matrices, which afford a means of attacking the problem of the atom, are matrices of this kind. In addition Heisenberg's matrices are infinite. As such, they may be associated with infinite quadratic forms, representing quadric surfaces situated in a space having an infinite number of imaginary dimensions (Hilbert space). The same problem of transforming these infinite quadratic forms into a sum of squares occurs as it did with real matrices. In particular, when this problem is investigated for the infinite Heisenberg matrices, we obtain, as noted previously, diagonal matrices the diagonal elements (eigenvalues) of which may form a discrete or a continuous spectrum or both.[*] In the continuous case the diagonal matrix becomes continuous, *i.e.*, there are elements at all points along the diagonal line.

A Geometric Interpretation of the Commutation Rules—We have said that the multiplication of matrices is not usually a commutative operation, but we mentioned that in special cases matrices do commute, *e.g.*, when the two matrices involved are diagonal. The general rule which determines under what conditions matrices will commute may be given a simple geometric interpretation when the matrices considered are real and symmetric, or complex and Hermitean.

Thus let A and B be two Hermitean matrices which we may suppose to be 3-dimensional. We associate with each matrix the corresponding quadric surface in a space of three dimensions. We shall assume that the quadric surfaces are ellipsoids. The geometric commutation rule then states:

> Two Hermitean matrices will always commute when the principal axes of their associated ellipsoids coincide. The matrices will not commute when these axes do not coincide.

This rule applies to all Hermitean matrices, whether finite or infinite.

On page 792 we said that two diagonal matrices always commute; that a diagonal matrix (which is not a scalar) does not commute with a non-diagonal one; and that a scalar matrix commutes with any matrix. Our geometric rule renders these statements intuitive in connection with Hermitean matrices. First, we recall that a diagonal matrix can be associated with a quadric surface, say, an ellipsoid, whose centre is at the origin and whose principal axes coincide with the axes of coordinates; in the particular case where the diagonal matrix is a scalar matrix, the three principal axes have the same length and the ellipsoid becomes a sphere.

Consider, then, two diagonal matrices. Since the principal axes of both ellipsoids coincide with the coordinate axes, these principal axes

---

[*] In this last case the sum of squares (61) is of type

$$\sum_n b_n x_n^* x_n + \int b(\lambda) x^*(\lambda) x(\lambda)\, d\lambda.$$

necessarily coincide two by two; and hence, according to our geometric rule, the two diagonal matrices commute.

Similarly let us show that a diagonal matrix which is not a scalar matrix cannot commute with an arbitrary matrix. A diagonal matrix of this type is represented by an ellipsoid of unequal principal axes, and these axes coincide with the coordinate axes. The principal axes of the other ellipsoid associated with the arbitrary matrix cannot possibly coincide with the coordinate axes, for, if they did, the arbitrary matrix would be a diagonal one. Hence the principal axes of the two ellipsoids cannot coincide. And so according to our geometric rule the two matrices cannot commute.

Finally, we may understand why a scalar matrix necessarily commutes with any arbitrary matrix. A scalar matrix is associated with a sphere having the origin as centre; and since the principal axes of a sphere are defined by any three mutually perpendicular diameters of the sphere, a certain triad of these mutually perpendicular diameters will always coincide with the principal axes of the arbitrary ellipsoid. Since the principal axes of the two surfaces always coincide, we conclude in accordance with our geometric rule that a scalar matrix commutes with any arbitrary matrix.

We may give an analytical presentation of our geometric rule. Thus, since two commuting Hermitean matrices are associated with ellipsoids whose principal axes coincide, it will always be possible to rotate our coordinate axes in such a way that they will be brought into coincidence with the principal axes of both ellipsoids. In other words, it will always be possible to apply an orthogonal transformation which will transform, at one and the same time, two commuting Hermitean matrices into their diagonal forms. Conversely, whenever a simultaneous transformation of this sort is possible, we may be certain that our Hermitean matrices commute. These statements constitute the analytic expression of our former geometric commutation rule.

On the other hand, if two Hermitean matrices do not commute, their simultaneous transformation into the diagonal form (by means of an orthogonal transformation) will be impossible; the most we shall be able to do in this case will be to transform one *or* the other of the matrices into the diagonal form. As will be seen later, this last statement constitutes the matrix expression of Heisenberg's Uncertainty Principle.

CHAPTER XXXVII

THE MATRIX METHOD

FROM a mathematical point of view, Heisenberg's Matrix Method and Wave Mechanics are equivalent; they permit the solution of the same atomic problems, and the same results are obtained. A choice between the two methods is largely a matter of convenience: some problems may be approached more easily by one method and others by the other. In general, wave mechanics is more adapted to the determination of the energy levels, whereas the matrix method is preferable when it is the intensities and polarizations of the radiated frequencies that we wish to consider. For this reason, in the solution of a problem the two methods are often combined.

Though the two methods are mathematically equivalent, they are conceptually entirely different. Wave mechanics stresses the wave aspect, whereas in the more symbolic matrix method this aspect is lost and can be revealed only indirectly. The fact that wave mechanics suggests a picture which may often be represented in 3-dimensional space, while the matrix method defies such a representation, might be thought to give wave mechanics the advantage. But this conclusion would be too hasty, because in the majority of cases wave mechanics can be pictured only in a hyperspace of n dimensions, and in such cases the picture it suggests loses its intuitive appeal. Still more important is the fact that for a picture to have any physical value, it must describe actual physical occurrences. We have seen, however, that the waves of Schrödinger's theory are regarded today as mere mathematical symbols, so that wave pictures cannot claim any physical reality. Moreover, even the formal analogy between the de Broglie waves in 3-dimensional space and the waves in material media must not be pressed too far, for it would lead to erroneous conclusions. For instance, with elastic waves, such as sound waves, the absolute amplitude of the waves determines the intensity of the musical note and is thus a significant feature of the waves. But with the de Broglie waves, the absolute amplitude is of no interest; only the relative amplitude from place to place is of importance in determining the probability.

The conclusion to be drawn from the foregoing discussion is that the wave treatment must not mislead us into believing that atomic processes can be interpreted by familiar pictures. In this respect the matrix method has the advantage, for it eliminates the possibility of any such misconception from the outset. The procedure of the matrix method is entirely mathematical; it makes no appeal to familiar modes of representation. Its technique is difficult, and at no stage in theoretical physics has so great a demand been made on the mathematical knowledge and on the powers of abstraction of the investigator. The mathematical trend of modern theoretical physics first made its appearance with the theory of relativity and its 4-dimensional non-Euclidean world. This trend is still more accentuated in the matrix method. The result has been that the modern theoretical physicist, whether he like it or not, must necessarily have a far greater knowledge of advanced mathematics than was required of his predecessors. In this connection it is significant to note that the recent rapid progress in the new theories of quantum mechanics has been attributed to Hilbert's publication in 1904 of a book in which many of the mathematical problems that have since confronted the quantum theorists were discussed.*

A matter of considerable interest is the philosophy underlying the matrix method. When we say that theoretical physics must necessarily yield conclusions that can be observed, we are stating a truism, because if the conclusions are beyond the scope of observation, we cannot test a theory. But the matrix method professes to go further; it attempts to exclude from its equations all magnitudes which cannot be observed. Obviously, we have here a drastic extension of the general requirements just noted for any theory of mathematical physics; for not only must the conclusions now be observable, but so also must all the intermediary magnitudes. That these stringent requirements are not really necessary for a theory of mathematical physics to furnish important results is suggested, however, by wave mechanics. Wave mechanics, especially in its incipient stages, was not restricted by the previous limitations. De Broglie was not concerned with the question of deciding whether his waves were observable or not; his only reason for introducing them was the hope that they would control in an indirect way phenomena which were observable.

Before examining the matrix method in detail, let us consider the various magnitudes utilized in Bohr's theory and see which ones Heisenberg claims to be observable and which ones he rejects as unobservable. The position, the orbit, and the motion of an electron in the atom are assumed to be unobservable, and so no use is made of such magnitudes in

---

* "Methoden der mathematischen Physick" by Courant and Hilbert.

the theory. On the other hand, the frequencies, intensities, and polarizations of the radiations emitted by the atom, as also the energy levels, are claimed to be observable. This classification indicates clearly that the word "observable" must not be construed here with its commonplace meaning. How, indeed, could ultra-violet radiations, which are invisible to the human eye, be called observable if the commonplace meaning were credited to the word? Ultra-violet radiations are in all truth observable only indirectly by the effects that they produce, and in order to connect the effects actually observed with the invisible ultra-violet radiation, which is assumed to be the cause of these effects, we must construct a theory. Thus, only in virtue of a theory can many magnitudes be classed as observable; and this implies that a subsequent modification of our theory may cause us to modify our list of observable magnitudes.

Let us, however, accept the highly sophisticated meaning attributed to the word "observable" in the new quantum theory. We must then inquire whether Heisenberg has been successful in adhering consistently to his program. His success in this matter appears to be far from complete. The phases of the vibrations, for instance, play a most important rôle in the equations of the theory, and yet these phases do not come under the category of things observable. As Bohr himself stated at the Solvay Congress of 1927:

> "Although the notion of a phase is indispensable in the calculations, it can scarcely be said that it introduces itself in the interpretation of our observations."

Elsewhere Bohr makes the following significant admission:

> "The matrix theory has often been called a calculus with directly observable quantities. It must be remembered, however, that the procedure described is limited just to those problems in which in applying the quantum postulate the space-time description may largely be disregarded, and the question of observation in the proper sense therefore placed in the background." *

These quotations show that Heisenberg was not altogether successful in his attempt to devise a purely phenomenological theory of the atom. Heisenberg's partial failure does not appear to reflect, however, on the soundness of his main contention, namely, that, in the construction of an atomic theory, the position, orbit, and motion of the electron in the atom should not be considered. The fact is that these magnitudes, being merely inferred, are necessarily uncertain, whereas the frequencies, intensities, and polarizations of the radiations can be measured with pre-

---

* Atomic Theory and the Description of Nature, page 72.

cision; and since furthermore Bohr's theory shows that the motion of the electron plays no direct part in the phenomenon of radiation, it will obviously be to our advantage to construct a theory of the atom in which the motion of the electron is disregarded entirely. Thus quite independently of the attitude we may adopt towards Heisenberg's phenomenological philosophy, we must agree that the magnitudes he rules out on account of their unobservability are precisely those which we should rule out, if possible, for reasons of safety.

But the most convincing argument in favor of Heisenberg's matrix method, is that this theory furnishes correct information on the frequencies, intensities, and polarizations of the radiations, and is thus far superior to Bohr's theory. We shall therefore examine the matrix method without further concern for the philosophy which it may embody.

Classical Preliminaries—If a point P is describing a circle with uniform speed, the projection P' of the point P on any diameter executes a vibratory motion which is called *simple harmonic*.

The motion repeats itself periodically and for this reason is called periodic; the number of vibrations occurring each second is the frequency ν of the vibration. The distance from the centre of oscillation O to the extreme points A and B (*i.e.*, the radius of the circle) is the amplitude of the oscillation.

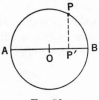

Fig. 56

We call q the distance OP' at any instant, and R the radius of the circle and hence also the amplitude of the oscillation. The variation in the value of q with time may be represented by

(1) $\qquad q = R \cos 2\pi\nu t \quad \text{or} \quad R \sin 2\pi\nu t.$

The first expression implies that the vibrating point P' is situated at B at the initial instant, $t = 0$. The second expression implies that, at the initial instant, the vibrating point is at the centre of oscillation O and is proceeding towards the right. We note that the second motion lags behind the first by one quarter of the complete oscillation. This fact is expressed by the statement that the two vibrations, though of the same frequency ν and amplitude R, differ in *phase* by a quarter period.

In many cases it is necessary to represent harmonic vibrations in which the vibrating point P' is situated at any arbitrary point between A and B

at the initial instant. This may be done by combining the two expressions (1). Thus

$$(2) \qquad q = a \cos 2\pi\nu t + b \sin 2\pi\nu t,$$

in which a and b are constants, defines a vibration of frequency ν and of amplitude

$$(3) \qquad \sqrt{a^2 + b^2}.$$

At the initial instant $t = 0$, the elongation q of the vibrating point from the centre of oscillation is measured by a. By attributing appropriate values to the constants a and b, we can make the expression (2) represent a harmonic vibration of frequency ν, which has any amplitude we choose and in which, at the initial instant, the vibrating point occupies any specified position between its limits of oscillation.

The expression (2) is sometimes written in a different, though equivalent, form: We introduce the imaginary magnitude $i = \sqrt{-1}$, and when this is done, we obtain

$$(4) \qquad q = q_1 e^{2\pi i\nu t} + q_{-1} e^{-2\pi i\nu t},$$

where q_1 and q_{-1} are two constants. If (4) is to represent exactly the same motion as (2), the new constants q_1 and q_{-1} must be connected with our former ones, a and b, by the relations

$$(5) \qquad \begin{cases} q_1 = \dfrac{a - ib}{2} \\[2ex] q_{-1} = \dfrac{a + ib}{2}. \end{cases}$$

From this we see that the two constants q_1 and q_{-1} are conjugate complex.[†] If we represent the conjugate complex of a magnitude by placing an asterisk over it, we may write

$$(6) \qquad q_1 = q_{-1}^*, \text{ or } q_1^* = q_{-1}.$$

The amplitude of our vibration in terms of the new constants may be deduced from (3) and (5). This amplitude is

$$(7) \qquad 2\sqrt{q_1 q_{-1}}, \text{ or } 2\sqrt{q_1 q_1^*}.$$

In the expression (4), we have assumed that the position of the vibrating point is measured from the centre of oscillation. More generally we may take any point on the line of oscillation as origin. Thus,

---

† The two coefficients q_1 and q_{-1} are called the complex amplitudes because, as is seen from (7), the amplitude of the motion is 2 mod q_1, or 2 mod q_{-1}.

if q_0 measures the distance from our new origin to the centre of oscillation, the expression (4) must be replaced by

$$(8) \qquad q = q_0 + q_1 e^{2\pi i \nu t} + q_{-1} e^{-2\pi i \nu t}.$$

Note that q_0, in contradistinction to q_1 and to q_{-1}, is always real.

Harmonic vibrations play an important part in physics, because the harmonic vibrations of the electromagnetic field and of the air give rise to monochromatic radiations and to pure musical notes. When the vibrations are not harmonic, more complicated effects are produced. In mechanics, a particle executes a harmonic vibration along a straight line when it is attracted to a point O on the line of oscillation by a force which is proportional to the distance (elastic force). The centre of attraction O is then the centre of the harmonic oscillation. A mechanical system of this sort is called a *linear harmonic oscillator*. If we consider a definite harmonic oscillator in which the vibrating particle is attracted to the centre of oscillation by a given restoring force, the amplitude of the vibration can be shown to increase with the energy of the motion. But the frequency of the vibration is characteristic of the oscillating system; it remains the same regardless of the amplitude and hence of the energy. In the particular case where the vibrating particle is an electron, the laws of classical electrodynamics require that a monochromatic radiation be emitted from the oscillator, and that its frequency be precisely that of the vibrating electron. These laws also require that the intensity of the radiation (energy radiated per second) be proportional to the square of the amplitude of the vibration, and hence to $q_1 q_1^*$ (if the form (4) or (8) is adopted).

We next consider the case where the motion is periodic, but no longer harmonic. It is then called nonharmonic, or anharmonic. We assume that the motion is repeated ν times every second, so that ν is the frequency of the vibration. Fourier's theorem shows that, at any instant t, the distance q of the vibrating particle from some point O selected on the line of oscillation is expressed by a Fourier series

$$(10) \quad \begin{cases} q = q_0 + q_1 e^{2\pi i \nu t} + q_2 e^{2\pi i (2\nu) t} + q_3 e^{2\pi i (3\nu) t} + \dots \\ \qquad + q_{-1} e^{-2\pi i \nu t} + q_{-2} e^{-2\pi i (2\nu) t} + q_{-3} e^{-2\pi i (3\nu) t} + \dots, \end{cases}$$

where q_0, q_n, q_{-n} $(n = 1, 2, 3 \dots)$ are constants. The coefficient q_0 is real; it measures the average distance of the vibrating particle from the origin O. As for q_n and q_{-n}, they are conjugate complex. Thus

$$(11) \qquad q_n = q_{-n}^* \quad \text{or} \quad q_n^* = q_{-n} \qquad (n = 1, 2, 3 \dots).$$

When we compare the nonharmonic periodic motion (10) with the harmonic one (4), we see that (10) may be viewed as resulting from the superposition of harmonic vibrations of type

(12) $q_n e^{2\pi i(n\nu)t} + q_{-n}e^{-2\pi i(n\nu)t}$.

The vibratory motion (12) constitutes a harmonic vibration of frequency $n\nu$ and of amplitude $2\sqrt{q_n q_{-n}}$, or $2\sqrt{q_n q_n^*}$. Hence we may say that the nonharmonic motion (10) results from an appropriate superposition of harmonic vibrations having the frequencies

(13) $\nu, 2\nu, 3\nu, \ldots \ldots n\nu, \ldots \ldots$,

appropriate amplitudes

(14) $2\sqrt{q_n q_n^*}$,

and appropriate phases. These statements illustrate the essence of Fourier's discovery.

Nonharmonic oscillations occur in mechanics when a particle is attracted to a fixed point on the line of oscillation by a force which is not proportional to the distance. A mechanical system of this sort is called an "anharmonic oscillator." Let us assume that in such an oscillator, the oscillations of which are supposed to be periodic, the vibrating particle is an electron. Since the total motion is equivalent to an appropriate superposition of harmonic vibrations having the frequencies (13) and the amplitudes (14), we infer that the oscillator will emit radiations of frequencies (13) simultaneously. The intensities of the radiations will be proportional to the squares of the corresponding amplitudes, and therefore to expressions of type $q_n q_n^*$. These are the classical conclusions, and we may summarize them by saying that the radiated frequencies are the mechanical frequencies of the motion, and that the intensities of the radiation are proportional to the squares of the mechanical amplitudes.

There is an important difference between the anharmonic, and the harmonic, oscillators. In a given harmonic oscillator, the amplitude of the swing increases with the energy of the motion, but the frequency of the vibration remains the same. In an anharmonic oscillator, if we increase the energy of the motion, not only are the amplitudes of the component vibrations increased but in addition the frequencies are changed. In particular, the fundamental frequency ν becomes, say ν';

and in the corresponding Fourier development of the motion, the frequencies are given no longer by (13) but by

(15) ν', $2\nu'$, $3\nu'$ . . . $n\nu'$, ...

In our new Fourier development the coefficients q_n and q_{-n}, which are connected with the partial amplitudes, must of course also be changed. We may represent the new coefficients by q'_n and q'_{-n}. The frequencies radiated (in the event of the vibrating particle being an electron) are the new mechanical frequencies (15), and the intensities of these radiations are proportional to the new magnitudes $q'_n q'_{-n}$, or $q'_n q'^*_n$.

Now, we know that the classical anticipations for the radiated frequencies are incorrect and that Bohr's theory was devised for the purpose of correcting and refining the classical theory. According to Bohr, the anharmonic oscillator can vibrate only with certain energies. Furthermore, when the oscillator is vibrating with one of the permissible energies, the frequencies radiated are derived from the frequency condition, and they are not the mechanical frequencies of the motion. We might treat the problem of the anharmonic oscillator by Bohr's methods, improving thereby on the classical treatment; but we know that Bohr's theory itself does not always lead to correct results, so that a further refinement is required. Heisenberg's Matrix Method proposes to furnish the more refined theory.

The Heisenberg Arrays—The Matrix Method, being a refinement of the classical theory and of Bohr's, is most easily approached when we start from classical considerations and refine them progressively. Thus, suppose we are dealing with an anharmonic oscillator. In the classical theory, the oscillator may have any energy, but we shall here avail ourselves of Bohr's restrictions, which require that only certain energy values be stable. For the present we are not concerned with the exact values of the stable energy levels and shall merely denote them by $E^{(1)}$, $E^{(2)}$, $E^{(3)}$, ... In the order written, each energy level is assumed to be higher than its predecessor.

We now propose to list all the frequencies which may be radiated by the anharmonic oscillator when it vibrates with one or another of the possible energies. We proceed classically. If, for example, the oscillator has the energy $E^{(1)}$, the fundamental frequency of the vibration has some corresponding value $\nu^{(1)}$. The motion of the electron is given by a Fourier series of type (10) in which the various terms are associated with the frequencies $\nu^{(1)}$, $2\nu^{(1)}$, $3\nu^{(1)}$, ... $n\nu^{(1)}$, . . . , i.e., with the fundamental frequency $\nu^{(1)}$ and its harmonics. Let us then call $\nu^{(1)}$, $\nu^{(2)}$, $\nu^{(3)}$,

.... $\nu^{(n)}$, ..., respectively, the fundamental frequencies of the oscillator's motion when its energy has the corresponding possible values $E^{(1)}, E^{(2)}, E^{(3)}, \ldots E^{(n)} \ldots$. The various terms of the Fourier series which correspond to the energies $E^{(1)}, E^{(2)}, E^{(3)}, \ldots$ are listed in the table below.

$$
\begin{cases}
q_0^{(1)} & q_1^{(1)}e^{2\pi i\nu^{(1)}t} & q_2^{(1)}e^{2\pi i(2\nu^{(1)})t} & q_3^{(1)}e^{2\pi i(3\nu^{(1)})t} & \cdots \\[4pt]
 & q_{-1}^{(1)}e^{-2\pi i\nu^{(1)}t} & q_{-2}^{(1)}e^{-2\pi i(2\nu^{(1)})t} & q_{-3}^{(1)}e^{-2\pi i(3\nu^{(1)})t} & \cdots \\[8pt]
q_0^{(2)} & q_1^{(2)}e^{2\pi i\nu^{(2)}t} & q_2^{(2)}e^{2\pi i(2\nu^{(2)})t} & q_3^{(2)}e^{2\pi i(3\nu^{(2)})t} & \cdots \\[4pt]
 & q_{-1}^{(2)}e^{-2\pi i\nu^{(2)}t} & q_{-2}^{(2)}e^{-2\pi i(2\nu^{(2)})t} & q_{-3}^{(2)}e^{-2\pi i(3\nu^{(2)})t} & \cdots \\[8pt]
q_0^{(3)} & \cdots & & & \\
\cdots & & \cdots & & \cdots
\end{cases}
$$

Instead of writing the terms in this order, we shall agree to place them in an array, as indicated in the table (16):

$$
(16)\begin{cases}
q_0^{(1)} & q_{-1}^{(1)}e^{-2\pi i\nu^{(1)}t} & q_{-2}^{(1)}e^{-2\pi i(2\nu^{(1)})t} & \cdots \\[6pt]
q_1^{(1)}e^{2\pi i\nu^{(1)}t} & q_0^{(2)} & q_{-1}^{(2)}e^{-2\pi i\nu^{(2)}t} & \cdots \\[6pt]
q_2^{(1)}e^{2\pi i(2\nu^{(1)})t} & q_1^{(2)}e^{2\pi i\nu^{(2)}t} & q_0^{(3)} & \cdots \\[6pt]
q_3^{(1)}e^{2\pi i(3\nu^{(1)})t} & q_2^{(2)}e^{2\pi i(2\nu^{(2)})t} & q_1^{(3)}e^{2\pi i\nu^{(3)}t} & \cdots \\[6pt]
\cdots & \cdots & \cdots & \cdots
\end{cases}
$$

In the arrangement (16) the constant terms $q_0^{(1)}$, $q_0^{(2)}$, ... of the various Fourier series are placed along the diagonal, and the terms of the Fourier series associated with, say, the energy $E^{(n)}$ are placed below the diagonal term $q_0^{(n)}$ and also on the horizontal to the right of it. The array gives at a glance the various mechanical frequencies of the oscillator when the successive energy values are assumed. These frequencies, according to classical theory, are also the radiated frequencies. As for the classical intensity of the radiation of frequency, say, $s\nu^{(n)}$, it is proportional to $q_s^{(n)}q_{-s}^{(n)}$, or $q_s^{(n)}q_s^{(n)*}$.

Now, we know that even if the energy values $E^{(1)}, E^{(2)}, E^{(3)}, \ldots$ were the correct quantum energy levels of our oscillator, the radiation frequencies and intensities expressed in the array would be wrong. For instance, we know that if the mechanical frequency $\nu^{(1)}$ should happen to be a possible quantum frequency for the radiation, then $2\nu^{(1)}$, $3\nu^{(1)}$, ... would not in general be possible radiation frequencies. Heisenberg's aim is to retain the general form of the array (16) but to replace the classical magnitudes which it contains by corrected quantum magnitudes. If this can be done in any particular problem, we shall have an array exhibiting all the quantum frequencies of the emitted radiation and also the corresponding intensities. It must be emphasized that, at this stage of the discussion, we have no idea of the numerical values of the quantum magnitudes in any particular problem. Our immediate concern does not lie, however, in this direction. All that we wish to do for the present is to devise a notational scheme for the quantum analogues of the classical amplitudes $q_s^{(n)}$ and $q_{-s}^{(n)}$ and of the classical frequencies $s\nu^{(n)}$.

To understand the new notation, we shall restrict our attention to those terms of the array (16) which are the Fourier terms pertaining to the nth energy level. The constant term of this Fourier series is the term $q_0^{(n)}$. It is situated on the diagonal of the array (16), and it constitutes the nth term counted along the diagonal from the upper left-hand corner. The other Fourier terms pertaining to this series are situated in the column below the diagonal term $q_0^{(n)}$ and on the horizontal line to the right of it. The distribution of these Fourier terms in the array (16) is illustrated in the table (17)

$$(17) \begin{cases} q_0^{(n)} \qquad q_{-1}^{(n)}e^{-2\pi i\nu^{(n)}t} \quad . \ . \ . \quad q_{-s}^{(n)}e^{-2\pi i(s\nu^{(n)})t} \qquad . \ . \ . \\[2em] q_1^{(n)}e^{2\pi i\nu^{(n)}t} \\[1em] \vdots \\[2em] q_s^{(n)}e^{2\pi i(s\nu^{(n)})t} \\[1em] \vdots \end{cases}$$

The classical frequencies are thus

$$(18) \qquad \nu^{(n)}, \ 2\nu^{(n)}, \ . \ . \ . \ s\nu^{(n)}, \ . \ . \ . \ ,$$

and the corresponding classical intensities are proportional to

$$(19) \qquad q_1^{(n)}q_{-1}^{(n)}, \quad q_2^{(n)}q_{-2}^{(n)}, \ . \ . \ . \ q_s^{(n)}q_{-s}^{(n)}. \ . \ . \ .$$

The correspondence principle guides us in our selection of a system of symbols whereby to represent the corrected quantum values of the frequencies and of the amplitudes. Thus we know that, if the energy value $E^{(n)}$ is high and if s is a small number, the classical frequencies (18) tend to become equal to the quantum frequencies which are generated when the system drops from the level $E^{(n)}$ to the levels $E^{(n-s)}$ ($s = 1, 2, 3 \ldots$), or also from the levels $E^{(n+s)}$ to $E^{(n)}$.

Let us, then, represent by $\nu(n + s, n)$ the quantum frequency which is emitted when the system drops from the level $E^{(n+s)}$ to the level $E^{(n)}$. When the conditions of the correspondence principle are realized, we have

(20) $$s\nu^{(n)} = \nu(n + s, n).$$

The notation for the frequency suggests that we represent the correct quantum value of the amplitude $q_s^{(n)}$ by $q(n + s, n)$. When the conditions of the correspondence principle are realized, we may therefore set

(21) $$q_s^{(n)} = q(n + s, n).$$

Also, since $q_s^{(n)}$ and $q_{-s}^{(n)}$ are conjugate complex, we have

(22) $$q_{-s}^{(n)} = q_s^{(n)*} = q^*(n + s, n).$$

From (21) we see that the quantum value of $q_0^{(n)}$ is written $q(n, n)$. Hence, when the conditions of the correspondence principle are satisfied, we obtain

(23) $$q_0^{(n)} = q(n, n).$$

Taking all these results into consideration, we have, under the limiting conditions of the correspondence principle,

(24) $$\begin{aligned} q_0^{(n)} &= q(n, n) \\ q_s^{(n)} e^{2\pi i (s\nu^{(n)})t} &= q(n + s, n) e^{2\pi i \nu(n+s, n)t} \\ q_{-s}^{(n)} e^{-2\pi i (s\nu^{(n)})t} &= q^*(n + s, n) e^{-2\pi i \nu(n+s, n)t}. \end{aligned}$$

We may modify our notation as a result of the following considerations:

We have agreed that when the system drops from the higher level $(n + s)$ to the lower level n, the frequency it emits is to be represented by $\nu(n + s, n)$. Suppose, then, the reverse transition takes place, i.e., from the lower level n to the higher one $(n + s)$. According to our notations we should associate this transition with a radiation frequency $\nu(n, n + s)$. Now, during the reverse transition, the system absorbs exactly the same radiation it emitted when it dropped from the higher level $(n + s)$ to the lower one n. Hence the two frequencies $\nu(n + s, n)$ and $\nu(n, n + s)$ are exactly the same in value; the only difference is that the former refers to an emission and the latter to an absorption. In a purely formal way, we may regard an absorption as a negative emission, and, under these conditions, we may set

$$(25) \qquad \nu(n, n+s) = -\nu(n+s, n).$$

The last term in (24) may then be written

$$q^*(n+s, n)e^{2\pi i\nu(n, n+s)t};$$

and, so as to preserve the symmetry of the notation, we shall write this

$$(26) \qquad q(n, n+s)e^{2\pi i\nu(n, n+s)t},$$

with the understanding that

$$(27) \quad q^*(n+s, n) = q(n, n+s), \quad \text{or} \quad q(n+s, n) = q^*(n, n+s).$$

With this change in notation, the terms in the table (24) become:

$$(28) \quad \begin{cases} q_0^{(n)} = q(n, n) \\ q_s^{(n)}e^{2\pi i(s\nu^{(n)})t} = q(n+s, n)e^{2\pi i\nu(n+s, n)t} \\ q_{-s}^{(n)}e^{-2\pi i(s\nu^{(n)})t} = q(n, n+s)e^{2\pi i\nu(n, n+s)t}. \end{cases}$$

Under the limiting conditions of the correspondence principle, namely, n large and s small, the classical terms on the left are equal to the quantum terms on the right. But suppose that n is small or s high (or both). The conditions of the correspondence principle are no longer fulfilled, and the quantum terms on the right of (28) will differ from the classical ones on the left. Nevertheless, we shall agree in all cases to represent the corrected values of the classical magnitudes in (28) by the corresponding quantum magnitudes of the same table.

Effecting the required quantum substitutions in the array (16), we obtain the quantum array

$$(29) \quad \begin{cases} q(1,1) & q(1,2)e^{2\pi i\nu(1,2)t} & q(1,3)e^{2\pi i\nu(1,3)t} & \cdots \\ q(2,1)e^{2\pi i\nu(2,1)t} & q(2,2) & q(2,3)e^{2\pi i\nu(2,3)t} & \cdots \\ q(3,1)e^{2\pi i\nu(3,1)t} & q(3,2)e^{2\pi i\nu(3,2)t} & q(3,3) & \cdots, \\ \cdots & \cdots & \cdots & \cdots \\ \cdots & \cdots & \cdots & \cdots \end{cases}$$

where

$$(30) \qquad q(n, k) = q^*(k, n) \quad \text{and} \quad \nu(n, k) = -\nu(k, n).$$

We shall call the array (29) a "Heisenberg array."

For the present this array is merely an aggregate of symbols; the precise numerical values of its terms are unknown. But if by some means or other the correct numerical values of the various terms could be inserted for any particular problem, the frequencies listed in the array would represent the possible quantum frequencies of the radiations. Also, the intensity of the frequency $\nu(n, k)$ would be proportional to

$$(31) \qquad q(n, k)q(k, n), \ i.e., \ q(n, k)q^*(n, k).†$$

All that we have done up to this point has been to explain the symbolism that will be adopted in the Heisenberg arrays. But it must be emphasized that when we are dealing with any particular problem, the numerical values to be attributed to the various terms of the array are completely unknown, so that, for the present at least, the array is a mere blank.

The Coordinate q—When we give the correct values to the terms of the Heisenberg array which corresponds to some particular physical system, we obtain the frequencies and intensities of all the radiations that the system can emit. The method whereby the correct values for the terms are secured will be considered later. For the present we shall be concerned with other features. Let us suppose, for instance, that we are

---

† The classical intensity of the radiation of frequency $s\nu^{(n)}$ emitted by a vibrating electric charge e is

$$(32) \qquad \frac{(2\pi s\nu^{(n)})^4}{3c^3} \times 4e^2 q_s{}^{(n)}q_s{}^{(n)*}.$$

We know that when the limiting conditions of the correspondence principle are realized, the classical expression (32) gives the correct value for the intensity. But we also know that when these limiting conditions are departed from, the value determined by (32) ceases to be correct. Now the general significance of the correspondence principle is to suggest that classical formulae are usually correct, and that their failure to yield correct results under all conditions is due to the erroneous values credited by the classical theory to the magnitudes entering in these formulae. Guided by these considerations, we may infer that the substitution in (32) of the corrected magnitudes for $s\nu^{(n)}$, $q_s{}^{(n)}$ and $q_s{}^{(n)*}$ will yield the correct expression of the intensity under all conditions. The corrected magnitudes being represented by $\nu(n + s, n)$, $q(n + s, n)$ and $q^*(n + s, n)$, respectively, we obtain in place of (32) the expression

$$(33) \qquad \frac{[2\pi\nu(n + s, n)]^4}{3c^3} \times 4e^2 \times q(n + s, n)q^*(n + s, n);$$

and this expression is assumed to yield the correct intensity under all conditions.

Since the intensity radiated is equal to the number of quanta of energy $h\nu(n + s, n)$ emitted every second, we have only to divide (33) by this quantum of energy and we obtain the *a priori* probability of the drop from the level $(n + s)$ to the level n. This magnitude is the Einstein probability coefficient of spontaneous emission, $A_{n+s,n}$.

dealing with the anharmonic oscillator and with its Heisenberg array. Now, we note that the array does not yield any information on the motion of the electron. Of course, if the terms in the array were the classical ones, we might sum up those terms situated on a horizontal and on the associated vertical, and thereby reconstruct the Fourier series which defines the motion of the electron when the energy has a corresponding value. By this means we could deduce from the array the various quantized motions which the electron of our oscillator might assume. But we have seen that the correct values of the terms in the Heisenberg array are not the Fourier values, so that even if the preceding summations were performed, they would not yield the electron's motions. It would thus appear that our present theory disregards the motion of the electron entirely and concentrates solely on the radiations emitted.

An important consequence of these considerations is that the coordinate q of the electron (and also its momentum p) appears to play no part in the theory. However, as subsequent results will show, this conclusion is only partly true, for we shall find that the Heisenberg array of a given problem (say, of the anharmonic oscillator), though it primarily determines the radiations emitted by the oscillator, also plays the part of the classical coordinate q of the vibrating electron at each instant t. This array, viewed as forming one unit, constitutes, therefore, Heisenberg's version of the classical coordinate q. But then it follows that Heisenberg's coordinate must be identified with all the electromagnetic radiations which the anharmonic oscillator may emit when it undergoes all possible energy drops. Obviously, only by a stretch of the imagination can we recognize in Heisenberg's coordinate q the slightest similarity to the classical conception of the electron's coordinate. In addition to the glaring physical difference in the two conceptions of a coordinate, a mathematical difference must also be mentioned: in all departments of mathematics and physics, a coordinate has heretofore been viewed as a mere number, whereas a Heisenberg array is not a number. Heisenberg's identification of a coordinate with an array is therefore revolutionary and, indeed, utterly incomprehensible—at least so it would appear at this stage. The justification for Heisenberg's views will be vindicated, however, by subsequent results. The identification of the coordinate q of a mechanical problem with a Heisenberg array leads, as we shall see, to a similar identification for the momentum p which is connected with q. The net result is that in the matrix method the variables q and p of classical mechanics will be replaced by Heisenberg arrays.

Now, in classical dynamics, all the mechanical magnitudes (energy, angular momentum, etc.) can be expressed by suitable combinations of the q's and p's of the mechanical system considered. The same combinations are retained in the matrix method, the only difference is that the q's and p's will now be Heisenberg arrays instead of ordinary numbers. But this difference complicates the situation considerably, for whereas the rules of addition, subtraction, multiplication, and division for ordinary numbers are well established, the corresponding rules for the Heisenberg arrays are as yet unknown. For the present we are in complete ignorance of the mathematical properties of these arrays; the arrays are mere empty symbols and cannot be utilized. The immediate problem is thus to establish their mathematical significance and their rules of combination. In this attempt Heisenberg, Born, and Dirac were guided by several clues. We shall first enumerate these clues and then examine how they were utilized.

The Clues to the Matrix Theory—The clues to the matrix theory may be classed in three groups.

(a) A first clue is Ritz's combination principle, which refers to the empirically observed regularities in the distribution of the spectral lines. Ritz's principle is accepted as accurate by Heisenberg. We obtain another clue by noting that although classical anticipations are incorrect in the subatomic world, yet, under the limiting conditions of the correspondence principle, they tend to furnish correct results. This suggests that the Heisenberg array and the Fourier series, which represent the coordinate q in Heisenberg's scheme and in classical science respectively, should have many properties in common. As will be explained presently, these two clues suffice to show that the Heisenberg arrays are infinite Hermitean matrices. At the same time we are able to assign similar matrix forms to the momenta p. The rules of addition and of subtraction for the Heisenberg arrays are thus secured; they are the rules that hold for matrices.

(b) In discussing matrices we mentioned that they do not usually exhibit the commutative property for multiplication. This peculiarity introduces a vagueness in calculations where matrices must be combined, and until this vagueness is removed, no progress can be made. Here another clue suggests itself; it is Bohr's quantizing condition. It is true that Bohr's quantizing condition often gives wrong results, $e.g.$, it does not even suggest the half-quantum numbers and the more complicated relations which are sometimes found to be necessary in the expression of the energy levels. Nevertheless Bohr's theory may be regarded as a first approximation, and on this basis we must expect that when the limiting conditions of the correspondence principle are approached, Heisenberg's more refined theory must pass over into Bohr's. Finally, when the conditions of the correspondence principle are fully satisfied, both theories must tend to merge into the classical theory. Inasmuch as the gradual passage to the limiting conditions is realized when we assume that Planck's constant h becomes smaller and smaller, we see that we have here an illustration of the progressive merging of quantum science into classical science, which accompanies the hypothetical decrease in the value of h. Considerations of this sort enabled Heisenberg, Born, and Dirac to obtain the refined quantizing condition which would have to replace Bohr's less refined condition. The new quantizing condition plays, however, a dual rôle. In the first place, its significance is physical since it imposes restrictions on the possible energy values of a system. But, in addition, it serves a purely mathematical purpose, for it establishes the exact commutation rules (for multiplication) which must hold between the Heisenberg matrices, and it thereby removes from our calculations the vagueness we mentioned previously. Combinations of matrices now yield determinate results.

(c) Thus far we have been concerned with the mathematical aspect of the Heisenberg arrays and with their rules of combination. Other points, however, must be clarified. In the classical treatment and in Bohr's theory, the mechanical motion is obtained by integrating the equations of dynamics, e.g., Hamilton's equations. But in Heisenberg's treatment, where coordinates are represented by matrices, we cannot assert a priori that the same dynamical equations will be valid. At this point the correspondence principle once more affords a clue; for since, under the limiting conditions of the correspondence principle, the classical mechanical equations yield correct results, we must assume that these classical equations express the laws to which the quantum laws will tend when the limiting conditions are neared. This suggests that the correct quantum laws controlling the radiation from mechanical systems will be obtained, not by rejecting the classical equations of mechanics, but merely by replacing in these equations the classical magnitudes q and p by the corresponding matrices q and p. In this way classical mechanics serves as a guide to the matrix method.

We shall now supplement the foregoing general considerations by examining the successive steps in the development of the theory.

Ritz's Combination Principle—The first clue in establishing the nature of a Heisenberg array is afforded by Ritz's combination principle. According to this principle (explained in Chapter XXV), the various frequencies which an atomic system radiates are connected in such a way that they may all be expressed by the differences in the various terms of some appropriate aggregate. In Bohr's theory, the frequency condition shows that the Ritz terms are the energy levels divided by h. But here we are not availing ourselves of Bohr's theory, and so this last feature will be disregarded.

In a one-dimensional system (such as the anharmonic oscillator) the Ritz terms form a single sequence and may be represented by the terms

$$(34) \qquad T_1, T_2, T_3, \ldots T_n, \ldots$$

We shall assume that these terms have increasing values from left to right. According to Ritz's rule, the frequencies radiated will be of type

$$(35) \qquad \nu(n, k) = T_n - T_k,$$

where $n > k$.

The name "combination principle" arises from the fact that if we add two radiated frequencies, such as $\nu(n, k)$ and $\nu(k, s)$, we obtain another possible radiated frequency, $\nu(n, s)$. This results from the relations

$$(36) \quad \nu(n, k) + \nu(k, s) = T_n - T_k + T_k - T_s = T_n - T_s = \nu(n, s).$$

We also see that

$$(37) \qquad \nu(n, k) = T_n - T_k = -(T_k - T_n) = -\nu(k, n).$$

(That $\nu(n, k)$ and $\nu(k, n)$ are equal, but of opposite sign, has been mentioned on a previous page.)

If the numerical values of the Ritz terms T were known, we should know the exact radiated frequencies. Of course, Ritz's principle does not give the values of these terms; it merely states that a peculiar relationship connects the frequencies, so that combinations of the type illustrated in (36) hold.

If we assume the correctness of Ritz's principle, we are assured that, regardless of the particular problem we may be considering, the frequencies in Heisenberg's array (29) must satisfy Ritz's relations (36). We have also seen that the coefficients $q(n, k)$ and $q(k, n)$ must be complex conjugate.

Heisenberg Matrices—A further point of importance concerning the nature of the array (29) is elucidated when we attempt to preserve as much as possible the analogy between the new coordinate q and the classical one. Thus in the classical theory, if a particle is vibrating with a periodic motion, the classical coordinate q of the particle may be expressed by means of a Fourier series of type (10). The square of the coordinate q is then expressed by a different Fourier series, but the new series contains exactly the same mechanical frequencies as the original one. Since, under the limiting conditions of the correspondence principle, Heisenberg's theory must yield the same results as the classical theory (for the latter is correct in this case), we extrapolate and assume that, under all conditions the square of a Heisenberg array q must be a Heisenberg array containing exactly the same frequencies.

If this requirement is imposed, it can be shown that, in view of Ritz's combination principle linking the frequencies, the rule for multiplying a Heisenberg array by itself is precisely the same rule that holds for matrices. We conclude that a Heisenberg array is a matrix. We also note that the terms symmetrically placed with respect to the diagonal are conjugate complex. Thus

$$q(n, k) e^{2\pi i \nu(n,k)t} \text{ and } q(k, n) e^{2\pi i \nu(k,n)t}$$

are conjugate complex, for

(38) $q(k, n) = q^*(n, k)$ and $\nu(k, n) = -\nu(n, k)$.

A Heisenberg matrix is therefore of the type called *Hermitean*.[†] Furthermore, considerations which will be developed presently show that the Heisenberg matrices contain an infinite number of lines and columns.[‡] Thus, they are infinite matrices. This feature appears plausible even at the present stage of the discussion, for the Heisenberg array (29) is a mere refinement of the Fourier array (16), and the Fourier terms may be infinite in number.

The mathematical status of the Heisenberg arrays is now clarified: these arrays are infinite Hermitean matrices; and so we know how to add and subtract them. As we explained in Chapter XXXVI, in order to add or subtract two matrices, we merely add or subtract their corresponding elements, or terms. The new terms form a matrix which represents the sum or the difference of the former two matrices.

The Heisenberg matrices contain t and hence vary with time. The only terms that are always independent of time are the diagonal terms; these are real constants. There is no difficulty in obtaining the derivative of a matrix q with respect to time.[¶] We have merely to differentiate each

---

[†] See page 806.

[‡] See note page 832.

[¶] If $q(n, k) e^{2\pi i \nu(n,k)t}$ is the general term of the matrix q, the general term of the derivative $\dot{q}$ is

$$2\pi i \nu(n, k) q(n, k) e^{2\pi i \nu(n,k)t}.$$

individual term; the constant terms on the diagonal vanish under the derivation, since they are independent of time. The new matrix we thus obtain from the matrix q is represented by $\dot{q}$, *i.e.*, by placing a dot over q.

The Conjugate Matrix p—In classical (or relativistic) dynamics, a system of f degrees of freedom is a system whose configuration at any instant is determined by the numerical values of f coordinates,

$$q_1, q_2, q_3, \ldots q_f.$$

If we know how these coordinates vary with time in any given problem, we know how the system evolves. A magnitude which varies with time is a function of time (*i.e.*, of t). Hence we may say that when the f functions

$$q_1(t), q_2(t), \ldots q_f(t)$$

are known, the motion is determined. For this reason these functions are sometimes called the *solution functions*.

Corresponding to each solution function $q_r(t)$ there is a function represented by $p_r(t)$. This latter function represents the variation through time of a momentum or of an angular momentum associated with that part of the system whose position at any instant t *is defined by* $q_r(t)$. In any given problem, if the f solution functions $q(t)$ are known, the corresponding f functions $p(t)$ may be deduced,* so that, in a certain sense, to express them is redundant. However, we shall retain the functions $p(t)$. Under these conditions the f functions

$$q_1(t), \quad q_2(t), \ldots q_f(t)$$

and the f corresponding functions

$$p_1(t), \quad p_2(t), \ldots p_f(t)$$

may be regarded as defining jointly the solution functions of the system. The functions $q(t)$ determine the configuration of the system at any instant, and the functions $p(t)$ determine its momentum or angular momentum.

In Chapter XXX, we mentioned that in a system having several degrees of freedom, the corresponding solution functions, *e.g.*, $q_r(t)$ and $p_r(t)$, considered two by two are said to be *canonically conjugate*, or,

---

* If the Hamiltonian function is expressed in terms of the coordinates q and of their derivatives $\dot{q}$, we have the relations

$$p_r = \frac{\partial H\ (q,\ \dot{q})}{\partial \dot{q}_r} \qquad\qquad (r = 1, 2, 3 \ldots f).$$

more simply to be *conjugate*. If our system has only one degree of freedom, there is only one solution function $q(t)$ and only one solution function $p(t)$. These two solution functions are then necessarily conjugate. As a special case of a problem of one degree of freedom, we may consider the linear oscillator. Here the solution function $q(t)$ defines the position of the vibrating particle at all instants during the motion, and $p(t)$ defines its momentum. Since momentum is mass × velocity, the two solution functions are connected by the relation

$$(39) \qquad\qquad p(t) = \mu\dot{q}(t),$$

where μ is the mass of the particle and $\dot{q}(t)$ its velocity at any instant.

Let us now consider the matrix theory. The Heisenberg matrices q and p contain the variable t (time), so that they vary with time. These matrices are thus the analogues of the coordinates q and p not at one specific instant, but throughout time. When therefore the correct matrices q and p are obtained for a given problem, these matrices are the analogues of the solution functions $q(t)$ and $p(t)$ of the classical theory. For this reason they may be called the solution matrices of the problem.

In the case of the oscillator and of similar systems, Heisenberg carries over the classical relation (39) into his matrix theory. Thus, if q is the solution matrix of the problem, the conjugate matrix p is defined by the matrix (39). The matrix p is itself a Heisenberg matrix, and it contains exactly the same frequencies as the matrix $\dot{q}$ and therefore also as the matrix q. By analogy with the classical solution functions $q(t)$ and $p(t)$, we may say that the two solution matrices q and p are canonically conjugate. For short, we speak of these magnitudes as *conjugate* without further specification.

The Commutation Rules—We shall restrict our attention for the present to systems of one degree of freedom, so that we have only one solution matrix q and only one momentum matrix p. In classical (or (relativistic) dynamics, the two conjugate functions $q(t)$ and $p(t)$ are mere numbers, and we may express their product by qp or pq indifferently. But now that q and p are viewed as matrices, we cannot accept the commutative property for multiplication without further inquiry. From what we know of matrices, commutability is a highly exceptional condition, and we may reasonably suspect that the two solution matrices q and p will not commute. Closer inspection shows that these two matrices cannot commute, and we must therefore determine the nature of the matrix which defines the difference between pq and qp. Unless this information is forthcoming, we are unable to proceed. Till now we know so little about our matrices that we cannot, on the basis of mathematical considerations alone, determine the value of $pq - qp$. Heisenberg and Born were thus compelled to resort to other methods of approach; they were guided by Bohr's quantizing condition.

We consider a mechanical system of one degree of freedom the motion of which is periodic. If we call $q(t)$ and $p(t)$ the conjugate solution functions of classical mechanics, which determine the motion, the Maupertuis action over a complete oscillation is

$$\int_0 pdq,$$

where $\int_0$ expresses that the integration is taken over a complete oscillation.

In classical (or relativistic) theory, no restriction is placed on the numerical value of this action.

Bohr's quantizing condition consists in restricting the value of the action by stipulating that it must always be some multiple of the quantum of action h. We thus obtain

(40) $$\int_0 pdq = nh \qquad (n = 1, 2, 3, \ldots).$$

This quantum restriction imposes a physical restriction on the possible motions, and hence also a mathematical restriction on the two solution functions $q(t)$ and $p(t)$.

For our present purpose, we find it advantageous to write Bohr's quantizing condition in a slightly different form. Thus suppose we consider the successive permissible values of the action (40) when n receives successive integral values. The difference in two consecutive values of the action is then

(41) $$\left[\int_0 pdq \right]_{n+1} - \left[\int_0 pdq \right]_n = h.$$

We may view the relation (41) as a consequence of Bohr's quantizing condition.

Let us return to the Heisenberg solution matrices q and p. Any given mechanical problem will be solved in the matrix theory if we can determine the solution matrices q and p. Now, we have just recalled that in Bohr's quantum theory the solution functions $q(t)$ and $p(t)$ must be submitted to the additional restriction of satisfying the quantizing condition (40), or (41). We infer that some similar restriction will have to be imposed when we follow the matrix treatment and consider the solution matrices q and p. The problem is to determine the nature of this restriction.

The only information we have to guide us is Bohr's quantizing condition itself. But it must be emphasized that this condition can only serve as a clue, for we know that it does not always yield the correct energy levels (*e.g.*, in the cases of the oscillator and the rotator), and hence cannot be accurate. However, Bohr's quantizing condition increases in accuracy when the energy levels are high, and as a result the clue which it affords to the matrix treatment becomes increasingly reliable in these circumstances. Incidentally, we recall that for extremely high values of the energy we find ourselves in the limiting conditions of the correspondence principle. Bohr's theory then merges into classical theory, and his quantizing condition ceases to impose any restriction on the solution functions.*

Since Bohr's quantizing condition becomes increasingly accurate as the energy increases, we conclude that the restriction which will play the part of the quantizing condition, in connection with the solution matrices q and p, must merge into Bohr's quantizing condition, *e.g.*, into (41), when the energy is high (*i.e.*, n large).

Now, calculation shows that when n is high, the matrix analogue of Bohr's quantizing condition (41) is expressed by the statement that $2\pi i$ times the nth diagonal term of the matrix $pq - qp$ must have the value h. Thus, the higher diagonal terms of this matrix must be equal to $\dfrac{h}{2\pi i}$. On the other hand, Bohr's quantizing condition fails to throw any light on the values of the lower diagonal terms or on the non-diagonal terms of the matrix $pq - qp$, so that the exact nature of this matrix is not

---

* This statement may be verified by supposing that, in (40), n is a very large number, *e.g.*, $n = 1000$. The consecutive permissible values of nh are then

(42) $1000h,\ 1001h,\ 1002h,\ \ldots\ .$

Each value differs from its predecessor by the same amount h, but the change in value, when we pass from one term to the next in (42), is insignificant when contrasted with the value of the term itself. The net result is that the discrete sequence (42) tends to become indistinguishable from a continuous sequence. Reverting then to the quantizing condition (41), we see that when n is very high, the integral may, for all practical purposes, be assumed to have any value within a continuous range, so that in effect no restriction is placed on its value. The quantizing condition thus ceases to impose any restriction, and Bohr's theory passes over into the classical one. It will be noted that this same situation would occur for all values of n provided h were infinitesimal. The successive values of nh in (42) would differ as before by h; but, with h now assumed to be exceedingly small, the sequence (42) would always approximate to a continuous succession of values. In short, if h were infinitesimal, Bohr's quantum theory would degenerate into classical science—a feature which is already known to us.

established. The information gathered on the higher diagonal terms was, however, taken as a clue by Born, who accordingly supposed that all the diagonal terms should have the same value $\dfrac{h}{2\pi i}$. Born further surmized that all the non-diagonal terms should vanish. He was guided to this assumption by the desire to simplify as much as possible the commutative properties of the matrices q and p. His step was of course speculative, but it has since been justified by results. Born's assumptions are summarized by the statement that the matrix $pq - qp$ is equal to the scalar matrix $1\dfrac{h}{2\pi i}$ (where 1 is the unit matrix). We thus obtain the following fundamental relation, usually called the "commutation relation," or the "commutation rule:"

$$(43) \quad pq - qp = \frac{h}{2\pi i}1, \quad i.e., \quad \left\{ \begin{array}{ccccc} \dfrac{h}{2\pi i} & 0 & 0 & 0 & .\ . \\[2mm] 0 & \dfrac{h}{2\pi i} & 0 & 0 & .\ . \\[2mm] 0 & 0 & \dfrac{h}{2\pi i} & 0 & .\ . \\[2mm] . & . & . & . & . \end{array} \right\}^{*}.$$

It is important to understand the significance of the relation (43). Primarily, it defines a mathematical restriction which must be imposed on the solution matrices q and p of any one-dimensional mechanical system.[†] As such, it informs us of the difference in the values of pq and of qp, and thereby enables us to formulate definite algebraic rules of combination for the solution matrices q and p. But the relation (43) has also a physical significance, for its mode of derivation implies clearly that it is the matrix analogue of Bohr's quantizing condition. Besides, the presence of Planck's physical constant h in (43) shows that physical restrictions are involved. The dual rôle played by the relation (43)

---

* The relation (43) is possible only if q and p are infinite matrices, so that, if (43) is accepted, the Heisenberg matrices are necessarily infinite—a fact mentioned previously.

† If q and p are solution matrices of two different problems, there is no reason why the commutation relation (43) should apply. The commutation relation has therefore no bearing on an arbitrary matrix q and an arbitrary matrix p, but solely on two solution matrices of the same problem. We shall see presently that the commutation relation is connected with the fact that the two solution matrices are conjugate.

justifies our referring to it indifferently as the "commutation relation" or as the "quantizing condition." At the same time we see how closely the mathematics of the theory is interwoven with its physical significance. No such situation arose in Bohr's theory.

Although Born's quantizing condition (43) passes over into Bohr's for high quantum values, it yields different results when the energy values are low; and we shall find that the new condition, when applied to specific problems, gives the correct energy levels, which Bohr's theory was not always able to furnish. Under the limiting conditions of the correspondence principle, when the energy is very high, Heisenberg's theory, like Bohr's, merges into classical science, so that q and p commute. To verify this important point, we note that the limiting conditions are realized when the relative importance of h decreases, and hence when we assume that in our formulae the value of h becomes infinitesimal. If now in the relation (43), we take h to be as small as we choose, $i.e.$, zero at the limit, we obtain

$$pq - qp = 0, \text{ or } pq = qp,$$

so that q and p commute. Thus, under the limiting conditions of the correspondence principle, one of the distinguishing features of Heisenberg's magnitudes q and p ($i.e.$, their non-commutability under multiplication) tends to become obliterated, and the matrices q and p behave like the familiar magnitudes of classical science. In other words, the matrix theory passes over into classical theory.

So far we have considered systems of one degree of freedom; and in such systems there is only one solution matrix q and only one corresponding matrix p, and the two matrices are necessarily conjugate. But we are often required to deal with systems having several degrees of freedom. We mentioned that in the classical treatment the solution of a problem of f degrees of freedom necessitates our obtaining the solution functions

$$q_1(t), q_2(t), \ldots \ldots q_f(t)$$

and

$$p_1(t), p_2(t) \ldots \ldots p_f(t).$$

We also mentioned that two solution functions such as $q_r(t)$ and $p_r(t)$ are canonically conjugate.

These results may be carried over into the matrix theory. All we

need do is to view the various magnitudes q and p as so many matrices.*
We retain the conception of conjugate magnitudes as before, except that
we are now dealing with matrices. Dirac established the commutation
relations for multidimensional solution matrices. He showed that,
for the Heisenberg method to yield Bohr's quantizing conditions in the
case of high quantum numbers, we must assume that the rth solution
matrix q_r commutes with the non-conjugate solution matrices q and p,
but that it does not commute with the conjugate momentum matrix p_r.
Similarly, the solution matrix p_r will commute with all the other matrices
except with q_r.

Expressed analytically, these commutation rules are

$$(46) \quad \begin{cases} q_r q_s - q_s q_r = 0 & (s = r \text{ or } s \neq r) \\ p_r p_s - p_s p_r = 0 & (s = r \text{ or } s \neq r) \\ p_r q_s - q_s p_r = 0 & (s \neq r) \\ p_r q_r - q_r p_r = \dfrac{h}{2\pi i} 1. \end{cases}$$

An additional commutation rule imposes itself when we apply the
relativistic refinements. Thus, in classical science, time may be treated
formally as a fourth dimension of space and therefore the time variable t
may be represented by a q, e.g., q_0. The momentum p_0 conjugate to
q_0 is found to be $-E$, where E is the total energy of the system. This
assimilation of time to a fourth dimension, though purely formal in
classical science, is seen to be necessary in the theory of relativity. Hence
when we adapt the matrix method to the requirements of the relativity

---

* In a system of one degree of freedom, a matrix q is 2-dimensional, and its
general term is of type

$$q(n, k) e^{2\pi i \nu(n,k)t}.$$

But if we are dealing with a system of f degrees of freedom, the matrices are
$2f$-dimensional. The general term, say of the rth matrix, is of form

$$(44) \quad q_r(n_1, n_2, \ldots n_f; k_1, k_2 \ldots k_f) e^{2\pi i \nu(n_1,n_2, \ldots n_f; k_1,k_2, \ldots k_f)t}.$$

Here the frequency $\nu(n_1, n_2, \ldots n_f; k_1, k_2 \ldots k_f)$ is connected with a drop in the
energy of the system from a quantum state $(n_1, n_2, \ldots n_f)$ to a state $(k_1, k_2, \ldots k_f)$.
The intensity radiated in connection with the coordinate q_r is proportional to

$$(45) \quad q_r(n_1, n_2, \ldots n_f; k_1, k_2 \ldots k_f) q_r^*(n_1, n_2, \ldots n_f; k_1, k_2 \ldots k_f).$$

theory, we must operate as though q_0 and p_0 were conjugate matrices and we must therefore extend the commutation rules (46) to them. When this is done, we get

(47)
$$\begin{cases} p_0 q_0 - q_0 p_0, \quad i.e., \quad -Et + tE = \dfrac{h}{2\pi i}\,1 \\[2mm] q_r t - t q_r = 0 \\[2mm] p_r t - t p_r = 0 \qquad (r \neq 0) \\[2mm] -E q_r + q_r E = 0 \qquad (r \neq 0) \\[2mm] -E p_r + p_r E = 0 \end{cases}$$

Dirac's Generalized Commutation Rules—Suppose we are dealing with a system of f degrees of freedom in classical dynamics, and let $F(q, p)$ and $G(q, p)$ represent two arbitrary functions of the solution functions $q_1(t)$, $q_2(t)$, . . . $q_f(t)$, $p_1(t)$, $p_2(t)$ . . . $p_f(t)$. If, in F and G, we replace the various functions $q(t)$ and $p(t)$ by the solution matrices, then F and G themselves become matrices, and according to whether these two matrices commute or not, the matrix $FG - GF$ will be vanishing or non-vanishing. Dirac, guided by the knowledge that for high energy values the matrix theory must pass over into Bohr's quantum theory and yield Bohr's quantizing conditions, established the following results:

If q_1, q_2 . . . . q_f, p_1, p_2 . . . p_f are the solution matrices of any problem, and if F and G are functions of these solution matrices, we have the matrix equation

(48)
$$FG - GF = \dfrac{h}{2\pi i}[F, G],$$

where $[F, G]$ represents what is known as the Poisson bracket expression of the functions F and G with respect to the variables q and p.* If this bracket happens to vanish, our two matrices F and G commute.

---

* By definition, the Poisson bracket $[F , G]$ of two functions F and G of the coordinates q and p is

$$\sum_r \left(\frac{\partial F}{\partial p_r} \frac{\partial G}{\partial q_r} - \frac{\partial F}{\partial q_r} \frac{\partial G}{\partial p_r} \right),$$

where the summation extends from $r = 1$ to $r = f$; i.e., over all the coordinates involved. Poisson brackets are of frequent occurrence in pure mathematics and in mechanics. Poisson introduced them in mechanics for the purpose of obtaining first integrals of Hamilton's equations.

When we replace in (48) F and G by the various individual q's and p's, the relations (46) are seen to be verified.

q-Numbers and c-Numbers—In the matrix method our concern is with the solution matrices q and p of the problem of interest, and also with combinations of these elementary matrices. Let us call $A(q, p)$, $B(q, p)$, $C(q, p)$, . . . any functions of the solution matrices. These functions A, B, and C are therefore matrices also. The combination rules we have established for such matrices are then

(49)
$$
\begin{cases}
A + B = B + A \\
A(B + C) = AB + AC \\
A(BC) = (AB)C \\
(A + B) + C = A + (B + C) \\
AB - BA = \dfrac{h}{2\pi i}[A, B] \quad \text{(see (48)).}
\end{cases}
$$

All these relations except the last are imposed by the fact that we are dealing with matrices. The last relation on the other hand cannot be derived from the properties of matrices alone. It is true that the multiplication of matrices is usually non-commutative, so that in general $AB - BA$ will not vanish; however, only when we supplement our mathematical information with physical considerations (*i.e.*, Bohr's quantizing condition) is the above precise value of $AB - BA$ obtained.

An alternative interpretation of the relations (49) is due to Dirac. Provisionally, let us dismiss matrices and let us view the relations (49) as defining a postulate system establishing the combination rules for numbers A, B, C . . . . These rules are not those which apply to the common numbers of arithmetic (for if A, B, C . . . were common numbers, the last relation (49) would be replaced by $AB - BA = 0$). We must therefore regard the numbers A, B, C . . . which satisfy the relations (49) as numbers of a novel kind. Dirac calls them q-numbers.*
The common numbers of arithmetic are then called c-numbers. Dirac

---

* The name q-numbers is taken as the contracted form of quantum-numbers. The shorter appellation is preferred because the new numbers have nothing in common with the quantum numbers n, l, j, m of Bohr's theory.

developed the mathematical analysis that holds for *q*-numbers. Fourier's decomposition was found to be valid and, indeed, except where the non-commutative property comes into play, the analysis for *q*-numbers bears a marked resemblance to the classical analysis that is valid for ordinary *c*-numbers. Although the algebraic combination rules (49), which characterize *q*-numbers, are those that apply to Heisenberg matrices, these matrices are but special illustrations of *q*-numbers. As Dirac has shown, many mechanical magnitudes which cannot be expressed by matrices may nevertheless be represented by *q*-numbers. To this greater generality of the *q*-numbers is due their utility in quantum mechanics.

The essence of these considerations is to show that in the study of microscopic processes (*e.g.*, in the atom), a new kind of algebra determined by the relations (49) is required. Under the limiting conditions of the correspondence principle, where h may be viewed as vanishing, the last relation (49) passes over into $AB - BA = 0$, so that our *q*-numbers A, B, C commute and may thus be identified with ordinary *c*-numbers. This passage from an unfamiliar to a familiar mathematics, accompanying the progressive realization of a certain limiting condition, also occurs in the theory of relativity. In the latter theory, Minkowski's strange 4-dimensional non-Euclidean geometry is the one best suited for the interpretation of physical phenomena; but when the limiting condition of low velocities is neared, the strange geometry passes over into the commonplace 3-dimensional Euclidean variety, and the fourth dimension, *i.e.*, time, becomes separate. Dirac's non-commutative algebra is thus the analogue of Minkowski's geometry. It is significant that both the commonplace algebra and the commonplace geometry are amply sufficient to cope with situations on the level of commonplace experience. Presumably for this reason, commonplace algebra and geometry were devised earlier and appear more familiar.

A further point of interest is that, so long as we are considering merely the algebraic rules of combination (46) for the *q*-numbers q and p, we need not be concerned with the exact significance of these *q*-numbers, for any magnitudes which satisfy the rules of combination (46) will perform the same rôle in our calculations. For instance, the pairs of operators

$$(50) \qquad q_r \quad \text{and} \quad \frac{h}{2\pi i} \frac{\partial}{\partial q_r} \qquad (r = 1, 2, 3 \ldots f),$$

have been shown by Schrödinger to satisfy the commutation relations

(46) of the matrices q and p.† The significance of Schrödinger's operators (50) will be explained on page 873. For the present we are merely drawing attention to the fact that magnitudes q and p, which Heisenberg interprets as matrices, may receive other interpretations.

Diagonal Matrices—Let us revert to the original conception of the magnitudes q and p as matrices. We shall confine ourselves to periodic systems of one degree of freedom, so that only one q and only one p will be involved. A first point to be clarified is the connection between a solution matrix q and the solution function $q(t)$, which in classical mechanics would represent the position of the oscillating particle at any instant.

Suppose we have obtained the matrix q which corresponds to the classical solution function $q(t)$ of some definite problem. Our matrix is of the general form (29), and it usually contains a large number of different frequencies $\nu(n, k)$; the latter represent the frequencies of the light that may be radiated from the system. The solution matrix also contains complex amplitudes $q(n, k)$, which serve to define the intensities of the radiated frequencies. It would therefore seem that all the magnitudes contained in the matrix refer to the radiations, and that the mechanical motion is disregarded entirely. As a matter of fact, this

---

† We may verify this for a one-dimensional system. With the matrices q and p, we have $pq - qp = \frac{h}{2\pi i}1$, where 1 is the unit matrix. If, then, any matrix ψ is multiplied (before) by $pq - qp$, the effect will merely be to multiply ψ by $\frac{h}{2\pi i}$. Let us now replace the matrix q by an ordinary variable and the matrix p by the operator $\frac{h}{2\pi i} \frac{\partial}{\partial q}$ and then let us form the operator which is the analogue of $pq - qp$. This operator is

$$\frac{h}{2\pi i} \frac{\partial}{\partial q}q - q\frac{h}{2\pi i} \frac{\partial}{\partial q}.$$

With this operator, we operate on a function $\psi(q)$. We get

$$(51) \quad \left[\frac{h}{2\pi i} \frac{\partial}{\partial q}q - q\frac{h}{2\pi i} \frac{\partial}{\partial q}\right]\psi(q) = \frac{h}{2\pi i}\left[\frac{\partial}{\partial q}(q\psi) - q\frac{\partial\psi}{\partial q}\right]$$
$$= \frac{h}{2\pi i}\left[\psi + q\frac{\partial\psi}{\partial q} - q\frac{\partial\psi}{\partial q}\right] = \frac{h}{2\pi i}\psi.$$

Hence the effect of our operation has been merely to multiply $\psi(q)$ by $\frac{h}{2\pi i}$. We are thus led to the result we obtained previously when, in place of the operators q and $\frac{h}{2\pi i} \frac{\partial}{\partial q}$, we considered the matrices q and p.

conclusion is erroneous, but even if it were correct, the matrix theory would not be inconsistent on this score, for we started out with the explicit understanding that the mechanical motions in the atom, being unobservable, would be ignored.

Let us show that the solution matrix does, however, give some information on the mechanical motion. To see this, suppose we are dealing with a problem of one degree of freedom in classical mechanics. We shall assume that the correct, quantized energy values E_1, E_2, ... of the system are known, and we consider the motion of the system when it has one or another of these energy values. The motion corresponding, say, to the energy value E_n is expressed by a Fourier series, and the constant term $q_0^{(n)}$ of this series defines the mean position of the vibrating particle over a long period of time. We then arrange the various terms of the different Fourier series into an array of type (16), placing the constant terms of the different series along the diagonal line.

The solution matrix of the same problem is given by an array of type (29), which has the same form as the Fourier array though its terms differ numerically from those of the latter array, since the Fourier developments do not give correct results for the radiations. The numerical differences between the corresponding terms of the two arrays tend, however, to disappear when we consider terms near the diagonal and far removed from the upper left-hand corner. This gradual merging of the two arrays is imposed by the correspondence principle, according to which for high energy values the lower terms of the various Fourier series give the correct frequencies and intensities.

Suppose, then, that E_n is a high energy value. In this case the diagonal Fourier term $q_0^{(n)}$ and the corresponding matrix one $q(n, n)$ will have the same value, and hence $q(n, n)$, like $q_0^{(n)}$, must measure the average value of the coordinate q of the vibrating particle, taken over a long period of time. We now extend these conclusions to *all* the diagonal terms of the Heisenberg array and assert:

Each diagonal term of the solution matrix q defines the average value of the classical coordinate q when the energy of the system has one quantum value or another.

Let us, then, strike out all the non-diagonal terms from the solution matrix q. We shall be left with a diagonal matrix, the diagonal terms of which define the various average values of the coordinate q of the vibrating particle when its energy has one or another of the permissible quantum values. We may speak of this matrix as the matrix of the average values of the coordinate q. Our procedure of striking out all the non-diagonal terms of a matrix may appear artificial; however, we may obtain the same diagonal matrix in a less arbitrary way. Thus let us take the average values of all the terms of the solution matrix q over a long period of time; all the non-diagonal terms now vanish, and we are left with the diagonal terms alone.* Hence our diagonal matrix, which determines the average positions of the coordinate q, is seen to result from an averaging of the solution matrix over a long period of time.

More generally, let $F(q, p)$ be some function of the solution matrices q and p of some particular problem. Then F is itself a matrix, and usually it will be of non-diagonal form. This matrix determines the manner in which the magnitude F varies when the system evolves. The diagonal terms of the matrix F (like those of any Hermitean matrix) are necessarily constants; they define the average values of the magnitude F over a long period of time when the system is in one or another of the energy states.

As a particular case the matrix F may happen to be a diagonal one. Since a diagonal matrix does not involve time, the magnitude F cannot involve time either, and so its average values are also its actual values. The diagonal terms of the diagonal matrix F will thus define the actual fixed values which the magnitude F will have when the system is moving with one or another of the quantized energies.

The foregoing analysis shows that, in Heisenberg's theory, a diagonal matrix is analogous to a constant of the classical theory, or rather to an aggregate of different constants.

The Solutions of Definite Problems—We are now in a position to understand how the Heisenberg matrices may be utilized to furnish the radiations that will be emitted from an atomic system. For simplicity we shall suppose that we are dealing with a one-dimensional system, *e.g.*, an anharmonic oscillator in which an electron vibrates back and forth along a straight line under the action of some force. Inasmuch as the matrix treatment of the problem utilizes the classical theory as a clue, we shall first recall the classical procedure.

According to classical electrodynamics, the frequencies of the radiations emitted by the system should be the mechanical frequencies of the

---

* We are assuming that all the non-diagonal terms are of form $q(n, k)e^{2\pi i\nu(n,k)t}$. If, in some of these terms, $\nu(n, k) = 0$, the terms in question reduce to $q(n, k)$ and they no longer vanish in the averaging process. The average matrix is no longer a diagonal one. We shall exclude a consideration of this situation; it involves degeneracy.

SOLUTIONS OF DEFINITE PROBLEMS 841

electron's motion; and the intensities of these radiations should be proportional to the squared amplitudes of the mechanical vibrations. The classical treatment thus compels us to determine the motion of the electron, and hence to obtain the solution function $q(t)$ of Hamilton's equations

$$(52) \qquad \dot{q} = \frac{\partial H(q,\, p)}{\partial p}, \quad \dot{p} = -\frac{\partial H(q,\, p)}{\partial q},$$

in which $H(q, p)$ is the Hamiltonian function corresponding to our problem.

Now we know that the classical theory furnishes incorrect results, since the frequencies of the radiations usually differ from the mechanical frequencies of the electron. But we also know that under the limiting conditions of the correspondence principle, the classical anticipations become increasingly correct, so that in this case the solution $q(t)$ of Hamilton's equations furnishes the peculiarities of the radiation (frequency, intensity, polarization) with increasing accuracy. Thus when we are operating under the limiting conditions of the correspondence principle, Hamilton's equations, which primarily control the mechanical motion, are seen also to control the radiations. We may express this situation more concisely by saying that, under the limiting conditions, Hamilton's mechanical equations tend to become *radiation equations*.

Let us now pass to the matrix method. In this method we are not concerned with the mechanical motion, but only with the radiations. All the peculiarities of the emitted radiation will be known if we can secure the solution matrix q corresponding to our problem, the various terms of this matrix furnishing the frequencies and intensities of the radiations. Since this solution matrix q, and also the conjugate matrix p, necessarily differ from one problem to another, they must be restricted by some condition which characterizes the problem we are considering; in other words they must be the solutions of matrix equations which play the part of radiation equations for the problem of interest. Our aim is therefore to discover these matrix equations which control the radiation.

It is here that correspondence considerations guide us. Thus we have seen that under the limiting conditions of the correspondence principle, the classical theory yields correct results; hence we must suppose that, under these limiting conditions, the matrix theory merges into the classical one and that the solution matrix q furnishes the same information as the classical solution function $q(t)$. As a result, the radiation equations of the matrix theory must merge into those of the classical theory, *i.e.*, into Hamilton's equations. We conclude that, under the limiting conditions of the correspondence principle, the radiation equations in the matrix theory will be matrix equations having the same form as Hamilton's equations, and the solution matrices q and p will be solutions of these matrix equations.

It should be observed that we have as yet no knowledge of the form of the matrix radiation equations when we are no longer operating under the limiting conditions of the correspondence principle. However, in view of our success in extrapolating correspondence considerations in Bohr's theory, it seems permissible to pursue the same course here and to say that under all conditions Hamilton's equations, in matrix form, will constitute the correct radiation equations of the matrix method. These preliminaries explain why Hamilton's mechanical equations constitute the radiation equations of the matrix theory, even though the mechanical motion exhibits no direct connection with the radiation and is completely disregarded in the matrix treatment.

Let us summarize. In any given problem we retain the classical Hamiltonian of the problem and also Hamilton's equations (52). The only difference is that now q and p are to be viewed as *matrices* and no longer as ordinary numerical magnitudes. As a result the Hamiltonian function will be represented by a matrix, and Hamilton's equations will be equations between matrices. Our procedure will now be to determine the matrices q and p which satisfy these matrix equations. These matrices q and p will then be the solution matrices.

Matrix equations are more difficult to solve than ordinary equations, for each matrix equation is equivalent to as many ordinary equations as there are terms in the matrix. $\Big($Hamilton's first equation, for example, implies that each term of the matrix $\dot{q}$ is equal to the corresponding term of the matrix $\dfrac{\partial H}{\partial p}\Big)$. Since the Heisenberg matrices are infinite, each one of Hamilton's two equations is equivalent to an infinite number of ordinary equations. This fact in itself indicates how complicated is the solution of a problem in the matrix method.

In addition to being solutions of Hamilton's equations, the two solution matrices q and p must satisfy the general commutation relation which holds between all conjugate matrices. Only then are the two matrices truly solution matrices. The commutation relation, *i.e.*,

$$(54) \qquad\qquad pq - qp = \frac{h}{2\pi i}1$$

is itself a matrix equation. The restriction (54) is not redundant, for we can obtain matrices which are solutions of Hamilton's equations and yet which do not satisfy (54).*

Suppose, then, we have secured matrices q and p satisfying all these conditions. These matrices will constitute the solution matrices,† and our problem will be solved. The solution matrix q in particular gives all the possible radiated frequencies and their intensities. We need not appeal to Bohr's frequency condition to derive the frequencies from the energy levels; the solution matrix yields these frequencies directly. Nor need we be guided by the correspondence principle so as to rule out some of the theoretically possible frequencies and to compute the intensities.‡ All this information is given by the solution matrix, for the amplitudes of the vibrations in the matrix determine the intensities correctly, and hence the forbidden frequencies of Bohr are those which are associated with vanishing amplitudes in the matrix.

Nothing has as yet been said of the energy levels, but the classical analogy shows how we may proceed. Suppose that $q(t)$ and $p(t)$ are the solution functions of some mechanical problem treated classically. Let $F(q, p)$ be an arbitrary function of q and p. The value of $F(q, p)$

---

* At this point an interesting comparison may be made with the classical treatment. In classical mechanics, we may obtain solutions of Hamilton's equations without having to consider Bohr's quantum condition. The quantum condition merely imposes a restriction which is found to be necessary for the interpretation of atomic phenomena. It would therefore appear permissible to assume that, in the matrix method also, we might omit the commutation relation (54), which plays the part of Bohr's quantum condition. We should then obtain the matrix analogue of classical mechanics unhampered by quantum restrictions. But this course is impossible, for the commutation relation (54) plays, as we have seen, a dual rôle. Not only is it the analogue of Bohr's physical quantum condition, which in a certain sense is imposed as an afterthought, but it also defines the algebraic rules of combinations which our matrices must satisfy. A failure to impose the commutation rule would therefore render all calculations with our matrices indeterminate. From this standpoint, the commutation relation represents the revised form of the similar classical relation $pq - qp = 0$, which is assumed tacitly in classical mechanics and which merely expresses that q and p commute as do ordinary numbers. The last condition is of course never written out explicitly in classical problems, because the commutative property for multiplication in classical mechanics is taken for granted.

† The matrices will be well determined except that the phases of the vibrations will be arbitrary. This indeterminateness has its exact analogue in classical mechanics when action and angle variables are used; the phases connected with the latter are likewise arbitrary.

‡ In deriving the intensities of the radiation from the amplitude functions in the solution matrix, it is nevertheless necessary to appeal to correspondence considerations; for we obtain the intensities by applying the formula (33); and this formula is merely the classical expression adapted to matrices.

at any instant during the motion is obtained when we replace q and p in this function by the solution functions $q(t)$ and $p(t)$. If F is an arbitrary function, it will usually vary in value during the motion; special functions, however, may retain fixed values. As a particular example, let us operate with the Hamiltonian function $H(q, p)$. If our mechanical system is conservative, we shall find that H remains constant during the motion. Since $H(q, p)$ represents the total energy of the mechanical system, the constancy of H expresses the conservation of energy. Of course, according to the initial conditions stipulated for the solution functions, H may assume one constant value or another. This merely implies that different initial conditions impose different energy values on the system.

Let us now proceed in the same way with the solution matrices. Owing to correspondence considerations, we may retain for our matrix magnitudes the same mechanical significance that the corresponding functions have in classical science. The Hamiltonian matrix $H(q, p)$ in particular represents the energy matrix. Now, it can be shown that if in this Hamiltonian matrix we substitute the solution matrices q and p, we obtain a diagonal matrix.* We shall represent this diagonal matrix by the letter E. A diagonal matrix is the matrix analogue of a constant, or rather of an aggregate of different constants represented by the diagonal terms of the matrix.† The fact that H is transformed into a diagonal matrix E, when the solution matrices q and p are substituted, implies therefore that the energy retains a constant value during the motion, and that the permissible values of the energy are given by the diagonal terms

$$E_1, E_2, E_3 \ldots \ldots$$

of the matrix E. We have thus proved that the conservation of energy holds in matrix mechanics, and, in addition, we have indicated how the various energy levels of a system can be obtained.

It can also be shown that the frequencies which appear in the solution matrix are connected with the energy levels by Bohr's frequency condition. Thus, Bohr's frequency condition is itself a consequence of the theory.

---

* This theorem may be proved without our having to know the precise expression of the solution matrices. The sole information utilized is that q and p are the solution matrices, *i.e.*, are solutions of Hamilton's matrix equations and also satisfy the commutation rule.

† See page 840.

The Linear Harmonic Oscillator—We shall illustrate these general considerations by examining the problem of the linear harmonic oscillator. Here the classical Hamiltonian is

(55)
$$H(q, p) = \frac{p^2}{2\mu} + 2\pi^2\nu_0^2\mu q^2,$$

where μ is the mass of the vibrating particle, and ν_0 is the mechanical frequency of the motion. The coordinate q of the particle is measured from the centre of oscillation.

If we proceed by Bohr's method, imposing Bohr's quantizing conditions, we find that the energy levels are

(56)
$$0, \ h\nu_0, \ 2h\nu_0, \ \ldots \ nh\nu_0 \ \ldots \ \ldots$$

Bohr's frequency condition then indicates that the frequencies radiated may be all integral multiples of the mechanical frequency ν_0. When we apply the correspondence principle, many of these theoretically possible frequencies are ruled out, and we find that drops between contiguous energy levels are the only ones possible. This indicates that ν_0 is the only possible frequency for the radiation. The calculation of the intensities also requires that we appeal to the correspondence principle.

Let us now indicate briefly the results obtained when the matrix method is applied. Proceeding along the lines explained previously, we find that the solution matrix q of the Hamilton equations is

(57)

$$\left\{ \begin{array}{ccccc} 0 & q(0,1)e^{2\pi i\nu(0,1)t} & 0 & 0 & \cdots \\ q(1,0)e^{2\pi i\nu(1,0)t} & 0 & q(1,2)e^{2\pi i\nu(1,2)t} & 0 & \cdots \\ 0 & q(2,1)e^{2\pi i\nu(2,1)t} & 0 & q(2,3)e^{2\pi i\nu(2,3)t} \\ 0 & 0 & q(3,2)e^{2\pi i\nu(3,2)t} & 0 & \cdots \\ \cdots & \cdots & \cdots & \cdots & \cdots \end{array} \right.$$

All the diagonal terms vanish, and the only other terms which do not vanish are those symmetrically situated next to the diagonal.

The frequencies in this table all have the same value ν_0. The only complex amplitudes to be considered are those of type $q(n+1, n)$ and $q(n, n+1)$; all others vanish. Calculation gives

(58)
$$q(n+1, n) = \sqrt{\frac{(n+1)h}{8\pi^2\mu\nu_0}} \ e^{i\delta(n+1,n)},$$

where $\delta(n+1, n)$ is real but of arbitrary value.† A complex amplitude of type $q(n, n+1)$, being the conjugate complex of (58) has exactly the same expression except for a change in the sign of the imaginary number i. The arbitrariness in the phases $\delta(n+1, n)$ is of no importance for our present purpose, because in the calculation of the intensities only the expressions of type $q(n+1, n)q^*(n+1, n)$ need be considered, and in such expressions the factor $e^{i\delta(n+1,n)}$ is cancelled.

Altogether then, the solution matrix (57) shows that the oscillator can emit only the radiation frequency ν_0. The intensities of the radiations are determined from the coefficients (58). All this information, be it noted, has been obtained without explicit reference to the correspondence principle. Finally, the fact that all the diagonal terms of the solution matrix q vanish implies that the average position of the vibrating particle over an extended period of time is the centre of oscillation. This conclusion is in keeping with the classical theory, for regardless of the energy of the motion, the vibrating particle always spends as much time on one side of the centre of oscillation as on the other.

The momentum solution matrix p is μ times the solution matrix q differentiated with respect to time. It is thus easily obtained. When these two matrices are substituted in the Hamiltonian matrix (55), it becomes a diagonal matrix E with diagonal terms

$$(59) \qquad \frac{h\nu_0}{2}, \quad \frac{h\nu_0}{2} + h\nu_0, \quad \ldots \frac{h\nu_0}{2} + nh\nu_0, \quad \ldots \; .$$

These, then, are the energy levels, and we see that they differ from the Bohr levels (56). The levels (59) are the correct levels, for they are the ones which appear to be required by experiment. Wave mechanics also furnishes the levels (59), so that the matrix method and wave mechanics both yield the same results. This agreement between the two methods holds also for the intensities.

More General Systems—The linear oscillator constitutes a system having only one degree of freedom, but in theory the matrix method may be extended to systems having any number of degrees of freedom. The difficulty of the matrix method increases with the number of degrees of freedom, so that we must be satisfied with general results. Some of these we shall now examine. Suppose we have an atomic system represented by several electrons circling around a nucleus. We assume that the system is acted upon by external forces which are distributed symmetrically about some axis, e.g., the Oz axis. To discuss a specific example,

---

† It is this arbitrariness which entails that of the phases.

we shall suppose that the forces are generated by a uniform magnetic field parallel to the Oz axis. This is the situation which occurs in the Zeeman effect.

In the classical treatment, the motion of the system is determined if we can obtain the solution functions

$$x(t),\ y(t),\ z(t),\ p_x(t),\ p_y(t),\ p_z(t)$$

for each individual electron. The total angular momentum of the system and its projection M_z on the Oz axis are defined at any instant by appropriate mathematical combinations of the various solution functions. However, it is not necessary to obtain the solution functions in order to prove that both of these angular momenta will remain constant in magnitude during the motion; this constancy follows from general mechanical considerations.

In Bohr's theory, quantum conditions must be imposed, and according to Bohr's original treatment the total angular momentum M can assume only the values

$$j\frac{h}{2\pi},$$

where j is a positive integer or zero. We also find that the projected angular momentum M_z must assume the values

$$m\frac{h}{2\pi},$$

where m is any integer between $-j$ and $+j$. But Bohr's original treatment gives wrong results. To obtain correct ones, we must suppose that j and m may also assume half-integral values. Furthermore, the total angular momentum should be expressed not by $j\dfrac{h}{2\pi}$, but by

$$\sqrt{j(j+1)}\,\frac{h}{2\pi}.$$

We recall that it was the empirical study of the anomalous Zeeman effect, leading as it did to Landé's g-formula, that was responsible for this modification.

We now treat the same problem by the matrix method. The classical expressions of the angular momenta in terms of the solution functions are retained. This course is justified by the correspondence principle, which shows that under the limiting conditions the classical expressions become correct. But in the matrix treatment the solution matrices, not the solution functions, must be substituted in the expressions of M and of

M_z, with the result that these angular momenta will now be matrices. We do not have to determine the precise solution matrices to show on general principles that, when the substitution is performed, the two matrices M and M_z become diagonal. This circumstance implies that in the matrix treatment, as in classical science, the above angular momenta remain constant in magnitude during the motion.

The diagonal terms of the diagonal matrices M and M_z define the permissible constant values that the angular momenta may assume during the motion. In particular, we find that the values of the total angular momentum M are

$$\sqrt{j(j+1)}\,\frac{h}{2\pi},$$

where j may be any positive integer or half-integer, or else zero. Also the values of M_z are found to be

$$m\frac{h}{2\pi},$$

where m may have integral or else half-integral values between $-j$ and $+j$. The results required by Landé are thus obtained.

We may go further and study the angular momentum of the optical electron. We find values

$$\sqrt{l(l+1)}\,\frac{h}{2\pi}.$$

To obtain Landé's g-formula, we must introduce the hypothesis of the spinning electron or some equivalent hypothesis. At the present stage of the theory, there is nothing to indicate the existence of any such spin. But if we accept the spin, we may give a matrix representation to it, and then Landé's g-formula can be accounted for. There is, however, no great advantage in stressing these matters, because the proper way to introduce the electron spin is entirely different; it results from Dirac's subsequent investigations.

The matrix method also allows us to obtain the selection rules and the polarizations of the emitted radiations. Thus the solution matrices of any one of the electrons are, say,

$$x,\ y,\ z.$$

These matrices are multidimensional, and their terms contain the various quantum numbers m, j, l just mentioned and also the main quantum number n. For instance in the solution matrix z, the term which is associated with a quantum transition from the state (n, l, j, m) to the state (n', l', j', m') will be of the form

$$z(n, l, j, m; \; n', l', j', m').$$

Calculation shows that these terms always vanish when $m' \neq m$, and hence when m changes in value during a quantum transition. As for the solution matrices x and y, their terms vanish when $m' \neq m \pm 1$, and hence except when m does not change by ± 1 during a transition. Now a vanishing term in a solution matrix signifies that the transition associated with this term cannot generate radiation: the transition of interest is then a "forbidden" one. We conclude that, insofar as the quantum number m is concerned, the only possible transitions are those in which m does not change or else changes by ± 1. In the former case, some of the other quantum numbers must obviously change, for otherwise there would be no energy drop. We are thus in possession of the selection rule for m. Calling Δm the change in the value of m during a transition, we must have

$$\Delta m = 0 \text{ or } \pm 1.$$

In a similar way we should find

$$\Delta j = 0 \text{ or } \pm 1; \text{ and } \Delta l = \pm 1.$$

These selection rules are precisely the ones we obtained in Bohr's theory when we applied the correspondence principle. The advantage of the matrix method is that the required results are secured directly.

The polarizations of the radiated frequencies can also be determined. The general rule which furnishes the polarizations is as follows:

If a solution matrix corresponds to a classical coordinate vibrating back and forth along a line, the frequencies in the solution matrix refer to waves that are plane polarized (the electric vector in the radiated wave is then parallel to the line of vibration of the classical coordinate).

Let us apply this rule to the present situation. We have seen that when m does not change during a quantum transition, the solution matrix z is the only matrix for which the corresponding term does not vanish. Consequently, according to our rule, we must assume that the radiation from the atom will be plane polarized and that the direction of vibration of the electric vector will be parallel to the Oz axis. Suppose now that m changes by ± 1. The matrices x and y are then the only ones to have non-vanishing terms, so that the vibration of the electromagnetic wave must be parallel to the xy plane and hence normal to the Oz axis. Calculation shows that the phases of the corresponding terms of the matrices

x and y differ by a quarter period; a fact which implies that the radiation will be circularly polarized. The matrix method thus accounts for the polarization effects noted in the normal Zeeman effect.

Summary of the Earlier Matrix Method —The matrix method developed to this stage may be referred to as the earlier method. In the next chapter a further development will be considered. To appreciate the scope of the earlier method, let us glance back at Bohr's theory.

Aside from its failure in the majority of cases to furnish the correct energy levels, Bohr's theory also failed to predict the forbidden transitions, the intensities, and the polarizations of the radiations. Bohr sought to overcome this defect by introducing the correspondence principle. We mentioned, however, that the correspondence principle did not permit an accurate computation of the intensities, so that a certain lack of precision was unavoidable. Besides, Bohr's appeal to the correspondence principle was in itself an admission of failure, for it implied that his theory was not self sufficient and that its application required the assistance of a foreign adjunct.

The matrix theory marks a great advance over Bohr's theory. Not only does it yield the correct energy levels and the various characteristics of the radiation, but it furnishes this information by its own methods without having to rely on the correspondence principle. This latter remark is not meant to imply that correspondence considerations play no part in the matrix method; indeed the contrary is true, for time and again we were guided by such considerations when we were constructing the theory. It should be observed, however, that these correspondence considerations served merely as clues in the building of the theory and played no further part once the theory was completed. Thus though it is true that correspondence considerations are interwoven in the very fabric of the matrix theory, they do not constitute corrective norms to be applied as and when needed. To this extent we are justified in saying that the matrix theory, in contradistinction to Bohr's, is self-sufficient.

To be accurate, however, we should add that, on certain occasions, correspondence considerations are invoked even after the basic tenets of the matrix theory are agreed upon: for instance, when we utilize Hamilton's equations as radiation equations, or apply the classical formula (32) to determine the intensity of the radiation (see note on page 822). The theory would have been more satisfactory had it been able to furnish these equations and this formula by an application of its own peculiar methods. However this may be, the intrusion of correspondence considerations is far less objectionable in the matrix theory than it was in Bohr's.

An obvious improvement on the matrix theory would be realized if, in the solution of problems, Hamilton's classical equations were replaced by their relativistic analogues. This improvement will be examined when we consider Dirac's theory of the electron. The theory of relativity was not applied in the earlier development of the matrix theory for two reasons. Firstly, the introduction of relativistic considerations entailed considerable technical difficulties. Secondly, relativistic science differs perceptibly from the classical theory only for extremely high velocities approximating that of light; and, in the problems with which the quantum theory is concerned, the velocities do not usually attain these high values. Besides, the modifications entailed in the classical notions by the quantum theory itself are of so much greater importance than those which would be introduced by relativistic refinements that, as a first approximation at least, we may disregard the relativity theory.

The success of the matrix theory is demonstrated by its results. Nevertheless it is strange to find that an abstract mathematical scheme, founded on such slender clues, should be capable of giving definite answers to physical problems. Lorentz at the Solvay Congress of 1927 expressed his surprise in the following words:

"I have been astonished to see that the matrices satisfy the equations of motion. Theoretically, this is very fine, but for my part it is a deep mystery, which, I hope will be clarified. I am told that these considerations have led to the formation of matrices which represent what can be observed of the atom, for instance the frequencies of the radiations emitted. Nevertheless the fact that the coordinates, the potential energy, etc., are now represented by matrices shows that these magnitudes have lost their original meaning, and that a tremendous step has been taken towards increasing abstraction.''

CHAPTER XXXVIII

THE MATRIX METHOD (*Continued*)

THE theoretical simplicity of the matrix method must not obscure the fact that, in the practical solution of any given problem, it involves very complicated mathematics. The Hamilton equations of classical mechanics now become equations between infinite matrices and are therefore extremely cumbersome. Even the problem of the harmonic oscillator, which in classical mechanics is one of the simplest, becomes complicated when treated by the matrix method. The difficulties are further aggravated when we deal with systems having several degrees of freedom. Attempts were therefore made to obtain a general method of solution which would involve a more familiar mathematical technique. Dirac's transformation theory, which we shall examine presently, is one of the most successful attempts in this direction.

First let us recall a theorem mentioned in the preceding chapter. In the case of a mechanical system having one degree of freedom, the theorem states:

The solution matrices q and p of Hamilton's matrix equations are infinite Hermitean matrices which satisfy the commutation relation and which, when they are inserted in the Hamiltonian function $H(q, p)$ of the problem, convert it into a diagonal matrix E. The diagonal terms of this diagonal matrix are then the energy levels.*

---

* Suppose the Hamiltonian contains a term p^2q. In classical science, where q and p are mere numbers and commutation always holds, the term p^2q may equivalently be written

(60) $$p^2q \text{ or } pqp \text{ or } qp^2.$$

But in the matrix method, where matrices must be substituted for q and p in the Hamiltonian, the different ways of writing the term p^2q will give rise to different matrices. It is therefore necessary to establish some definite rule which will state how a term such as p^2q must be written. This difficulty did not occur in the case of the oscillator, as may readily be understood from the form of its Hamiltonian (55).

The general rule to be observed is that a term such as (60) must be so written that the Hamiltonian matrix is Hermitean (*i.e.*, the terms symmetrically situated with respect to the diagonal must be conjugate complex. The matrices q and p are themselves Hermitean, but it can be shown that the Hamiltonian will be Hermitean only when its terms are symmetrized. This requires that we replace p^2q or qp^2 by pqp in the Hamiltonian.

This theorem may be expressed in another form. Let

(61) $q(n, k)e^{2\pi i\nu(n,k)t}$ and $p(n, k)e^{2\pi i\nu(n,k)t}$

be the general terms of the two solution matrices. We may easily verify that, if in these matrices (61) we change the frequencies $\nu(n, k)$ but in such a way that they remain compatible with Ritz's combination principle, the new matrices obtained will still satisfy the commutation relation. Furthermore, when these new matrices are inserted in the Hamiltonian function $H(q, p)$ they will convert it into exactly the same diagonal matrix as before. Since vanishing frequencies afford a particular illustration of frequencies satisfying Ritz's combination principle, we may, if we choose, assume that all the frequencies $\nu(n, k)$ in the solution matrices (61) have the value zero. The matrices (61) are thus reduced to their amplitude parts alone and become

$$
(62)\quad q = \begin{cases} q(11) & q(12) & q(13) \cdots \\ q(21) & q(22) & q(23) \cdots \\ q(31) & q(32) & q(33) \cdots \\ \vdots & \vdots & \vdots \end{cases} \; ; \; p = \begin{cases} p(11) & p(12) & p(13) \cdots \\ p(21) & p(22) & p(23) \cdots \\ p(31) & p(32) & p(33) \cdots \\ \vdots & \vdots & \vdots \end{cases}
$$

These matrices will be referred to as the solution matrices *without time terms*. According to our previous theorem, the matrices (62) still satisfy the commutation rule; they convert the Hamiltonian function into a diagonal matrix; and the diagonal terms of this diagonal matrix will be the energy levels.

The Converse Theorem—The converse of this theorem is also true. Thus suppose we can obtain two infinite Hermitean matrices having the general form (62) and satisfying the commutation rule. If, when we substitute these two matrices in the Hamiltonian function $H(q, p)$ of a given problem of one degree of freedom, a diagonal matrix E is obtained, the converse theorem states that our two matrices are necessarily the solution matrices without time terms.* Consequently, the diagonal elements

(63) $E_1, E_2, E_3, \ldots \ldots$

---

* More generally, any two Hermitean matrices of form (61) will have the same amplitude parts as the solution matrices provided their frequencies satisfy Ritz's combination principle; provided the matrices satisfy the commutation rule; and provided their substitution in the Hamiltonian function converts it into a diagonal matrix. (The Hermitean nature of the matrices must be imposed since the matrices are to represent real magnitudes.)

of the diagonal matrix E, into which the Hamiltonian function has been converted, define the energy levels of the system.

Suppose, then, that by some means or other we have discovered two matrices q and p of form (62) satisfying the conditions mentioned in the converse theorem. These two matrices define, as we know, the amplitude parts of the solution matrices. It is now an easy matter to obtain the solution matrices themselves. We first substitute our two matrices q and p in the Hamiltonian function and obtain thereby a diagonal matrix E, whose diagonal elements E_1, E_2, E_3 . . . are the energy levels of the problem. Having done this, we utilize Bohr's frequency condition (which is a consequence of the matrix theory), and we thus obtain the correct frequencies

$$(64) \qquad \nu(n, k) = \frac{E_n - E_k}{h},$$

which must appear in the solution matrices. To obtain the solution matrices, we have now but to adjoin the time terms

$$(65) \qquad e^{\frac{2\pi i}{h}(E_n - E_k)t}, \text{ or } e^{2\pi i \nu(n,k)t}$$

to the corresponding terms of our matrices q and p.

The Transformation Theory—With this preliminary information disposed of, we may examine one of the general methods evolved for the solving of a matrix problem. For simplicity, we confine our attention to problems of one degree of freedom.

Instead of attempting to obtain the solution matrices of Hamilton's equations by a direct integration of these equations, we proceed indirectly by noting:

1. The solution matrices q and p must be infinite Hermitean matrices and must be conjugate (*i.e.*, must satisfy the commutation rule).

2. When inserted in the Hamiltonian function of the problem, they must transform it into a diagonal matrix.

These two requirements suffice to determine the solution matrices (without time terms) of any particular problem.

Let us, then, start with any two arbitrary Hermitean and conjugate matrices (without time terms). We call these two matrices q_0 and p_0. Two such matrices are furnished by the solution matrices of any problem of one degree of freedom; and since we have already solved the problem of the linear harmonic oscillator, we have merely to take its solution matrices (without time terms) and we obtain in this way two conjugate Hermitean matrices q_0 and p_0. Let $H(q, p)$ be the Hamiltonian function of the problem we are now considering. If in this Hamiltonian we substitute the matrices q_0 and p_0, we obtain a matrix $H(q_0, p_0)$. The matrix $H(q_0, p_0)$ is certainly not a diagonal one, for, if it were, the matrices q_0 and p_0 would be the solution matrices of our present problem; and this is impossible since they are the solution matrices of the oscillator, and hence of a different problem. But we may utilize the conjugate matrices q_0 and p_0 as a starting point for the discovery of the solution matrices of the problem of interest. The following geometric considerations suggest how the required solution matrices q and p can be derived from q_0 and p_0.

When discussing quadratic forms, we mentioned that an infinite Hermitean matrix may be viewed as the matrix of an infinite quadratic form whose coefficients a_{ik} and a_{ki} are conjugate complex. Let us, then, associate the two solution matrices of the linear oscillator with quadratic forms of this type. Calling $q_0(n, k)$ and $p_0(n, k)$ the general terms of the matrices q_0 and p_0, we have the quadratic forms

$$(66) \qquad \sum_{nk} q_0(n, k) x_n^* x_k \quad \text{and} \quad \sum_{nk} p_0(n, k) x_n^* x_k .$$

If we equate each of these two quadratic forms to some real constant, say 1, we obtain the equations of two quadric surfaces having the origin as centre, situated in the Hilbert space, and referred to some Cartesian coordinate system in this space. Thus,

$$(67) \qquad \sum_{nk} q_0(n, k) x_n^* x_k = 1 \quad \text{and} \quad \sum_{nk} p_0(n, k) x_n^* x_k = 1$$

define these two quadric surfaces.

We now substitute the two initial matrices q_0 and p_0 into the Hamiltonian H (symmetrized) of our problem. We obtain a Hermitean matrix $H(q_0, p_0)$ having as general term, say $H_0(n, k)$. Proceeding as before, we associate an infinite quadratic form with this matrix, and if we equate this form to 1, we obtain

$$(68) \qquad \sum_{nk} H_0(n, k) x_n^* x_k = 1.$$

This is the equation of some quadric surface having the origin as centre, situated in the same Hilbert space, and referred to the same Cartesian axes as are the two surfaces (67).

Let us examine the relative positions of the principal axes of these three quadric surfaces. The matrices, q_0 and p_0, being conjugate, satisfy the commutation rule

$$p_0 q_0 - q_0 p_0 = \frac{h}{2\pi i} 1,$$

and hence do not commute. According to the considerations developed on pages 808 and 809, this implies that the principal axes of the two quadric surfaces (67) do not coincide. Furthermore, since neither q_0 nor p_0 is a diagonal matrix, we are certain that the principal axes of neither of the two surfaces (67) can coincide with the axes of coordinates.

Next we consider the quadric surface (68). As $H(q_0, p_0)$ is not a diagonal matrix, the principal axes of the surface (68) cannot coincide with the axes of coordinates. We may also show that the principal axes of the surfaces H_0 and q_0, or of H_0 and p_0, will not in general coincide.*

A graphical description of these results will be helpful. Since the Cartesian coordinate system in the Hilbert space has an infinite number

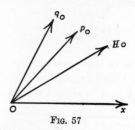

FIG. 57

of axes, we cannot represent all these axes in a drawing. But we may agree to represent the infinity of coordinate axes by a single line Ox. Similarly, the lay of the principal axes (infinite in number) of the quadric surface q_0 will be represented by a line Oq_0. The same procedure is followed for the principal axes of the quadric surfaces p_0 and H_0. We thus obtain the schematic representation illustrated in the figure. The various lines Oq_0, Op_0, and OH_0 are traced as distinct, because the principal axes of the corresponding quadric surfaces do not coincide. Similarly, since none of these quadric

---

* When we apply Dirac's Poisson-bracket relation to the matrices H_0 and p_0, we obtain the relation

$$H_0 p_0 - p_0 H_0 = \frac{h}{2\pi i}[H_0, p_0] = -\frac{h}{2\pi i}\frac{\partial H_0}{\partial q_0}.$$

Hence H_0 and p_0 commute only if the Hamiltonian is not a function of q_0. We may disregard this possibility, because, with our present Cartesian coordinates, the Hamiltonian function for a particle moving under a force always contains q_0. Consequently, H_0 and p_0 do not commute, and therefore the principal axes of the corresponding quadric surfaces do not coincide. A similar argument shows that the principal axes of the surfaces H_0 and q_0 will coincide only when the Hamiltonian does not contain p_0. Inasmuch as this situation does not arise, we conclude that the principal axes of the surfaces H_0 and q_0 do not coincide.

surfaces has principal axes which coincide with the coordinate axes, none of the three lines Oq_0, Op_0, OH_0 will coincide with Ox.

Suppose now we rotate the coordinate axes to any new position. The equations of the three quadric surfaces in the new axes will be modified, and so they will be associated with three new matrices, which we shall call q_1, p_1, and H_1. It can be proved that the new matrices q_1, p_1, and H_1 are still Hermitean, and that the commutation rule holds between q_1 and p_1. It can also be shown that the matrix H_1, into which H_0 (*i.e.*, $H(q_0, p_0)$) is transformed, is equal to $H(q_1, p_1)$; and hence H_1 is the matrix obtained when we substitute q_1 and p_1 in the Hamiltonian of the problem.

If we call S the matrix of the orthogonal transformation, or rotation of axes, the following relations hold for the various matrices: †

$$(69) \quad \begin{cases} q_1 = \tilde{S}^* q_0 S = S^{-1} q_0 S \\[4pt] p_1 = \tilde{S}^* p_0 S = S^{-1} p_0 S \\[4pt] H_1 = H(q_1, p_1) = \tilde{S}^* H_0 S = S^{-1} H_0 S. \end{cases}$$

These considerations suggest how the problem may be solved. Instead of performing an arbitrary rotation of axes (arbitrary orthogonal transformation), we rotate these axes so that they coincide with the principal axes OH_0 of the quadric surface H_0. In the figure, the line Ox will be pivoted so as to coincide with OH_0. The equation of the quadric surface H_0 is now given by an infinite sum of squares (equated to 1), and hence the matrix H_0 is transformed into a matrix H_1 which is diagonal. As before, the matrix H_1 (now diagonal) is the same as the matrix that is obtained when we substitute the matrices q_1 and p_1 in the Hamiltonian.

Thus, as a result, of our orthogonal transformation S, the original matrices q_0 and p_0 are transformed into matrices q_1 and p_1 which are Hermitean, which satisfy the commutation rule, and which, when inserted in the Hamiltonian function of the problem, convert it into a diagonal matrix. But then we know from the converse theorem of page 853 that the matrices q_1 and p_1 are none other than the solution matrices (without time terms) of our problem, and that the diagonal terms of the diagonal matrix H_1 are the energy levels. We conclude that the problem is solved if we can obtain the orthogonal matrix S which transforms the Hamiltonian matrix $H(q_0, p_0)$, or H_0, into a diagonal matrix.

So as to stress that the matrices q_1 and p_1, derived from q_0 and p_0 as a result of the transformation S, are the solution matrices (without time

---

† See page 807.

terms), we shall represent them by the capital letters Q and P. Also the matrix H_1, which is now diagonal, will be represented by E, and its diagonal terms, which are the energy levels, will be written

(70) $E_1, E_2, E_3, \ldots \ldots$

The relations (69) may then be written

(71) $Q = \tilde{S}^* q_0 S = S^{-1} q_0 S; \quad P = S^* p_0 S = S^{-1} p_0 S;$

and

(72) $E = \tilde{S}^* H_0 S = S^{-1} H_0 S$, or equivalently $H_0 S - SE = 0$.

We have also said that E is the matrix that results when Q and P are substituted in the Hamiltonian. Thus

$$E = H(Q, P).$$

It is now a simple matter for us to complete the solution of the problem by adjoining the correct time terms to the matrices Q and P. As explained previously, the frequencies $\nu(n, k)$ entering into the expression of the matrices Q and P are derived from Bohr's frequency condition

(73) $\nu(n, k) = \dfrac{E_n - E_k}{h},$

where the energy levels E_n and E_k, being diagonal terms of the diagonal matrix E, are known. If, then, we call

(74) $Q(n, k)$ and $P(n, k)$

the general terms of the matrices Q and P, the general terms of the full solution matrices are

(75) $Q(n, k) e^{\frac{2\pi i}{h}(E_n - E_k)t}$ and $P(n, k) e^{\frac{2\pi i}{h}(E_n - E_k)t}$.

This method of solution may be summarized by the statement:

1. We start from any arbitrary pair of conjugate Hermitean matrices q_0 and p_0 without time terms. (These matrices may always be obtained.)

2. We substitute the matrices q_0 and p_0 in the classical Hamiltonian $H(q, p)$ of the problem. If the matrix obtained is a diagonal one, our problem is solved, for q_0 and p_0 are then the solution matrices (without time terms), and the diagonal terms of the matrix H are the energy levels. In the general case, however, this peculiar situation does not occur.

3. When $H(q_0, p_0)$ does not happen to be a diagonal matrix, we seek an orthogonal matrix S which, when applied to the matrix $H(q_0, p_0)$, transforms it into a diagonal matrix E. The diagonal elements of E are

the energy levels. As for the solution matrices Q and P (without time terms), they are given by the relations (71); and the full solution matrices (with time terms) are defined by (75).

The method just described is the crux of the transformation theory. We shall therefore mention different interpretations that may be placed on it.

In the language of matrices, the problem of obtaining the solution matrices of Hamilton's equations and of deriving the energy levels is equivalent to that of transforming, by means of an orthogonal matrix, a given infinite Hermitean matrix (the Hamiltonian matrix) into its diagonal form. The eigenvalues of the Hamiltonian matrix (i.e., the diagonal elements of the diagonal matrix) are then the energy levels of the problem.

From the algebraic standpoint, the method is equivalent to that of transforming, by means of an orthogonal transformation, a given infinite quadratic form into an infinite sum of squares. The coefficients of the terms in the sum of squares, namely, the eigenvalues, are the energy levels of the problem.

Finally, we may adopt a more geometric presentation and say that the method is equivalent to rotating the coordinate axes in such a way as to bring them into coincidence with the principal axes of a given quadric surface (the quadric surface of the Hamiltonian). The lengths of the principal axes are connected with the eigenvalues, or energy levels as was explained in Chapter XXXVI.

The transformation method we have outlined is simple in theory and would be simple in practice if we were dealing with finite matrices or finite quadratic forms. The eigenvalues (or energy levels) in particular would then be given by the roots of an algebraic equation of finite degree. But we must remember that our present matrices or quadratic forms are infinite; and at the time Heisenberg was developing his matrix theory, a general method of transforming such infinite matrices into the diagonal form had not been elaborated. Special cases, however, had been investigated by Hilbert. In their attempt to transform the matrices of the matrix theory, mathematical physicists were thus thrown back on their own resources. Lanczos and then Dirac devised similar methods by connecting the problem with the solution of an "integral equation." In this chapter Dirac's method alone will be examined.

Even before general methods of transforming Heisenberg matrices into the diagonal form had been perfected, methods involving successive approximations were applied. Thus, let us suppose that we are dealing with a problem which differs but slightly from one which we know how to solve. As an example, we may consider an anharmonic oscillator which differs from a given harmonic oscillator only as a result of a small additional force acting on the vibrating particle. The problem of the harmonic oscillator can be solved. It is referred to as the "unperturbed problem." The problem of the anharmonic oscillator, which we propose to solve, is called the "perturbed problem," because the motion of this latter oscillator may be viewed as due to a small perturbation acting on the harmonic oscillator. The Hamiltonian functions of the two problems are necessarily different, but the difference is slight owing to the smallness of the perturbing force.

Let us substitute the solution matrices of the harmonic oscillator into the Hamiltonian function of the anharmonic oscillator. The Hamiltonian cannot of course become a diagonal matrix, for the solution matrices of the two oscillators cannot be the same. In other words, the coordinate axes, which now coincide with the principal axes of the Hamiltonian matrix for the unperturbed problem, will not coincide with the principal axes of the Hamiltonian for the perturbed problem. Nevertheless, in view of the small numerical differences between the two problems, the non-diagonal terms of the matrix of the perturbed Hamiltonian will be small, and a small rotation of axes will suffice to transform this matrix into the diagonal form. The exact rotation required to bring about this result is not determined immediately in the present method, but it may be calculated by a series of successive approximations. The method is modelled closely on the perturbation theory of celestial mechanics, as was also the case for the perturbation theory of wave mechanics. It was by means of the perturbation theory of the matrix method that Heisenberg was led to the discovery of resonance degeneracy, a phenomenon which accounts for the forbidden transitions between the "ortho" and the "para," levels in the helium atom. In Chapter XXXV the nature of this degeneracy was investigated by the medium of wave mechanics, but a point of historical interest is that Heisenberg obtained his original results by applying the method of perturbations of the matrix theory.

Continuous Matrices—In Bohr's treatment of atomic phenomenon, the only mechanical motions which had to be quantized were the periodic or multiply periodic ones; the aperiodic motions were not subjected to quantization. Incidentally, the freedom of aperiodic motions from quantum restrictions does not constitute a separate hypothesis in Bohr's theory. Thus an aperiodic motion may, if we choose, be viewed as the limiting case of a periodic motion with an extremely long period; and calculation shows that if a motion of this kind is quantized, the successive quantized values differ so little that they appear to form a continuous sequence. All energy values then become permissible, and the discreteness characteristic of quantized energy levels ceases to manifest itself. This result is an immediate consequence of the general rule whereby Bohr's theory passes over into classical science whenever the relative importance of h is small.

Let us apply these considerations to the hydrogen atom in Bohr's theory. The elliptical orbits are periodic and must be quantized. The hyperbolic orbits, being aperiodic, are not affected. According to the usual conventions, the potential energy is assumed to vanish when the electron is at infinity. As a result of this convention, the total energy in an elliptical orbit is always negative, and in a hyperbolic orbit always positive. The zero value for the total energy is associated with the intermediary type of orbit, namely, the parabolic orbit. We represent the discrete values of the energy for the elliptical orbits by the discrete sequence of negative numbers $E_1, E_2, \ldots E_n, \ldots 0$. The value E_1 is the largest of these negative numbers, and hence it corresponds to the lowest energy level. As for the energy levels of the hyperbolic orbits, they form a continuous sequence of positive values extending from zero to $+\infty$. The complete list of possible energy values is thus expressed by

$$(76) \qquad E_1, E_2, \ldots E_n, \ldots 0 \ldots E \ldots +\infty.$$

$$\underbrace{\qquad\qquad\qquad\qquad}_{\substack{\text{discrete aggregate} \\ \text{of negative values}}} \underbrace{\qquad\qquad\qquad}_{\substack{\text{continuous aggregate} \\ \text{of positive values}}}$$

The spectral lines are supposed to be generated by drops from one energy level to another; and the frequencies radiated are determined by the frequency condition

$$\nu = \frac{\text{drop in energy}}{h}.$$

We conclude that the radiations from a large aggregate of variously excited hydrogen atoms should form a discrete spectrum, as a result of drops between the quantized elliptical orbits; and also a continuous spectrum, generated by drops from hyperbolic orbits to elliptical ones or to other hyperbolic orbits. We know that these expectations are in agreement with observation, so that there is no conflict between theory and experiment on this score. If, in place of the hydrogen atom, we consider the linear oscillator, where the motion is always periodic and hence must always be quantized, we shall obtain only a discrete spectrum; the continuous one will be lacking.

Wave mechanics likewise, by a totally different method, predicts discrete and continuous aggregates of energy values for the hydrogen atom. We recall that in wave mechanics the discrete set of negative energy values for hydrogen and the continuous range of positive values are an immediate consequence of the general conditions imposed on the solution function ψ of Schrödinger's wave equation. More generally, the nature of the aggregates of energy levels to be expected in any particular problem results from the mathematical properties of the wave equation.

We have now to examine how the matrix method can cope with continuous ranges of energy levels. Heisenberg, when he created the matrix theory, was concerned with the discrete states of an atomic system. For instance, he utilized Ritz's combination principle, which controls the frequencies of the discrete spectral lines but which conveys no information on continuous spectra. Now an aggregate of discrete states is represented by a discrete matrix, and, accordingly, all the matrices we have dealt with up to this point are of the discrete variety. To use a mechanical analogy, discrete matrices are connected with periodic motions. In view, however, of the demonstrated existence of continuous spectra, we must demand of the matrix theory that it furnish continuous ranges of energy levels when such continuous ranges are known to be required. We may readily verify that an extension of this kind will involve the introduction of continuous matrices.

Let us revert to the discrete and continuous ranges of energy levels exhibited in (76). According to the transformation theory, the energy levels of a problem are the diagonal elements, or eigenvalues, of the diagonal energy-matrix E. Hence we must assume that in the case of the energy levels (76) our diagonal matrix has the following appearance:

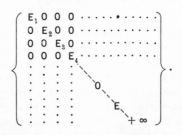

This matrix is discrete from the upper left-hand corner to the term 0 on the diagonal, but, from then on, the diagonal terms form a continuous and no longer a discrete range. The present matrix is thus partly discrete and partly continuous. If all the energy levels were situated over a continuous range, the matrix would be totally continuous.

The physical existence of continuous energy levels in the atom has prompted us to consider continuous and partly continuous diagonal energy matrices. But, as the following argument shows, continuous energy matrices in turn entail continuous solution matrices. Suppose, for instance, we are dealing with the hyperbolic orbits of the hydrogen atom. The energy levels now form a continuous range, and the application of Bohr's frequency condition (which is a consequence of the matrix theory) shows that the radiated frequencies will likewise be spread over a continuous range. The various terms of the solution matrix Q furnish, as we know, the possible frequencies; and since the frequencies are spread continuously, the same continuity must hold for the terms of the solution matrix. The solution is thus itself continuous in this case (though not diagonal); and so we must expect our theory to involve matrices with continuous rows and columns.

These considerations show the necessity of introducing into the matrix method continuous matrices side by side with the discrete ones. We might suppose that Heisenberg, by developing from the start a theory of discrete matrices, had precluded the introduction of continuous matrices, and that a complete remodelling of his theory would be required. But this is not so, for, Hilbert proved that in certain cases the matrix of an infinite quadratic form (of the discrete type) is transformed by an orthogonal transformation into a diagonal matrix which' may have continuous diagonal elements. From a mathematical standpoint, therefore, discrete (infinite) and continuous matrices are not mutually exclusive; and so the immediate aim is not to remodel Heisenberg's theory but to extend it by investigating the properties of continuous matrices.

Dirac's Continuous Matrices—When dealing with discrete matrices, we had to define the unit matrix 1. This matrix is scalar and is represented by a discrete diagonal matrix, all the diagonal terms of which have the value 1, whereas the non-diagonal terms are zero. We shall represent the term situated in the nth row and kth column of this matrix by the symbol $\delta_{n,k}$ or $\delta(n, k)$. The diagonal terms are obtained when we set $k = n$. The values of all the terms of this unit matrix are thus defined by

$$(77) \qquad \delta(n, k) = \begin{cases} 0, & \text{if } k \neq n \\ 1, & \text{if } k = n. \end{cases}$$

If we add all the terms situated on the nth row, we have

$$\delta(n, 1) + \delta(n, 2) + \delta(n, 3) + \ldots , \text{ or } \sum_k \delta(n, k).$$

The value of this infinite sum is 1, for all the terms vanish, with the exception of the diagonal term $\delta(n, n)$, which has the value 1.

Thus

(78)
$$\sum_k \delta(n, k) = 1.$$

Similarly, if we add all the terms of the kth column, we have

(79)
$$\sum_n \delta(n, k) = 1.$$

We wish to obtain the analogue of this unit matrix in the case of continuous matrices. Let us first agree on a general symbolism to represent the various terms of a continuous matrix. As before, a term is specified by the position it occupies in the matrix and hence by the row and column at the intersection of which it is situated. The only difference between the continuous case and the discrete one is that, in a continuous matrix, the rows and columns cannot be numbered by means of integers; instead they must be associated with continuously varying indices.

Let us agree to number the rows by a variable number b' and the columns by a variable number b''. We assume that b' increases progressively in value as we move downward, and that b'' increases progressively as we move to the right (see figure). With this convention, the term situated at the intersection of the row of number b' and of the column of number b'' is represented by (b', b''). If the matrix refers to some magnitude x, the term (b', b'') of this particular matrix is written $x(b', b'')$. We observe

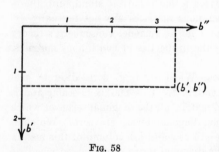

Fig. 58

that if we keep b' fixed and allow b'' to vary continuously, we shall be describing the row b' of the matrix. If, on the other hand, we keep b'' fixed and make b' vary, we describe the column b''. Since a diagonal term is always at the intersection of a row and of a column which are associated with the same number, all diagonal terms of our matrix x

are of type $x(b', b')$. If, in $x(b', b')$, we make b' vary continuously, we shall describe the diagonal line of the matrix. It may happen that the terms situated along the vertical columns of the matrix form continuous ranges, but that the terms situated on the horizontal rows are discrete. The matrix in this case will be continuous in the vertical direction and discrete in the horizontal one. When we represent by $x(b', b'')$ the terms of a matrix of this type, the letter b' will vary continuously whereas b'' will assume only discrete values.

We may now obtain the unit matrix in the continuous case. We represent its terms by

$$\delta(b', b''), \text{ or by } \delta(b' - b'').$$

If we use the second form of symbolism, any diagonal term $\delta(b' - b')$ may be written $\delta(0)$. By analogy with the unit matrix in the discrete case, we might suppose that the continuous unit matrix would be represented by a continuous diagonal line each term of which would have the value 1, whereas the non-diagonal terms of the matrix would vanish. This assumption would imply $\delta(b' - b'') = 1$ or 0 according to whether $b'' = b'$ or $b'' \neq b'$. But, as we shall now see, the first part of our assumption must be rejected. Thus, since the continuous matrix we are considering is to play the part of a unit matrix, it must satisfy relations analogous to (78) and (79). Now, a summation in the discrete case passes over into an integration in the continuous one. Hence the continuous analogues of (78) and (79) are

$$(80) \qquad\qquad \int \delta(b' - b'') db'' = 1$$

and

$$(81) \qquad\qquad \int \delta(b' - b'') db' = 1.$$

These relations must therefore be satisfied by the unit continuous matrix. In (80) we are integrating along the horizontal row b', and in (81) along the column $b.''$

Suppose, then, we accept our former tentative definition of the unit continuous matrix. The two integrals in (80) and (81) will be found to vanish; and the relations (80) and (81), which must be satisfied by our matrix, will be impossible. To remedy this difficulty, we must assume that the non-diagonal terms of the unit continuous matrix vanish, but that, in the immediate neighborhood of the diagonal, the values of the terms tend to increase rapidly, so that a term actually on the diagonal is infinite. When we introduce a matrix of this kind, the relations (80) and (81) can be satisfied.* Accordingly, we define the unit continuous matrix as a matrix whose non-diagonal terms $\delta(b' - b'')$ always vanish except when b' and b'' differ only infinitesimally; in this latter case, δ tends to become infinite. In addition, the unit continuous matrix satisfies the relations (80) and (81).

The Diagonal Matrix q—An infinite Hermitean matrix, whether discrete or continuous, may be associated with an infinite quadratic form which is also discrete or continuous. By equating the quadratic form to some real constant, we obtain the equation of a quadric surface in the Hilbert space of an infinite number of dimensions. If our matrix is not a diagonal one, the quadratic form will not be a sum of squares,† and the principal axes of the corresponding quadric surface will not coincide with the coordinate axes. However, in theory we can always rotate the coordinate axes in such a way that they will coincide with the principal axes; and in the new coordinate system the quadratic form is of course a sum of squares (discrete or continuous), and the matrix becomes diagonal.

We have already applied these considerations to the Hamiltonian matrix, but they hold quite generally for all Hermitean matrices, e.g.,

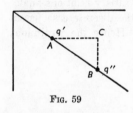

Fig. 59

for the matrix q. Let us, then, consider the coordinate system in which this matrix is diagonal. We shall assume that the matrix is continuous. As in all diagonal matrices, the non-diagonal terms will vanish, the only non-vanishing terms being situated along the diagonal. Suppose we proceed from the upper left-hand corner of the matrix downwards along the diagonal line. The successive terms we shall

encounter may be assumed to have continuously increasing values. Let q' and q'' be the values of the two terms situated, say at A and B in the

---

* Let us integrate along the row b'. The terms along this row are given by $\delta(b' - b'')$ in which b' is fixed and b'' varies. A graphical representation of the integral (80) may be obtained by considering two Cartesian axes b'' and y and the curve $y(b'')$ defined by $y = \delta(b' - b'')$. The area comprised between this curve and the horizontal axis b'' then measures the value of our integral. (The area is shaded in the figure.) Now we wish this area to have the value 1. Since the curve is to have a zero elevation except in the immediate vicinity of the diagonal element, where $b'' = b'$, we must assume that, as b'' tends to the value b', the value of $\delta(b' - b'')$ increases indefinitely. It would be of no use to suppose that $\delta(b' - b'')$ would vanish except when $b'' = b'$ and that its value would then be infinite; for this would not prevent the integrals (80) and (81) from vanishing as before.

Fig. 60

† By sum of squares we wish to imply either a discrete sum or an integral.

figure. A row and a column of the matrix pass through the point A, and similarly for the point B. We shall agree to call the row and the column which pass through A the row q' and the column q' of the matrix. Similarly, the row and the column passing through B are called the row q'' and the column q''. A non-diagonal term such as the one at the point C in the figure, being situated at the intersection of the row q' and of the column q'', is designated by $q(q', q'')$.

Now in the discrete case, if we have a discrete diagonal matrix with diagonal terms $D_1, D_2, \ldots D_n, \ldots$, the vanishing term on the nth row and kth column may be represented by $D_n\delta(n, k)$, where $\delta(n, k)$ is the corresponding term of the discrete unit matrix. Keeping n fixed and allowing k to assume the various integral values, we obtain in succession all the terms on this nth row. In particular, the diagonal term D_n is obtained by setting $k = n$, for we then get $D_n\delta(n, n)$, i.e., D_n (since $\delta(n, n) = 1$).

Let us adopt a similar notation for the terms of the continuous diagonal matrix q. Previously, the term situated at the intersection of the row q' and of the column q'' of this matrix was represented by $q(q', q'')$. But we wish now to utilize the unit continuous matrix in this expression. To do so, we shall suppose that the rows and columns of the unit continuous matrix are numbered in the same way as the rows and columns of the matrix q. For instance, in the unit continuous matrix we shall represent by $1(q', q'')$, or $\delta(q' - q'')$, the term which is situated at the intersection of the row q' and of the column q''. If, then, we follow the system of notation adopted in the discrete case, the term $q(q', q'')$ of the diagonal matrix will be represented by

$$(82) \qquad q(q', q'') \equiv q'\delta(q' - q'').$$

If we keep q' fixed and allow q'' to vary continuously, we shall describe the row q' of the matrix q. In a similar way, all the terms of the column q' of the matrix q are expressed by

$$(83) \qquad q(q'', q'), \text{ or } q'\delta(q'' - q'),$$

where q' is fixed and q'' assumes all values.*

---

* The same peculiarity we noted in connection with the unit continuous matrix is also present here. Thus, strictly speaking, we cannot say that the diagonal term on the row q' has the value q', for this diagonal term is represented by

$$q'\delta(q' - q'), \text{ or } q'\delta(0),$$

the value of which is infinite (since $\delta(0)$ is infinite). If, however, we integrate over as small a distance as we choose, through the diagonal, the value q' is obtained. This statement results from the properties of the unit matrix δ. We thus have

$$\int q'\delta(q' - q'')dq'' = q' \qquad \text{(integration along the row } q')$$

$$\int q'\delta(q'' - q')dq'' = q' \qquad \text{(integration along the column } q')$$

The matrix just considered is the continuous diagonal matrix q. But quite generally, if x is some continuous diagonal matrix whose diagonal terms are represented by x', x'', x''', . . . , we shall agree to number the rows and columns of the matrix by the values x', x'', x''' . . . of the diagonal terms. Thus the term at the intersection of the row x' and the column x'' will be represented by $x(x', x'')$, and, by following the same argument as for the matrix q, we have

$$x(x', x'') \equiv x' \delta(x' - x'').$$

The coordinate system in which an arbitrary continuous matrix x is diagonal is a very special system, for if we rotate our axes, x will necessarily cease to be diagonal.

Suppose, now, we are dealing with two or more continuous matrices $x, y, z, \ldots$, all of which are expressed in the same coordinate system. In this coordinate system let x be diagonal, but y and z non-diagonal. The coordinate system is characterized by the fact that, in it, the matrix x is diagonal, and for this reason it is often referred to as the coordinate system x. A convention which proves of advantage is that the terms of all non-diagonal matrices y and z, when expressed in this coordinate system x, are to be numbered in the same way as the terms of the diagonal matrix x. Thus, if y is a non-diagonal matrix expressed in the coordinate system x, that term of y, which occupies the same relative position as the term $x(x', x'')$ of x, is represented by

$$y(x', x'').$$

The Conjugate Continuous Matrix p and the Commutation Rules for Continuous Matrices—In the theory of discrete matrices, we saw that two conjugate matrices q and p must satisfy the relation

$$(84) \qquad pq - qp = \frac{h}{2\pi i} 1.$$

Thus, these matrices do not commute. The equation (84), being a matrix equation, is satisfied when the corresponding terms of the two matrices are equal in pairs. If, then, we denote as usual by (n, k) the terms in the nth row and the kth column of a discrete matrix, the matrix equation (84) can equivalently be written

$$(85) \qquad (pq - qp)(n, k) = \frac{h}{2\pi i} 1(n, k) = \frac{h}{2\pi i} \delta(n, k),$$

for all integers n and k.

These considerations may be extended to continuous matrices. Suppose the continuous matrix q is represented in the coordinate system in which it is diagonal (the coordinate system q), As in the discrete case, there will be a continuous matrix p which is conjugate to q. Now conjugate matrices are necessarily expressed in the same coordinate system, and since q and p, whether discrete or continuous, do not commute, we may be certain that the matrix p, when expressed in the coordinate system q, cannot be diagonal. For the present we do not know what appearance the conjugate matrix p will have, but we may obtain the required information by noting that q and p must be linked by the commutation rule.

We are thus led to determine the commutation rule for continuous matrices. It will be the analogue of (85) adapted to the peculiarities of continuous matrices. The coordinate system is taken to be the one in which q is diagonal, and hence the terms of our matrices will be represented by the notation (q', q''). A general term of the matrix $pq - qp$ may then be written

$$(pq - qp)(q', q'').$$

The analogue of the equation (85) becomes

$$(86) \qquad (pq - qp)(q', q'') = \frac{h}{2\pi i}\delta(q' - q'')$$

(for all values of q' and q'').

We observe that in (85), if we set $k = n$ we obtain

$$(87) \qquad (pq - qp)(n, n) = \frac{h}{2\pi i}\delta(n, n) = \frac{h}{2\pi i}.$$

This relation expresses that, in the discrete case, all the diagonal elements of the matrix $(pq - qp)$ have the value $\frac{h}{2\pi i}$. But if we follow the same procedure in the continuous case and set $q'' = q'$, the relation (86) becomes

$$(88) \qquad (pq - qp)(q', q') = \frac{h}{2\pi i}\delta(q' - q') = \frac{h}{2\pi i} \times \infty.$$

The infinite value we here obtain for the diagonal elements of the continuous matrix $pq - qp$ is due to the more complicated nature of the continuous unit matrix $\delta(q' - q'')$. However, we may remedy the situation by the usual method of integration. Thus, if we integrate (86) along the row q', we get

$$\int (pq - qp)(q', q'')dq'' = \frac{h}{2\pi i}\int \delta(q' - q'')dq'' = \frac{h}{2\pi i}.$$

Hence, thanks to the integration, we obtain the same result as in the discrete case (84).

To obtain the continuous matrix p which is conjugate to the continuous matrix q and is expressed in the coordinate system in which q is diagonal, we must discover a Hermitean matrix p which is a solution of the matrix equation (86). Dirac gives a solution for the term (q', q'') of this conjugate matrix p. It is

$$(89) \qquad p(q', q'') = \frac{h}{2\pi i} \dot{\delta}(q' - q''),$$

where $\dot{\delta}(q' - q'')$ means the derivative of $\delta(q' - q'')$ with respect to $(q' - q'')$. As pointed out previously, the conjugate matrix p expressed in the coordinate system in which q is diagonal cannot itself be diagonal, for this would be possible only if q and p were to commute, and such is not the case.

Matrices as Operators—Consider any continuous function $f(q', y')$ of two variables q' and y'. We assume that the two variables vary continuously and independently. By giving all pairs of values to q' and y', we obtain a doubly infinite aggregate of values for the function. These values may be represented by the terms of an infinite continuous matrix. The columns of this matrix are secured by giving various fixed values $y', y'', y''', \ldots$ to y' and allowing q' to vary; and the rows are obtained by giving fixed values $q', q'', q''', \ldots$ to q' and allowing y' to vary. The general nature of the matrix may be understood from the figure. The arrows pointing downwards mean that q' is to be given all

$$(90) \qquad f = \left\{ \begin{array}{ccc} f(q', y') & f(q', y'') & f(q', y''') \\ \Big\downarrow & \Big\downarrow & \Big\downarrow \end{array} \right.$$

values while y', y'', and y''' are kept fixed. We call this aggregate the matrix f.

Suppose now that we wish to form the matrix qf, where q is the continuous diagonal matrix. The rules of multiplication for continuous matrices are the same as those for the discrete ones except that integrations replace summations. In particular, the term, say in the row q''' and the column y'' of the matrix qf, is given by

$$(91) \int q(q''', q'') f(q'', y'') dq'' = \int q''' \delta(q''' - q'') f(q'', y'') dq''$$
$$= q''' f(q''', y'').$$

This shows that all the terms of the column y'' of the matrix qf are expressed by

(92) $$q'f(q', y''),$$

where y'' is kept fixed and q' assumes all values. The same conclusions follow for the other columns of the matrix qf, so that this matrix is represented by

(93) $$qf = \left\{ \begin{array}{ccc} q'f(q', y') & q'f(q', y'') & q'f(q', y''') \ \ldots \\ \downarrow & \downarrow & \downarrow \end{array} \right.$$

Proceeding in the same way, we examine the matrix pf, in which p is the continuous matrix (89) conjugate to q. Calculation yields the matrix

(94) $$pf = \left\{ \begin{array}{ccc} \dfrac{h}{2\pi i}\dfrac{\partial f(q', y')}{\partial q'} & \dfrac{h}{2\pi i}\dfrac{\partial f(q', y'')}{\partial q'} & \dfrac{h}{2\pi i}\dfrac{\partial f(q', y''')}{\partial q'} \\ \downarrow & \downarrow & \downarrow \end{array} \right.$$

In this matrix, the column, say y'', is given by $\dfrac{h}{2\pi i}\dfrac{\partial f(q', y'')}{\partial q'}$ where y'' is kept fixed while q' receives all continuous values.

These results may be generalized. Thus let $F(q, p)$ be any continuous function of our two continuous matrices q and p. F is then itself a continuous matrix. We consider the matrix Ff resulting from the multiplication of the matrix f by the matrix F. If $F(q, p)$ is an algebraic function of the matrix p, it can be shown that the matrix Ff will have the form

(95) $$Ff = \left\{ \begin{array}{cc} F\!\left(q', \dfrac{h}{2\pi i}\dfrac{\partial}{\partial q'}\right)f(q', y') & F\!\left(q', \dfrac{h}{2\pi i}\dfrac{\partial}{\partial q'}\right)f(q', y'') \ \ldots \\ \downarrow & \downarrow \end{array} \right.$$

where the arrows mean that the columns of the matrix are obtained by allowing q' to vary continuously while y' and y'' are kept fixed. The matrix Ff is seen to differ from f in that a column such as $f(q', y'')$ of the matrix f is replaced by a column

$$F\left(q', \frac{h}{2\pi i}\frac{\partial}{\partial q'}\right)f(q', y'').$$

As an illustration, let us suppose that $F(q, p)$ stands for $pq - qp$. The column y'' of the matrix Ff is then

$$(96) \qquad \left(\frac{h}{2\pi i}\frac{\partial}{\partial q'}\cdot q' - q'\frac{h}{2\pi i}\frac{\partial}{\partial q'}\right)f(q', y'').$$

On performing the calculations, we get

$$(97) \qquad \frac{h}{2\pi i}f(q', y'').$$

This expression was to be expected, for, according to the commutation relation, the matrix $pq - qp$ is $\frac{h}{2\pi i}1$.

We now propose to examine a different interpretation of these results. In the columns of our foregoing matrices, the variable q' is allowed to assume all values while the second variable (of value y' or y'' or y''' ...) is kept fixed. These different columns represent therefore the successive values of different continuous functions of the single variable q'. The operation we represented by qf thus replaces the continuous functions $f(q', y')$, $f(q', y'')$, ... of the matrix f by the corresponding continuous functions $q'f(q', y')$, $q'f(q', y'')$ ... of the matrix qf. Similarly, the operation pf replaces the continuous functions $f(q', y'), f(q', y''), \ldots$ by the continuous functions

$$\frac{h}{2\pi i}\frac{\partial f(q', y')}{\partial q'}, \qquad \frac{h}{2\pi i}\frac{\partial f(q', y'')}{\partial q'}, \ldots$$

Since no restriction has been set on the nature of the continuous functions $f(q', y')$, $f(q', y'')$, ..., we may generalize the foregoing conclusions and say: If $f(q')$ is any continuous function of q', we have

$$qf(q') = q'f(q')$$

and

$$pf(q') = \frac{h}{2\pi i}\frac{\partial f(q')}{\partial q'}.$$

Furthermore, when $F(q, p)$ is a continuous function of q and p (algebraic in p), we have

$$F(q, p)f(q') = F\left(q', \frac{h}{2\pi i}\frac{\partial}{\partial q'}\right)f(q').$$

The conclusions to be drawn from this discussion are:

Dirac's continuous conjugate matrices q and p may, if we choose, be viewed as operators

$$(98) \qquad q' \quad \text{and} \quad \frac{h}{2\pi i} \frac{\partial}{\partial q'}$$

operating on an arbitrary function of q'. Also, an arbitrary continuous function $F(q, p)$ (algebraic in p) of Dirac's matrices q and p may be treated as the operator

$$(99) \qquad F\!\left(q', \frac{h}{2\pi i} \frac{\partial}{\partial q'}\right).$$

In particular, $pq - qp$ may be viewed as the operator

$$\frac{h}{2\pi i} \frac{\partial}{\partial q'} \cdot q' - q' \cdot \frac{h}{2\pi i} \frac{\partial}{\partial q'}$$

and hence as the operator $\dfrac{h}{2\pi i}$ (see 97). The last result shows that the two operators (98) constitute mathematical entities which satisfy the commutation rule; they thus afford illustrations of q-numbers.* This

---

* The association of operators with the matrices which represent a coordinate q and the momentum p may be extended to any number of dimensions. Thus in 3-dimensional space, if x, y, z are Cartesian coordinates, we have to consider the three continuous matrices which represent these coordinates. In the Hilbert space we take a coordinate system in which the continuous matrices x, y, and z are diagonal; their various diagonal elements are then represented by x', y', and z' respectively. The momenta p_x, p_y, and p_z conjugate to x, y, and z are also expressed by continuous matrices, and in the foregoing coordinate system they may be assimilated to the three operators

$$\frac{h}{2\pi i} \frac{\partial}{\partial x'}, \quad \frac{h}{2\pi i} \frac{\partial}{\partial y'}, \quad \frac{h}{2\pi i} \frac{\partial}{\partial z'}.$$

In short, the six continuous matrices x, y, z, p_x, p_y, and p_z give rise to the six operators

$$(98') \qquad x', y', z', \quad \frac{h}{2\pi i} \frac{\partial}{\partial x'}, \quad \frac{h}{2\pi i} \frac{\partial}{\partial y'}, \quad \frac{h}{2\pi i} \frac{\partial}{\partial z'}.$$

These results may be generalized. Thus, if we treat time as a fourth dimension, the three coordinates x, y, z must be supplemented by the fourth coordinate t. The magnitude conjugate to t is found to be $-E$, where E is the total energy. If then we view t as a continuous matrix and represent it in the coordinate system in which it is diagonal, a general term on the diagonal may be represented by t'. Proceeding as before, we find that the conjugate continuous matrix $-E$ may be viewed as the operator $\dfrac{h}{2\pi i} \dfrac{\partial}{\partial t'}$. Hence in addition to the operators (98'), which have the commutative properties of q-numbers, we may adjoin the operators

$$(98'') \qquad t' \text{ and } \frac{h}{2\pi i} \frac{\partial}{\partial t'}.$$

important discovery was made by Schrödinger,* but when Schrödinger first stated his results, the connection between the operators and the matrices was less clear.† Dirac's investigations throw full light on the matter, for they show that Schrödinger's operators are equivalent to Dirac's continuous conjugate matrices q and p expressed in the coordinate system in which q is diagonal; as a result, these operators satisfy the same commutation rules as the matrices.

The Equivalence of the Matrix Method and of Wave Mechanics —When discussing discontinuous matrices on an earlier page, we mentioned that the solution of a mechanical problem associated with a Hamiltonian function $H(q, p)$ may be conducted in the following way. First we substitute in the Hamiltonian (symmetrized) any two arbitrary conjugate Hermitean matrices q_0 and p_0 (without time terms). A Hermitean matrix $H(q_0, p_0)$ is thus obtained. We then seek to transform this matrix into its diagonal form E by means of an orthogonal transformation. If we call S the matrix of the orthogonal transformation, the equation which our problem requires us to solve is the matrix equation

$$(100) \qquad H(q_0, p_0)S - SE = 0.$$

In this equation there are two unknowns: the orthogonal matrix S and the diagonal matrix E into which H is transformed. We have mentioned that the solution of the problem is relatively simple when the matrices are finite, but that it is difficult in the case of infinite matrices. At all events, whether the problem can be solved or not, we know that the diagonal elements of the diagonal matrix E (also called the eigenvalues) are the energy levels. The diagonal elements are usually denoted by $E_1, E_2, \ldots E_n, \ldots$ when they are discrete, and by $E', E'', \ldots$ when they form a continuous range.

In our former treatment we assumed that the original conjugate matrices q_0 and p_0 were discrete. And since the solution matrices (without time terms) of the oscillator were known, we selected these latter to play the part of the matrices q_0 and p_0. But the same treatment would have been valid had we selected any other pair of conjugate matrices

---

* See page 837.

† On pages 708 to 710 we explained how Schrödinger utilized the operators (98′) in deriving his wave equation from the equation of energy. We saw that the momenta p_x, p_y, and p_z were replaced in the energy equation by the corresponding operators of (98′). The equation we called the generalized wave equation was obtained by having recourse to the former substitutions and to the additional one indicated in (98″).

Thus $-E$ in the energy equation was replaced by the operator $\dfrac{h}{2\pi i}\dfrac{\partial}{\partial t'}$.

q_0 and p_0. In particular, continuous conjugate matrices might have been taken. In the earlier development of the theory, however, continuous matrices were unknown; but now that Dirac has shown how to obtain continuous conjugate matrices q and p (*i.e.*, (83) and (89)), we may avail ourselves of these matrices. This utilization of continuous matrices presents considerable advantages, as will be understood presently.

Suppose, then, that in (100) we substitute Dirac's continuous matrices q and p. As these matrices are expressed in the coordinate system in which q is diagonal, the matrix $H(q, p)$, resulting from the substitution of q and p in the Hamiltonian, will itself be a continuous matrix expressed in the same coordinate system q. This Hamiltonian matrix $H(q, p)$ cannot be diagonal; for if the matrices H and q were diagonal in the same coordinate system, they would commute; and this is impossible owing to the presence of p in the function $H(q, p)$.† Our problem consists, as before, in rotating the coordinate axes in such a way that the matrix $H(q, p)$ assumes its diagonal form E. In more geometric language, our coordinate axes, which initially coincide with the principal axes of the quadric surface q (since q is diagonal), must be rotated so as to coincide with the principal axes of the quadric surface attached to the Hamiltonian H.

In the new coordinate system, the matrix q, which is no longer diagonal, now represents the amplitude part of the solution matrix of the problem. Let us suppose that the rotation has been performed. The Hamiltonian matrix H becomes a diagonal matrix E with diagonal terms

$$(101) \quad \underbrace{E_1, E_2, \ldots E_n, \ldots}_{\text{discrete}} \quad \Big| \quad \underbrace{E', E'', \ldots}_{\text{continuous}}$$

We are here assuming for greater generality that the diagonal elements, or eigenvalues, of E (the energy levels) form a discrete and also a continuous range. The diagonal matrix E is then partly discrete and partly continuous in this case.

We now examine the orthogonal matrix S which determines the required rotation of axes in the infinite Hilbert space. This matrix transforms from the coordinate system q in which q is diagonal, to the system E in which H is diagonal. As such, this orthogonal matrix is represented neither in the coordinate system q nor in the coordinate system E. In contradistinction to our former matrices, which were expressed in a single coordinate system, the matrix S pertains to both coordinate systems. This feature justifies a kind of dual notation for the elements

---

† See note page 856.

of the matrix S and we shall therefore represent a general element by the symbol $S(q', E')$. In particular, the elements, or terms, in the column E_n or in the column E' will be represented by

$$(102) \qquad\qquad S(q', E_n) \text{ or } S(q', E'),$$

where E_n or E' is kept fixed while q' is allowed to vary continuously. The general appearance of the matrix S is then

$$(103) \quad \left\{ \begin{array}{cccc} S(q', E_1) & S(q', E_2) \dots & \Big| & S(q', E') \quad S(q', E'') \dots \\[1em] \downarrow & \downarrow & & \downarrow \qquad\qquad \downarrow \end{array} \right.$$

The columns $S(q', E_1)$, $S(q', E_2) \dots$ correspond to the discrete diagonal terms, or energy levels, which we have assumed for the matrix E; and the columns $S(q', E')$, $S(q', E'') \dots$ are associated with the continuous range of diagonal terms. All the columns are continuous, but, as is seen in the table, if we follow a horizontal row, we shall encounter discrete elements in the left half of the table and continuous elements in the right. Thus, in contradistinction to the columns, the rows are partly discrete and partly continuous. In the general case they may be discrete, continuous, or both, according to the discrete or continuous nature of the diagonal matrix E.

Finally, since S must be an orthogonal matrix, it must satisfy the matrix equations

$$(104) \qquad\qquad \tilde{S}^*S = S\tilde{S}^* = 1, \text{ or } \tilde{S}^* = S^{-1}.$$

We have also seen that the solution matrices Q and P (without time terms) are given by

$$(105) \qquad \left\{ \begin{array}{l} Q = \tilde{S}^* \, qS = S^{-1}qS \\[0.5em] P = \tilde{S}^* \, pS = S^{-1}pS, \end{array} \right.$$

where q and p, being the matrices whence we started, are here the continuous matrices of Dirac. We note that these solution matrices Q and P are merely the original matrices q and p expressed in the rotated coordinate system E in which the Hamiltonian matrix (i.e. the energy matrix) is diagonal.

Our problem is, then, to discover the diagonal matrix E and also the orthogonal transformation matrix S. This will be done if we can solve the matrix equation (100). To simplify the discussion, let us write this matrix equation as

(106) $$HS = SE.$$

It is satisfied if each and every term (q', E') of the matrix HS is equal to the corresponding term of the matrix SE.

Let us consider the matrix SE. The matrix E is as yet unknown; all we can say is that it is some diagonal matrix. Aside from this, we do not know whether its diagonal terms are discrete or continuous, but we may suppose for the present that they exhibit both features and hence are represented by some aggregate of type (101). The matrix S is also unknown. We have already agreed to represent the terms of its column E' by $S(q', E')$ where E' is fixed and q' may vary continuously. These conventions enable us to write in a purely formal way the terms of the column E' of the matrix SE.† They are

(107) $$S(q', E') \times E', \quad \text{or} \quad E' \times S(q', E'),$$

where q' may vary continuously.

We now pass to the corresponding column of the matrix HS. Since we are dealing with continuous matrices, integrations must take the place of summations. The terms of the column E' of HS are thus defined by

(108) $$\int H(q', q'') S(q'', E') dq'',$$

where q' may have all continuous values.

Our matrix equation (106) is thus equivalent to

(109) $$\int H(q', q'') S(q'', E') dq'' = E' \cdot S(q', E'),$$

for all continuous values of q'. In this equation, E' is a parameter of unknown value, and $S(q', E')$ is the unknown function of q', corresponding to the value E' of the parameter. In addition, we shall have to stipulate that the functions $S(q', E')$ must be the columns of an orthogonal matrix; but this last point may be left in abeyance for the present. An equation of type (109) is called an integral equation, and its study pertains to a highly developed department of mathematical analysis. Dirac's method has thus enabled us to transform the original matrix equation

---

† The term (q', E') of the matrix SE is given, according to the rules, by

$$\int S(q', E'') E'' \delta(E'' - E') dE'' \equiv S(q', E') E'.$$

(100), or (106), into an integral equation and it has thereby considerably advanced our understanding of the problem.*

In the present problem, however, we may dispense with the theory of integral equations, for the Hamiltonian function H is in nearly every case an algebraic function of the momentum p. We may then apply Dirac's result (95), according to which the column E' of the matrix HS is expressed by the operation

$$(110) \qquad H\left(q', \frac{h}{2\pi i}\frac{\partial}{\partial q'}\right)S(q', E').$$

The matrix equation (106), or (100), thus becomes

$$(111) \qquad \left[H\left(q', \frac{h}{2\pi i}\frac{\partial}{\partial q'}\right) - E'\right]S(q', E') = 0.$$

We have here a partial differential equation containing a parameter E', so that in effect we have a different equation for each value we may see fit to attribute to this parameter.

Dirac's partial differential equation (111) is already familiar to us, for it is none other than Schrödinger's wave equation. This statement will be substantiated presently. Furthermore, we shall find that the restrictive conditions imposed in Dirac's method on the solution functions $S(q', E')$ are precisely the restrictions that were prescribed by Schrödinger for his wave functions ψ. The equations being the same in the two methods and the restrictions placed on the solutions being also the same, Dirac's solution functions $S(q', E')$, which represent the columns of the orthogonal transformation matrix S, are also Schrödinger's wave functions $\psi_{E'}(q)$, associated with the values E' of the energy.

Let us first verify that Dirac's equation (111) is the same as Schrödinger's wave equation. We recall that Schrödinger's wave equation may be written,†

$$(112) \qquad \left[H\left(q, \frac{h}{2\pi i}\frac{\partial}{\partial q}\right) - E\right]\psi(q) = 0.$$

In this equation the symbol $H\left(q, \frac{h}{2\pi i}\frac{\partial}{\partial q}\right)$ represents the operator we

---

* Integral equations have been studied by some of the foremost mathematicians. Abel is one of the first to have met with an equation of this kind. He solved it by an artifice. More general methods of solutions were investigated by Neumann and by Liouville. But only in recent times has the theory of integral equations been developed systematically. The whole theory is intimately connected with the works of Volterra, Fredholm, Hilbert, Schmidt, and others. The integral equation in the text is of the so-called homogenous type. The function of two variables $H(q', q'')$ is called the "kernel," and in the present case this kernel is Hermitean, i.e., $H(q', q'') = H^*(q'', q')$ (for H is a Hermitean matrix). This special class of integral equations was extensively studied by Hilbert and Schmidt.

† See page 708.

obtain when we replace p by $\dfrac{h}{2\pi i}\dfrac{\partial}{\partial q}$ in the Hamiltonian function $H(q, p)$ of the problem of interest. As for E, it is a parameter of unspecified value, which measures the energy of the system. Thus, Schrödinger's wave equation (112) and Dirac's equation (111) are the same except for the notation: Dirac writes q' and E' where Schrödinger sets q and E, and Dirac represents the solution function by a different letter. Let us examine the differences in the notation for the solution functions. When the value, say, E_n is given to the parameter E in Schrödinger's equation (112), Schrödinger denotes the solution by

$$\psi(q, E_n), \quad \text{or} \quad \psi_{E_n}(q), \quad \text{or} \quad \psi_n(q).$$

Dirac, on the other hand, writes it

$$S(q', E_n).$$

We now propose to show that the solution functions of Schrödinger and of Dirac are the same. This point will be verified if we can show that the restrictions which Schrödinger and Dirac, respectively, impose on the solution functions are the same. Let us, then, establish the equivalence of the restrictions imposed by these two physicists. We recall that Schrödinger's restrictions are:

(113)
1. The solution functions must be one-valued.

2. They must be continuous together with their first and second derivatives.

3. They must tend to a finite limit over the boundary of the q-space. We shall assume that this boundary is at infinity.

Now, we have seen in Chapter XXXII that, in the general case, Schrödinger's wave equation has no solutions which satisfy the foregoing conditions. But we have also seen that such solutions will exist whenever the parameter E in the equation is credited with privileged values (the eigenvalues), which may form a discrete set or a continuous set or both. To simplify our exposition, we shall confine our attention to the permissible discrete values of E. Let E_n be one of them. If we set this value for E in (112), there may be one or more corresponding solution functions, or eigenfunctions, $\psi_{E_n}(q)$. Any one of these eigenfunctions defines a possible distribution of the amplitudes of the de Broglie waves when E_n is the energy of the system.

We also know that the Schrödinger solution functions which are associated with different eigenvalues are orthogonal. We recall the meaning of this mathematical term. Let E_n and E_k be two different eigenvalues pertaining to a discrete set. The eigenfunctions corresponding to E_n and to E_k, respectively, may be written $\psi(q, E_n)$ and $\psi(q, E_k)$. The property of orthogonality, which these solution functions betray, is expressed by the relation

$$(114) \qquad \int \psi^*(q, E_n)\psi(q, E_k)dq = 0 \qquad\qquad (k \neq n).$$

The integral is taken over the entire range of q, from $-\infty$ to $+\infty$. If $k = n$, we obtain the integral.

$$(115) \qquad \int \psi^*(q, E_n)\psi(q, E_n)dq.$$

This integral converges to a finite limit, and hence we may always select our eigenfunctions in such a way that it has the value 1, *i.e.*,

$$(116) \qquad \int \psi^*(q, E_n)\psi(q, E_n)dq = 1.$$

The eigenfunctions are then said to be normalized to 1. Schrödinger imposes this normalization in order that the total electric charge of his charge cloud, which he assumes to be of density

$$(117) \qquad -e\psi^*(q, E_n)\psi(q, E_n),$$

should have the value $-e$ (the charge of an electron).

Now, the possibility of normalizing the continuous eigenfunctions implies that integrals of type (115) are finite, and therefore that the eigenfunctions vanish at infinity. Hence the normalizing condition (116) may take the place of Schrödinger's third condition, being even more drastic. The net result is that Schrödinger's solution functions for his wave equation satisfy the two first conditions (113) and the normalizing condition (116). In addition these eigenfunctions are automatically orthogonal and hence satisfy (114).

We now pass to the solution functions $S(q', E')$ of Dirac's equation (111). Since $S(q', E')$, in which E' is fixed and q' varies continuously, represents a column of an orthogonal matrix S, the function $S(q', E')$ viewed as a function of q' is necessarily one-valued and continuous. Thus, Schrödinger's first two conditions (113) are also imposed on Dirac's functions. Furthermore, since S must be an orthogonal matrix, the relation (104), *i.e.*,

$$(118) \qquad \tilde{S}^*S = 1$$

must be satisfied. If, as before, we assume discrete eigenvalues E_n and E_k, this relation (118) may be written

$$(119) \quad \text{or} \quad \begin{aligned} \int \tilde{S}^*(E_n, q')S(q', E_k)dq' &= \begin{cases} 0 \text{ if } k \neq n \\ 1 \text{ if } k = n \end{cases} \\[2ex] \int S^*(q', E_n)S(q', E_k)dq' &= \begin{cases} 0 \text{ if } k \neq n \\ 1 \text{ if } k = n. \end{cases} \end{aligned}$$

But (119) is the same as (114) and (116). Hence we conclude that the Dirac functions are orthogonal and are normalized to 1. In short, the restrictions imposed on the solution functions $S(q', E')$ of Dirac are exactly the same as those imposed on Schrödinger's normalized eigenfunctions; and since both sets of functions are solutions of the same partial differential equation, they are identical.

Let us drop the dash over q in Dirac's function, since we are now running no risk of confusing q' with the matrix q. We may then set

(120) $S(q, E_n) \equiv \psi(q, E_n)$, or $\psi_n(q)$.

The identity of Dirac's and of Schrödinger's functions shows that the latter, which represent the amplitude distributions of the de Broglie waves in the stationary states of the atom, are none other than the columns of the orthogonal matrix which transforms from the coordinate system in which Dirac's matrix q is diagonal to the coordinate system in which the energy matrix is diagonal. An alternative interpretation may also be given. Thus the columns $S(q, E_n)$ of an orthogonal matrix define, in the Hilbert space, the cosines of the angles which the various principal axes of the quadric surface q make with the principal axis E_n of the quadric surface connected with the energy. The same interpretation may then be extended to Schrödinger's eigenfunctions $\psi_n(q)$.

It is remarkable to find that Schrödinger's wave equation, derived from the assumption of underlying waves, should reveal itself as the equation to which we are led by totally different considerations in which waves play no part at all. We have here a further argument pointing to the symbolic nature of the de Broglie waves. Another point of interest is that, thanks to the introduction of Dirac's continuous matrices, Schrödinger's seemingly arbitrary procedure of replacing the momentum p in the Hamiltonian by the operator $\dfrac{h}{2\pi i}\dfrac{\partial}{\partial q}$ now shows itself to be a natural development.

Dirac's method furnishes immediately the connection between the wave functions and Heisenberg's solution matrices. Thus, the solution matrices (without time terms) of a problem are merely Dirac's matrices q and p expressed in the coordinate system in which the Hamiltonian matrix (i.e., the energy matrix) is diagonal. The solution matrices are then given by the expressions (105).

If we assume that the energy levels are discrete, the term (n, k) of the solution matrix (i.e., the term associated with the transition from the energy level E_n to the energy level E_k) is

$$Q(n, k) = (\tilde{S}^*qS)(n, k) = \iint \tilde{S}^*(E_n, q')\, q'\delta(q' - q'')S(q'', E_k)\, dq'dq''$$
$$= \int \tilde{S}^*(E_n, q')q'S(q', E_k)\, dq' = \int S^*(q', E_n)q'S(q', E_k)\, dq'.$$

Since there is no danger of confusing the ordinary variable q' with the matrix q, we may write the above relation

$$(121) \qquad Q(n, k) = \int S^*(q, E_n) q S(q, E_k) dq.$$

Similarly, for the solution matrix P, we have

$$P(n, k) = (\tilde{S}^* p S)(n, k) = \iint \tilde{S}^*(E_n, q') \stackrel{h}{} \dot{\delta}(q' - q'') S(q'', E_k) dq' dq'';$$

and this is found to be

$$(122) \qquad P(n, k) = \frac{h}{2\pi i} \int S^*(q, E_n) \frac{\partial S(q, E_k)}{\partial q} dq,$$

where q, as before, is an ordinary variable.

The formulae (121) and (122), connecting the wave functions of a problem and the solution matrices, were first obtained by Schrödinger, but the method just outlined (which is due to Dirac) establishes the same results more simply.

If, instead of confining our attention to the amplitude parts of the solution matrices, we wish to obtain the complete solution matrices with time terms, we must supplement the functions $S(q, E_n)$ by adjoining the vibrational term $e^{-\frac{2\pi i}{h} E_n t}$. Thus, if we assume that $S(q, E_n)$ is replaced by

$$(123) \qquad S(q, E_n) e^{-\frac{2\pi i}{h} E_n t}$$

in the relations (121) and (122), and similarly for $S(q, E_k)$, we obtain

$$(124) \qquad Q(n, k) e^{\frac{2\pi i}{h}(E_n - E_k)t}$$

and

$$(125) \qquad P(n, k) e^{\frac{2\pi i}{h}(E_n - E_k)t},$$

which represent the general terms of the complete solution matrices.

Dirac's treatment is more comprehensive than Schrödinger's on several points. The essence of Dirac's treatment is to transform a continuous non-diagonal Hermitean matrix $F(q, p)$ into its diagonal form by means of a rotation of axes. Thus let $F(q, p)$ be a Hermitean matrix in which q and p are Dirac's continuous matrices. The matrix F being expressed in the coordinate system in which the matrix q is diagonal, cannot itself be diagonal, for if it were, it would commute with the matrix q; and this is impossible since F contains p.† We wish now to rotate our axes so

---

† See note page 856.

that F will assume its diagonal form. The matrix F will then be represented by some diagonal matrix D. The diagonal terms of this diagonal matrix are, say, D', D'', ... If $F(q, p)$ is algebraic in p, we are led to the partial differential equation

$$(126) \qquad \left\{ F\left(q', \frac{h}{2\pi i} \frac{\partial}{\partial q'} \right) - D' \right\} S(q', D') = 0.$$

The solution functions $S(q', D')$ that satisfy the conditions required of the columns of an orthogonal matrix are then the columns of the matrix S which transforms from the coordinate system q, in which q is diagonal, to the coordinate system D, in which F is diagonal. As for the eigenvalues D', D'', . . . , they are the diagonal elements of the diagonal matrix D.*

The equation just written is Schrödinger's wave equation provided the function $F(q, p)$ be the Hamiltonian, or the energy function, of the problem. In this event the eigenvalues D' are the energy levels E', and the solution functions $S(q', D')$ are Schrödinger's amplitude functions. But in the general case where F is not the Hamiltonian, the solution functions $S(q', D')$ are not Schrödinger's functions, and the eigenvalues D' are not the energy levels. From this standpoint Schrödinger's wave equation shows itself to be a particular case of Dirac's more general equation (126). Another advantage of Dirac's treatment is that, even when the Hamiltonian $H(q, p)$ is not algebraic in p, so that Schrödinger's partial differential equation cannot be obtained, we may still express our problem by means of an integral equation.

---

* As an illustration, let us determine the orthogonal matrix S which transforms from the coordinate system q to the coordinate system p. To solve the problem, we note that in the coordinate system q Dirac's continuous matrix p is not diagonal, whereas it is of course diagonal in the coordinate system p. Our problem is thus to transform by means of an orthogonal matrix S the non-diagonal matrix p into its diagonal form. This problem is a special case of the more general one considered in the text. We have but to replace the matrix $F(q, p)$ by Dirac's matrix p (in the system q), and the diagonal matrix D by the diagonal matrix p (in the system p). If we set p' for any diagonal term of the continuous diagonal matrix p, the partial differential equation (126) of the text becomes

$$\left\{ \frac{h}{2\pi i} \frac{\partial}{\partial q'} - p' \right\} S(q', p') = 0.$$

The normalized solution functions $S(q', p')$, i.e., the columns of the transformation matrix S, are then found to be

$$(127) \qquad S(q', p') = \frac{1}{\sqrt{h}} e^{\frac{2\pi i}{h} q', p'}$$

The Physical Significance of the Transformation Functions
$S(q', E')$—The columns $S(q', E')$ and $S(q, E_n)$ of the transformation matrix S have an important physical significance. We shall restrict our attention to the case of discrete energy levels. Let us consider the significance of the function $S(q, E_n)$, for example, which defines the column E_n of the transformation matrix S.

In formula (121), which gives the amplitude parts of the solution matrix Q, let us set $k = n$. We then obtain the diagonal elements $Q(n, n)$ of this non-diagonal matrix, *i.e.*,

$$(128) \qquad Q(n, n) = \int q S^*(q, E_n) S(q, E_n) dq.$$

We mentioned on page 839 that the diagonal elements of the solution matrix (here called Q) represent the mean values of the classical coordinate Q over a long period of time when the system is in one energy state or another. In particular, $Q(n, n)$ in (128) gives the mean value of the coordinate Q of the particle when the system is in the nth energy state E_n. The formula (128) is consistent with the calculus of probabilities if we assume that

$$(129) \qquad S^*(q, E_n) S(q, E_n) dq$$

represents the probability that the particle is situated between a point q and a point $q + dq$ at any instant of time when the energy is E_n.† We may also say that

$$(130) \qquad S^*(q, E_n) S(q, E_n)$$

---

† Suppose we know that a particle is situated somewhere on a straight line. Let $f(q) dq$ define the probability that the particle is situated between the points q and $q + dq$. Since the particle must be somewhere on the line, the integral of the probability taken along this line must have the value 1. The probability function must therefore satisfy the condition

$$(131') \qquad \int f(q) dq = 1.$$

The average position $\bar{q}$ of the particle is then determined, according to the calculus of probabilities, by

$$(131'') \qquad \bar{q} = \int q f(q) dq.$$

We may verify that $S^*(q, E_n) S(q, E_n)$ satisfies the condition (131') imposed on the probability function $f(q)$, for since the functions $S(q, E_n)$ are normalized, we have

$$\int S^*(q, E_n) S(q, E_n) dq = 1,$$

which is the analogue of (131'). The average value of q is thus

$$(132) \qquad \bar{q} = \int q S^*(q, E_n) S(q, E_n) dq;$$

and this is (128).

measures the density of the probability that the particle will occupy a position q when the energy of the system is known to be E_n. Owing to this physical rôle played by the functions $S(q, E_n)$, these functions are often referred to as *probability amplitudes*. We must remember, however, that it is not $S(q, E_n)$ itself, but $S(q, E_n)$ multiplied by the conjugate complex magnitude, which expresses a probability.

We have considered the case of one dimension for simplicity, but exactly the same conclusions may be extended to any number of dimensions. For instance, in the hydrogen atom, where the electron has three degrees of freedom, the transformation function is of type

$$(133) \qquad S(x, y, z, E_{n,l,m}),$$

and therefore

$$(134) \qquad S^*(x, y, z, E_{n,l,m}) S(x, y, z, E_{n,l,m}) dx dy dz$$

defines the probability that, at any instant, the electron will be found in the volume $dx dy dz$ about the point x, y, z when the atom is known to be in the energy state $E_{n,l,m}$.

These results agree with those obtained by Schrödinger. Thus Schrödinger assumed that when his eigenfunctions were normalized,

$$(135) \qquad \psi^*_{n,l,m}(x, y, z) \psi_{n,l,m}(x, y, z) dx dy dz$$

defined the probability we have just mentioned, or that

$$(136) \qquad \psi^*_{n,l,m}(x, y, z) \psi_{n,l,m}(x, y, z)$$

measured the density of the probability of the electron's position. Inasmuch as Schrödinger's normalized eigenfunctions $\psi_{n,l,m}(x, y, z)$ are none other than Dirac's functions $S(x, y, z, E_{n,l,m})$, the two theories yield exactly the same results.

Dirac's treatment is obviously more satisfactory, for his results are obtained without necessitating the introduction of additional assumptions. Schrödinger, on the other hand, had to postulate that the intensity (136) of the de Broglie waves measured the density of a probability.

More General Probability Amplitudes—A magnitude which in classical science is represented by an ordinary number (c-number) is usually expressed by a q-number in the Heisenberg-Dirac theory. In some cases a q-number may be written as a matrix. Now if the matrix is in a non-diagonal form, it does not furnish the precise numerical values which the physical magnitude may assume. Average values are then the nearest that can be obtained. To secure accurate information, we must convert the matrix into its diagonal form (by means of an orthogonal transformation); the diagonal terms of the diagonal matrix thus obtained are c-numbers which define the possible numerical values of the physical magnitude. An illustration of this situation occurred in connection with the Hamiltonian matrix. The Hamiltonian function represents the energy of the system, and the Hamiltonian matrix, or the energy matrix, is the matrix representation of this energy. But when we sought the precise numerical values of the energy, we had to transform the energy matrix into its diagonal form. The diagonal terms then defined the energy levels. We may liken a non-diagonal matrix to the image of an object which is out of focus; the corresponding diagonal matrix represents the same image, but now perfectly focussed. Diagonal matrices thus play an important part when definite numerical answers to quantum problems are required.

Suppose, then, we have two physical magnitudes, A and B, represented by matrices. Let the following discrete terms

$$\left\{\begin{array}{l} a_1, \ a_2, \ a_3, \ \ldots \ldots \ a_n, \ \ldots \ldots \\ \text{and} \\ b_1, \ b_2, \ b_3, \ \ldots \ldots \ b_n, \ \ldots \ldots \end{array}\right.$$

be the diagonal terms of the matrices A and B, respectively, when these matrices are rendered diagonal. If the matrices A and B commute, it will be possible to obtain both matrices in the diagonal form simultaneously by means of an appropriate rotation of axes. In this case it will be possible to determine the values a and b of A and B simultaneously. If on the other hand A and B do not commute, only one of the two matrices at a time will be obtainable in the diagonal form. Suppose B is made diagonal, then A will remain non-diagonal. In this case, if the value of B is known to be b_k, that of A may be a_1 or a_2 or a_3, etc., and hence will be uncertain. Thus the possibility of obtaining simultaneous precise values for two physical magnitudes depends on whether or not the matrices representing these magnitudes commute. Nevertheless even when the two matrices do not commute, we may determine the probability that, say, the first magnitude will have one value or another when the value of the second magnitude is known, and vice versa. Let us determine this probability.

We call S the orthogonal matrix which transforms from the coordinate system A to the coordinate system B. Let us consider the term $S(a_n, b_k)$ of this matrix S. Then, according to the transformation theory,

(137) $S^*(a_n, \ b_k) \ S(a_n, \ b_k)$

is the probability that the physical magnitude A will have the numerical value a_n when the magnitude B is known to have the value b_k. Conversely, (137) also expresses the probability that B will have the value b_k when A is known to have the value a_n.†

A geometric interpretation may be given to the probability (137). Thus $S(a_n, b_k)$ measures in the Hilbert space the cosine of the imaginary angle which is formed by the nth principal axis of the quadric surface associated with the matrix A and by the kth principal axis of the quadric surface B. Hence (137) represents the squared value of this cosine.

The probability (137) is more general than the one we considered earlier, for our former transformation function $S(q', E_n)$ transformed solely from the coordinate system q, in which q was diagonal, to the one in which the energy matrix was diagonal. In (137), on the other hand, the magnitudes A and B need not refer to q and to the energy, though as a particular case they may do so.

Heisenberg's Uncertainty Relations—Let us apply the foregoing considerations to the matrices which represent the position and momentum of a particle moving in space.

For a system of several degrees of freedom, the relations (46) of the previous chapter indicate which of the q-numbers, q and p, pertaining to a mechanical system commute. All these magnitudes commute except those that are conjugate. Thus q_1 commutes with q_2 and with p_2, but not with p_1. If, then, we are dealing with a particle which is moving in space and which has for its position and momentum coordinates x, y, z, p_x, p_y, p_z, we should be able to determine simultaneously, say, x, y, and z; or x, p_y, and p_z; but not x and p_x. The commutation rules (46) thus account in a general way for the uncertainties expressed in Heisenberg's uncertainty relations. Moreover, we may show that our present theory furnishes exactly the same quantitative values for the uncertainties as were given previously by Heisenberg.

To prove this, we confine our attention to a system of one degree of freedom, having therefore a single coordinate q and a single momentum p. The general term of the transformation matrix which transforms from the coordinate system q, in which the q-number q is diagonal, to the coordinate system p, in which p is diagonal, is found to be

$$(138) \qquad S(q', p') = \frac{1}{\sqrt{h}} e^{\frac{2\pi i}{h} q', p'} \qquad (\text{see } (127)).$$

---

† In (137) we have a probability and not a density of a probability as in (130). The reason is that the diagonal terms a_n and b_k in (137) are assumed discrete.

In this expression q' and p' are typical diagonal elements of the matrices q and p expressed in diagonal form. The density of the probability that the value of the momentum will be p' when the coordinate of position is q', or vice versa, is given by

$$(139) \qquad\qquad S^*(q', p')\, S(q', p'),\, i.e.,\, \frac{1}{h}.$$

Since this density is constant, the momentum may have any value when the position is accurately known, and the position may have any value when the momentum is accurately known. We are here in perfect agreement with the Heisenberg uncertainty relations.

Let us now suppose that the position q' is not accurately known, and that Δq measures the probable error in our determination of the value of q' at the instant of interest. Calculation then shows that the value of p' is no longer totally uncertain: Under the most favorable conditions * there will be an uncertainty Δp; it satisfies the relation

$$(140) \qquad\qquad \Delta q.\, \Delta p \sim h;$$

and this is one of Heisenberg's uncertainty relations.

In Chapter XXX we also mentioned an uncertainty relation between time and energy. According to our present explanations, this latter uncertainty relation must be ascribed to the fact that time and energy, when viewed as q-numbers, do not commute (see formula (47) of the last chapter). The transformation theory also yields this uncertainty relation; it is expressed by

$$(141) \qquad\qquad \Delta t,\, \Delta E \sim h.$$

The foregoing analysis brings out an important point. We saw in Chapter XXX that Heisenberg's uncertainty relations could be derived theoretically from the properties of wave packets. Now, had the wave treatment afforded the only theoretical means of obtaining the uncertainty relations, we might plausibly have attributed the uncertainties to the inadequacy of the wave form of representation. It is thus of interest to find that the transformation theory, which follows an entirely different method of approach, leads to exactly the same uncertainty relations.

The matrix theory and the transformation theory show that the uncertainty relations arise because the numbers representing certain physical magnitudes do not commute. In these commutation relations

---

* Calculation shows that the most favorable conditions are realized when the uncertainty in the value of q' is expressed by Gauss's error curve.

therefore must be sought the source of the uncertainty relations. Now the commutation relations are refined expressions of Bohr's quantum conditions, so that we sense the connection between the uncertainty relations and the quantum states (which are characteristic of quantum science).

If all the quantum magnitudes were to commute, there would be no uncertainty relations. This is precisely the situation that occurs when we consider the relationships between the momentum matrices p_x, p_y, p_z of a particle and its energy E viewed as a q-number. All these matrices commute, as is seen from the formulae (46) and (47). Hence no theoretical objection interferes with our assigning precise values simultaneously to the momentum and to the energy of a particle or to the different components of momentum.

The various theoretical methods of obtaining the uncertainty relations shed no light on their physical origin. But we explained at length how Heisenberg traced the source of the uncertainties to the quantum disturbances generated by measurements.

We may give a more graphical representation of the disturbances generated by measurements. Suppose we have a particle restricted to move along a line so that only one coordinate q and one momentum p come into consideration. A position measurement performed with precision requires that the matrix q be diagonal, and this implies that the coordinate axes must coincide with the principal axes of the quadric surface attached to the matrix q. A position measurement will thus rotate the coordinate axes Ox from some initial position and bring them into coincidence with the principal axes of q. A momentum measurement will set the coordinate axes into coincidence with the principal axes of the matrix p. Suppose, then, we perform a position measurement. The coordinate axes now coincide with the principal axes of q, and the position q' of the particle may be measured with precision. Since the principal axes of q and p do not coincide, our coordinate axes cannot be

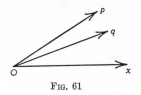

Fig. 61

in coincidence with the principal axes of p, and, as a result, the value of the particle's momentum is uncertain; or, if we prefer, the matrix p is out of focus. Next, suppose we perform a momentum measurement. The coordinate axes are thereby rotated so as to coincide with the principal axes of p; they no longer coincide therefore with the principal axes of q, so that q passes out of focus. In this situation the momentum

p' of the particle can be obtained with accuracy, while its position q' is uncertain.

In the present scheme, the quantum disturbances, entailed by measurements, are associated with rotations of axes in the Hilbert space and with an attendant blurring of the eigenvalues pertaining to the q-numbers of interest. We may contrast this interpretation with that of wave mechanics, in which the quantum disturbances are illustrated by disturbances in the probability waves.

As a further illustration of the transformation theory let us suppose that a particle has at the initial instant, $t = 0$, a position q'_0 and that it is moving freely in space. At a subsequent instant t, it occupies some position q'. (We are here using primed letters to indicate that we are dealing with ordinary coordinates and not with matrices).

According to classical mechanics, the position of the particle at time t is

$$(142a) \qquad q' = q'_0 + \frac{p'_0}{m}t,$$

where p'_0 is the initial momentum of the particle and m is its mass. Consequently, according to classical mechanics, the position of the particle is known at any instant t if we have determined its initial position q'_0 and its initial momentum p'_0. But we know that Heisenberg's uncertainty relations prohibit a simultaneous determination of q'_0 and p'_0, so that at time t the position q' of the particle will necessarily be uncertain. Inasmuch as the present transformation theory embodies Heisenberg's uncertainty relations we may rest assured that the foregoing conclusions will be consistent with the transformation theory. However, it is of interest to examine the matter more closely.

We call q_0, p_0, and q the matrices having as diagonal elements (when the matrices are in diagonal form) the ordinary numerical magnitudes q'_0, p'_0, p'. In the mechanical problem we are considering here, the three matrices q_0, p_0, and q are connected by a matrix equation modeled on the classical mechanical relation (142a). This matrix equation is

$$(142b) \qquad q = q_0 + \frac{p_0}{m}t.$$

Let us verify that, in full agreement with the uncertainty relations, the positions q' and q'_0 of the particle, at time t and at the initial instant, cannot both be accurately known. From the standpoint of matrices this requirement implies that the matrices q and q_0 cannot assume the diagonal form simultaneously and hence do not commute. Now we may easily verify that q and q_0 do not commute.† At first sight this result may appear strange for both matrices represent position coordinates, and up to this point all the position coordinates, which we have been concerned with, commuted. We must remember, however, that q_0 and q refer to different instants of time, so that our two matrices which coincided at the initial instant $t = 0$ (because q was then q_0) are no longer the same after an interval of time t. There is no particular reason therefore why the principal axes of the two matrices should coincide at any instant t after the initial instant. We may picture the situation by supposing that the principal axes of the matrix q, though coinciding with those of q_0 at the initial instant, undergo a gradual rotation, as a result of which they diverge more and more from those of q_0 as time passes.

Since, then, q and q_0 do not commute, there will exist an orthogonal matrix $S(q_0', q')$ transforming from the system q_0 (in which q_0 is diagonal) to the system q (in which q is diagonal) ; and since q varies with time, the matrix S will also vary with time.

Let us suppose that the initial position q_0' is accurately known, and let us calculate the density of the probability that at a specified instant t, the particle will occupy a position q'. This probability, as we know, is given by

(143) $$S^*(q_0', q')\ S(q_0', q').$$

If we calculate this expression, we find that its value is a constant, a fact which implies that the density of the probability is the same at all points, so that at time t the particle may be anywhere. Its position is thus completely uncertain. This conclusion is in agreement with the uncertainty relations, because an accurate knowledge of q_0' entails complete ignorance of the initial momentum p_0'.

Next let us assume that the initial position q_0' is not accurately known. In this case the uncertainty relations show that p_0' need no longer be completely unknown and hence that q' will no longer be completely uncertain. The transformation theory bears out these expectations.

Thus let

(144) $$f^*(q_0')f(q_0')$$

---

† Let us write (142b) as $q_0 = q - \dfrac{p_0}{m}t$. Now q_0 commutes with itself and hence with $q - \dfrac{p_0}{m}t$. If, then, q_0 were to commute with q, it would necessarily commute with p_0 and we know that this is not the case.

represent the density of the probability that the initial position is q'_0. The average value of the initial momentum can then be determined. Under these conditions the density of the probability that, at time t, the particle will be at q' can be shown to be

$$(145) \qquad \phi^*(q')\phi(q'),$$

where

$$(146) \qquad \phi(q') = \int f(q'_0) \, S(q'_0, q') \, dq'_0.$$

The density (145), in contradistinction to (143), is not a constant. Its value depends on the value attributed to q', so that the probability of finding the particle at one point or another at time t varies from point to point. A definite limited range within which the particle will most probably be situated at time t is thus determined. Calculation also shows that the uncertainty in the particle's position, as embodied in the probability (145), increases with time. Furthermore, the manner in which the region of maximum probability spreads as it advances is in all wise similar to the spreading of a wave packet in the wave theory. The same quantitative results are thus obtained by wave mechanics and by our present transformation theory.

In addition to the probability amplitudes of the type discussed, there are amplitudes connected with the probabilities of the transitions of the atom from one energy state to another. Dirac has developed a theory of emission and of absorption of radiation by atoms, in which the defect noted in connection with Schrödinger's treatment no longer occurs. We recall that, in Schrödinger's wave-mechanical method, the probability of energy drops on the part of a large number of atoms was found to depend not only on the number of atoms in the higher level, but also on the number in the lower level.[†] This result was in conflict with Bohr's original assumptions; and, besides, it appears to be refuted by experiment. Dirac's theory overcomes this difficulty; the higher energy level alone now enters into consideration.

---

† See page 740.

CHAPTER XXXIX

DIRAC'S THEORY OF THE ELECTRON

DIRAC's theory of the electron is an abstract mathematical theory, which combines wave mechanics and the theory of relativity. The major achievement of the new theory is that it establishes the existence of an electron spin, in full accord with the hypothesis of Uhlenbeck and Goudsmit.

We have seen that Uhlenbeck and Goudsmit introduced the hypothesis of the spinning electron in Bohr's theory for the purpose of dispelling certain inconsistencies in the interpretation of spectra. No theoretical argument, however, seemed to demand this hypothesis, so that had it not been for the spectroscopic evidence, there would have been no reason to anticipate the existence of the spin. The earlier wave theory and the matrix theory registered no better success in this connection. These considerations prompt us to examine how the electron spin may be expected to manifest itself in the wave theory.

We recall that Schrödinger originally utilized the wave equation derived from classical mechanics, and then refined his theory by considering the relativistic wave equation. The latter equation was found to be a considerable improvement over the classical one, and in the case of the hydrogen atom it furnished some of the features of the fine structure of the spectral lines. Neither wave equation, however, afforded any indication of the electron's spin. As a result, whenever the spin was taken into account in the wave treatment (*e.g.*, in the wave theory of the helium atom), it was merely grafted on the existing theory as a separate assumption.

The first attempt to devise a wave equation which would be consistent with the electron spin was undertaken by Pauli. Pauli observed that the spin of a rotating body could be represented by a matrix, and he therefore introduced suitable spin matrices into the classical wave equation. The effect of these matrices was to split the wave equation into a system of two simultaneous equations, so that the solutions of the system appeared in pairs. The two solutions of a pair were the representations of an electron spinning in one sense or the other. Equivalently, we may say that the amplitude function ψ of a wave was no longer represented by a scalar but by a magnitude having two components, and that the two components represented the opposite spins of the electron.

Properly speaking the electron spin pertains to the particle picture, and so we inquire: What is the wave representation of the spin? Pauli's treatment suggests the answer to this question, for it shows that the amplitude ψ of the wave must be defined by two components instead of by a single scalar. Since this is precisely the condition that is required in the representation of polarized waves, we may assume that the wave representation of a spinning electron is a polarized wave. C. G. Darwin therefore postulated that the de Broglie waves associated with an electron manifested polarization, and on this basis he was led to a wave equation which furnished the correct fine structure of the hydrogen spectrum.

The procedures of Pauli and Darwin, however, are both open to the objection that they introduce the spin of the electron (or the polarization of the waves) in an artificial way, without justifying this introduction by any theoretical argument.

At this juncture Dirac maintained that the problem of the spin had not been attacked properly, and he then devised the new method of approach which we shall now consider. But first let us recall certain notions on the wave equation.

At the end of Chapter XXXII we said that there were two different kinds of equations commonly referred to as wave equations. One of these was the amplitude equation, which was utilized in Schrödinger's theory of the hydrogen atom; the other wave equation was the generalized equation. The amplitude equation only furnishes the amplitudes of the wave functions, whereas the generalized equation yields the wave functions themselves, and its solutions may represent a superposition of different monochromatic waves forming a wave packet. Obviously, the generalized equation constitutes the true wave equation, and we shall therefore refer to it in future as the "wave equation" without further specification.

Now we have seen that the amplitude equation and the wave equation may be derived from the energy equation of mechanics. Thus let us suppose that we are dealing with an electron moving in a conservative field of force. The classical equation of energy is represented by

$$(1) \qquad H(q, p) - E = 0,$$

where $H(q, p)$ is the Hamiltonian function, and E the total energy of the system. If we consider the relativistic energy equation instead of the classical one, we must replace the classical energy E by the relativistic energy W, and take the relativistic Hamiltonian in place of the classical one.

If we assume that no force is acting on the electron, the classical energy equation is

$$(2) \qquad \frac{1}{2m}(p_x^2 + p_y^2 + p_z^2) - E = 0,$$

and the relativistic one may be written

$$(3) \qquad p_x^2 + p_y^2 + p_z^2 + m_0^2 c^2 - \frac{W^2}{c^2} = 0.$$

The classical wave equation is derived from (2) by replacing p_x, p_y, p_z, and $-E$ by

$$(4) \qquad \frac{h}{2\pi i}\frac{\partial}{\partial x}, \quad \frac{h}{2\pi i}\frac{\partial}{\partial y}, \quad \frac{h}{2\pi i}\frac{\partial}{\partial z}, \quad \frac{h}{2\pi i}\frac{\partial}{\partial t},$$

and then causing the operator thus obtained to operate on a function $u(x, y, z, t)$.

The relativistic wave equation is secured by proceeding in the same way with the relativistic energy equation (3), except that now it is $-W$ that must be replaced by the last operator (4).

The expressions of the two wave equations for a particle m were given in Chapter XXXII. Formula (40) gave the classical wave equation, though in it we must now set $E_{pot} = 0$ since we are assuming that the particle is moving in free space. Formula (45) furnished the relativistic wave equation (no field). It will be noted that, whereas the classical wave equation (40) involves the first derivative $\frac{\partial u}{\partial t}$ of the wave function u, the relativistic equation (45) involves the second derivative $\frac{\partial^2 u}{\partial t^2}$.

Now according to Dirac, the calculation of the various probabilities, which are of such importance in the quantum theory, requires that the wave equation should contain $\frac{\partial}{\partial t}$ but not $\frac{\partial^2}{\partial t^2}$. Thus whereas the classical wave equation furnishes the probabilities, the relativistic wave equation is ineffective. This situation appeared strange, for the relativity theory is a mere refinement of the classical theory. Dirac assumed therefore that the derivation of the relativistic wave equation was faulty, and that the correct relativistic equation, like the classical one, should involve only $\frac{\partial}{\partial t}$. Inasmuch as the presence of $\frac{\partial^2}{\partial t^2}$ in the relativistic wave equation is due to the presence of the squared energy, W^2, in the relativistic energy equation (3), Dirac sought to replace this latter equation by one which would be linear in W and would thus resemble the classical energy equation.

In order to obtain a relativistic energy equation of the required form, Dirac split the left hand side of the equation (3) into a product of two factors, both of which were linear in W. Either one of these two factors, when equated to zero, furnished an energy equation which satisfied Dirac's demands. It remained to be proved that these two equations satisfied the conditions of invariance of the special theory of relativity, *i.e.*, that they were invariant under the Lorentz transformation. Dirac, having verified this invariance, concluded that either one of the two equations could be taken to represent the revised energy equation for an electron moving in free space. The two equations being essentially the same, Dirac selected one and discarded the other. The revised relativistic equation was thus found to be

$$(5) \qquad \rho_1 \sum_{x,y,z} \sigma_x p_x + \rho_3 m_0 c + \frac{W}{c} = 0,$$

where ρ_1, ρ_3, σ_x, σ_y, and σ_z are 4-dimensional matrices involving only constant terms and satisfying certain relations. (The matrices σ are equivalent to Pauli's spin matrices.)

The equation (5) is linear in W, as required, but the presence of the matrices ρ and σ introduces a novel element. Since these matrices involve only constants, they must refer to some characteristic of the electron itself. Dirac, commenting on their presence in the energy equation of an electron, writes:

"They must therefore denote some quite new dynamical variables, which may be pictured as describing some internal motion in the electron. We shall see later that they just describe the spin of the electron." *

If we perform Schrödinger's substitutions (4) in the revised relativistic energy equation (5) and then operate on $u(x, y, z, t)$ we obtain Dirac's revised relativistic wave equation for a free particle.

Let us now consider the case of an electron of mass m_0 and charge $-e$ moving in an electromagnetic field. The field can be represented by a so-called vector potential of components A_x, A_y, A_z, and by a scalar potential ϕ. According to the classical theory of relativity, the energy equation in the presence of an electromagnetic field is derived from the

---

* Dirac. The Principles of Quantum Mechanics; Oxford; 1930; p. 241.

energy equation in the absence of the field by replacing in the latter equation p_x, p_y, p_z, and $\dfrac{W}{c}$ by

$$p_x + \frac{e}{c}A_x, \quad p_y + \frac{e}{c}A_y, \quad p_z + \frac{e}{c}A_z, \quad \text{and} \quad \frac{W}{c} + \frac{e\phi}{c},$$

respectively. Dirac followed this procedure, so that his energy equation in the presence of a field became

$$(6) \qquad \rho_1 \underset{xyz}{\Sigma} \sigma_x \left(p_x + \frac{e}{c}A_x \right) + \rho_3 m_0 c + \left(\frac{W + e\phi}{c} \right) = 0.$$

The relativistic wave equation in the presence of the electromagnetic field was then obtained by utilizing Schrödinger's substitution operators (4).

Dirac showed that the new wave equation entailed the following important consequences:

The electron according to this wave equation will have a magnetic moment of one magneton $\left(i.e., \dfrac{eh}{4\pi m_0 c} \right)$, **Furthermore, if we consider an electron moving in a central field of force, we find that the orbital angular momentum of the electron does not remain constant during the motion, so that conservation of angular momentum is not satisfied by the orbital motion alone. But if we credit the electron with an additional angular momentum,** $\dfrac{1}{2} \cdot \dfrac{h}{2\pi}$ **due to a half-quantum spin, the validity of the conservation principle is restored. Dirac's wave equation thus leads to the same assumptions that were made by Uhlenbeck and Goudsmit when they introduced the hypothesis of the spinning electron in Bohr's atom.** Thus, the hypothesis of the spinning electron, which proved so great an advance in our understanding of spectroscopic phenomena, has at last received a theoretical explanation and is no longer a hypothesis *ad hoc* introduced for reasons of expediency. The spin and the attendant magnetic moment now reveal themselves as relativistic effects.

Finally, Dirac on applying his revised wave theory to the hydrogen atom found that the peculiarities of the hydrogen spectrum were accounted for. He also showed that his wave equation for the hydrogen atom coincided with the one which Darwin had obtained previously. But Darwin had introduced the assumption that the waves were polarized, an assumption which we have seen to be equivalent to postulating the electron spin. Dirac's treatment has the advantage of making no *a priori* assumptions of this kind.

The Negative Energy Solutions—Dirac's relativistic wave equation of an electron in an electromagnetic field is compatible with two kinds of solutions: those in which the kinetic energy of the electron is positive and those in which it is negative. These latter solutions are called the negative energy solutions. From a theoretical standpoint both kinds of solutions have the same measure of validity and should therefore describe possible situations in Nature. But the difficulty is that from a physical standpoint the negative energy solutions appear to be impossible, for a negative kinetic energy has no physical significance. Leaving aside this objection for the present, Dirac showed, however, that a negative-energy solution (*i.e.*, an electron in a negative-energy state) could be interpreted as referring to a particle in motion having a positive charge (equal and opposite to the charge of an electron) and having the same mass as an electron. The existence of a positive electron was thus suggested by Dirac's wave equation. But at the time the only particle known to carry a positive charge was the proton, its charge being equal and opposite to that of the electron, as required, but its mass being considerably greater. In spite of this disparity in the mass, Dirac at first assumed that his negative-energy solutions referred to protons. Subsequently, the positive electron was discovered experimentally and was called the "positron." Thus Dirac's wave equation had in effect predicted the existence of the positron.

There remained the difficulty of interpreting the negative kinetic energy. Dirac assumed that in a perfect vacuum, all the negative-energy states were occupied by electrons, one electron occupying each state, as required by Pauli's Exclusion Principle. Furthermore, all the positive energy states remained vacant. Inasmuch as a perfect vacuum connotes the absence of all matter, the electrons in negative-energy states did not imply the existence of any effective negative kinetic energy. Consider next a region of space in which the physicist would recognize the presence of electrons. Such a region would be represented, as before, by occupied negative states, but now some of the positive states would likewise be occupied by electrons. These latter electrons could never fall into negative-energy states, as a result of quantum transitions, since these negative-energy states were already occupied. From a physical standpoint, this would mean that an electron could not suddenly vanish into nothingness.

Next, suppose that one of the negative states is unoccupied by its electron. An unoccupied negative-energy state will appear as something with a positive charge and with positive energy, since to fill the state we should have to add to it an electron with negative energy. Dirac therefore assumes that an unoccupied negative-energy state represents a positron. Now when we have an unoccupied negative-energy state, and when one or more positive-energy states are occupied by electrons, one of the latter electrons may quite well drop into the now vacant negative state and fill it. The result will be two-fold: in the first place, the dropping away of the electron from the positive-energy state will betray itself by the disappearance of an electron; and in the second place, the filling of the negative-energy state will result in the disappearance of a positron. Thus an electron and a positron will disappear simultaneously, their charges cancelling each other. Their energies should appear in a chargeless form, *viz* as light.

This mutual cancellation of an electron and positron, with an attendant transformation of their energies into light has since been verified experimentally, *viz* an electron and a positron vanish, and two photons take their place. The converse of the above phenomenon would be the conversion of light-energy into an electron and a positron. In Dirac's representation, such a transformation would correspond to an electron passing from a negative-energy level into a positive one.

The wave equation that we have discussed in this chapter is the one connected with an electron (or a positron); but the same general results would have been obtained had we considered a proton. We should have found that the proton, also, must be credited with a spin and a magnetic moment.* However, on account of the much greater mass of the proton, the magnetic moment will be considerably less than for the electron. If we assume a spin for the proton and hence for the nucleus of the hydrogen atom, we find that the wave theory indicates the existence of two different kinds of hydrogen molecules. Each kind is formed of two atoms; but in one kind of molecule the axes of spin of the two proton-nuclei are parallel, whereas in the other they are antiparallel. We have here a situation analogous to the one discussed in connection with the symmetric and the antisymmetric states of the helium atom. Experiment has since confirmed these anticipations.

---

* Theory indicates that photons, likewise, must be credited with spins. Polarized light would then be represented by a swarm of photons spinning in the same direction. Recent experiment has established the existence of these photon-spins.

FURTHER DEVELOPMENTS

In the earlier theory of wave-mechanics, a de Broglie wave in ordinary 3-dimensional space was assumed to constitute the wave picture of a particle. The probability of locating the particle in the neighborhood of a point x, y, z at an instant t was taken to be proportional to the intensity of the wave at this point at the instant t. As for the probable momentum and energy of the particle, they were connected with the wave length and frequency of the wave in accordance with Born's assumptions. The wave could also be associated with a swarm of similar non-interacting particles. In this event the intensity of the wave at a point was proportional to the density of the cluster of particles around the point. In all rigor, a swarm of non-interacting particles is a myth; but if the interactions are of slight importance (as would arise in a swarm of electrons sparsely scattered), it would still seem permissible, as a first approximation, to represent the swarm by a wave in 3-dimensional space. We may recall that this representation proved successful in the interpretation of the diffraction experiments on electrons (*e.g.*, those of Davisson and Germer).

On the other hand, when more than one particle is involved and when the interactions can no longer be neglected, we must represent the waves in a multidimensional configuration space. An instance of this complication was noted in connection with the helium atom. In the helium atom, we had two electrons exerting mutual repulsive influences, and we saw that the waves should be pictured in a 6-dimensional configuration space. Quite generally, if there are N interacting particles and we assume that each particle has three degrees of freedom, the configuration space has $3N$ dimensions.

Attempts were made to rid the theory of the multidimensional configuration spaces and to represent the waves in ordinary 3-dimensional space in all cases. To understand how this representation was secured, we shall first assume the presence of only one particle, *e.g.*, an electron. According to Schrödinger's method of treatment, the single electron circling in the hydrogen atom on a definite orbit is represented by a vibrational condition distributed throughout space. The intensity of the vibration at a point determines the electric density of the electrified cloud which takes the place of the corpuscular electron. In the particle picture (in which the nucleus is regarded as fixed), there is no other particle on which the electron may act, and the only potential energy that enters into consideration is the potential energy of the electron in the field developed by the nucleus. From the standpoint of the charge cloud, this situation implies that no account must be taken of the potential energy arising from the interactions of the various parts of the cloud. But suppose that we have N electrons and that we cannot afford to neglect their mutual repulsions. These mutual repulsions constitute a new source of potential energy, and the additional potential energy must be represented in the wave picture. Now we may retain the 3-dimensional wave picture if we ascribe this additional potential energy to the interactions of the various parts of the charge cloud. Also, since the charge cloud represents the N electrons, we must impose the condition that the total charge of the cloud be the same as that of these N electrons, viz., $-Ne$. Let us note, however, that the reason we impose this restriction on the total charge is that we happen to know from the empirical evidence that electric charge is atomic, and can exist only in discrete amounts e and not in fractions thereof. It would be more satisfying if our theory itself furnished this conclusion, for then atomicity would rest on a theoretical basis.

Jordan, Pauli, Heisenberg, and others set themselves the problem of modifying the wave theory so that it would automatically entail the atomicity of electric charge. Their attempts must be regarded as highly speculative. The essence of their procedure is to view the wave functions no longer as mere numbers, but as q-numbers satisfying appropriate noncommutative relations. As a result of these commutation relations (in which Planck's ubiquitous constant h appears), the new theories indicate that the total electric charge of a charge cloud must always be some multiple of the fundamental unit of charge $-e$. The existence of the electron is thus accounted for.

The characteristic trait of these more recent investigations is their extremely symbolic nature and the unfamiliar mathematical world to which they pertain. Even ordinary numbers, such as are used to specify the number of electrons in a given volume, appear as the eigenvalues of strange q-numbers. Practically all physical magnitudes lose their familiar associations, so that we seem to be penetrating into a new world whose abstruseness baffles the imagination. No logical conflict with the more classical notions is suggested, however; for, as in all cases, the new mathematics passes over into the more familiar kind when the relative importance of h decreases and hence when the level of commonplace experience is approximated to.

CHAPTER XL

THE NEW STATISTICS

In Chapter XXII we discussed the classical statistics of gases. We shall now elaborate certain points so as to understand the nature of the modifications introduced by the quantum theory.

The gas laws (for a perfect gas), as obtained from experiment and when supplemented by Avogadro's hypothesis, may be expressed by

$$(1) \qquad PV = NkT,$$

where P and V are the pressure and the volume of the gas, k is the gas constant, T the absolute temperature, and N the number of molecules present in the gas. When we are dealing with a gram-molecule of the gas, N is Avogadro's number, and Nk is then written R. Thus, for a gram-molecule, the gas laws, also called the equation of state, may be written

$$(2) \qquad PV = RT.$$

According to thermodynamics, the entropy S of a perfect gas (monoatomic) containing N molecules is defined by

$$(3) \qquad S = kN \log (VT^{\frac{3}{2}}) + \text{constant.}^*$$

The value of the constant in this expression is undetermined, so that the absolute value of the entropy eludes us. Thus the first two principles of thermodynamics, on which the formula (3) is based, yield no information on the absolute entropy of a gas. Accordingly physicists reconciled

---

* For a gram-molecule, thermodynamics gives the value

$$S = c_v \log T + R \log V + \text{constant,}$$

where c_v is the specific heat at constant volume. Direct measurement shows that c_v is determined by

$$c_v = \frac{3}{2}R \text{ (for a monoatomic gas),}$$

a result that may also be deduced from the kinetic theory. If we accept this value for the specific heat, the expression (3) of the text is obtained for the entropy of a perfect monoatomic gas containing an arbitrary number N of molecules.

themselves to the thought that absolute values of the entropy could not be obtained.

The discovery of the Third Principle of Thermodynamics, however, shed a new light on the situation. This principle, we recall, states that the entropy of a crystalline body vanishes at the absolute zero of temperature, and it thus affords a rational point of departure for the measurement of entropy. For example, if we start with a crystal at the absolute zero and then vaporize it, the resulting vapor, or gas, at stipulated temperature and pressure, must necessarily have a well-defined entropy. The problem confronting physicists was therefore to devise a theoretical means of determining the value of the entropy in any particular space.

We might suppose that Boltzmann's kinetic theory would furnish the entropy of a perfect gas. But, for reasons that will appear presently, the kinetic theory yields no information unless certain assumptions are made. In particular, restrictions must be placed on the partitioning of the phase space. Planck observed that the quantum theory could be used to secure these restrictions, and so he superimposed the quantum theory on the Boltzmann statistics. In this way he obtained an absolute value for the entropy of a gas. Subsequent developments have shown that Planck's method of imposing quantum restrictions on the kinetic theory was correct, but that his expression of the entropy was only approximate because the Boltzmann statistics, on which he had based his calculations, was itself a mere approximation. The Boltzmann statistics has since been modified into the Bose-Einstein statistics, thanks to which the correct expression of the entropy has been obtained.

In this chapter we shall be concerned with the successive alterations which the Boltzmann statistics has undergone.

The Classical, or Boltzmann, Statistics—So as to avoid constant reference to Chapter XXII, we recall some of the fundamental points of the kinetic theory. This theory proposes to interpret the behavior of gases by assuming that perfectly elastic molecules are rushing hither and thither at random. In what follows we shall consider only the perfect gases. This restriction requires that the molecules should exert no mutual action (except when collisions occur). Furthermore, we shall suppose that the molecules are monoatomic, similar to tiny elastic spheres. Each molecule will be taken to have only three degrees of freedom, *i.e.*, the translational degrees of freedom; rotations will be disregarded.

First we must define what we mean when we say that a gas is in a given *state*. We must here make a distinction between a *microscopic* and a *macroscopic* state. To say that a gas is in a given microscopic state implies that its various molecules occupy definite positions and have definite momenta (and hence velocities), the momenta being specified in direction as well as in magnitude. If, then, we define a point of space by x, y, z and the components of a molecule's momentum by p_x, p_y, p_z, the exact position and motion of a molecule of mass m will be determined by the specification of the six magnitudes

$$(4) \qquad x, y, z, p_x, p_y, p_z.$$

It is therefore convenient to introduce a fictitious 6-dimensional configuration space, of coordinates (4), and to represent the simultaneous position and momentum of a molecule by a single point having the six coordinates (4) in this 6-dimensional space. The 6-dimensional space is called the *phase space*, and the point which determines the position and the momentum of a molecule is named a *phase-point*. When we proceed in the manner stated, for all the N molecules, the instantaneous microscopic state of the gas is represented by N phase points appropriately distributed throughout the 6-dimensional phase space. Suppose that the position or the momentum of one of the molecules is changed slightly. The phase point of this molecule will undergo a displacement in the phase space, and in all truth the microscopic state of the gas will be modified. However, if the changes in the position and in the momentum of the molecule are so slight that even the most accurate experiment would be unable to reveal them in practice, we shall assume that no change in the microscopic state has occurred. These considerations must be extended to all the N molecules of the gas.

A graphical interpretation of the foregoing assumptions is easily obtained. We imagine the phase space to be partitioned into tiny juxtaposed 6-dimensional cubes called *cells*. These cells are all equal, the lengths of the six edges issuing from a corner of a cell being represented by the small, equal magnitudes:

$$(5) \qquad \Delta x, \ \Delta y, \ \Delta z, \ \Delta p_x, \ \Delta p_y, \ \Delta p_z.$$

The 6-dimensional volume of a cell is written ω; its value is

$$(6) \qquad \omega = \Delta x \Delta y \Delta z \Delta p_x \Delta p_y \Delta p_z.$$

The lengths of the edges (5) are assumed so small that, if a phase point originally situated at the cell's centre is displaced to any point within the cell, the changes in the position and in the momentum of the corresponding molecule will be too minute to be disclosed even by accurate measurement. Obviously, this criterion is vague, so that the exact dimensions and the volume of a cell are poorly determined.

Suppose then, we are dealing with a given microscopic state of the gas. The N phase points of the N molecules are distributed among the various cells. Let there be

(7)

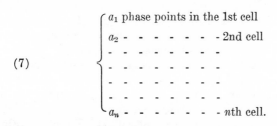

a_1 phase points in the 1st cell

a_2 - - - - - - - 2nd cell

- - - - - - - - -

- - - - - - - - -

- - - - - - - - -

- - - - - - - - -

a_n - - - - - - - nth cell.

If the position or the momentum (or both) of one of the molecules is changed, but so slightly that the corresponding phase point does not move out of the cell in which it was contained, we shall say that the gas is still in the same microscopic state. On the other hand, if the phase point passes into another cell, we shall agree that the microscopic state is no longer the same. These considerations may be extended to several molecules. Thus, if the positions and momenta of all the molecules are changed in such a way that all the phase points remain in their original cells, the microscopic state of the gas is not modified. The vagueness we mentioned in connection with the exact dimensions of the cells necessarily clouds our understanding of the identity of two microscopic states. For the present, however, this difficulty may be disregarded, and we shall merely assume that the cells are very small—physically infinitesimal.

Among the various microscopic states represented, for instance, by the distribution (7), a large number will be very much alike. As an example, if we exchange two representative points between two cells, there will still be the same numbers of points in each cell, the only difference being that the identity of the points is no longer the same. This situation corresponds to our exchanging the positions and motions of two of the molecules in the gas. Clearly, from the macroscopic standpoint, the situation will not have changed. Let us consider the different microscopic states in which the numbers of points per cell are not disturbed. Each one of these microscopic states is said to pertain to the same *macroscopic state* of the gas. The number of different microscopic states that are associated with a single macroscopic state is therefore the number of different ways in which the points may be interchanged among the cells (the number of points in each cell of course remains unchanged). In particular, if we assume the original distribution of the table (7), this number W of ways is found to be

$$(8) \qquad W = \frac{N!}{a_1!\, a_2!\, a_3!\, \ldots\, a_N!},$$

where N is the total number of molecules in the gas. In the course of time, as the molecules move, the various possible microscopic and also macroscopic states will follow in succession. By "possible" states we mean those in which the total number of molecules is the same number N and in which the sum total of the energy of all the molecules has the same value. This latter requirement is imposed by the principle of conservation of energy as applied to our mechanical gas system.

We now pass to the problem of the probabilities. When discussing the kinetic theory (Chapter XXII), we saw that, as a result of Gibbs's theorems and of the ergodic hypothesis, we could ascribe the same *a-priori* probability to each one of the possible microscopic states; and that in the course of time these states would be realized one after another with the same frequency (on an average). If the correctness of the foregoing conclusions is granted, the quantity (8) is necessarily proportional to the probability of the macroscopic state (7) being realized. The macroscopic state that will have the greatest probability of occurring at a given instant will then be that one (among the possible ones) for which the value of the expression (8) is the greatest. This particular macroscopic state, and those that differ only slightly from it, are so much more probable in their aggregate than the other states, that the gas is practically certain to be in one of these most probable states whenever we happen to observe it. For this reason the most probable macroscopic state will seem to endure permanently. This most probable macroscopic state is called the state of statistical equilibrium, and calculation shows that when it is realized, the theorem of equipartition of energy holds.

Inasmuch as the state of statistical equilibrium is the one to which the gas tends to pass, we must assume that it is the state of highest entropy. Maximum entropy and maximum probability thus go hand in hand. Also, since the probability of two situations occurring simultaneously is given by the product of the probabilities of the individual situations, and since the total entropy of the two systems is the sum of the entropies, Boltzmann assumed that the entropy of a state was proportional to the logarithm of its probability. Thus, if we call Π the probability of a given macroscopic state, the entropy S of this state, according to Boltzmann, is

$$(9) \qquad S = A \log \Pi,$$

where A is some constant to be determined.

At this point a difficulty occurs in connection with the correct expression to be given to the probability Π. The number W (defined by (8)) is the number of different microscopic distributions which are contained in a given macroscopic distribution, or state. The probability Π of the macroscopic state is proportional to W; and according to the usual definition of a probability, we have the relation $\Pi = \dfrac{W}{D}$, where D is the total number of different microscopic distributions comprised in all possible macroscopic states. There is, however, no particular reason to be bound by the usual definition of a probability; and so the safest course is to set $\Pi = BW$, where B is some positive constant whose value we leave undetermined for the present.

Setting $\Pi = BW$ in (9), we obtain for the entropy

$$S = A \log BW = A \log W + A \log B.$$

The entropy in the state of statistical equilibrium (*i.e.*, the entropy in the thermodynamical sense) is then secured when we give to W in the previous formula its greatest permissible value. Planck showed that when this was done and the thermodynamical relation between entropy, energy, and temperature was taken into account, the expression of the entropy became

$$(10) \qquad S = AN \log (VT^{\frac{3}{2}}) + AN \log \left(\frac{e^{\frac{5}{2}} (2\pi mk)^{\frac{3}{2}}}{N\omega} \right) + A \log B.*$$

In this formula, V is the volume of the gas, T the absolute temperature, e the number 2,87 . . . , m the mass of molecule, k the gas constant, ω the volume of a cell, and N the total number of molecules. We note that the last two terms of (10) are constants, and hence independent of the volume and of the temperature of the gas. The values of these constants are unknown, for the constants A, B, and ω are as yet undetermined.

A comparison of (10) with the thermodynamical formula (3) shows that the two formulae concur if we assume that the undetermined constant A of (10) is the gas constant k. We therefore set in (10)

$$A = k.$$

Even when A is replaced in (10) by the known magnitude k, the exact value of the entropy is undetermined, for the last two terms of (10) contain the unknown constants B and ω. Thus, the kinetic theory has not led us any further than thermodynamics: the absolute value of the entropy still eludes us.

---

* Planck's derivation of this formula was subsequently criticized by Ehrenfest.

To overcome this defect, Planck postulated that the value 1 should be assigned to the constant B. If this value is taken, log B vanishes; the entropy S is then expressed by

(11) $S = k \log W.$

The explicit expression of the entropy is now given by (10) from which the last term is deleted and in which A is replaced by the gas constant k. The result obtained may be written more compactly

(12) $S = kn \log\left(\dfrac{e^{\frac{5}{2}}(2\pi mkT)^{\frac{3}{2}}}{N\omega} V\right).$

It will be noted that, by setting $B = 1$, we are in effect assuming that the probability Π of a macroscopic state is defined by the number W of different microscopic states which are included in this macroscopic state. This assumption and also the identification of the constant A with the gas constant k will be retained in the new statistics to be discussed later.

Let us revert to the expression (12) of the entropy. It contains the volume ω of a cell and hence is indeterminate unless the magnitude of this volume is specified. In Boltzmann's day there was no theoretical reason to ascribe any specific value to the volume ω, and so the classical, or Boltzmann, statistics failed to yield an absolute value for the entropy.* But with the rise of the quantum theory and the discovery of Planck's constant h, more light was thrown on the situation. The manner in which Planck grafted the quantum theory on the kinetic theory becomes intuitive when we consider the explicit expression of the volume ω of a cell. This volume ω is defined by

(13) $\omega = \Delta x \Delta y \Delta z \Delta p_x \Delta p_y \Delta p_z.$

Now $\Delta x \Delta p_x$ has the dimensions of action, *i.e.*, length $\times$ momentum; and the same is of course true for the other pairs of terms in (13). And since Planck in his formulation of the quantum theory had found it necessary to postulate the existence of a quantum or atom of action h, he deemed it plausible to set

(14) $\Delta x \Delta p_x = \Delta y \Delta p_y = \Delta z \Delta p_z = h.$

---

* In the deduction of the most probable state of a gas, we need not specify the precise value of the volume ω; it is sufficient to assume that this volume is physically infinitesimal.

According to Planck's assumption, the volume ω of each cell in the phase space becomes

(15) $$\omega = h^3.$$

Planck therefore replaced ω by h^3 in the formula (12) and he thus obtained the following determinate expression for the entropy S of a monoatomic gas, *viz.*,

(16) $$S = kN \log \left(\frac{e^{\frac{5}{2}}(2\pi mkT)^{\frac{3}{2}}}{Nh^3}V \right).$$

Planck obtained this expression (16) of the entropy in 1921, but it had already been derived in a different way by Sackur and Tetrode. For this reason it is often called the *Sackur-Tetrode* expression.

Planck's method of calculating the entropy of a gas draws attention to the close relationship between the Third Principle of Thermodynamics and the quantum theory. The following analysis will make this point clear: We have seen that an absolute value for the entropy cannot be derived from the kinetic theory so long as the volumes of the cells are unspecified. On the other hand, the Third Principle of Thermodynamics, by giving physical significance to the absolute entropy indicates that the cells should have well-determined volumes. These volumes, which have the dimensions of the cube of an action, remained indeterminate in the classical treatment because classical theory viewed action as a magnitude which was susceptible of having any value. The quantum theory, by furnishing a universal unit of action, h, provided a rational measure for these volumes and hence an absolute value for the entropy, in accordance with the demands of the Third Principle. From this account we see that the Third Principle suggests, as it were, the existence of a unit of action. An earlier example of the relationship between the Third Principle and the quantum theory was mentioned in Chapter XXIV, when we discussed the decrease of the specific heats of solids at low temperatures.

Let us now examine some of the consequences of Planck's assumption (15). Whether we follow Boltzmann or Planck, the general significance of a cell is the same. The various phase points situated in the same cell represent various molecules, the spatial positions and also the momentum components of which are so nearly alike that they cannot be distinguished, even by microscopic observation. Equivalently, we may say that the precise position of a phase point within a given cell is fundamentally uncertain.* In both treatments therefore we may without loss of generality assume that all the phase points which are associated with a given cell are situated at the centre of the cell. Similarly, when we are dealing with the aggregate of molecules forming the gas, we may suppose that their phase points occupy the centres of the various cells, and hence that the molecules have the spatial positions and momenta defined by these centres. Thus far nothing essentially new is involved by Planck's assumption, but we shall now see that this assumption entails a novel result.

In the classical theory, where the cells were infinitesimal, the distance between the centres of two contiguous cells was itself infinitesimal. Consequently, the spatial positions and momentum components of the various gas molecules could differ by as little as we chose. But if we accept Planck's assumption, the centres of two adjacent cells are at a finite distance apart. Hence only a discrete set of positions and of momenta (or velocities) can be ascribed to the molecules. Thus, the permissible velocities of the molecules, differing as they do by finite amounts in which the constant h enters, may be said to be quantized. The significance of Planck's assumption now becomes clear. Its effect is to impose quantization on the molecular motions.

Having elucidated the significance of Planck's cells, we now inquire whether Planck was justified in postulating them and in quantizing thereby the molecular motions. Planck made his initial contributions to the statistical theory of gases in the early days of the quantum theory. At that time it was believed that quantum occurrences were connected solely with periodic motions. For example in Bohr's hydrogen atom, only the periodic elliptical orbits were quantized; the hyperbolic orbits, which are aperiodic, were not subjected to quantization. Now, in the molecular

---

* Planck's assumption of a volume h^3 for the cells reveals itself as a consequence of Heisenberg's uncertainty relations. Thus, according to the uncertainty relations, if the position of a particle is uncertain within a volume $\Delta x \Delta y \Delta z$, the momentum components of the particle are uncertain within a volume $\Delta p_x \Delta p_y \Delta p_z$ of the momentum space. And between these two volumes we have the relation

(17) $$\Delta x \Delta y \Delta z \, \Delta p_x \Delta p_y \Delta p_z \sim h^3.$$

Consequently, all particles situated in the spatial volume $\Delta x \Delta y \Delta z$ and having momenta within the volume $\Delta p_x \Delta p_y \Delta p_z$ are indistinguishable. This implies that the 6-dimensional phase-volume $\Delta x \Delta y \Delta z \Delta p_x \Delta p_y \Delta p_z$ of magnitude h^3 plays the part of a cell—and this is Planck's assumption. But of course when Planck first formulated his assumption, he could not justify it on the basis of the uncertainty relations, for these relations were discovered only later.

motions of a gas, we cannot well assume any trace of periodicity; and under these conditions it would seem that the molecular motions should remain unquantized and hence that Planck's assumption of the volume h^3 for the cells should not be considered. However this may be, Planck's intuition guided him correctly, though a vindication of his views had to await the development of the new methods of quantum mechanics. As we shall see later, wave mechanics leads quite naturally to Planck's assumption (15).

But even before the new quantum mechanics was formulated, Planck's quantization of the phase space into cells h^3 received support from another direction. In this connection we recall that Planck's expression (16) of the entropy, which he had derived, in 1921, from the assumption of cells h^3, had already been furnished in 1912 by Sackur and Tetrode. In 1913, Stern obtained this same expression of the entropy by a method which did not involve Planck's speculative quantization of the phase space. The fact that Stern's less speculative method of approach likewise yielded Planck's expression of the entropy, suggested the correctness of this expression, and hence the correctness of Planck's method of quantization. We shall now give a rapid sketch of Stern's derivation.

Consider the transformation of a crystal at temperature T_0 into a vapor at temperature T and pressure p. For simplicity, we shall assume that the vapor behaves like a perfect monoatomic gas. The formulae of thermodynamics show that the change in entropy which accompanies this transformation can be computed if we can determine the vapor pressure of the saturated vapor in contact with the crystal at temperature T_0. If the temperature T_0 is assumed to be low, the entropy of the crystal may be taken to be zero, in accordance with the Third Principle of Thermodynamics; and the above change in the entropy then represents the entropy of the gas itself.

Now the vapor pressure cannot be computed by the methods of thermodynamics because of its dependence on a constant of integration of undetermined value, which Nernst has called a "chemical constant." The chemical constant may be calculated, however, if adequate information is furnished on the behavior of the crystal, and if the ordinary statistical methods are then applied to determine the state of equilibrium between the crystal and the vapor. Stern assumed that the quantum theory of solids would determine the behavior of the crystal; he therefore quantized the vibrations of the crystal in the manner explained in Chapter XXIV when we discussed the specific heats of solids.

By proceeding along these lines, Stern obtained the expression of the chemical constant.* As will be seen in the note, the chemical constant involves Planck's constant h, this latter constant having been introduced in the formulae as a result of the application of the quantum theory to the crystal. The chemical constant being known, the expressions of the vapor pressure and of the entropy are derived immediately. They too involve Planck's constant. Stern was thus led to the familiar Sackur-Tetrode formula (16) for the entropy of the gas.

It will be observed that Stern obtained the Sackur-Tetrode expression by applying the quantum theory to the solid and not to the gas. Inasmuch as the quantum theory of solids had received such brilliant confirmation in connection with the falling off of the specific heats, Stern's derivation was thought to establish the correctness of the Sackur-Tetrode expression. Consequently, when Planck showed that this same expression of the entropy could be derived by applying the quantum theory to the gas directly, physicists came to regard Planck's quantization of the phase space as valid.

Today, as a result of the new statistical theories, we know that the Sackur-Tetrode expression of the entropy is only approximate.† Never-

---

* Stern's expression of the chemical constant is

$$(18) \qquad C = \log\left[\frac{(2\pi m)^{\frac{3}{2}} k^{\frac{5}{2}}}{h^3}\right].$$

This value is in very good agreement with measurements of vapor pressures.

† That the Sackur-Tetrode expression (16) was at best an approximation, might have been suspected for other reasons. Thus, if we take account of (16) and of the formula which expresses the energy of the gas, we find that the classical equation of state and the specific heat undergo no change however low the temperature and however high the pressure. Thus no degeneration is indicated, in spite of the fact that the Third Principle, according to Nernst, would require degeneration to set in at low temperatures. There is also another objection. The Sackur expression becomes negative at sufficiently low temperatures and high pressures (small volumes), whereas a negative entropy appears to be inconsistent with the Third Principle. If we were dealing with a real gas this objection might be dismissed, for a real gas would condense into a liquid and into a solid long before its entropy could become negative. But Planck, we recall, is dealing with a perfect gas, which satisfies the classical gas laws at all temperatures and for which a change of state does not come into consideration. However, neither of the two objections mentioned in this note is sufficient to invalidate the Sackur-Tetrode expression, for we may reasonably suppose that the implications of the Third Principle apply only to real existents and cannot be extended to such conceptual constructs as the perfect gas considered by Planck. Nevertheless, we shall see that the Bose-Einstein statistics accounts for the degeneration of a perfect gas and also for the non-occurrence of a negative entropy.

theless, as the more recent developments of wave mechanics have shown, Planck's method of quantization is correct, and the reason it led Planck to the incorrect expression of the entropy must be ascribed to the lack of validity of the Boltzmann statistics. These points will be understood later.

The Distribution of the Energy in the Boltzmann Statistics—In many cases it is of interest to determine how the total energy of a gas is distributed among the molecules in the state of statistical equilibrium. This information can be obtained when we pursue the calculations in accordance with the method previously indicated; but it can also be reached by a more direct procedure. We shall now examine this alternative procedure, because it enables us to understand more clearly in what respects the Boltzmann statistics differs from those to be discussed presently.

Let us consider one of the cells of the phase space. The coordinates of its centre are x, y, z, p_x, p_y, p_z. A phase point at the centre of the cell corresponds to a molecule situated at a point of space defined by the three first coordinates x, y, z of the cell's centre, and having components of momentum defined by the three other coordinates p_x, p_y, p_z. The molecule which corresponds to this phase point has a total momentum

$$(19) \qquad p = \sqrt{p_x^2 + p_y^2 + p_z^2}.$$

The molecule will also have a certain total energy ε. Since we are assuming that the molecules exert no interactions and that no external field of force is applied, the total energy of a molecule is also its kinetic energy. According to classical mechanics (which is a sufficient approximation for our present purpose), the kinetic energy, and hence the total energy, of the molecule is

$$(20) \qquad \varepsilon = \frac{p^2}{2m} = \frac{p_x^2 + p_y^2 + p_z^2}{2m}.$$

For reasons already explained, all molecules corresponding to phase points in the same cell may be represented by phase points situated at the centre of the cell. Consequently, if x, y, z, p_x, p_y, p_z are the coordinates of the centre point of some cell, we may attribute these same coordinates to any phase point situated within this cell. Any phase point in this cell corresponds therefore to a molecule which has the total momentum (19) and the total energy (20). In a purely figurative sense, the momentum and the energy common to all the molecules connected with a given cell may be associated with the cell itself. Thus a cell having as centre the point x, y, z, p_x, p_y, p_z will be said to be a cell of momentum p (defined by (19)) and of energy ε (defined by (20)).

Let us consider all those cells which are associated with the same values of the momentum and energy. As is seen from (19) and (20), the centres of these cells are defined by points of coordinates x, y, z, p_x, p_y, p_z satisfying the condition that $p_x^2 + p_y^2 + p_z^2$ should have the same value for all the cells. There will obviously be a large number of these cells, for the three first coordinates x, y, z of the cell are arbitrary, and the three last ones p_x, p_y, p_z are subjected to a single relation and are not otherwise specified. Calculation shows that all these cells lie over a cylindrical surface in the phase space.

Instead of grouping the cells which have exactly the same energy, let us consider those cells which are associated with energies lying within a small range of values, e.g., within the range ε and $\varepsilon + d\varepsilon$. There will then be, say,

(21)
$$
\begin{cases}
Z_0 \text{ cells associated with energies between 0 and } d\varepsilon \\
Z_1 \text{ - - - - - - - - - - - - - - } d\varepsilon \text{ and } 2d\varepsilon \\
Z_2 \text{ - - - - - - - - - - - - - - } 2d\varepsilon \text{ and } 3d\varepsilon \\
\text{- - - - - - - - - - - - - - - - -} \\
\text{- - - - - - - - - - - - - - - - -} \\
\text{- - - - - - - - - - - - - - - - -} \\
\text{- - - - - - - - - - - - - - - - -}
\end{cases}
$$

This tabulation may be expressed more concisely by saying that there are

(22) Z_i cells associated with energies between ε_i and $\varepsilon_i + d\varepsilon$,

where $\varepsilon_i = id\varepsilon$, $(i = 0, 1, 2, \ldots)$.

The numbers Z_i can be calculated once and for all, because they are independent of the manner in which the phase points may subsequently be distributed among the cells.

The aggregate of Z_i cells which are associated with the same range ε_i to $\varepsilon_i + d\varepsilon$ of the energy will be called the ith *layer* of cells ($i = 0, 1, 2, 3, \ldots$).

Consider now some given macroscopic state. It is determined by the numbers of molecules situated in the various cells. This macroscopic state also defines a certain energy distribution for the molecules. For example, there will be

$$\begin{cases} n_0 \text{ molecules having energies between 0 and } d\varepsilon \\ n_1 \text{ - - - - - - - - - - - - - - - } d\varepsilon \text{ and } 2d\varepsilon \\ n_2 \text{ - - - - - - - - - - - - - - } 2d\varepsilon \text{ and } 3d\varepsilon, \end{cases}$$

and so on. In more concise form, we may say that there are

n_i molecules having energies between ε_i and $\varepsilon_i + d\varepsilon$,

where $\varepsilon_i = id\varepsilon$ $(i = 0, 1, 2, \ldots\ldots)$.

A distribution of this kind implies that n_i phase points are distributed in any arbitrary way among the Z_i cells which have energies between ε_i and $\varepsilon_i + d\varepsilon$. In other words n_i phase points are situated in the ith layer of cells.

A state of the gas specified by the numbers of molecules that have energies comprised within consecutive small ranges may be called an "energy state" of the gas. Note that although a macroscopic state determines an energy state unambiguously, the reverse is not true, for any redistribution of the molecules would leave us with the same energy state provided the total numbers of molecules in the various layers were left unchanged, whereas a redistribution of this sort might alter the macroscopic state. We conclude that an energy state comprises a number of different macroscopic states and hence a still larger number of microscopic states. This conclusion becomes obvious when we note that in an energy state the spatial positions and the directions of motion of the molecules are unspecified.

We now calculate the number of different microscopic states that will enter into the constitution of a given energy state. This number is found to be

$$(23) \qquad P = \frac{N!}{n_1! \, n_2! \, n_3! \, \ldots\ldots} Z_1^{n_1} Z_2^{n_2} Z_3^{n_3} \cdots\cdots,$$

where n_i is the number of molecules distributed in any arbitrary way among the Z_i cells of the ith layer. Since the microscopic states are viewed as equally probable, the number P will be proportional to the probability of the energy state of interest. We may, however, proceed as we did in the case of W and assume that P is the probability itself instead of being merely proportional thereto. By giving different values to $n_1, n_2, \ldots$ (subject to the restriction that the total number of molecules and the total energy must remain unchanged), we obtain all the possible energy states. That state for which P has its greatest value will represent the energy state which is realized when the gas is in statistical equilibrium. We secure by this method a knowledge of the numbers n_1, n_2, ... of molecules having energies which lie in the successive energy ranges when the state of statistical equilibrium is attained. The state of statistical equilibrium here referred to is of course exactly the same state of statistical equilibrium that we derived from our former method of treatment, but in the present case only the energy distribution, not the positions and the momentum components of the molecules, are obtained.

If we assume, as we have done up to this point, that the molecules do not interact and are not acted upon by external forces (*e.g.*, gravitational), the energy of the molecules is purely kinetic, and a knowledge of the energy distribution entails a knowledge of the velocity distribution (except for direction).

We now come to the entropy S of the gas. Previously, we defined it by $S = k \log W$; but since W and P do not have the same numerical values, we cannot expect to obtain correct results if we set

$$(24) \qquad\qquad S = k \log P.$$

Nevertheless, calculation shows that, if it is only the entropy of the state of statistical equilibrium with which we are concerned, and if the numbers of molecules in the various cells are large, the two expressions of the entropy are equal. Thus, when in (24) we replace P by its greatest permissible value and take Planck's assumption of cells of volume h^3 into consideration, we obtain, as before, the Sackur-Tetrode expression (16) for the entropy of the gas in the state of statistical equilibrium.

The Boltzmann Statistics and Radiation—Planck's law of radiation gives the densities (or the intensities) of the radiations of different frequencies that will be found in a heated enclosure when the radiation is in equilibrium with matter. The last point requires elucidation. The walls of the enclosure are of course made of matter, but if these walls are perfectly reflecting, so that they can neither emit nor absorb radiation, and if there is no loose matter within the enclosure, the state of statistical equilibrium will be one affecting the radiation alone. In practice, however, the walls are never perfectly reflecting, and so they necessarily emit and absorb radiation. The state of statistical equilibrium will therefore be one that holds as between the radiation and the matter. In other words, there will be constant exchanges of energy, not only between the radiations but also between the radiations and the matter. Planck's law applies specifically to the case where these latter exchanges occur. All attempts to obtain Planck's radiation law from the statistics of electromagnetic waves in accordance with the classical ideas were failures, for they invariably led to Rayleigh's law of radiation, not to Planck's. We explained in Chapter XXIV that Planck, by postulating that the emission of radiation was a discontinuous process, was able to furnish a theoretical derivation of his law. In Planck's treatment, however, our attention was directed to the matter in the enclosure; the nature of the radiation was then obtained indirectly when we assumed the classical relationship between the mean energy of an oscillator and the density of the corresponding radiation. This appeal to classical considerations (which for the sake of consistency should have been barred) was regarded as unsatisfactory.

Shortly after Planck had formulated his theory, Einstein, in his analysis of the photo-electric effect, proposed to view radiation as corpuscular. According to Einstein's views, radiation of frequency ν was formed of corpuscles, or photons, having energy $h\nu$ and momentum $\dfrac{h\nu}{c}$ (in free space). If we avail ourselves of Einstein's corpuscular conception of radiation, we may investigate the problem of equilibrium radiation by another method. Thus we may suppose that the radiations in equilibrium in Planck's enclosure should be likened to a gas (the photon gas) in the state of statistical equilibrium at the temperature T of the enclosure—the onrushing photons of the photon gas playing the part of the molecules in an ordinary gas.* The pressure which the radiation exerts against the walls, and which in the electromagnetic theory is accounted for in terms of the electric and magnetic intensities, can now be attributed to the impacts of the photons. The pressure is thus inter-

---

* In point of fact, as the enclosure contains matter which emits and absorbs radiation, the state of equilibrium is not quite the same as in gas. Rather does it resemble the equilibrium of a vapor in contact with its solid or liquid phase (*e.g.*, water vapor in contact with ice or water). In a gas, the total number of molecules is fixed; and so in applying the statistical theory we must prescribe not only the total energy of the system but also the number N of molecules. On the other hand, in the equilibrium of a vapor with its solid, or liquid, phase, molecules are constantly passing from the vapor to the liquid or solid, and conversely. The total number of molecules forming the vapor is thus subject to variation. In the statistical treatment we must therefore refrain from stipulating that the number of molecules is fixed, though, as before, we must specify the constant value of the total energy. Since the equilibrium of radiation and matter in the enclosure is similar to that of a vapor with its condensed phase, we must not assume that the total number of photons in the enclosure is fixed. This is a minor consideration from our present standpoint, but it must be taken into account in the mathematical treatment of the problem.

preted by exactly the same mechanism for radiation as it was for ordinary gases.

Suppose, then, we apply the Boltzmann statistics and compute the numbers of photons which, in the state of statistical equilibrium at temperature T, have energies between $h\nu_i$ and $h(\nu_i + d\nu)$, where $i = 0, 1, 2, 3, \ldots$. When this is done, we may deduce immediately the density of the radiations whose frequencies are comprised between ν_i and $\nu_i + d\nu$. Since it is precisely this information which is furnished by a radiation law, we understand how we must proceed in order to obtain the radiation law *via* the photons.

To obtain the radiation law, we have but to determine the state of statistical equilibrium for a photon gas, just as we did for an ordinary gas at temperature T: Planck's law should result. Let us observe that the procedure we here propose to follow is purely mathematical; we are dealing with a mere problem of probabilities, and we do not have to assume that the photons (whatever their nature may be) will actually collide and rebound like elastic spheres. Indeed, even in the treatment of an ordinary gas, the mechanical analogy of perfectly elastic billiard balls colliding with one another was utilized by Boltzmann only to afford a physical basis for his theory. At all events, when we apply the Boltzmann statistics to the photon gas in equilibrium with matter and when we take into consideration the volume h^3 of the cells, we do not obtain the correct law, namely, Planck's. Instead, we are led to a law which has the same form as Wien's law of radiation, but in which the constant h appears. The intrusion of this constant is only natural since we have assumed that the photons have energy $h\nu$, and that the cells have the volume h^3. The mere presence of h in the present law does not, however, make our results correct, for the general form of Wien's law differs from Planck's.

Thus, when we treat the equilibrium problem by viewing the radiation as formed of electromagnetic waves for which the equipartition of energy.holds, we obtain Rayleigh's law; and when we treat the radiation as corpuscular on the basis of the quantum assumptions and then apply the Boltzmann statistics, Wien's law is obtained. Inasmuch as the correct law (Planck's) is a compromise between the two former ones, the results just mentioned were at one time thought to establish the dual nature of light; part wave and part corpuscle. Though the dual nature of light is recognized today, the aforementioned argument is of doubtful value in establishing this dualism. The fact is that when the correct statistics is utilized, Planck's law is obtained whether we choose to view radiation as wave-like or as corpuscular; and we do not have to suppose that the radiation manifests both appearances.

Bose's Statistics for Photons—We have just seen that the Boltzmann statistics, when applied to photons, furnishes Wien's radiation law and hence is incapable of yielding correct results. Nevertheless, the general idea behind the previous attempts cannot be entirely erroneous, for Planck's radiation law, which is the correct one, is similar to Wien's law and indeed passes over into it for high frequencies and low temperatures.* It would thus appear that our failure to obtain Planck's law must be ascribed to some lack of refinement in the statistics utilized rather than to the main ideas that have guided our investigations. Bose therefore retained Einstein's assumption that the photons have momentum $\frac{h\nu}{c}$ and energy $h\nu$. He also accepted Planck's assumption that the 6-dimensional phase space should be sub-divided into cells of volume h^3. But Bose rejected the Boltzmann statistics and formulated in its place a new statistics which he devised for the express purpose of securing Planck's law of radiation. We must take into consideration the semi-empirical nature of Bose's attempt, for otherwise his statistics would appear arbitrary. With the advent of quantum mechanics, the theoretical basis of Bose's new statistics became clear; but in 1924, when Bose communicated his paper to Einstein, quantum mechanics had not yet been invented and the only justification for the new statistics was that it led to Planck's law of radiation.

Let us revert to the 6-dimensional phase space with its cells of volume h^3. A point of coordinates x, y, z, p_x, p_y, p_z in this phase space represents a photon situated at the point x, y, z of ordinary space and having components of momentum defined by p_x, p_y, p_z. Suppose, then, we start with a distribution of phase points determined by the presence of a_1 phase points in the first cell, a_2 in the second cell, a_3 in the third, and so on. The exact positions of the phase points in their cells are disregarded as usual. Let us interchange the phase points in all manner of ways. After each interchange, the phase points in at least two of the cells cannot be the same phase points that were present before the interchange was performed. From this standpoint, therefore, the interchange has generated a change in the distribution of the phase points. On the other hand, the interchange does not modify the numbers a_1, a_2, a_3 . . . . .

---

* More generally, Planck's law passes over into Wien's if we assume that the constant h is decreased indefinitely in value. This implies that the two laws tend to coincide when the limiting conditions of the correspondence principle are neared.

of phase points in the various cells, so that from the latter standpoint no change in distribution has occurred. Thus, the question of deciding whether or not an interchange of phase points constitutes a change which has physical significance is equivalent to the problem of deciding whether or not the phase points (or rather the photons which the phase points represent) are to be regarded as physically distinct individuals.

It is here that the statistics of Boltzmann and of Bose differ. In the Boltzmann statistics the phase points were treated as distinct individuals, and hence an interchange of phase points was regarded as physically significant: each interchange was held to give rise to a new microscopic state. The different microscopic states obtained in this way were, however, regarded as indistinguishable from the macroscopic standpoint; and so Boltzmann assumed that any one of the microscopic states characterized by the same numbers a_1, a_2, a_3 . . . . . of phase points in the various cells would represent the same macroscopic state.

Bose does not proceed in this way. The essence of his statistics is to deny any individuality to the phase points, so that interchanges of phase points situated in different cells constitute no changes at all. Even in theory the phase points cannot be tagged. As a result, the different microscopic states of Boltzmann, in which the numbers a_1, a_2, a_3 . . . . . of phase points in the various cells remain the same, are viewed by Bose as forming a single microscopic state. Thus, we may say: A microscopic state of Bose is determined by the numbers a_1, a_2, a_3 . . . . . of phase points in the various cells, regardless of the individualities of the phase points. Obviously, a microscopic state for Bose coincides with a macroscopic state for Boltzmann. The novelty thus far is to deny all individuality, even in theory, to the phase points and hence to the photons. On the other hand, both statistics recognize that if the numbers of phase points in the cells are modified, a new microscopic distribution arises.

We must now consider the probabilities of the various microscopic states in the Bose statistics. The Boltzmann statistics was based on a mechanical model, and hence we could derive the probabilities of the microscopic states from the laws of dynamics (supplemented by the ergodic hypothesis). We remember that all the microscopic states were found to be equally probable. But the Bose statistics is for the present a mere mathematical scheme, and we cannot appeal to the dynamical laws to obtain any insight into the probabilities. It thus appears impossible to establish these probabilities. However this may be, Bose postulated the equiprobability of the microscopic states. Bose's ideas are justified by results, for, as we shall see, his statistics, supplemented by the assumption of equiprobability, leads to Planck's law of radiation.

Thus far we have discussed Bose's microscopic states. As these states are equally probable, they furnish no immediate means of determining the state of maximum probability for the aggregate of particles, *i.e.*, the state of statistical equilibrium. We must therefore proceed as we did in the Boltzmann statistics and introduce macroscopic states of unequal probability; the most probable of these macroscopic states will then be the state of statistical equilibrium. Of course the macroscopic states of Bose cannot possibly be the same as those of Boltzmann, for we have seen that a macroscopic state in the Boltzmann sense is what Bose calls a microscopic one. Bose takes for his macroscopic states the energy states to which we referred in the second method of treating the Boltzmann statistics. Thus, a macroscopic state for Bose is an energy state in which there are

$$(25) \quad \begin{cases} n_i \text{ particles distributed in one way or another among} \\ \text{the } Z_i \text{ cells of the energy layer } \varepsilon_i \ (i=0,1,2,\ldots). \end{cases}$$

The difference between the Bose and the Boltzmann statistics now appears more clearly. In both statistics, a given energy state is illustrated by any one of those distinct microscopic states in which the total number of particles in each energy layer remains unchanged. The two statistics differ only in their definitions of the microscopic states that must be treated as distinct, for in the Bose statistics we shall have to view as identical all those distinct microscopic states of Boltzmann which differ only by interchanges of particles. The net result is that a Bose energy state contains fewer microscopic states than does the corresponding energy state of Boltzmann. In the Boltzmann statistics, the number of distinct microscopic states corresponding to a given energy state was defined by P (see (23)). In the Bose statistics, the corresponding number is defined by the smaller magnitude

(26)

$$P_{Bose} = \frac{(Z_1+n_1-1)!}{n_1!(Z_1-1)!} \times \frac{(Z_2+n_2-1)!}{n_2!(Z_2-1)!} \times \frac{(Z_3+n_3-1)!}{(n_3!(Z_3-1)!} \times \cdots$$

Since the various microscopic states of Bose have been assumed equally probable, the probability of an energy state is proportional to the number of different microscopic states contained in this energy state. For example, the probability of the energy state defined by (25) is proportional to (26). The state of statistical equilibrium of the photon gas, being the most probable energy state, will be determined by that state for which (26) is a maximum. In the determination of an energy state and of the state of statistical equilibrium in particular, it is necessary to prescribe the total energy of the particles or photons, for this total energy inevitably affects the energies which the individual photons will have. On the other hand, as explained in the note on page 917, the presence of matter in Planck's enclosure prevents us from stipulating that the total number of photons is fixed.

We may now easily understand how the law of radiation is obtained. We calculate the number n_i of photons which have energies comprised in the range from ε_i to $\varepsilon_i + d\varepsilon$ ($i = 0, 1, 2 \ldots\ldots$) when the state of statistical equilibrium is attained in the enclosure. Since the energy ε_i of a photon is $h\nu_i$ (where ν_i is its frequency), we obtain the number n_i of photons having frequencies between ν_i and $\nu_i + d\nu$. The total energy of the radiations of frequencies between ν_i and $\nu_i + d\nu$ present in the enclosure is then given by the sum of the energies of the corresponding photons, i.e., by $n_i h\nu_i$. The energy density of this radiation ν_i (per unit volume) is deduced by dividing the foregoing total energy by the volume of the enclosure. In this way we can determine the densities of the energies of the various radiations in the enclosure at any given temperature T. This is precisely the information which the radiation law aims to furnish. The radiation law thus obtained by Bose coincides with Planck's law, as required.

Bose defines the entropy of the photon gas in the state of statistical equilibrium by the same formula as was used in (24), i.e., by

$$(27) \qquad S = k \log P_{Bose,}*$$

where P_{Bose} stands for the maximum value of (26). It will be noted that the new statistics furnishes no information on the spatial positions of the photons or on the directions in which they are moving. It merely determines the ranges of their energy (or frequency). This situation occurs because Bose starts from the probability P_{Bose} of an *energy state*. The same situation was noted in connection with the second method of treating the Boltzmann statistics.

The only argument in favor of Bose's statistics at the present stage is that it leads to the correct law of radiation. Not until we examine the implications of wave mechanics can any theoretical justification be obtained. The new statistics is exceedingly strange, and it seems to be incompatible with any familiar form of representation by means of

---

* This formula implies that, just as in the Boltzmann statistics, we are assuming P_{Bose} to define the probability itself and not merely a number proportional thereto.

particles. The difficulty is due to the lack of individuality we have ascribed to the photons. To understand this point let us suppose that we toss two coins simultaneously. We assume that for either coin the probability of heads or tails is the same. Let us represent heads by + and tails by −. When we toss the coins, the four different situations illustrated in the table may arise:

(a) + +

(b) − −

(c) + −

(d) − +

The probability of each one of these situations is ¼.

The four situations here considered may, for the purpose of discussion, be assimilated to four microscopic states in the Boltzmann sense. In accordance with this simile, the different macroscopic states of Boltzmann are three in number and are represented by (a), (b), and by (c) or (d). The last two microscopic states define a single macroscopic state, because from the macroscopic viewpoint they are indistinguishable. Now, since all the microscopic states are equally probable, the last macroscopic state, being realized by either one of two microscopic states, must be twice as probable as is each of the first two macroscopic states. In short, the situation in which one coin shows heads and the other shows tails is twice as probable as the situations in which heads appears for both coins or tails appears for both. This conclusion does not seem to be dependent on our ability to distinguish the two coins, for even if they were so nearly alike as to be indistinguishable under the microscope, our former results would still be valid.

In spite of the seemingly obvious nature of this conclusion, it is incompatible with Bose's statistics. According to Bose, we must view the situations (c) and (d) as identical and as representing only one microscopic state, which has the same probability as (a) and as (b). This would imply that when two coins are tossed, there is the same probability of heads appearing for one coin and tails for the other as there is of heads appearing for both coins (or of tails appearing for both). The difficulty might be overcome by supposing that when one coin shows heads, it acts so as to decrease the probability of the other coin showing tails. The foregoing illustration is intended only in a figurative sense, but it does illustrate the paradoxical consequences of Bose's statistics, at least when we apply this statistics to particles. An interpretation appears less remote if we abandon the idea of particles and assume waves, so that a photon is everywhere at the same time, so to speak.

Einstein's Gas Theory—The strangeness of the Bose statistics must not be attributed to its connection with mysterious photons. As we shall soon see, Bose's statistics is also believed to be the correct one for ordinary material gases. We have no reason to be surprised at this development, for we know that the seemingly contradictory wave-like and corpuscular properties of the mysterious photons are likewise characteristic of material particles.

Einstein therefore applied the Bose statistics to an ordinary monoatomic gas. He also retained Planck's assumption that the 6-dimensional cells have a volume h^3. The only difference between Einstein's treatment of a gas and the Boltzmann-Planck treatment is that Einstein utilizes Bose's probability P_{Bose}, given by (26), in place of Boltzmann's probability (23). Following the usual method for gases, we must consider the various energy states which involve the same total number N of molecules and the same total energy E, and then select from this aggregate of energy states the one which is the most probable. This particular state is that which makes the probability P_{Bose} of (26) a maximum; it defines the state of statistical equilibrium.

Calculation shows that the state of statistical equilibrium is characterized by the following features:

Calling n_i the number of molecules which in the state of statistical equilibrium have energies between ε_i and $\varepsilon_i + d\varepsilon$, we find that

$$(28) \qquad n_i = \frac{Z_i}{Ae^{\frac{\varepsilon_i}{kT}} - 1}.$$

In this expression k is the familiar gas constant, T the absolute temperature, and Z_i the number of cells of the ith energy layer, *i.e.*, the number of cells having energies between ε_i and $\varepsilon_i + d\varepsilon$.

The number Z_i is found to be

$$(29) \qquad Z_i = \frac{4\pi m V}{h^3}\sqrt{2m}\ \sqrt{\varepsilon_i}\ d\varepsilon,$$

where V is the volume of the enclosure, and m the mass of a molecule.

Finally, let us consider the magnitude A in (28). This magnitude is furnished by a complicated expression involving the temperature, the volume, and the mass of a molecule. When the temperature is not too low, and the number of molecules per unit volume (particle density) is not too great, and the mass m of the molecules not too small, the value of A increases to such an extent that it is permissible to disregard -1 in the denominator of (28). In this case the expression of A is found to approximate to

$$(30) \qquad A = \frac{V}{Nh^3}(2\pi m k T)^{\frac{3}{2}},$$

where N is the total number of molecules.

Had we followed the classical Boltzmann statistics, we should have obtained in place of (28) the expression

$$(31) \qquad n_i = \frac{2N\sqrt{\varepsilon_i}}{kT\sqrt{\pi kT}}\ \frac{1}{e^{\frac{\varepsilon_i}{kT}}}d\varepsilon.$$

It was this value (31) that defined the state of statistical equilibrium in the Boltzmann statistics and which entailed the theorem of equipartition of energy and the constancy of the specific heat.

We have said that when the temperature is not too low and the particle density $\frac{N}{V}$ of the gas not too great, and the mass of the gas particles not too small, the value of A is given by (30). If we replace A by this value in (28) and also take into account the value of Z_i, we find that (28) coincides with the classical expression (31). We conclude that the foregoing restrictions affecting the temperature and the particle density constitute the limiting conditions under which Einstein's gas theory passes over into Boltzmann's classical theory. Calculation shows that the limiting conditions are realized for the gases with which we deal in the laboratory at ordinary pressures and temperatures. Hence there is no conflict between the new statistics of gases and the Boltzmann statistics under those conditions where the latter statistics is known to give correct results. This of course might have been anticipated since the new statistics must be in agreement with experiment.

On the other hand, considerable discrepancies between the two statistics occur when the limiting conditions are not realized, e.g., when the temperature is extremely low, or when the pressure and hence the particle density become enormous. In such cases the equation of state of the gas differs considerably from the classical equation

$$(32) \qquad PV = RT.$$

This change in the form of the equation of state constitutes the phenomenon of gas degeneration. The new equation of state shows that when the gas is in a degenerate condition, it will be much more compressible than Boyle's law would indicate, the departure from Boyle's law increasing in importance with a lowering of the temperature or an increase of the pressure. Furthermore, the specific heat of the gas is seen to vanish at the absolute zero.

Finally, we have to examine what information the new statistics furnishes on the entropy. At the time the new statistics was formulated, the most accurate expression for the entropy of a monoatomic gas was the Sackur-Tetrode formula (16). Einstein, by applying the new statistics and by utilizing the formula $S = k \log P_{Bose}$, obtained an expression of the entropy which differed from the Sackur-Tetrode formula and which vanished at the temperature of the absolute zero. Einstein showed, however, that the new expression of the entropy passed over into the Sackur-Tetrode formula at ordinary temperature and pressure. This is precisely what might have been expected, for the Sackur formula has been verified by experiment under these conditions. At the same time we see that the Sackur formula is only approximate.

It will be observed from the approximate formula (30) that the quantity A would always be very great in a world where Planck's constant h happened to be infinitesimal (as was thought to be the case in classical science). Consequently, in such a world there would be no gas degeneration and the new statistics would be indistinguishable from the classical statistics of Maxwell and Boltzmann. This circumstance shows that the gas degeneration, which the new statistics requires, is essentially a quantum manifestation. As such, it may be referred to as "quantum degeneracy." Inasmuch as gas degeneration is an immediate consequence of the Third Principle of Thermodynamics, we see once again how intimate is the connection between the quantum theory and the Third Principle. At the same time the gradual merging of the new statistics into the classical one, when h is assumed to decrease, illustrates a feature ever recurrent in the quantum theory; namely, the quantum theory passes over into classical theory as the limiting conditions of the correspondence principle are neared.

The Fermi Statistics—Shortly after Bose had formulated his statistics, Fermi suggested an alternative one. The only difference between the two statistics is that some of the microscopic states that appear in Bose's statistics are ruled out as impossible by Fermi. But the interest of Fermi's statistics lies more particularly in the method whereby it was obtained. Fermi was guided by Pauli's exclusion principle, according to which no two electrons in an atom can have the same four quantum numbers n_i, l_i, j_i, m_{l_i}. Pauli's principle was arrived at through empirical considerations, and there is no obvious reason why it should be valid, even in a modified form, when we are no longer dealing with the electrons in an atom. Fermi, however, was of the opinion that the exclusion principle might have a more general significance, and that it might be valid for the particles of a gas.

An obvious objection to Fermi's views is that the motions of the particles of a gas exhibit no periodicity, and that as a result their motions cannot be associated with quantum numbers. Fermi overcame this difficulty by assuming that the gas particles were acted upon by an elastic force directed towards a fixed point. The vibratory motions set up by the force introduced periodicities and thereby justified the allocation of three quantum numbers to the translational motions of the particles. Fermi then supposed that no two particles which occupied approximately the same position * could be associated with the same three quantum numbers. This assumption implied that no two particles which occupied approximately the same position could have the same momentum, or velocity (in direction and in magnitude). From the standpoint of the phase space, Fermi's assumption requires that a cell should never contain more than one phase point. In Bose's statistics, on the other hand, a cell could contain any number of points. Apart from this difference, the statistics of Fermi and of Bose are the same.

At first sight the exclusion principle utilized by Fermi does not appear to be the same as Pauli's, for whereas in Fermi's principle only three quantum numbers are involved, in Pauli's there are four quantum numbers. This difference is, however, of minor importance, because if we assume that the particles have a spin, like the electrons in an atom, Fermi's exclusion principle also deals with four quantum numbers. Besides, when the problem is investigated by the methods of wave mechanics, the two exclusion principles are seen to be the same. Fermi's introduction of a hypothetical restoring force is obviously artificial. This defect will be overcome when we treat the problem by wave mechanics, so it need not detain us any further. We shall therefore examine the mathematical results which follow from Fermi's assumptions, leaving in abeyance for the present the question of deciding whether these assumptions are justified.

As we mentioned previously, the only difference between the statistics of Bose and of Fermi is that in the latter a cell can contain one phase

---

* The vagueness of this statement is overcome when the cells are introduced.

point at most. As before, the cells have the volume h^3, and the problem is to calculate the number of different ways in which N phase points can be distributed among the cells when interchanges of phase points are excluded as irrelevant, and when one phase point at most may be situated in a cell. If we consider an energy state in which there are n_i phase points in the Z_i cells associated with energies between ε_i and $\varepsilon_i + d\varepsilon$ ($i = 0, 1, 2, \ldots$), we find that the number of different microscopic distributions which define the same energy state is given by a number, P_{Fermi}, which is smaller than the corresponding number, P_{Bose}, of the Bose statistics. (This is necessarily so since Fermi excludes many of the microscopic distributions accepted by Bose.)

Calculation gives

(33)

$$P_{Fermi} = \frac{Z_1!}{n_1!(Z_1 - n_1)!} \times \frac{Z_2!}{n_2!(Z_2 - n_2)!} \times \frac{Z_3!}{n_3!(Z_3 - n_3)!} \times \ldots \ldots$$

We assume, as before, that all microscopic states have the same probability, so that the number (33) is proportional to the probability of the corresponding energy state. Following the usual procedure, we identify P_{Fermi} with the probability itself. In short, except for the change from the probability (26) of Bose to the new probability (33), we may proceed exactly as we did previously.

For a gas, Fermi obtains results very similar to Einstein's. Thus his formula for the state of statistical equilibrium is practically the same as the formula (28) given by Einstein; the only difference is that, in the denominator of (28), we must replace -1 by $+1$. The Fermi statistics also yields the same results as the classical statistics under the same limiting conditions of average temperatures, densities, and pressures, mentioned in connection with Einstein's treatment. In particular, Fermi's expression of the entropy passes over into Sackur's. As in the Bose-Einstein statistics, degeneracy (i.e., the modification of the equation of state) occurs when the limiting conditions are far from being realized.

Of course, in view of the difference between the statistics of Bose and of Fermi, the numerical results to which they lead cannot be quite the same. For instance, a perfect monoatomic gas, according to the Bose-Einstein statistics, is more compressible, and, according to the Fermi statistics, is less compressible than is required by the classical gas law.

Another peculiarity of Fermi's statistics, which is not present in the Bose-Einstein statistics, is that at the absolute zero of temperature a gas must still exert some pressure and retain some energy. The reason why the energy cannot vanish in Fermi's statistics may easily be understood. Thus there can be only one phase point in each cell, and hence the cells of zero energy cannot possibly contain all the phase points. Some phase points will always be situated in cells which have non-vanishing energies; and this implies that some of the molecules must have non-vanishing energies, so that the total energy of the gas cannot vanish.

The differences in the statistics of Boltzmann, Bose, and Fermi may be summarized as follows:

In the Boltzmann statistics interchanges of phase points are supposed to generate distinct microscopic states, whereas in the new statistics such interchanges do not count. Bose's statistics, like Boltzmann's, assumes that there may be more than one phase point in a cell; the Fermi statistics excludes this possibility.

The similarities in the results furnished by the two new statistics make it difficult to decide which of them should be applied in any particular case. For photons, the matter is settled beyond dispute because Bose's statistics alone yields Planck's law—Fermi's does not. On the other hand, Fermi's statistics is presumably the correct one in the case of the electron gas that exists in a metal, because the electrons in the atom are controlled by Pauli's exclusion principle, with which Fermi's statistics is closely related. In the case of ordinary material gases the choice between the two statistics is less certain. The consensus of opinion, however, is that the Bose-Einstein statistics must be applied to an aggregate of neutral atoms, and more generally to an aggregate of particles whose electric charges are any even multiple of the unit e. Thus, photons and neutral molecules, having no net charge, and alpha particles, which have a charge $+2e$, will be treated in accordance with the Bose statistics. If the charge is an odd multiple of the unit e, the Fermi statistics must be taken (this occurs for a gas formed of electrons or of protons).

In the previous paragraph we stated that the Bose statistics presumably applies to a material gas. It should be observed, however, that our discussion of the new statistics has been restricted to perfect gases, i.e., to gases whose molecules are treated as points which exert no forces on one another. This restriction prevents the results which we have just obtained for perfect gases from being of much use when we are dealing with real material gases, such as hydrogen or helium. It is true that, at ordinary temperatures and pressures, real gases approximate to perfect gases; but we have seen that the novel properties which perfect gases should exhibit, according to the new statistics, become noticeable only at extremely low temperatures and high pressures, and therefore under conditions for which the real gases are no longer even approximately perfect. For these reasons we cannot test by direct experiment the validity of the new statistics in its appication to real gases. Fortunately, difficulties of this sort are not encountered in connection with the photon gas or the electron gas. We have already seen that the Bose statistics furnishes correct results for the photon gas, and we shall now illustrate an application of the Fermi statistics to the electron gas.

Sommerfeld's Electronic Theory of Metals—In Chapter XXII we mentioned that the application of the classical kinetic theory to the atoms vibrating in a crystal failed to account for the phenomenon of degeneracy (*i.e.*, the progressive decrease in the specific heat, which accompanies a lowering of the temperature). In Chapter XXIV we saw that Einstein overcame this difficulty by quantizing the vibratory motions of the atoms. In Einstein's treatment, as also in the classical one, no attention was paid to the free electrons, forming an electron gas, which should be present in a metal crystal; and we explained on page 424 that a proper treatment would require that the degrees of freedom of these electrons be taken into consideration. Since the degrees of freedom of the atoms alone sufficed to account for the observed specific heat, it was recognized that the electrons in no wise participated in the partition of energy, even at ordinary temperatures. Presumably some quantum manifestation prevented their degrees of freedom from becoming effective. The earlier quantum theory, however, which Einstein had applied with success to the vibratory motions of the atoms, afforded no means of imposing quantum restrictions on the translational motions of the electrons. For this reason the problem of the specific heats did not receive a satisfactory solution.

The situation is clarified by the new statistics. Dealing as we are with an electron gas, we must apply Fermi's statistics. Sommerfeld followed this course. Let us first recall that in the Bose-Einstein statistics a magnitude A appeared in the formula (28). We mentioned that when conditions were such that this magnitude was large, Bose s statistics passed over into the classical one. In this case Bose's statistics led to the equipartition of energy and to the classical value for the specific heat of the gas. On the other hand, if conditions were such that A was very small, the influence of the quantum manifestations became pronounced and degeneracy appeared: the specific heat of the gas tended to vanish. Although our discussion of the constant A was given in connection with Bose's statistics, it plays exactly the same rôle in Fermi's, and its value under the limiting conditions is given as before by (30). Consequently, all that we said on the subject of degeneracy in Bose's statistics applies in equal measure to the statistics of Fermi, which we are here considering.

Now, the expression (30)* shows that even for ordinary values of the temperature T, the value of A will be small when the mass m of the particles is exceedingly small or when the particle-density $\frac{N}{V}$ is high. For ordinary gases, m is always relatively large, but for the electron gas m is exceedingly small, and at the same time the particle density is extremely high. Consequently, even at ordinary temperatures, the electron gas will be in a condition of quasi-complete degeneracy, with the result that its specific heat will be exceedingly small. This implies that the degrees of freedom of the electrons will absorb practically no heat energy when heat is conveyed to the metal; all the energy goes to the degrees of freedom of the atoms, as though the electrons were inexistent. The difficulty concerning the specific heats of metals in the electronic theory is thus explained.

The case of the electron gas is of interest because it furnishes an illustration of practically complete degeneracy. To obtain pronounced degeneracy with ordinary gases, we should have to operate near the absolute zero of temperature or under tremendous pressures which are beyond our present reach. With the electron gas, ordinary laboratory conditions suffice.

The New Statistics and Wave Mechanics—The first step in connecting the kinetic theory with the quantum theory was undertaken by Planck when he postulated that the 6-dimensional cells of the phase space should have the common volume h^3. We mentioned that Planck's procedure had no theoretical basis, and that quite generally it was difficult to see any possible connection between the quantum restrictions, which were thought to apply only to periodic motions, and the molecular motions, which are not periodic. Wave mechanics solves this difficulty by introducing periodicity through its waves.

According to de Broglie's original theory, a particle, of total energy W and of momentum p, is associated with a wave of frequency ν and wave length λ, where

$$(36) \qquad \nu = \frac{W}{h} \text{ and } \lambda = \frac{h}{p}.$$

---

* Although this expression of A is only the value to which A tends under the limiting conditions, it affords a general indication of the manner in which A depends on the volume, the temperature, and the mass of a particle.

The essence of wave mechanics is to establish the evolution of the wave by the methods of wave optics and to derive therefrom the probable behavior of the associated particle or particles. We recall that by utilizing the theory of wave mechanics, we were able to obtain a satisfactory interpretation of the experiments on the diffraction of electrons. In some situations, however, the methods of classical mechanics suffice to determine the particle's motion; in such cases we need not consider the wave For instance, if a particle is moving in empty space and no field of force is applied, the particle follows a straight course in any direction and with any velocity. De Broglie's association of a wave with the particle does not affect this classical conclusion perceptibly, so that in this instance the wave may be disregarded.

But suppose the particle is moving in an enclosure from the walls of which it may rebound without loss of energy. The situation is now entirely different, and we are compelled to take the wave into consideration. The associated wave is reflected back and forth from the walls of the enclosure, and the superpositions that ensue produce interference effects. Situations of this kind were discussed in Chapter XXXI. There we examined the interference effects due to the superposition of waves in elastic media. We saw that when boundary conditions were imposed, standing waves were formed and that these standing waves (also called the normal modes of vibration) were characterized by privileged wave lengths and frequencies. We also mentioned that the various waves could occur singly or simultaneously. Now, the generation of standing waves under the conditions specified illustrates a general wave phenomenon. This phenomenon arises with air waves in a rigid box and with electromagnetic waves in an enclosure which has reflecting surfaces; it must also be expected when de Broglie waves are considered.

Let us, then, revert to a de Broglie wave in an enclosure. A specific set of standing waves having definite wave lengths and frequencies will alone be possible. Consequently, if we imagine a wave which does not have one of the privileged values for the wave length and for the frequency, we may be certain that after successive reflections this wave will destroy itself. If now we apply Born's assumptions, the information obtained for the waves is extended to the associated particle. Thus let us assume that an impossible wave is generated. We know that it will soon destroy itself. In the particle picture, this means that a particle moving in the enclosure with the momentum and the energy corresponding to the wave is a physical impossibility. Next we assume that a permissible wave is generated, so that a standing wave is formed. According to Born's assumptions, the particle will most probably be found at those points where the intensity of the wave is the greatest; such points are distributed over appropriate surfaces. Furthermore, the momentum p and the energy W of the particle will be connected with the wave length λ and with the frequency ν of the wave by the fundamental formulae (36). Instead of supposing that only one of the standing waves is present, we may suppose that all the possible waves are vibrating simultaneously. In this event the momentum and the energy of the particle will be those corresponding to one of the waves. We may also suppose that a swarm of non-interacting particles is associated with the waves. The various particles will then have the momenta and the energies corresponding to the different possible waves.

To summarize these results: The particles of a swarm contained in a closed volume can have only certain momenta and energies; these permissible values of the momentum and energy form discrete sets, and they are derived from the wave lengths and frequencies of the possible standing waves, or normal modes of vibration. Thus, the quantization of the molecular motions, which Planck postulated when he introduced his finite cells, is seen to be an immediate consequence of the wave theory. Incidentally, let us observe that the quantization of the molecular motions occurs only in an enclosure, for only then do normal modes of vibration come into being. In an unlimited region of space devoid of obstacles, the motions would not be quantized.

We have yet to justify Planck's crediting of the precise volume h^3 to his cells. The first step in this direction consists in discovering the wave counterpart of a cell. As we shall see, each standing wave in the enclosure plays the part of a cell. We recall that a phase point situated in a given cell represents a particle whose spatial position and momentum components are defined by the coordinates x, y, z, p_x, p_y, p_z of the cell's centre. Thus, the cell determines the position, the total momentum p of the corresponding particle, and also its energy. Next let us pass to the wave picture. We shall suppose that only one of the standing waves of the enclosure is present. A particle associated with this wave, of wave length λ and of frequency ν, has a total momentum $p = \dfrac{h}{\lambda}$ and an energy $h\nu$. The total momentum and the energy of the particle are thus determined by the wave. As for the particle itself, it will most probably be found on the surfaces where the intensity of the wave is greatest. To some extent, therefore, the wave determines a position for the particle. It is true that this position is vague whereas the momentum and energy appear to be defined with precision; but some vagueness must be expected in any case in view of the uncertainty relations.

The foregoing considerations show that when account is taken of the unavoidable difference between a wave picture and a particle picture, a standing wave in the enclosure may be viewed as the wave counterpart of a cell in the phase space.

Let us now see how the wave picture can furnish the volumes of these cells. A formula established by Jeans, long before the present considerations arose, shows that the number of standing waves of wave lengths between λ and $\lambda + d\lambda$ that can exist in an enclosure of volume V is given by

$$(37) \qquad \frac{4\pi V}{\lambda^4}\, d\lambda.*$$

Jeans's formula applies to longitudinal waves. For transverse waves, *e.g.*, electromagnetic ones, the number (37) must be doubled on account of the two kinds of polarization that may arise.

Inasmuch as de Broglie waves are assumed to be determined by a scalar ψ and not by a vector, they exhibit no polarization and thus behave like longitudinal waves. The formula (37) may therefore be taken to define the number of distinct standing de Broglie waves, of wave lengths between λ and $\lambda + d\lambda$, that can exist in an enclosure of volume V. We may now readily derive the number of waves which are associated with momenta between p and $p + dp$ or with energies between W and $W + dW$. We shall consider the momenta. If in (37) we replace λ by its value (36), we see that

$$(39) \qquad \frac{4\pi V}{h^3} p^2 dp$$

is the number of de Broglie waves associated with particles having momenta between p and $p + dp$†.

---

* The number of standing waves having frequencies between ν and $\nu + d\nu$ is

$$(38) \qquad \frac{4\pi V \nu^2}{V'^2 U} d\nu,$$

where V' is the wave velocity and U the group velocity.

† When we neglect the relativistic refinements, we find that the number of de Broglie waves associated with energies between W_i *and* $W_i + dW$ (and hence, if the rest-energy $m_0 c^2$ is disregarded, associated with energies ϵ_i and $\epsilon_i + d\epsilon_i$, where $\epsilon_i = W_i - m_0 c^2$) is

$$(40) \qquad \frac{4\pi m_0 V}{h^3} \sqrt{2m_0\epsilon_i}\ d\epsilon.$$

This expression coincides with (29), which gives the number of cells of volume h^3, corresponding to energies between ϵ_i and $\epsilon_i + d\epsilon$. It thus affords an alternative justification of Planck's assumption for the volume of the cells, *i.e.*, $\omega = h^3$.

On the other hand, if we consider the tiny cells in the phase space and assume that they have the volume ω, we find that the number of these cells associated with momenta between p and $p + dp$ is

(41)
$$\frac{4\pi V}{\omega} p^2 dp.$$

Now we have agreed to identify the standing waves with the cells of the phase space. Hence we must suppose that the numbers (39) and (41) are equal. The equating of (39) to (41) then yields

(42)
$$\omega = h^3.$$

Thus, Planck's assumption for the volumes of the cells is justified by the wave treatment.

The Various Statistics in the Wave Picture—The standing de Broglie waves that can exist in an enclosure may be viewed as the solution functions, or eigenfunctions, of Schrödinger's wave equation adapted to the peculiar boundary conditions of the problem. For example, let us suppose that the enclosure contains but one particle. Schrödinger's wave equation admits acceptable solutions only when the energy of the system (i.e., the energy of the particle) has any one among a discrete set of values—the eigenvalues. The corresponding solution functions are the eigenfunctions and they represent the amplitude distributions of the possible waves. The energies associated with these waves are the eigenvalues, and the frequencies of the waves are deduced from these energies.

Let us suppose that the Schrödinger equation has been solved, so that all the possible standing waves are known. We recall that we are assuming a single particle. If several waves of different frequencies $\nu_1, \nu_2, \ldots$ are present, we are dealing with a superposition of states. The particle is then associated with one or another of these waves, and its energy is determined by the particular wave with which it happens to be associated. We may also say that the phase point of the particle is situated in one or another of the corresponding Planck cells. If only one wave is present, say the nth wave ψ_n of frequency ν_n, the particle is necessarily associated with this wave, and hence has the energy $\varepsilon_n = h\nu_n$. We shall refer to this particle as the particle a; and to stress the fact that it is attached to the wave ψ_n, we shall represent the wave function by

(43)
$$\psi_n^a.$$

Thus, when the phase point of the particle a is in the nth cell, the corresponding wave picture is afforded by the wave function, or eigenfunction, ψ_n^a.

Similarly, if instead of the particle a, we were dealing with another particle b, and if the phase point of this particle were situated in the kth cell, the wave picture would be represented by the eigenfunction

$$(44) \qquad \psi_k^b.$$

Next, let us suppose that both particles a and b are present simultaneously in the enclosure, their phase points being situated in the nth and the kth cell respectively. We shall assume that the particles do not interact. The wave representation of the foregoing situation is given by the product of the two eigenfunctions ψ_n^a and ψ_k^b, i.e., by

$$(45) \qquad \psi_n^a \psi_k^b.$$

To secure correct results, we must suppose that the wave distribution (45) is represented in a 6-dimensional configuration space (3 dimensions for each particle).

The wave function (45) is an eigenfunction of Schrödinger's wave equation, the equation being here adapted to the case of two particles present in the enclosure. The corresponding value of the energy of the system is the sum $\varepsilon_n + \varepsilon_k$ of the energies of the two particles. This value of the energy is one of the eigenvalues of Schrödinger's equation for two particles, and the function (45) is an eigenfunction connected with this eigenvalue.

Let us exchange the two particles, so that the phase point of the particle a is now in the kth cell and the phase point of the particle b is in the nth cell. The particle a is now associated with the wave ψ_k and the particle b with the wave ψ_n. The eigenfunction which expresses the new situation is derived from (45) by an interchange of the letters a and b. We thus obtain the eigenfunction

$$(46) \qquad \psi_n^b \psi_k^a.$$

It is important to observe that the two eigenfunctions (45) and (46) are not the same. This point was explained in Chapter XXXV in our discussion of the helium atom.* On the other hand, the energy of the system has the same value in the two cases, so that two different eigenfunctions are connected with the same value of the energy. This implies

* See page 759.

that the system is degenerate, the degeneracy being of the same resonance type that was discussed in Chapter XXXV.

Owing to the existence of two different eigenfunctions associated with the same value of the energy, the most general eigenfunction, or wave function, is obtained by taking any linear combination (with constant coefficients) of the two functions (45) and (46). Two combinations are particularly simple; they are given by the symmetric eigenfunction

$$(47) \qquad \psi_n^a \psi_k^b + \psi_n^b \psi_k^a$$

and by the antisymmetric one

$$(48) \qquad \psi_n^a \psi_k^b - \psi_n^b \psi_k^a.^*$$

The last eigenfunction may be written as a determinant, *i.e.,*

$$(49) \qquad \begin{vmatrix} \psi_n^a & \psi_n^b \\ \psi_k^a & \psi_k^b \end{vmatrix}.$$

The eigenfunctions (45) and (46) indicate clearly the cells in which the particles are situated. But, the eigenfunctions (47) and (48) confront us with the difficulty we discussed in Chapter XXXV; namely, either particle appears to be associated with the two cells simultaneously. We shall not insist on this point further, for we have already examined it. Suffice it to say, it illustrates the usual difficulty we experience when we attempt to incorporate the concept of "particle" into a wave picture. Although the eigenfunctions (47) and (48) suggest that each of the two particles occupies two cells (or is associated with two waves), we can transcribe into the wave picture some of the features we encountered when we were discussing the particle representation. For instance, an exchange in the positions of the two particles in the phase space is expressed in the eigenfunctions (47) and (48) by an exchange in the positions of the letters a and b.

If we make this exchange in (47), the eigenfunction is reproduced without modification; and for this reason the eigenfunction (47) is called symmetric. The same exchange in the eigenfunction (48) leaves the absolute value of the function unaffected but reverses its sign. Accordingly this eigenfunction is called antisymmetric. Finally, the effect of the exchange on the eigenfunction (45) is to transform it into the different

---

* If the functions $\psi_n^a, \psi_k^b, \psi_n^b, \psi_k^a$ are normalized to 1, the solutions (47) and (48) are normalized when multiplied by $\dfrac{1}{\sqrt{2}}$.

eigenfunction (46); and conversely, (46) is transformed into (45). The eigenfunctions (45) and (46) are thus neither symmetric nor antisymmetric.

Let us extend these results to N particles, which we denote by $A_1, A_2 \ldots \ldots A_N$; and let $C_1, C_2 \ldots \ldots C_N$ represent N different cells. We assume that the phase point of the particle A_1 is in the cell C_1, the phase point of the particle A_2 in the cell C_2, and so on. To simplify the exposition, we shall drop all reference to phase points and speak of a particle as being in a cell. The eigenfunction which corresponds to the distribution of particles just defined is the analogue of (45) extended to N particles. It is expressed by

$$(50) \qquad \psi_{C_1}^{A_1} \psi_{C_2}^{A_2} \ldots \ldots \psi_{C_N}^{A_N}.$$

Since there are N particles, each of which has three degrees of freedom, the eigenfunction (50) must be represented in a configuration space of $3N$-dimensions.

Suppose now we interchange the particles in some definite way, obtaining thereby a new distribution of particles among the same cells. The eigenfunction which corresponds to this new distribution is the eigenfunction (50) in which the A's are subjected to the same interchanges that were performed on the particles. For instance, if the interchange of particles consists in setting the particle A_2 in the cell C_1, the particle A_N in the cell C_2, and the particle A_1 in the cell C_N (whereas the other particles are left undisturbed), the corresponding eigenfunction will be

$$(51) \qquad \psi_{C_1}^{A_2} \psi_{C_2}^{A_N} \ldots \ldots \psi_{C_N}^{A_1}.$$

Since our eigenfunctions are represented in a hyperspace, the eigenfunctions (50) and (51) are different functions. We conclude that any interchange in the distribution of the N particles among the N cells entails a change in the eigenfunction. In the present case there are $N!$ different distributions (in each of which one particle occupies one of the cells $C_1, C_2, \ldots \ldots C_N$). Hence there will be $N!$ different eigenfunctions of type (50), which we may derive from (50) by permuting the letters $A_1, A_2, \ldots \ldots A_N$ in all manner of ways. One of these eigenfunctions is (50) itself, and one of the others is represented by (51). We shall call these $N!$ eigenfunctions the fundamental eigenfunctions; they are neither symmetric nor antisymmetric.

From these $N!$ fundamental eigenfunctions many different varieties of eigenfunctions may be built. The fact is that any linear combination (with constant coefficients) of the $N!$ fundamental eigenfunctions is itself an eigenfunction. Among the linear combinations that may be constructed, two stand out prominently on account of their simplicity. A symmetric eigenfunction is obtained by adding the $N!$ fundamental ones. We may represent it by

(52)
$$\Sigma\psi_{C_1}^{A_1}\psi_{C_2}^{A_2}\ldots\ldots\psi_{C_N}^{A_N}.$$

An antisymmetric eigenfunction is obtained by adding and subtracting in an appropriate way the various eigenfunctions of type (50). It may be expressed by the determinant

(53)
$$\begin{vmatrix} \psi_{C_1}^{A_1} & \psi_{C_1}^{A_2} & \cdots & \psi_{C_1}^{A_N} \\ \psi_{C_2}^{A_1} & \psi_{C_2}^{A_2} & \cdots & \psi_{C_2}^{A_N} \\ \cdots & \cdots & \cdots & \cdots \\ \psi_{C_N}^{A_1} & \psi_{C_N}^{A_2} & \cdots & \psi_{C_N}^{A_N} \end{vmatrix}.$$

If we assume that the partial eigenfunctions ψ are normalized to 1, the eigenfunctions (52) and (53) will be normalized when we multiply them by $\frac{1}{\sqrt{N!}}$. The appropriateness of the names "symmetric" and "antisymmetric" for the eigenfunctions (53) and (54) will be understood presently.

Now, an eigenfunction gives the wave picture of a definite distribution of the phase points among the cells. As such, it determines a microscopic state of the gas. Two eigenfunctions that are different (or the same) define therefore two microscopic states that will have to be regarded as different (or identical). We conclude that the peculiarities of the eigenfunctions determine the peculiarities of the various microscopic states. Hence we shall be led to different conceptions of the microscopic states, and thereby to different statistics, according to whether we select one or another of the families of eigenfunctions. Let us first examine the fundamental eigenfunctions (50).

To obtain a better understanding of the eigenfunctions (50), we shall suppose that the particles we have called A_1, A_2, A_3 are in the same cell C_1 and the remaining $(N-3)$ particles in the cell C_N. The eigenfunction of type (50) which describes this distribution is

(54)
$$\overbrace{\psi_{C_1}^{A_1}\psi_{C_1}^{A_2}\psi_{C_1}^{A_3}}\overbrace{\psi_{C_N}^{A_4}\ldots\ldots\psi_{C_N}^{A_N}.}$$

All other eigenfunctions which correspond to 3 particles in the cell C_1 and the remaining $(N-3)$ in the cell C_N are obtained by permuting in all manner of ways the letters A_1, A_2, A_3 . . . . . A_N in (54). But, in contradistinction to what occurred in (50), the various permutations will not always yield different eigenfunctions. For instance, if in (54) we permute the particles A_1, A_2 and A_3 among themselves or the particles A_4, A_5 . . . . . A_N among themselves, the expression (54) obviously remains unchanged, so that the eigenfunction (54) is not modified. For a permutation in (54) to yield a different eigenfunction, the particles that are interchanged must belong to the two different cells C_1 and C_N. Calculation then shows that the total number of different eigenfunctions which correspond to 3 particles in the cell C_1 and $(N-3)$ in the cell C_N is

$$(55) \qquad\qquad \frac{N!}{3!(N-3)!}.$$

Since different microscopic states are determined by different eigenfunctions, we conclude that the interchanges of particles situated in the same cell do not furnish different microscopic distributions, and that, only when particles are exchanged between different cells (or when a particle is removed from a cell to a different one), does a new microscopic state arise. As a result, the number (55) defines the number of different microscopic states in each of which 3 particles are situated in the cell C_1 and $(N-3)$ in the cell C_N. Finally, if we assume that the different eigenfunctions have the same *a-priori* probability, we must extend the same equiprobability to the different microscopic states. The number (55) then defines the probability that 3 particles will be in the cell C_1 and $N-3$ in the cell C_N; hence (55) defines the probability of the corresponding macroscopic state. The statistics which we thus obtain is obviously the statistics of Boltzmann. Thus, when we select eigenfunctions of type (50), Boltzmann's statistics is the natural outcome.

Next we consider the symmetric eigenfunction (52). It is composed of a sum of terms, and whereas in the first term the particles A_1 and A_2 are associated with the cells C_1 and C_2 respectively, we should find that in the remaining terms the particles A_1 and A_2 are associated in turn with all the other cells. The same situation is apparent in the antisymmetric eigenfunction (53). We cannot therefore connect the particle A_1 with any particular one of the N cells; the particle seems to be situated in all the cells at the same time. We have already commented on this situation in connection with two particles; it illustrates the vagueness which often arises when we attempt to translate the wave picture into the particle one.

This situation is brought to light in another way when we interchange particles. If we exchange, say, the two particles A_1 and A_2, the symmetric eigenfunction (52) undergoes no change whatsoever. Hence we must suppose that an exchange of particles does not generate a new microscopic distribution. More generally, all permutations of the particles leave the symmetric eigenfunction unaffected. For this reason the eigenfunction is called symmetric. At the same time, since an interchange never generates a change in the eigenfunction and hence in the microscopic state, we must conclude that the particles cannot be differentiated and cannot be credited with any individuality.

The same condition holds for the antisymmetric eigenfunctions. Thus in (53), if we exchange A_1 and A_2, we are exchanging two columns of the determinant. This change reproduces the same determinant, or eigenfunction, except that its sign is reversed. However, a change of sign is trivial and does not modify the nature of the eigenfunction, so that here also interchanges of particles do not generate different microscopic states. Thus, the particles lose their identity as before. More generally, we find that all even permutations leave the determinant (53) unaltered, whereas odd permutations change its sign. To this feature is due the name antisymmetric given to the eigenfunctions of type (53). Symmetric and antisymmetric eigenfunctions will thus furnish a form of statistics in which all interchanges among particles will have to be disregarded as meaningless.

Let us examine more closely the difference between these two kinds of eigenfunctions. In the symmetric eigenfunction, we shall show that two or more particles may be situated in the same cell. For instance, in (52), if we replace C_2, C_3, ... C_N by C_1, we are implying that all the N particles are situated in the cell C_1. Now, the original eigenfunction is changed thereby into a new one, but since it does not vanish, we conclude that the situation contemplated is a possible one. Suppose we follow the same procedure with the antisymmetric eigenfunction (53). All the rows of the determinant become identical, and as a result the determinant, or the eigenfunction, vanishes. A vanishing eigenfunction implies that the situation contemplated is impossible. Hence the particles cannot all be in the same cell when antisymmetric eigenfunctions are selected. In point of fact, we cannot have more than one particle in the same cell; for if in the determinant (53) we replace, say, C_2 by C_1, the two top rows become identical, and so the determinant or eigenfunction vanishes. The net result is that, in the statistics derived from the antisymmetric eigenfunctions, a cell can contain only one particle or none.

We may now easily verify that the symmetric eigenfunctions lead to the statistics of Bose and that the antisymmetric ones yield the statistics of Fermi. To verify this statement, we recall that the symmetric eigenfunctions entail the identity of those microscopic states which are derived from one another by mere interchanges of particles. We conclude that the statistics which ensues from the use of the symmetric eigenfunctions will differ from the statistics of Boltzmann in that the foregoing microscopic states will be regarded as identical. But this is precisely the characteristic of the Bose statistics. Hence the symmetric eigenfunctions lead to Bose's statistics.

On the other hand, when we utilize the antisymmetric eigenfunctions, not only do we have to view all the foregoing microscopic states as identical, but we must also regard as impossible the presence of more than one particle in a cell. And these restrictions are precisely those embodied in the statistics of Fermi. Thus, the antisymmetric eigenfunctions lead to the Fermi statistics. In short the statistics of Boltzmann, of Bose, and of Fermi appear to be various possible statistics derived from the wave treatment.

Other points should be mentioned in connection with Fermi's statistics. In Fermi's original treatment, an elastic force was introduced for the purpose of quantizing the translational motions of the molecules. In the wave treatment, this unsatisfactory feature is obviated, for the statistics follows directly from our choice of the eigenfunctions. We have also seen that Fermi was guided to his statistics by Pauli's exclusion principle, which states that no two electrons in an atom can be associated with the same four quantum numbers. The existence of four quantum numbers is due to the spin of the electron which adds a fourth degree of freedom to the three translational ones. If, then, we wish to view the Fermi statistics as a mere consequence of Pauli's principle, we must suppose that the particles are spinning as well as undergoing translational motions. This supposition does not introduce any new assumption (at least in the case of electrons and protons), for we have seen that Dirac's relativistic wave equation entails the existence of a spin. The only complication caused by the spin is that the eigenfunctions utilized will have to incorporate this spin.

Summarizing the wave treatment of the various statistics, we conclude that when no interaction is assumed among the particles, the statistics of Boltzmann, Bose, and Fermi appear to be equally justified. In addition to these three statistics, others may be considered, for the symmetric and the antisymmetric eigenfunctions are only some among those that we might have constructed. The wave treatment does not therefore impose any particular form of statistics; it merely suggests different possibilities.

The wave treatment does, however, show that transitions cannot occur from the symmetric to the antisymmetric states, and vice-versa. We explained this restriction on page 766 in connection with the helium atom. In the present case it implies that the Bose and the Fermi statistics are mutually exclusive, so that if one of the two statistics is valid at the start, it will remain so throughout time. But in the final analysis, the choice of the statistics in any particular case is not dictated by the wave theory; it is imposed by considerations of an empirical nature. So far as can be determined, the Bose and the Fermi statistics are the only two that are possible. The Bose statistics applies when the particles carry an even multiple of the fundamental charge e (or no charge at all, $e.g.$, photons), whereas the Fermi statistics is the correct one when the charge of each particle is an odd multiple of the fundamental charge e.

Our conclusions up to this point are based on the assumption that the particles do not interact. If there are small mutual actions, we may proceed to calculate the eigenfunctions by the method of successive approximations, as was explained in connection with the helium atom (Chapter XXXV). In this event the eigenfunctions of type (50) and with them the Boltzmann statistics are ruled out as impossible. On the other hand, the eigenfunctions (52) and (53) are imposed as first approximations. Further approximations do not affect the characteristic features of these eigenfunctions, namely, their symmetric or antisymmetric properties. Since it is these properties which lead to the statistics of Bose and of Fermi, we conclude that in the event of interactions the last two statistics are the ones to be expected.

CHAPTER XLI

QUANTUM MECHANICS AND CAUSALITY

CLASSICAL physicists believed that the evolution of any physical system could be represented by a continuous chain of events causally related. But when Planck advanced his quantum theory of equilibrium radiation, science was confronted with a doctrine in which the succession of events was not continuous, and in which causal relations between successive events were not specifically asserted. The highly speculative nature of Planck's ideas prevented them from receiving immediate acceptance; but around the year 1907, the remarkable successes of Planck's theory in other realms of physics silenced all opposition, and from then on serious doubts were cast on the validity of rigorous causality in physical science. Poincaré, after his return from the Solvay Congress of 1911, gave expression to these doubts in the following passage:

"Newton realized (or thought he realized—today we are beginning to wonder) that the state of a mobile system, or more generally of the universe, depended only on its immediately preceding state; and that all change in Nature occurred continuously. . . . Well, it is this fundamental belief that is being questioned today."

The doubts alluded to by Poincaré did not, however, stimulate much discussion at the time, for the quantum theory was still in its infancy, and the prevalent feeling was that a deeper study of quantum processes might eventually reveal an underlying continuous causal scheme.

In 1912 and in the following years Bohr developed his quantum theory of the atom. In spite of its many triumphs, this theory was unable to foretell the durations of the various excited states, and to predict therefore at what exact instant a given energy drop should occur and the corresponding radiation be emitted. With the failure of Bohr's theory to give information on individual processes, statistical methods and hence probability considerations appeared to offer the only avenue of advance. The systematic introduction of probability coefficients in the study of Bohr's atom was undertaken by Einstein in 1917 (see Chapter XXVIII).

Now, the intrusion of probability coefficients in a theory of mathematical physics does not necessarily connote the absence of underlying rigorous laws. For instance, the classical kinetic theory of gases is a statistical theory, and, as such, it appeals to probability considerations. Nevertheless the theory is based on the assumption that the rigorous laws of mechanics control the motions of the gas molecules. Indeed, the only reason for introducing probabilities in the study of gases is to obviate the insuperable mathematical difficulties which a direct application of the mechanical laws would involve. The mechanical laws, however, are utilized indirectly, for it is by their means (and also by accepting the ergodic hypothesis) that the probabilities are computed.

But in the quantum theory, the situation is entirely different. For example, when we consider the probabilities of the quantum drops, the underlying rigorous laws (if there be any) which control these drops are unknown, and so we cannot proceed, as we did in the kinetic theory, by calculating the probabilities from the laws. Einstein, we recall, obtained some insight into the numerical values of these quantum probabilities by pursuing a semi-empirical course; namely, by selecting the *a-priori* probabilities of emission and of absorption in such a way that Planck's statistical radiation law would be satisfied. It is true that thanks to the more modern methods of quantum mechanics, these probabilities may be computed theoretically; but, even so, their computation is not based on the assumption of rigorous laws controlling individual processes. A decision on whether such laws truly exist is thus a mere matter of opinion.

Despite the difficulty of arriving at a definite conclusion, Bohr at one time maintained that the cumulative evidence drawn from the quantum theory pointed to the absence of rigorous laws for individual atomic processes. He denied, in particular, any absolute validity to the mechanical laws of conservation of energy and of momentum. The semblance of rigor, which these laws seemed to betray in many phenomena, was ascribed by Bohr to a statistical effect generated by the large number of individual processes taking place. According to this view, the laws of conservation joined the law of entropy in having only a statistical validity.

Bohr's views may be illustrated in connection with the Compton effect. A beam of X-rays is directed against matter containing loosely bound electrons. The X-rays are found to be deflected with loss of energy (decreased frequency) in various directions, and some of the electrons are ejected from the matter. The original interpretation of this effect was furnished by Compton himself. Compton assumed that the radiation was formed of onrushing corpuscles (photons); the incident photons collided with the electrons, rebounding with decreased energies, and at the same time the electrons were driven out of the matter by the impacts. The impacts were supposed to be perfectly elastic and hence of the same kind that occur when perfectly elastic billiard balls collide. By applying to the collisions the mechanical laws of conservation of energy and of momentum, Compton was able to give a good quantitative account of the deflections actually observed.

In Compton's interpretation, the laws of conservation establish a precise causal connection between the impact of a photon and the ejection of an electron. But in Bohr's interpretation, where the laws of conservation are viewed as statistical, the precise causal connection no longer exists. We must now suppose that, under the influence of the incident radiation, the atoms of the matter pass over into a state conducive to the emission of electrons and then eject the electrons at random. At the same time the photons are deflected with decreased energies, also at random. In short, there is no immediate causal connection between the arrival of a photon and the ejection of an electron. The semblance of a causal connection is due to a general statistical effect.

With the idea of testing the merits of the two interpretations, Bothe and Geiger devised an experiment. The ejected electrons and the deviated photons were received into suitably disposed ionization chambers, where their entrance could be detected. If Compton were right, the penetrations of an electron and of a photon in the respective ionization chambers should be simultaneous; whereas if Bohr's interpretation were correct, the penetrations should occur independently and at random. The experiment showed that simultaneous penetrations occurred frequently, far more frequently than would be expected on the basis of the laws of probabilities. Indeed, when account was taken of parasitical effects which could not easily be avoided, simultaneous penetrations were found to be the rule. Thus, the experiment established the existence of rigorous causal connections for the individual atomic processes; and so, Bohr recognized his error.

When Heisenberg discovered his uncertainty relations, in 1927, a new element was injected into the discussions; and doubts on the existence of rigorous causality became widespread. But before considering the conclusions of the quantum theorists, we must recall the more important features of the classical causal doctrine.

Commonplace experience shows that inorganic phenomena appear to be causally related; and from this elementary observation arose a belief in rigorous causality. The causal doctrine may be compressed into the statement: The evolution of any self-contained, or isolated, system is determined by the initial state of the system.

Now, in physical science, all prospective laws, or doctrines should be submitted to experimental tests. Only thus can any assurance of their accuracy be obtained.*   Classical physicists realized, however, that for a number of reasons the doctrine of causality could not be tested rigorously: firstly, because the inaccuracies attendant on all human measurements precluded the possibility of accurate observation; secondly, because perfectly isolated systems were unattainable idealizations.   They argued that the only truly isolated system was the system represented by the entire universe, and that, in view of the tremendous number of cross influences that would have to be taken into account in so complicated a system, the existence of perfectly rigorous causal connections would be impossible to verify.

Classical physicists also recognized a logical difficulty which has since been stressed by the quantum theorists.   This difficulty results from the fact that an observation requires an observer and a system that is observed; in other words, there must be a subject and an object.   If this division is ignored, circularity cannot be avoided.   Let us suppose, then, that we have at our disposal an ideally isolated system which does not comprise the entire universe; and that we ourselves, as observers, do not form a part of the system, so that the necessary distinction between subject and object is respected.   To observe the state of the system and its subsequent evolution, we must perform measurements on its internal processes.   But measurements issue from interactions between the processes to be measured and our measuring instruments, which are foreign to the system.   Measurements are therefore incompatible with the isolation of the system.   Thus, we cannot observe the internal processes and hence we cannot test the presence of rigorous causal connections.

To overcome this difficulty, we might include ourselves and our measuring instrument in the system to be observed.   We should then have a single isolated system of which we ourselves would form a part.   Measurements, being now internal to the system, could be made without destroying its isolation.   But here the confusion between the subject and the object would render observation impossible, for the sum total of the processes taking place in the isolated system would include not only the processes originally contemplated, but also the processes involved in the workings of our brain and in the cognitive act generally.   Apart from any qualms we might feel in extending rigorous causality to vital activities, the circularity of the situation would of itself preclude the possibility of observation.

* The general objection that no test or series of tests can demonstrate the universality of a law is disregarded, for if tests are rejected as useless, no definite conclusions can ever be drawn.

From this analysis we might suppose that, even if perfectly isolated systems of manageable proportions were to exist in this universe, we should still be unable to observe their evolution with absolute rigor. A rigorous test of causality being unattainable, a blind adherence to the causal doctrine would seem to be a questionable procedure. Classical scientists, however, dismissed these elementary difficulties for the following reasons:

(a) They claimed that, though in practice there were no perfectly isolated systems (other than the entire universe), some systems were more nearly isolated than others, and that in view of the varying degrees of isolation, there was no objection to our imagining a perfectly isolated system.

(b) They conceded that the measurements performed from without on the internal processes of an isolated system would disturb these processes and would thereby destroy the system's isolation. But, as against this, they argued that the disturbances were contingent, being mere effects of human clumsiness; that by exercising sufficient care we could reduce these disturbances indefinitely; and that a super-experimenter could measure without disturbing. For these reasons they concluded that the principle of strict causality, though it eluded an experimental test in practice, might be submitted to such a test in theory.

Of course the classical physicists recognized that the foregoing arguments could not establish the validity of the rigorous causal doctrine, for the bare theoretical possibility of testing a doctrine could not of itself ensure the doctrine's validity, and so long as no practical test was possible the truth of a doctrine necessarily remained in doubt. But they argued that a doctrine which could be tested in theory was on safer ground than one which could not be so tested. To this extent therefore, the classical arguments served to dispel any *a-priori* objections against the doctrine of rigorous causality.

The essence of the classical arguments (a) and (b) is the belief that we may appeal to a kind of limiting process of thought, thanks to which, disturbances and extraneous influences may progressively be reduced to the vanishing point. By applying this limiting process to approximate causality (which was known to be valid), classical physicists inferred the existence of rigorous causality. As we shall see, it is the foregoing limiting process of thought that is ruled out as unjustified in the new quantum theory—and this leads us to Heisenberg's uncertainty relations.

According to the uncertainty relations, we cannot obtain simultaneous accurate information on the position of a particle at a given instant and on its momentum and energy. Since in mechanics, the position and the momentum of a particle determine its state, we conclude that the state of a particle cannot be determined. Consequently, we cannot decide whether or not a given initial state determines the subsequent states; and, as a result, the existence of rigorous causal connections and of rigorous laws cannot be tested in mechanics. Inasmuch as similar uncertainty relations hold in other fields of physical science, the conclusions which we have arrived at in mechanics must be extended to the physical world in general.

In Chapter XXX we mentioned that the uncertainties must be attributed to the disturbances which our measurements generate. We also saw that these disturbances must not be confused with the contingent ones which accompany all human observations. In contradistinction to the latter, the quantum disturbances are *essential*; they owe their existence to the non-vanishing value of Planck's constant h. For this reason even a super-experimenter could not avoid them.

Let us now revert to the classical argument which defended the theoretical possibility of verifying the causal doctrine by experiment. This argument maintained that, by appealing to a limiting process of thought, we could imagine perfectly isolated systems, the internal processes of which could be observed accurately without being subjected to disturbances. This classical argument now becomes untenable, for the limiting process of thought, though conceivable of course in a purely formal sense, cannot be utilized in the construction of a philosophy of physical Nature. The limiting process is now claimed to break down eventually because Nature, through the quantum of action h, imposes a boundary beyond which the process cannot be extended. Beyond this boundary nothing but vagueness and uncertainty would be found. To appeal to the limiting process would thus be to reason on a world which is not the one in which we live. The difference between the classical and the modern stand may be expressed in the statement: Whereas classical science believed that the sequence of successive approximations in our measurements converged to a limit, the uncertainty relations show that there is no such convergence. Thus, not only is a test of rigorous causality excluded for *practical* reasons, but it is also seen to be impossible *in theory*—at least in this world of ours. Let us observe that the refutation of the classical argument does not reflect on the logic of the earlier physicists; it issues solely from our increased knowledge of natural processes.

Many thinkers will contend that the new views have not settled the problem of rigorous causality in its major aspects; they will maintain that the theoretical impossibility of an empirical test does not of itself invalidate a doctrine. Be this as it may, the uncertainty relations have uncovered limitations heretofore unsuspected.

As we may well imagine, the novel situation created by the quantum theory has led to various interpretations. We shall first examine the views of the quantum theorists, Bohr, Heisenberg, and Dirac. According to these thinkers, the classical doctrine of causality for individual processes must be abandoned and a modified form of the causal doctrine accepted in its stead. But it would be incorrect to suppose that the rejection of the causal doctrine was undertaken solely on the basis of the arguments previously listed; the whole body of facts encompassed by the quantum theory was also given careful consideration.

Bohr's Principle of Complementarity—We mentioned that Bohr at one time believed that the mechanical laws of conservation of momentum and energy had only a statistical validity for individual processes. Subsequently, the Bothe-Geiger experiments showed that Bohr was mistaken, and that rigorous conservation held for the individual impacts between photons and electrons. Since the conservation laws illustrate causal relations, we may say that the foregoing experiments established the existence of rigorous causality for individual processes. At first sight this conclusion appears incompatible with the philosophy we have credited to the quantum theorists. But Bohr, in his principle of complementarity, has shown that the incompatibility is more apparent than real. Bohr's principle is today generally accepted by the leading quantum theorists. Strictly speaking, it does not embody a thoroughgoing rejection of the causal doctrine; rather does it exhibit a compromise between rigorous causality and complete indeterminism.

The significance of Bohr's principle is readily understood when we adopt a relativistic interpretation for Heisenberg's uncertainty relations. In Minkowski's 4-dimensional space-time, a measurement of position and of time may be represented by a point which, on being joined to the origin, determines a 4-dimensional vector called a space-time vector. A measurement of momentum and of energy determines another vector called the momentum-energy vector. Heisenberg's uncertainty relations, which connote the impossibility of obtaining a simultaneous knowledge of the space-time position of a particle and of its momentum and energy, may therefore be expressed by the statement: The space-time vector and the energy-momentum vector cannot be focussed simultaneously; the better the one is focussed, the more blurred does the other become. Now, the rigorous conservation of momentum and of energy has a meaning only insofar as momentum and energy can be accurately defined, and this is possible only when the momentum-energy vector is rigorously focussed. The space-time position then becomes utterly uncertain, so that the rigorous laws of conservation are incompatible with an accurate localization in space-time. Thus, the rigorous conservation laws, which illustrate rigorous causal connections, may be retained but only when an accurate space-time localization is relinquished. Conversely, when an accurate localization is secured, rigorous conservation laws cease to hold. These results may be generalized into the statement: A rigorous space-time description and a rigorous causal sequence for individual processes cannot be realized simultaneously—the one or the other must be sacrificed. This statement expresses Bohr's Principle of Complementarity.

A few illustrations will be helpful. Suppose two elastic particles enter into collision. If we determine with accuracy the momenta and the energies of the two particles before and after the collision, the laws of conservation may be verified. But, according to the uncertainty relations, the space-time positions of the particles before and after the collision are now uncertain; the particles may be anywhere. Though energy and momentum are conserved, we cannot say *where* and *when* this conservation is realized, so that to this extent the space-time description eludes us. The other extreme case occurs when the space-time positions of the two particles are accurately determined. The momenta and the energies are then uncertain, and rigorous conservation cannot be asserted. When we recall that the conservation laws for particles are special examples of causal relations for individual processes, we see that the foregoing illustrations exhibit the incompatibility between strict causality and an accurate space-time description. Bohr's principle is thus verified. In practice neither of the two extreme cases just considered is realized. Usually we know more or less where the particles are situated and we also have some knowledge of their momenta and energies. In practice therefore the space-time description is not accurate and conservation is not rigorous.

The processes occurring in an isolated system furnish another example of Bohr's principle. We first assume that the observer is exterior to the system. In this case a space-time observation of the internal processes of the system is precluded by the system's isolation. Consequently Bohr's principle is consistent with the assumption that the internal processes are controlled by rigorous laws.* Since, however, the only truly isolated

---

* An atom which is not radiating furnishes an example of an isolated system. We know that the internal energy state of the atom remains unchanged, so that conservation of energy is satisfied. The causal principle is thus verified in this case.

system is represented by the entire universe, we may dismiss the foregoing hypothetical situation and restrict our attention to the case where the observer and his measuring devices are contained within the system (*i.e.*, the universe). Here the distinction between subject and object becomes confused, and so, as before, observation is impossible. Accordingly, Bohr's principle allows us to postulate rigorous causal connections. Needless to say, however, no use can be made of these connections. Heisenberg, commenting on these considerations, writes:

"The chain of cause and effect could be quantitatively verified only if the whole universe were considered as a single system—but then physics has vanished, and only a mathematical scheme remains. The partition of the world into observing and observed systems prevents a sharp formulation of the law of cause and effect." *

Bohr has interpreted his principle of complementarity as implying that an exact localization in space-time on the one hand, and rigorous causal relations on the other, illustrate two different aspects of reality. Reality itself is not depicted correctly by either one of these two modes of representation considered singly. The two aspects are viewed as "complementary"; and it was to stress this feature that the name "principle of complementarity" was given by Bohr to his principle.

In Chapter XXX we explained Heisenberg's and Bohr's interpretation of a wave picture. We may readily verify that this interpretation and the principle of complementarity express the same philosophy. For example, when the momentum and energy of an electron are measured with a high degree of accuracy, with the result that the space-time position of the electron becomes uncertain, Heisenberg and Bohr would say that the electron has become diffused throughout a certain volume of space-time. Now, to say that an electron has become diffused implies that it can no longer be located at a definite space-time point. This circumstance in turn may be ascribed to a breakdown in the space-time form of representation; and we may say that space-time becomes progressively blurred as the momentum and energy measurement is performed with increasing accuracy. We are thus led to the conclusions expressed in Bohr's principle of complementarity.

We mentioned that the attitude of Heisenberg and Bohr removed many contradictions from the wave picture, and that for this reason among others it appeared to be justified. In view of the intimate connection between the ideas of Heisenberg-Bohr and the principle of complementarity, the arguments listed in previous chapters in support of

* The Physical Principles of the Quantum Theory, p. 58.

these ideas furnish an equal measure of support to the principle. Thus, in addition to the uncertainty relations on which the principle of complementarity is based, this principle is rendered plausible by important indirect evidence. On a later page some of the indirect evidence will be reviewed.

Bohr's principle and its implications embody the philosophy which, according to the quantum theorists, should replace the classical doctrine of causal relations for individual processes. It is therefore of interest to determine on what points the two philosophies differ. In discussing the classical doctrine, we did not emphasize the fact that the events forming a causal chain were assumed to be susceptible of localization in space-time. This peculiarity was not mentioned because it was taken for granted. For instance, when the classical physicist stated that the laws of conservation (and hence rigorous causal connections) were verified in the collision of two particles, he also claimed that it was possible to specify the exact space-time point at which the collision and the causal exchange of momentum and energy had taken place.

Now, the principle of complementarity does not require that, in all cases, strict causal connections for individual processes be impossible. The principle merely denies the possibility of our picturing with accuracy a sequence of causally related events in space-time. We may, if we choose, consider a sequence of causally related events, but we shall then be unable to secure a precise space-time representation of this sequence, and vice-versa. These considerations exhibit one of the differences between Bohr's principle and the classical doctrine; at the same time they justify our former statement that the principle of complementarity is a compromise between classical causality and indeterminism.

Another difference in the two philosophies concerns the distinction between subject and object. In the classical scheme a system could, in theory, be observed without suffering any disturbance from the observation; and as a result the system was assumed to evolve in exactly the same way even when it was not observed. We could thus conceive of an impersonal outside world (the object), of which the observer (the subject) would become aware. In other words, a clear-cut separation between subject and object was deemed to be justified. But in the modern quantum theory, the observer, by changing the nature of his observations, may destroy the space-time form of representation or else the causal chain. The outside world is thus deeply affected by the actions of the observer. A clear-cut distinction between the knowing subject and the passive object ceases to be possible. To this extent a subjective tinge colors the new philosophy. However, we must remember that it is not the cognitive act, as such, which causes the disturbances; it is the physical measurements performed with a view of rendering the cognitive act possible. The disturbances are thus of physical and not of psychic origin. To be sure, the cognitive act by supplementing the physical measurements does affect the mathematical representation, whether the latter be expressed by means of waves or of quadric surfaces in the Hilbert space. But no mystery is attached to this additional modification, for it is of the same kind that occurs in our daily life when the probability of a certain event is modified by an increase in the information at our disposal.

Lastly, let us inquire whether there is any conflict between the principle of complementarity and the possibility of obtaining a space-time representation for causal processes on the commonplace level of experience. We have already answered this question in the negative in the course of the present work. Thus we have seen that the principle of complementarity issues from the uncertainty relations and we know that the importance of the uncertainties tends to vanish when the action involved in the phenomenon of interest is large compared with the quantum of action h. Since this is precisely the situation that is realized when we are dealing with bodies of average mass, we may reconcile Bohr's principle with classical causality by saying: In all truth, rigorous causal connections and an accurate space-time representation cannot be focussed simultaneously and hence are mutually incompatible, but a simultaneous focussing comes closer to realization when we approach the macroscopic level of ordinary experience.

Thus far, in our analysis of the differences between the classical and the new school of thought, we have been concerned with the doctrines of rigorous causality for individual processes. Let us now pass to statistical processes. Where statistical processes are involved, the new quantum philosophy retains the classical belief in strict causal relations accompanying the possibility of an accurate space-time description. The principle of complementarity thus ceases to play any part. Let us note, however, that statistical processes connote probabilities, so that the vagueness which was inherent in the principle of complementarity is merely replaced by the vagueness that surrounds probabilities.

As an example, let us consider the evolution of a wave picture in wave mechanics. The waves, being controlled by rigorous mathematical laws, represent causally related events unfolding themselves in space-time. But we have seen that these waves do not represent individual physical processes; they are symbolic and serve to measure the probabilities of individual processes. In short, it is only the probabilities of individual processes (and not the individual processes themselves) that are causally related and that can be pictured in space-time. The statistical method of approach thus appears to be fundamental in quantum mechanics.

These considerations entail interesting consequences. In classical physics, the probabilities, which were introduced in many theoretical discussions, were supposed to be derivable from underlying rigorous laws controlling individual processes. The classical probabilities were thus mere makeshifts, and for a statistical theory to rest on solid ground, the probabilities utilized had to be calculated rigorously. We recall in this connection that Boltzmann assumed the equiprobability of the microscopic states in the kinetic theory, and that this assumption could be justified only in part on theoretical grounds. A complete justification required the acceptance of the ergodic hypothesis, a hypothesis which could receive no theoretical support. For this reason Kelvin rejected Boltzmann's kinetic theory. But in the quantum statistics of Bose and of Fermi, where the equiprobability of the microscopic states is also assumed, the situation is entirely different and Kelvin's objection would lose its force. For now the probabilities are themselves fundamental and cannot even in principle be traced to underlying rigorous laws since the latter are inexistent. We are therefore justified in postulating whatever probabilities appear to yield results consistent with the facts of observation. Thus, the new outlook removes the type of difficulty noted in connection with the ergodic hypothesis.

The Controversies on Quantum Mechanics—The new quantum theory indicates that we cannot, even in theory, test the classical doctrine of causality. On the strength of these findings, the quantum theorists have rejected classical causality, adopting the principle of complementarity in its place.

Those who are opposed to the revolutionary implications of the principle will presumably claim that the impossibility (even theoretical) of testing a doctrine does not necessarily prove that the doctrine is erroneous; and on this basis they might propose to retain the classical philosophy. But the weakness of this claim is that if we accept doctrines which cannot be verified in principle, no restriction is placed on caprice; anything may then be postulated with impunity.

But there is another way of contesting the philosophy of the quantum theorists. As a preliminary let us consider an example drawn from classical mechanics. According to mechanics the evolution of a self-contained conservative mechanical system is determined by the initial positions and momenta of its various parts. If, then, two perfectly elastic billiard balls collide, we might suppose that a knowledge of the initial positions and momenta of the two balls should apprise us of their behavior after the collision. Yet such is not the case; only statistical results can be obtained from the mechanical laws. Here, our inability to predict precise results is due, not to any inherent indeterminism in Nature, but to the insufficiency of the information expressed in the initial conditions. If this information is supplemented by a knowledge of the diameters of the balls, our predictions become determinate.* Thus, the original semblance of indeterminism in our collision problem was caused by the incompleteness of our information.

In other cases, a failure to predict future events with accuracy from a knowledge of initial conditions is due to the incompleteness of the laws on which our reasonings are based. For instance, Lorentz's equations, which regulate the behaviour of electromagnetic fields in the presence of free electrons, constitute incomplete laws. An additional equation would be required to secure completeness.

More generally, the predictions that can be derived from a theory are restricted by the concepts which the theory accepts as valid. If, then, a theory is deficient in relevant concepts, it may be unable to predict certain events with precision, even though these events are not meaningless and even though they occur in a perfectly determinate way.* A theory of this latter type may be called *incomplete*. Thermodynamics furnishes an illustration: by restricting its field to macroscopic concepts and magnitudes, thermodynamics can give no information on microscopic occurrences. Yet this does not mean that microscopic occurrences are meaningless and cannot affect observable ones. Indeed, in the kinetic theory, these microscopic occurrences are made the basis of our speculations, and as a result many phenomena which thermodynamics was unable to anticipate can be predicted. At the same time the macroscopic concepts of thermodynamics, such as temperature, pressure, and entropy, are replaced by mechanical concepts which apply to the individual processes.

In view of the foregoing illustrations it is conceivable that the vagueness of quantum mechanics (as is evidenced in the uncertainty relations) may be due to the incompleteness of the theory. If this be the case, no fundamental vagueness need be postulated in Nature. The claim that quantum mechanics is incomplete has been made by Einstein and by

---

* The problem would also become determinate if the diameters of the balls were unknown, but if, as against this, the behavior of one of the balls *after the collision* was revealed to us. In this case, however, the problem would assume a teleological aspect.

† Compare with Planck's statement on page 89.

Planck. Einstein's arguments are highly technical, dealing with the mathematics of the theory, and so we shall not consider them. In the following pages we shall confine ourselves to an elementary discussion.

We recall that in the theoretical derivation of the uncertainty relations from quantum mechanics, the uncertainties define a boundary beyond which the theory can give no information. We are thus prompted to adopt one of the following alternatives:

(a) The uncertainties are due to the incompleteness of quantum mechanics.

(b) They express a fundamental vagueness in Nature.

The quantum theorists, as we know, have accepted the second alternative. Dirac writes:

"When an observation is made on any atomic system which has been prepared in a given way and is thus in a given state, the result will not in general be determinate, *i.e.*, if the experiment be repeated several times under identical conditions several different results may be obtained." *

An obvious objection to Dirac's statement might be expressed as follows:

What requirements must be realized for us to be justified in asserting that the conditions under which successive experiments are performed are identical? Until a rigorous method of determining the identity of such conditions is given, we shall always be free to suppose that differences in the observed results arise precisely because the conditions are not identical.

Let us consider, for instance, a collision between an electron and a photon, such as occurs in the experiment with Heisenberg's microscope. If we wish to repeat the experiment under identical conditions, we must operate each time with the same electron and the same photon. But even then, how can we be certain that the electron and the photon have remained unaffected by the first collision and that they will react in the same way during the following collisions? Furthermore, what right have we to assert that each successive collision always occurs under exactly the same conditions? As an example, two billiard balls may suffer a more or less glancing collision, and the behavior of the balls after the collision will differ accordingly. In short, it is easy enough to suggest any number of influences which might differ from one experiment to another, and which might thereby be responsible for the possibly erroneous belief in the fundamental indeterminism of Nature.

* The Principles of Quantum Mechanics, p. 10.

However, we must grant that such criticisms are of so elementary a nature that even a child might suggest them; and we cannot expect them to have been overlooked by such deep thinkers as Heisenberg, Born, Dirac, and Bohr. Besides, the possibility that the quantum uncertainties are due to ignorance is so obvious that it was one of the first avenues to be explored. Born and Heisenberg, commenting on the quantum uncertainties at the Solvay Congress of 1927, stated:

"It was thought at first that there was here a gap which would be filled when the theory was further investigated. Soon, however, it was recognized that this is not so, and that we are here faced with a deficiency of an essential kind, deeply anchored in the very nature of our ability to understand physical phenomena."

Let us, then, examine the evidence step by step and see what light can be thrown on the problem. The uncertainty relations are usually derived theoretically from quantum mechanics. But they may also be derived by following a different procedure—we refer to their derivation from the analysis of physical measurements. When we follow this latter procedure, only elementary quantum concepts need be utilized, the advanced methods of quantum mechanics being disregarded. In particular, the only reference to the quantum theory consists in the assumption of a corpuscular structure for light and in the crediting of the value $\frac{h\nu}{c}$ to the photon's momentum. This value of the momentum is the one originally postulated by Einstein in the early days of the quantum theory. Now, when the uncertainty relations are derived from an analysis of measurements, they might appear at first sight to be due to our ignorance of details affecting the processes measured. For instance in Heisenberg's microscope, an electron is illuminated by γ-rays, so that its position is accurately known. The collision of the incident photon with the electron gives a kick to the latter and changes its momentum by an unknown amount. In the elementary treatment of this problem, we proceed exactly as we would if two elastic billiard balls were to collide; the laws of conservation of momentum and energy are applied, and the uncertainty in the recoil of the billiard ball or electron arises only from the fact that we are not apprised of the angle under which the collision takes place.

From this illustration we might infer that the uncertainties which are introduced into the analysis of measurements do not betray any fundamental vagueness in Nature; they merely represent the price we have to pay for our neglect in stipulating all the data necessary for an accurate answer. But what makes the situation mysterious is that exactly the same numerical values for the uncertainties are obtained when we vary the experiments in which the measurements are performed. Furthermore, we are again led to the same numerical values when we derive the uncertainty relations theoretically from the mathematics of quantum mechanics. It is difficult to understand how the uncertainties, if they are contingent, should always turn out to have the same values regardless of the manner in which we derive them. We can scarcely claim that the agreement is due to chance.

The most plausible explanation is to suppose that the reason the uncertainties are always found to have the same values is that they betray some fundamental characteristics of Nature. Under this view the vagueness of quantum mechanics cannot be ascribed to any incompleteness of the theory; it must be traced to a deep-seated vagueness in natural processes.

In spite of the cogency of the preceding argument, we doubt whether the quantum theorists would have adopted their indeterministic philosophy had there not been further evidence pointing in the same direction. We should be deceiving ourselves therefore if we were to pass on the new philosophy until all the evidence was heard and understood. As we shall see, no single argument by itself is convincing; it is the aggregate of a number of different arguments that gives weight to the new philosophy.

A first argument issues from the mode of derivation of the uncertainty relations. If we assume that the uncertainties are due to the incompleteness of quantum mechanics, we must view them as expressing a statistics of individual processes, these latter processes being presumably controlled by rigorous laws. We shall now see that this solution does not appear possible.

Thus the uncertainty relations may be derived theoretically from the matrix method or from wave mechanics. We prefer the former because the significance of the waves is as mysterious as that of the uncertainty relations themselves. The essence of the matrix method is to replace the familiar coordinates of position and of momentum of classical mechanics by matrices. The commutation rules, which render calculations with matrices determinate, are mere refinements of Bohr's quantizing conditions, which determine the stable energy levels in atomic systems. Since these commutation rules entail the uncertainty relations, we must suppose that the latter are closely connected with the existence of stable energy levels. Now the energy levels are not statistical manifestations; they are revealed in individual atoms. Under these conditions it would seem that the uncertainty relations in turn cannot be of a statistical nature, but that they must have the same fundamental significance as the energy levels. In short, the uncertainty relations cannot result from any looseness in our treatment of physical problems. We may grant, however, that the preceding argument is suggestive rather than convincing, for the introduction of matrices may be responsible for statistical features being incorporated unawares. Let us, then, pass to a second argument.

Bohr's frequency condition, which is assumed to illustrate individual processes, is closely allied with one of the uncertainty relations. If we call ΔE the drop in energy and ν the frequency radiated, Bohr's frequency condition may be written

$$\nu = \frac{\Delta E}{h} \, .$$

But to recognize the frequency of a vibration, we must allow the lapse of sufficient time for one complete vibration to be performed. This interval of time is the period of the vibration and is defined by the inverse of the frequency. Consequently, we cannot state at what precise instant the radiation is emitted; all we may assert is that the emission of the radiation covers at least one period of the vibration. If we call Δt the interval of time which represents a period, Bohr's frequency condition becomes

$$\Delta t \cdot \Delta E = h \, ;$$

and it is thus seen to be Heisenberg's uncertainty relation between energy and time. In it ΔE represents the uncertainty in the exact value of the energy of the atom while it is dropping from one level to another, and Δt measures the uncertainty in the instant at which the photon is emitted. This illustration shows that the uncertainty relation, which is here derived from a phenomenon involving an individual process, cannot be claimed to owe its vagueness to a statistical outlook.

A further argument which lends support to the fundamental nature of the uncertainties is found in the new statistics of gases. The partitioning of the 6-dimensional phase space into cells of volume h^3 was originally devised by Planck in order to incorporate the ubiquitous quantum of action h into the gas theory. Planck recognized that by this means he could assign an absolute value to the entropy—a situation demanded by the third principle of thermodynamics. Planck's partitioning of the phase space was therefore believed to illustrate something fundamental. Now, in the previous chapter we mentioned that this partitioning is a direct consequence of the uncertainty relations. We conclude that these relations must likewise have a fundamental significance. Indeed, the very presence of the constant h in the expression of the uncertainty relations points to the same conclusion. Moreover, we must remember that quantum mechanics may also be developed in the form of wave mechanics. Now, when de Broglie proposed his wave mechanics, his aim was to establish a rigorous theory and not a mere statistical one. The fact that his theory eventually led to the uncertainty relations cannot therefore be traced to an unwarranted neglect of detail; rather does it show that these relations are inevitable and hence fundamental.

More striking perhaps is the part the uncertainty relations play in accounting for the dual nature of light and matter. The appearance of light, now as a corpuscle and now as a wave, but never as both simultaneously, cannot be due to a statistical effect. It is therefore remarkable to find that quantum mechanics accounts for the mutually exclusive features of the corpuscle and of the wave, and that furthermore it allows us to anticipate, in any given situation, which of the two aspects will be manifested. Such precise information would seem strange on the part of an incomplete theory. Finally, we may recall that no inconsistency is involved between the vagueness of the uncertainty relations and the extreme accuracy with which rigorous causal relations may be represented in space-time on the macroscopic level.

Many other arguments might be advanced. As we warned the reader, no one of these arguments by itself is absolutely convincing, but in their aggregate they do compel the recognition that more lies behind the philosophy of the quantum theorists than might appear at first sight.

Einstein and Planck, however, have expressed the opinion that quantum mechanics is incomplete and that the introduction of new concepts may reinstate rigorous determinism. We cannot prophesy what the future may hold in store, but we may safely assert that at the present writing the new concepts have not been discovered, and that, if ever they are found, they will differ considerably from any we know of today. Thermodynamics was refined by the introduction of mechanical concepts, yet neither mechanical concepts nor the field categories can assist us in passing beyond quantum mechanics. Entirely new concepts will have to be devised.

MAX WIEN (1866-1938)

MAX PLANCK (1858-1938)

ALBERT EINSTEIN (1879-   )

HENRY MOSELEY (1887-1915)

WALTHER RITZ (1878-1909)

NIELS BOHR (1885-   )

ARNOLD SOMMERFELD (1868-)

WOLFGANG PAULI (1900-

LOUIS VICTOR DE BROGLIE (1892-   )

ERWIN SCHRODINGER (1887-   )

WERNER HEISENBERG (1901-   )

PAUL ADRIEN MAURICE DIRAC (1902-   )

INDEX

Ritz's Combination Principle, 481, *482*,
483, 824, *826*, 853, 862
Rosseland, 589
Rotator
in earlier quantum theory and in Bohr's
theory, *465-467*, *505*, *506*
in de Broglie's wave theory, 626
in Schrödinger's wave theory, *696*, 721,
722
Rumford, 333
Rupp, 635
Russell (Bertrand), 200, 208, 209
Rutherford, 473, 474, 529
Rutherford's atom, *474*, 476, 480, 488
Rydberg constant, *478*, *482*
its expression in Bohr's theory, *497*

S

Sackur, O., 909, 911
Sackur-Tetrode expression of entropy;
see under Entropy
Sainte-Claire Deville, 367
Scalar (mathematics), 71
Schmidt, 878
Schrödinger, 100, 121, 628, 630, 638, 639,
Chapters XXXII, XXXIII,
XXXIV, 754, 837, 874, 879, 892,
893
Schrödinger's Wave Mechanics, 100, 444,
Chapters XXXII-XXXV
charge cloud; *see* Charge Cloud
connection between waves and matrices,
881, *882*
electron in Schrödinger's theory, 638,
639, *725-728*
operators, *708*, *710*, 837, 838, 873, 874,
895
quantizing conditions, *691*
radiation emission, *711-723*, 727-729,
765, 766
superposition of states, *716*, 717-719,
725, 729, 765-767, 935
Born's interpretation, *735-740*
wave equation (classical approximation), 689, *690*, 707, *708-710*, 878,
895

Schrödinger's Wave Mechanics, *cont'd*
wave equation (classical approximation), *cont'd*
for hydrogen atom, 121, *697*; *see
also under* Hydrogen atom, Eigenfunctions, Eigenvalues
wave equation (relativistic), 690, 704,
708, *710*, 895
for hydrogen atom, 704
(*See also* Helium atom by wave mechanics, Wave pictures)
Schwartzschild, 177
Screening-doublet levels (in hydrogen
atom), *583*, 602, 630, 705
Selection rules (for quantum transitions)
in Bohr's theory (*see* Bohr's atom)
in matrix method, *843*, *849*
in Schrödinger's wave mechanics, *720-
722*, 765, 766, *767*, 774, 943
Series (mathematics), *124-126*, 154, 247,
248
Fourier series; *see* Fourier series
Lindstedt series (celestial mechanics),
248
power or Taylor series
complex variables, *143-146*, 248,
249
real variables, *126-133*, 248, 249
Silberstein, L., 359
Simplicity (in theoretical physics),
Chapter IV
Simultaneity (in theory of relativity),
435, 436
Sinusoid, *129*, 273, 274
de Sitter, 108
Smoluchowski, 416
Soddy, 475, 477
Sodium (spectral series), *539*, *540*, *550*
Sommerfeld, 41, 425, 445, 509-514, 522,
551-553, 930
his application of Fermi's statistics to
electron gas, 425, *930*, *931*
application of relativity to Bohr's
atom (fine structure of lines), *509-
514*, 583
quantum number *j* in Bohr's atom,
551-555

Velocities of waves
 group velocity (or amplitude velocity),
  291, *292*, 294, *296*, 299, 934
 wave velocity (or phase velocity), 269,
  273, *275*, 283, 294, *296*, 299
Velocity of a propagation (propagation
 velocity, *291*
Vibrations
 harmonic, *272*, 484, *813-815*
  amplitude, *272*, 713, 813-815
  frequency, *272*, 713
  phase, *272*
 non-harmonic, 485, *815*, *816*
  fundamental frequency, *485*
  harmonics, *485*
 of membranes, *680*
  nodal lines, *680*
 of solids, *680-686*; see also Eigenfunctions *and* Eigenvalues
  degeneracy, *685*
  fundamental frequency, *684*
  modes, *684*, 685
  nodal surfaces, *684*, 685
 of strings, 133, 134, 181, 182, *674-680*;
  see also Eigenfunctions and
  Eigenvalues
  fundamental frequency, *674*
  fundamental mode, *674*
  harmonics, *675*
  modes of vibration, 675
  nodal points, *675*
Viscosity of gases, 28, *403*
Volta, 12
Volterra, 118, 140, 149, 184, 878

W

Wave equation; *see under* d'Alembert,
 Schrödinger, Dirac
Wave Mechanics; *see under* Hamilton, de
 Broglie, Schrödinger
Wave Optics; *see under* Optics
Wave packets, *294-299*; *see also under* de
 Broglie waves
Wave pictures (wave mechanics), *640*,
 642, 644-648, 649, *652*, 654, *666-*
 669, *725*, *726*, 735, *741-750*

Wave theory of light; *see* Light
Wave velocity; *see under* Velocity
Waves, *Chapter XIX*
 coherence of waves, *303*, 304, 305
 electromagnetic; *see* Electromagnetic
  Waves
 equiphase surfaces, *275*, 276, 282; *see*
  also under de Broglie waves
  their construction; *see* Huyghens's
  construction
 frequency of waves, *277*, 284
 group of waves, *291;* see also Wave
  packets
 group velocity; *see under* Velocities of
  waves
 longitudinal waves, *270*, 271, 277
 phase of waves, *273*, 275
 plane waves, *270*, 276
 polarization of waves, *272*
 sinusoidal waves, 273-275, 306
 spherical waves, *269*, 276, 281-283
 standing, *271*
 transverse, *270*, 271, 277
 wave front, 269, *275*, 276
 wave length, *273*, 280, *284*, 285
 wave motions, *270*
 wave number, *290*, *291*
 wave velocity; *see under* Velocities of
  Waves
Weierstrass, 117, 136, 143, 151, 201, 241,
 268, 317
Weierstrass curve, 136, 137, 166
Weyl, H., 85, 110, 111, 191, 199, 202, 207-
 209
Whittaker, 242, 249
Wiedemann-Frantz (law of), 424
Wien, 453, 592
Wien's radiation law, *457*, 460, 467
 obtained with photons, *918*, 919
Wien's relation (Displacement law), 94,
 453, *454*, 456
Work, *216*, 219, 220, *336*, *337*, 338-344

X

X-rays, 469, 478, 585, *634*
 diffraction of, *634*
X-ray spectra, *478*, 546, 548, *574*

A CATALOGUE OF
SELECTED DOVER BOOKS
IN ALL FIELDS OF INTEREST

A CATALOGUE OF SELECTED DOVER
BOOKS IN ALL FIELDS OF INTEREST

RACKHAM'S COLOR ILLUSTRATIONS FOR WAGNER'S RING. Rackham's finest mature work—all 64 full-color watercolors in a faithful and lush interpretation of the *Ring*. Full-sized plates on coated stock of the paintings used by opera companies for authentic staging of Wagner. Captions aid in following complete Ring cycle. Introduction. 64 illustrations plus vignettes. 72pp. 8⅝ x 11¼.   23779-6 Pa. $6.00

CONTEMPORARY POLISH POSTERS IN FULL COLOR, edited by Joseph Czestochowski. 46 full-color examples of brilliant school of Polish graphic design, selected from world's first museum (near Warsaw) dedicated to poster art. Posters on circuses, films, plays, concerts all show cosmopolitan influences, free imagination. Introduction. 48pp. 9⅜ x 12¼.
23780-X Pa. $6.00

GRAPHIC WORKS OF EDVARD MUNCH, Edvard Munch. 90 haunting, evocative prints by first major Expressionist artist and one of the greatest graphic artists of his time: *The Scream, Anxiety, Death Chamber, The Kiss, Madonna*, etc. Introduction by Alfred Werner. 90pp. 9 x 12.
23765-6 Pa. $5.00

THE GOLDEN AGE OF THE POSTER, Hayward and Blanche Cirker. 70 extraordinary posters in full colors, from Maitres de l'Affiche, Mucha, Lautrec, Bradley, Cheret, Beardsley, many others. Total of 78pp. 9⅜ x 12¼.   22753-7 Pa. $6.95

THE NOTEBOOKS OF LEONARDO DA VINCI, edited by J. P. Richter. Extracts from manuscripts reveal great genius; on painting, sculpture, anatomy, sciences, geography, etc. Both Italian and English. 186 ms. pages reproduced, plus 500 additional drawings, including studies for *Last Supper*, Sforza monument, etc. 860pp. 7⅞ x 10¾. (Available in U.S. only)
22572-0, 22573-9 Pa., Two-vol. set $19.90

THE CODEX NUTTALL, as first edited by Zelia Nuttall. Only inexpensive edition, in full color, of a pre-Columbian Mexican (Mixtec) book. 88 color plates show kings, gods, heroes, temples, sacrifices. New explanatory, historical introduction by Arthur G. Miller. 96pp. 11⅜ x 8½. (Available in U.S. only)   23168-2 Pa. $7.95

UNE SEMAINE DE BONTÉ, A SURREALISTIC NOVEL IN COLLAGE, Max Ernst. Masterpiece created out of 19th-century periodical illustrations, explores worlds of terror and surprise. Some consider this Ernst's greatest work. 208pp. 8⅛ x 11.   23252-2 Pa. $6.00

DRAWINGS OF WILLIAM BLAKE, William Blake. 92 plates from Book of Job, *Divine Comedy, Paradise Lost,* visionary heads, mythological figures, Laocoon, etc. Selection, introduction, commentary by Sir Geoffrey Keynes. 178pp. 8⅛ x 11. 22303-5 Pa. $5.00

ENGRAVINGS OF HOGARTH, William Hogarth. 101 of Hogarth's greatest works: *Rake's Progress, Harlot's Progress, Illustrations for Hudibras, Before and After, Beer Street and Gin Lane,* many more. Full commentary. 256pp. 11 x 13¾. 22479-1 Pa. $12.95

DAUMIER: 120 GREAT LITHOGRAPHS, Honore Daumier. Wide-ranging collection of lithographs by the greatest caricaturist of the 19th century. Concentrates on eternally popular series on lawyers, on married life, on liberated women, etc. Selection, introduction, and notes on plates by Charles F. Ramus. Total of 158pp. 9⅜ x 12¼. 23512-2 Pa. $6.00

DRAWINGS OF MUCHA, Alphonse Maria Mucha. Work reveals draftsman of highest caliber: studies for famous posters and paintings, renderings for book illustrations and ads, etc. 70 works, 9 in color; including 6 items not drawings. Introduction. List of illustrations. 72pp. 9⅜ x 12¼. (Available in U.S. only) 23672-2 Pa. $4.50

GIOVANNI BATTISTA PIRANESI: DRAWINGS IN THE PIERPONT MORGAN LIBRARY, Giovanni Battista Piranesi. For first time ever all of Morgan Library's collection, world's largest. 167 illustrations of rare Piranesi drawings—archeological, architectural, decorative and visionary. Essay, detailed list of drawings, chronology, captions. Edited by Felice Stampfle. 144pp. 9⅜ x 12¼. 23714-1 Pa. $7.50

NEW YORK ETCHINGS (1905-1949), John Sloan. All of important American artist's N.Y. life etchings. 67 works include some of his best art; also lively historical record—Greenwich Village, tenement scenes. Edited by Sloan's widow. Introduction and captions. 79pp. 8⅜ x 11¼. 23651-X Pa. $5.00

CHINESE PAINTING AND CALLIGRAPHY: A PICTORIAL SURVEY, Wan-go Weng. 69 fine examples from John M. Crawford's matchless private collection: landscapes, birds, flowers, human figures, etc., plus calligraphy. Every basic form included: hanging scrolls, handscrolls, album leaves, fans, etc. 109 illustrations. Introduction. Captions. 192pp. 8⅞ x 11¾. 23707-9 Pa. $7.95

DRAWINGS OF REMBRANDT, edited by Seymour Slive. Updated Lippmann, Hofstede de Groot edition, with definitive scholarly apparatus. All portraits, biblical sketches, landscapes, nudes, Oriental figures, classical studies, together with selection of work by followers. 550 illustrations. Total of 630pp. 9⅛ x 12¼. 21485-0, 21486-9 Pa., Two-vol. set $17.90

THE DISASTERS OF WAR, Francisco Goya. 83 etchings record horrors of Napoleonic wars in Spain and war in general. Reprint of 1st edition, plus 3 additional plates. Introduction by Philip Hofer. 97pp. 9⅜ x 8¼. 21872-4 Pa. $4.50

CATALOGUE OF DOVER BOOKS

THE EARLY WORK OF AUBREY BEARDSLEY, Aubrey Beardsley. 157 plates, 2 in color: *Manon Lescaut, Madame Bovary, Morte Darthur, Salome,* other. Introduction by H. Marillier. 182pp. 8⅛ x 11. 21816-3 Pa. $6.50

THE LATER WORK OF AUBREY BEARDSLEY, Aubrey Beardsley. Exotic masterpieces of full maturity: *Venus and Tannhauser, Lysistrata, Rape of the Lock, Volpone,* Savoy material, etc. 174 plates, 2 in color. 186pp. 8⅛ x 11. 21817-1 Pa. $5.95

THOMAS NAST'S CHRISTMAS DRAWINGS, Thomas Nast. Almost all Christmas drawings by creator of image of Santa Claus as we know it, and one of America's foremost illustrators and political cartoonists. 66 illustrations. 3 illustrations in color on covers. 96pp. 8⅜ x 11¼. 23660-9 Pa. $3.50

THE DORÉ ILLUSTRATIONS FOR DANTE'S DIVINE COMEDY, Gustave Doré. All 135 plates from Inferno, Purgatory, Paradise; fantastic tortures, infernal landscapes, celestial wonders. Each plate with appropriate (translated) verses. 141pp. 9 x 12. 23231-X Pa. $5.00

DORÉ'S ILLUSTRATIONS FOR RABELAIS, Gustave Doré. 252 striking illustrations of *Gargantua and Pantagruel* books by foremost 19th-century illustrator. Including 60 plates, 192 delightful smaller illustrations. 153pp. 9 x 12. 23656-0 Pa. $6.00

LONDON: A PILGRIMAGE, Gustave Doré, Blanchard Jerrold. Squalor, riches, misery, beauty of mid-Victorian metropolis; 55 wonderful plates, 125 other illustrations, full social, cultural text by Jerrold. 191pp. of text. 9⅜ x 12¼. 22306-X Pa. $7.00

THE RIME OF THE ANCIENT MARINER, Gustave Doré, S. T. Coleridge. Dore's finest work, 34 plates capture moods, subtleties of poem. Full text. Introduction by Millicent Rose. 77pp. 9¼ x 12. 22305-1 Pa. $4.50

THE DORE BIBLE ILLUSTRATIONS, Gustave Doré. All wonderful, detailed plates: Adam and Eve, Flood, Babylon, Life of Jesus, etc. Brief King James text with each plate. Introduction by Millicent Rose. 241 plates. 241pp. 9 x 12. 23004-X Pa. $6.95

THE COMPLETE ENGRAVINGS, ETCHINGS AND DRYPOINTS OF ALBRECHT DURER. "Knight, Death and Devil"; "Melencolia," and more—all Dürer's known works in all three media, including 6 works formerly attributed to him. 120 plates. 235pp. 8⅜ x 11¼. 22851-7 Pa. $7.50

MECHANICK EXERCISES ON THE WHOLE ART OF PRINTING, Joseph Moxon. First complete book (1683-4) ever written about typography, a compendium of everything known about printing at the latter part of 17th century. Reprint of 2nd (1962) Oxford Univ. Press edition. 74 illustrations. Total of 550pp. 6⅛ x 9¼. 23617-X Pa. $7.95

THE COMPLETE WOODCUTS OF ALBRECHT DURER, edited by Dr. W. Kurth. 346 in all: "Old Testament," "St. Jerome," "Passion," "Life of Virgin," Apocalypse," many others. Introduction by Campbell Dodgson. 285pp. 8½ x 12¼. 21097-9 Pa. $7.50

DRAWINGS OF ALBRECHT DURER, edited by Heinrich Wolfflin. 81 plates show development from youth to full style. Many favorites; many new. Introduction by Alfred Werner. 96pp. 8⅛ x 11. 22352-3 Pa. $6.00

THE HUMAN FIGURE, Albrecht Dürer. Experiments in various techniques—stereometric, progressive proportional, and others. Also life studies that rank among finest ever done. Complete reprinting of *Dresden Sketchbook*. 170 plates. 355pp. 8⅜ x 11¼. 21042-1 Pa. $7.95

OF THE JUST SHAPING OF LETTERS, Albrecht Dürer. Renaissance artist explains design of Roman majuscules by geometry, also Gothic lower and capitals. Grolier Club edition. 43pp. 7⅞ x 10¾ 21306-4 Pa. $3.00

TEN BOOKS ON ARCHITECTURE, Vitruvius. The most important book ever written on architecture. Early Roman aesthetics, technology, classical orders, site selection, all other aspects. Stands behind everything since. Morgan translation. 331pp. 5⅜ x 8½. 20645-9 Pa. $5.00

THE FOUR BOOKS OF ARCHITECTURE, Andrea Palladio. 16th-century classic responsible for Palladian movement and style. Covers classical architectural remains, Renaissance revivals, classical orders, etc. 1738 Ware English edition. Introduction by A. Placzek. 216 plates. 110pp. of text. 9½ x 12¾. 21308-0 Pa. $10.00

HORIZONS, Norman Bel Geddes. Great industrialist stage designer, "father of streamlining," on application of aesthetics to transportation, amusement, architecture, etc. 1932 prophetic account; function, theory, specific projects. 222 illustrations. 312pp. 7⅞ x 10¾. 23514-9 Pa. $6.95

FRANK LLOYD WRIGHT'S FALLINGWATER, Donald Hoffmann. Full, illustrated story of conception and building of Wright's masterwork at Bear Run, Pa. 100 photographs of site, construction, and details of completed structure. 112pp. 9¼ x 10. 23671-4 Pa. $5.95

THE ELEMENTS OF DRAWING, John Ruskin. Timeless classic by great Viltorian; starts with basic ideas, works through more difficult. Many practical exercises. 48 illustrations. Introduction by Lawrence Campbell. 228pp. 5⅜ x 8½. 22730-8 Pa. $3.75

GIST OF ART, John Sloan. Greatest modern American teacher, Art Students League, offers innumerable hints, instructions, guided comments to help you in painting. Not a formal course. 46 illustrations. Introduction by Helen Sloan. 200pp. 5⅜ x 8½. 23435-5 Pa. $4.00

THE ANATOMY OF THE HORSE, George Stubbs. Often considered the great masterpiece of animal anatomy. Full reproduction of 1766 edition, plus prospectus; original text and modernized text. 36 plates. Introduction by Eleanor Garvey. 121pp. 11 x 14¾.   23402-9 Pa. $8.95

BRIDGMAN'S LIFE DRAWING, George B. Bridgman. More than 500 illustrative drawings and text teach you to abstract the body into its major masses, use light and shade, proportion; as well as specific areas of anatomy, of which Bridgman is master. 192pp. 6½ x 9¼. (Available in U.S. only)
22710-3 Pa. $4.50

ART NOUVEAU DESIGNS IN COLOR, Alphonse Mucha, Maurice Verneuil, Georges Auriol. Full-color reproduction of *Combinaisons ornementales* (c. 1900) by Art Nouveau masters. Floral, animal, geometric, interlacings, swashes—borders, frames, spots—all incredibly beautiful. 60 plates, hundreds of designs. 9⅜ x 8-1/16.   22885-1 Pa. $4.50

FULL-COLOR FLORAL DESIGNS IN THE ART NOUVEAU STYLE, E. A. Seguy. 166 motifs, on 40 plates, from *Les fleurs et leurs applications decoratives* (1902): borders, circular designs, repeats, allovers, "spots." All in authentic Art Nouveau colors. 48pp. 9⅜ x 12¼.
23439-8 Pa. $5.00

A DIDEROT PICTORIAL ENCYCLOPEDIA OF TRADES AND IN-DUSTRY, edited by Charles C. Gillispie. 485 most interesting plates from the great French Encyclopedia of the 18th century show hundreds of working figures, artifacts, process, land and cityscapes; glassmaking, papermaking, metal extraction, construction, weaving, making furniture, clothing, wigs, dozens of other activities. Plates fully explained. 920pp. 9 x 12.
22284-5, 22285-3 Clothbd., Two-vol. set $40.00

HANDBOOK OF EARLY ADVERTISING ART, Clarence P. Hornung. Largest collection of copyright-free early and antique advertising art ever compiled. Over 6,000 illustrations, from Franklin's time to the 1890's for special effects, novelty. Valuable source, almost inexhaustible.
Pictorial Volume. Agriculture, the zodiac, animals, autos, birds, Christmas, fire engines, flowers, trees, musical instruments, ships, games and sports, much more. Arranged by subject matter and use. 237 plates. 288pp. 9 x 12.
20122-8 Clothbd. $15.00

Typographical Volume. Roman and Gothic faces ranging from 10 point to 300 point, "Barnum," German and Old English faces, script, logotypes, scrolls and flourishes, 1115 ornamental initials, 67 complete alphabets, more. 310 plates. 320pp. 9 x 12.   20123-6 Clothbd. $15.00

CALLIGRAPHY (CALLIGRAPHIA LATINA), J. G. Schwandner. High point of 18th-century ornamental calligraphy. Very ornate initials, scrolls, borders, cherubs, birds, lettered examples. 172pp. 9 x 13.
20475-8 Pa. $7.95

ART FORMS IN NATURE, Ernst Haeckel. Multitude of strangely beautiful natural forms: Radiolaria, Foraminifera, jellyfishes, fungi, turtles, bats, etc. All 100 plates of the 19th-century evolutionist's *Kunstformen der Natur* (1904). 100pp. 9⅜ x 12¼.   22987-4 Pa. $5.00

CHILDREN: A PICTORIAL ARCHIVE FROM NINETEENTH-CENTURY SOURCES, edited by Carol Belanger Grafton. 242 rare, copyright-free wood engravings for artists and designers. Widest such selection available. All illustrations in line. 119pp. 8⅜ x 11¼.
                                                    23694-3 Pa. $4.00

WOMEN: A PICTORIAL ARCHIVE FROM NINETEENTH-CENTURY SOURCES, edited by Jim Harter. 391 copyright-free wood engravings for artists and designers selected from rare periodicals. Most extensive such collection available. All illustrations in line. 128pp. 9 x 12.
                                                    23703-6 Pa. $4.95

ARABIC ART IN COLOR, Prisse d'Avennes. From the greatest ornamentalists of all time—50 plates in color, rarely seen outside the Near East, rich in suggestion and stimulus. Includes 4 plates on covers. 46pp. 9⅜ x 12¼.   23658-7 Pa. $6.00

AUTHENTIC ALGERIAN CARPET DESIGNS AND MOTIFS, edited by June Beveridge. Algerian carpets are world famous. Dozens of geometrical motifs are charted on grids, color-coded, for weavers, needleworkers, craftsmen, designers. 53 illustrations plus 4 in color. 48pp. 8¼ x 11. (Available in U.S. only)   23650-1 Pa. $1.75

DICTIONARY OF AMERICAN PORTRAITS, edited by Hayward and Blanche Cirker. 4000 important Americans, earliest times to 1905, mostly in clear line. Politicians, writers, soldiers, scientists, inventors, industrialists, Indians, Blacks, women, outlaws, etc. Identificatory information. 756pp. 9¼ x 12¾.   21823-6 Clothbd. $65.00

HOW THE OTHER HALF LIVES, Jacob A. Riis. Journalistic record of filth, degradation, upward drive in New York immigrant slums, shops, around 1900. New edition includes 100 original Riis photos, monuments of early photography. 233pp. 10 x 7⅞.   22012-5 Pa. $7.00

NEW YORK IN THE THIRTIES, Berenice Abbott. Noted photographer's fascinating study of city shows new buildings that have become famous and old sights that have disappeared forever. Insightful commentary. 97 photographs. 97pp. 11⅜ x 10.   22967-X Pa. $6.00

MEN AT WORK, Lewis W. Hine. Famous photographic studies of construction workers, railroad men, factory workers and coal miners. New supplement of 18 photos on Empire State building construction. New introduction by Jonathan L. Doherty. Total of 69 photos. 63pp. 8 x 10¾.
                                                    23475-4 Pa. $4.00

THE DEPRESSION YEARS AS PHOTOGRAPHED BY ARTHUR ROTH-STEIN, Arthur Rothstein. First collection devoted entirely to the work of outstanding 1930s photographer: famous dust storm photo, ragged children, unemployed, etc. 120 photographs. Captions. 119pp. 9¼ x 10¾.
23590-4 Pa. $5.95

CAMERA WORK: A PICTORIAL GUIDE, Alfred Stieglitz. All 559 illustrations and plates from the most important periodical in the history of art photography, Camera Work (1903-17). Presented four to a page, reduced in size but still clear, in strict chronological order, with complete captions. Three indexes. Glossary. Bibliography. 176pp. 8⅜ x 11¼.
23591-2 Pa. $6.95

ALVIN LANGDON COBURN, PHOTOGRAPHER, Alvin L. Coburn. Revealing autobiography by one of greatest photographers of 20th century gives insider's version of Photo-Secession, plus comments on his own work. 77 photographs by Coburn. Edited by Helmut and Alison Gernsheim. 160pp. 8⅛ x 11.
23685-4 Pa. $6.00

NEW YORK IN THE FORTIES, Andreas Feininger. 162 brilliant photographs by the well-known photographer, formerly with Life magazine, show commuters, shoppers, Times Square at night, Harlem nightclub, Lower East Side, etc. Introduction and full captions by John von Hartz. 181pp. 9¼ x 10¾.
23585-8 Pa. $6.95

GREAT NEWS PHOTOS AND THE STORIES BEHIND THEM, John Faber. Dramatic volume of 140 great news photos, 1855 through 1976, and revealing stories behind them, with both historical and technical information. Hindenburg disaster, shooting of Oswald, nomination of Jimmy Carter, etc. 160pp. 8¼ x 11.
23667-6 Pa. $6.00

THE ART OF THE CINEMATOGRAPHER, Leonard Maltin. Survey of American cinematography history and anecdotal interviews with 5 masters— Arthur Miller, Hal Mohr, Hal Rosson, Lucien Ballard, and Conrad Hall. Very large selection of behind-the-scenes production photos. 105 photographs. Filmographies. Index. Originally Behind the Camera. 144pp. 8¼ x 11.
23686-2 Pa. $5.00

DESIGNS FOR THE THREE-CORNERED HAT (LE TRICORNE), Pablo Picasso. 32 fabulously rare drawings—including 31 color illustrations of costumes and accessories—for 1919 production of famous ballet. Edited by Parmenia Migel, who has written new introduction. 48pp. 9⅜ x 12¼. (Available in U.S. only)
23709-5 Pa. $5.00

NOTES OF A FILM DIRECTOR, Sergei Eisenstein. Greatest Russian filmmaker explains montage, making of Alexander Nevsky, aesthetics; comments on self, associates, great rivals (Chaplin), similar material. 78 illustrations. 240pp. 5⅜ x 8½.
22392-2 Pa. $7.00

HOLLYWOOD GLAMOUR PORTRAITS, edited by John Kobal. 145 photos capture the stars from 1926-49, the high point in portrait photography. Gable, Harlow, Bogart, Bacall, Hedy Lamarr, Marlene Dietrich, Robert Montgomery, Marlon Brando, Veronica Lake; 94 stars in all. Full background on photographers, technical aspects, much more. Total of 160pp. 8⅜ x 11¼. 23352-9 Pa. $6.95

THE NEW YORK STAGE: FAMOUS PRODUCTIONS IN PHOTO-GRAPHS, edited by Stanley Appelbaum. 148 photographs from Museum of City of New York show 142 plays, 1883-1939. *Peter Pan, The Front Page, Dead End, Our Town,* O'Neill, hundreds of actors and actresses, etc. Full indexes. 154pp. 9½ x 10. 23241-7 Pa. $6.00

DIALOGUES CONCERNING TWO NEW SCIENCES, Galileo Galilei. Encompassing 30 years of experiment and thought, these dialogues deal with geometric demonstrations of fracture of solid bodies, cohesion, leverage, speed of light and sound, pendulums, falling bodies, accelerated motion, etc. 300pp. 5⅜ x 8½. 60099-8 Pa. $5.50

THE GREAT OPERA STARS IN HISTORIC PHOTOGRAPHS, edited by James Camner. 343 portraits from the 1850s to the 1940s: Tamburini, Mario, Caliapin, Jeritza, Melchior, Melba, Patti, Pinza, Schipa, Caruso, Farrar, Steber, Gobbi, and many more—270 performers in all. Index. 199pp. 8⅜ x 11¼. 23575-0 Pa. $7.50

J. S. BACH, Albert Schweitzer. Great full-length study of Bach, life, background to music, music, by foremost modern scholar. Ernest Newman translation. 650 musical examples. Total of 928pp. 5⅜ x 8½. (Available in U.S. only) 21631-4, 21632-2 Pa., Two-vol. set **$12.00**

COMPLETE PIANO SONATAS, Ludwig van Beethoven. All sonatas in the fine Schenker edition, with fingering, analytical material. One of best modern editions. Total of 615pp. 9 x 12. (Available in U.S. only) 23134-8, 23135-6 Pa., Two-vol. set **$17.90**

KEYBOARD MUSIC, J. S. Bach. Bach-Gesellschaft edition. For harpsichord, piano, other keyboard instruments. English Suites, French Suites, Six Partitas, Goldberg Variations, Two-Part Inventions, Three-Part Sinfonias. 312pp. 8⅛ x 11. (Available in U.S. only) 22360-4 Pa. **$7.95**

FOUR SYMPHONIES IN FULL SCORE, Franz Schubert. Schubert's four most popular symphonies: No. 4 in C Minor ("Tragic"); No. 5 in B-flat Major; No. 8 in B Minor ("Unfinished"); No. 9 in C Major ("Great"). Breitkopf & Hartel edition. Study score. 261pp. 9⅜ x 12¼. 23681-1 Pa. $8.95

THE AUTHENTIC GILBERT & SULLIVAN SONGBOOK, W. S. Gilbert, A. S. Sullivan. Largest selection available; 92 songs, uncut, original keys, in piano rendering approved by Sullivan. Favorites and lesser-known fine numbers. Edited with plot synopses by James Spero. 3 illustrations. 399pp. 9 x 12. 23482-7 Pa.$10.95

PRINCIPLES OF ORCHESTRATION, Nikolay Rimsky-Korsakov. Great classical orchestrator provides fundamentals of tonal resonance, progression of parts, voice and orchestra, tutti effects, much else in major document. 330pp. of musical excerpts. 489pp. 6½ x 9¼.   21266-1 Pa. $7.50

TRISTAN UND ISOLDE, Richard Wagner. Full orchestral score with complete instrumentation. Do not confuse with piano reduction. Commentary by Felix Mottl, great Wagnerian conductor and scholar. Study score. 655pp. 8⅛ x 11.   22915-7 Pa. $13.95

REQUIEM IN FULL SCORE, Giuseppe Verdi. Immensely popular with choral groups and music lovers. Republication of edition published by C. F. Peters, Leipzig, n. d. German frontmaker in English translation. Glossary. Text in Latin. Study score. 204pp. 9⅜ x 12¼.
                 23682-X Pa. $6.50

COMPLETE CHAMBER MUSIC FOR STRINGS, Felix Mendelssohn. All of Mendelssohn's chamber music: Octet, 2 Quintets, 6 Quartets, and Four Pieces for String Quartet. (Nothing with piano is included). Complete works edition (1874-7). Study score. 283 pp. 9⅜ x 12¼.
                 23679-X Pa. $7.50

POPULAR SONGS OF NINETEENTH-CENTURY AMERICA, edited by Richard Jackson. 64 most important songs: "Old Oaken Bucket," "Arkansas Traveler," "Yellow Rose of Texas," etc. Authentic original sheet music, full introduction and commentaries. 290pp. 9 x 12.   23270-0 Pa. $7.95

COLLECTED PIANO WORKS, Scott Joplin. Edited by Vera Brodsky Lawrence. Practically all of Joplin's piano works—rags, two-steps, marches, waltzes, etc., 51 works in all. Extensive introduction by Rudi Blesh. Total of 345pp. 9 x 12.   23106-2 Pa. $15.95

BASIC PRINCIPLES OF CLASSICAL BALLET, Agrippina Vaganova. Great Russian theoretician, teacher explains methods for teaching classical ballet; incorporates best from French, Italian, Russian schools. 118 illustrations. 175pp. 5⅜ x 8½.   22036-2 Pa. $2.75

CHINESE CHARACTERS, L. Wieger. Rich analysis of 2300 characters according to traditional systems into primitives. Historical-semantic analysis to phonetics (Classical Mandarin) and radicals. 820pp. 6⅛ x 9¼.
                 21321-8 Pa. $12.50

THE WARES OF THE MING DYNASTY, R. L. Hobson. Foremost scholar examines and illustrates many varieties of Ming (1368-1644). Famous blue and white, polychrome, lesser-known styles and shapes. 117 illustrations, 9 full color, of outstanding pieces. Total of 263pp. 6⅛ x 9¼. (Available in U.S. only)   23652-8 Pa. $6.00

AN ETYMOLOGICAL DICTIONARY OF MODERN ENGLISH, Ernest Weekley. Richest, fullest work, by foremost British lexicographer. Detailed word histories. Inexhaustible. Do not confuse this with Concise Etymological Dictionary, which is abridged. Total of 856pp. 6½ x 9¼.
                 21873-2, 21874-0 Pa., Two-vol. set $13.00

A MAYA GRAMMAR, Alfred M. Tozzer. Practical, useful English-language grammar by the Harvard anthropologist who was one of the three greatest American scholars in the area of Maya culture. Phonetics, grammatical processes, syntax, more. 301pp. 5⅜ x 8½. 23465-7 Pa. $4.00

THE JOURNAL OF HENRY D. THOREAU, edited by Bradford Torrey, F. H. Allen. Complete reprinting of 14 volumes, 1837-61, over two million words; the sourcebooks for *Walden*, etc. Definitive. All original sketches, plus 75 photographs. Introduction by Walter Harding. Total of 1804pp. 8½ x 12¼. 20312-3, 20313-1 Clothbd., Two-vol. set $80.00

CLASSIC GHOST STORIES, Charles Dickens and others. 18 wonderful stories you've wanted to reread: "The Monkey's Paw," "The House and the Brain," "The Upper Berth," "The Signalman," "Dracula's Guest," "The Tapestried Chamber," etc. Dickens, Scott, Mary Shelley, Stoker, etc. 330pp. 5⅜ x 8½. 20735-8 Pa. $4.50

SEVEN SCIENCE FICTION NOVELS, H. G. Wells. Full novels. *First Men in the Moon, Island of Dr. Moreau, War of the Worlds, Food of the Gods, Invisible Man, Time Machine, In the Days of the Comet.* A basic science-fiction library. 1015pp. 5⅜ x 8½. (Available in U.S. only) 20264-X Clothbd.$15.00

ARMADALE, Wilkie Collins. Third great mystery novel by the author of *The Woman in White* and *The Moonstone*. Ingeniously plotted narrative shows an exceptional command of character, incident and mood. Original magazine version with 40 illustrations. 597pp. 5⅜ x 8½. 23429-0 Pa. $7.95

FLATLAND, E. A. Abbott.)Science-fiction classic explores life of 2-D being in 3-D world. Read also as introduction to thought about hyperspace. Introduction by Banesh Hoffmann. 16 illustrations. 103pp. 5⅜ x 8½. 20001-9 Pa. $2..75

AYESHA: THE RETURN OF "SHE," H. Rider Haggard. Virtuoso sequel featuring the great mythic creation, Ayesha, in an adventure that is fully as good as the first book, *She*. Original magazine version, with 47 original illustrations by Maurice Greiffenhagen. 189pp. 6½ x 9¼. 23649-8 Pa. $3.50

ORIENTAL RUGS, ANTIQUE AND MODERN, Walter A. Hawley. Persia, Turkey, Caucasus, Central Asia, China, other traditions. Best general survey of all aspects: styles and periods, manufacture, uses, symbols and their interpretation, and identification. 96 illustrations, 11 in color. 320pp. 6⅛ x 9¼. 22366-3 Pa. $6.95

CHINESE POTTERY AND PORCELAIN, R. L. Hobson. Detailed descriptions and analyses by former Keeper of the Department of Oriental Antiquities and Ethnography at the British Museum. Covers hundreds of pieces from primitive times to 1915. Still the standard text for most periods. 136 plates, 40 in full color. Total of 750pp. 5⅜ x 8½. 23253-0 Pa. $10.00

UNCLE SILAS, J. Sheridan LeFanu. Victorian Gothic mystery novel, considered by many best of period, even better than Collins or Dickens. Wonderful psychological terror. Introduction by Frederick Shroyer. 436pp. 5⅜ x 8½. 21715-9 Pa. $6.95

JURGEN, James Branch Cabell. The great erotic fantasy of the 1920's that delighted thousands, shocked thousands more. Full final text, Lane edition with 13 plates by Frank Pape. 346pp. 5⅜ x 8½. 23507-6 Pa. $4.50

THE CLAVERINGS, Anthony Trollope. Major novel, chronicling aspects of British Victorian society, personalities. Reprint of Cornhill serialization, 16 plates by M. Edwards; first reprint of full text. Introduction by Norman Donaldson. 412pp. 5⅜ x 8½. 23464-9 Pa. $5.00

KEPT IN THE DARK, Anthony Trollope. Unusual short novel about Victorian morality and abnormal psychology by the great English author. Probably the first American publication. Frontispiece by Sir John Millais. 92pp. 6½ x 9¼. 23609-9 Pa. $2.50

RALPH THE HEIR, Anthony Trollope. Forgotten tale of illegitimacy, inheritance. Master novel of Trollope's later years. Victorian country estates, clubs, Parliament, fox hunting, world of fully realized characters. Reprint of 1871 edition. 12 illustrations by F. A. Faser. 434pp. of text. 5⅜ x 8½. 23642-0 Pa. $6.50

YEKL and THE IMPORTED BRIDEGROOM AND OTHER STORIES OF THE NEW YORK GHETTO, Abraham Cahan. Film *Hester Street* based on *Yekl* (1896). Novel, other stories among first about Jewish immigrants of N.Y.'s East Side. Highly praised by W. D. Howells—Cahan "a new star of realism." New introduction by Bernard G. Richards. 240pp. 5⅜ x 8½. 22427-9 Pa. $3.50

THE HIGH PLACE, James Branch Cabell. Great fantasy writer's enchanting comedy of disenchantment set in 18th-century France. Considered by some critics to be even better than his famous *Jurgen*. 10 illustrations and numerous vignettes by noted fantasy artist Frank C. Pape. 320pp. 5⅜ x 8½. 23670-6 Pa. $4.00

ALICE'S ADVENTURES UNDER GROUND, Lewis Carroll. Facsimile of ms. Carroll gave Alice Liddell in 1864. Different in many ways from final Alice. Handlettered, illustrated by Carroll. Introduction by Martin Gardner. 128pp. 5⅜ x 8½. 21482-6 Pa. $2.50

FAVORITE ANDREW LANG FAIRY TALE BOOKS IN MANY COLORS, Andrew Lang. The four Lang favorites in a boxed set—the complete *Red, Green, Yellow* and *Blue* Fairy Books. 164 stories; 439 illustrations by Lancelot Speed, Henry Ford and G. P. Jacomb Hood. Total of about 1500pp. 5⅜ x 8½. 23407-X Boxed set, Pa. $16.95

HOUSEHOLD STORIES BY THE BROTHERS GRIMM. All the great Grimm stories: "Rumpelstiltskin," "Snow White," "Hansel and Gretel," etc., with 114 illustrations by Walter Crane. 269pp. 5⅜ x 8½.
21080-4 Pa. $3.50

SLEEPING BEAUTY, illustrated by Arthur Rackham. Perhaps the fullest, most delightful version ever, told by C. S. Evans. Rackham's best work. 49 illustrations. 110pp. 7⅞ x 10¾. 22756-1 Pa. $2.95

AMERICAN FAIRY TALES, L. Frank Baum. Young cowboy lassoes Father Time; dummy in Mr. Floman's department store window comes to life; and 10 other fairy tales. 41 illustrations by N. P. Hall, Harry Kennedy, Ike Morgan, and Ralph Gardner. 209pp. 5⅜ x 8½. 23643-9 Pa. $3.00

THE WONDERFUL WIZARD OF OZ, L. Frank Baum. Facsimile in full color of America's finest children's classic. Introduction by Martin Gardner. 143 illustrations by W. W. Denslow. 267pp. 5⅜ x 8½.
20691-2 Pa. $4.50

THE TALE OF PETER RABBIT, Beatrix Potter. The inimitable Peter's terrifying adventure in Mr. McGregor's garden, with all 27 wonderful, full-color Potter illustrations. 55pp. 4¼ x 5½. (Available in U.S. only)
22827-4 Pa. $1.50

THE STORY OF KING ARTHUR AND HIS KNIGHTS, Howard Pyle. Finest children's version of life of King Arthur. 48 illustrations by Pyle. 131pp. 6⅛ x 9¼. 21445-1 Pa. $5.95

CARUSO'S CARICATURES, Enrico Caruso. Great tenor's remarkable caricatures of self, fellow musicians, composers, others. Toscanini, Puccini, Farrar, etc. Impish, cutting, insightful. 473 illustrations. Preface by M. Sisca. 217pp. 8⅜ x 11¼. 23528-9 Pa. $6.95

PERSONAL NARRATIVE OF A PILGRIMAGE TO ALMADINAH AND MECCAH, Richard Burton. Great travel classic by remarkably colorful personality. Burton, disguised as a Moroccan, visited sacred shrines of Islam, narrowly escaping death. Wonderful observations of Islamic life, customs, personalities. 47 illustrations. Total of 959pp. 5⅜ x 8½.
21217-3, 21218-1 Pa., Two-vol. set $14.00

INCIDENTS OF TRAVEL IN YUCATAN, John L. Stephens. Classic (1843) exploration of jungles of Yucatan, looking for evidences of Maya civilization. Travel adventures, Mexican and Indian culture, etc. Total of 669pp. 5⅜ x 8½. 20926-1, 20927-X Pa., Two-vol. set $7.90

AMERICAN LITERARY AUTOGRAPHS FROM WASHINGTON IRVING TO HENRY JAMES, Herbert Cahoon, et al. Letters, poems, manuscripts of Hawthorne, Thoreau, Twain, Alcott, Whitman, 67 other prominent American authors. Reproductions, full transcripts and commentary. Plus checklist of all American Literary Autographs in The Pierpont Morgan Library. Printed on exceptionally high-quality paper. 136 illustrations. 212pp. 9⅛ x 12¼. 23548-3 Pa. $12.50

AN AUTOBIOGRAPHY, Margaret Sanger. Exciting personal account of hard-fought battle for woman's right to birth control, against prejudice, church, law. Foremost feminist document. 504pp. 5⅜ x 8½.

20470-7 Pa. $7.50

MY BONDAGE AND MY FREEDOM, Frederick Douglass. Born as a slave, Douglass became outspoken force in antislavery movement. The best of Douglass's autobiographies. Graphic description of slave life. Introduction by P. Foner. 464pp. 5⅜ x 8½.

22457-0 Pa. $6.50

LIVING MY LIFE, Emma Goldman. Candid, no holds barred account by foremost American anarchist: her own life, anarchist movement, famous contemporaries, ideas and their impact. Struggles and confrontations in America, plus deportation to U.S.S.R. Shocking inside account of persecution of anarchists under Lenin. 13 plates. Total of 944pp. 5⅜ x 8½.

22543-7, 22544-5 Pa., Two-vol. set $12.00

LETTERS AND NOTES ON THE MANNERS, CUSTOMS AND CONDITIONS OF THE NORTH AMERICAN INDIANS, George Catlin. Classic account of life among Plains Indians: ceremonies, hunt, warfare, etc. Dover edition reproduces for first time all original paintings. 312 plates. 572pp. of text. 6⅛ x 9¼.

22118-0, 22119-9 Pa.. Two-vol. set $12.00

THE MAYA AND THEIR NEIGHBORS, edited by Clarence L. Hay, others. Synoptic view of Maya civilization in broadest sense, together with Northern, Southern neighbors. Integrates much background, valuable detail not elsewhere. Prepared by greatest scholars: Kroeber, Morley, Thompson, Spinden, Vaillant, many others. Sometimes called Tozzer Memorial Volume. 60 illustrations, linguistic map. 634pp. 5⅜ x 8½.

23510-6 Pa. $10.00

HANDBOOK OF THE INDIANS OF CALIFORNIA, A. L. Kroeber. Foremost American anthropologist offers complete ethnographic study of each group. Monumental classic. 459 illustrations, maps. 995pp. 5⅜ x 8½.

23368-5 Pa. $13.00

SHAKTI AND SHAKTA, Arthur Avalon. First book to give clear, cohesive analysis of Shakta doctrine, Shakta ritual and Kundalini Shakti (yoga). Important work by one of world's foremost students of Shaktic and Tantric thought. 732pp. 5⅜ x 8½. (Available in U.S. only)

23645-5 Pa. $7.95

AN INTRODUCTION TO THE STUDY OF THE MAYA HIEROGLYPHS, Syvanus Griswold Morley. Classic study by one of the truly great figures in hieroglyph research. Still the best introduction for the student for reading Maya hieroglyphs. New introduction by J. Eric S. Thompson. 117 illustrations. 284pp. 5⅜ x 8½.

23108-9 Pa. $4.00

A STUDY OF MAYA ART, Herbert J. Spinden. Landmark classic interprets Maya symbolism, estimates styles, covers ceramics, architecture, murals, stone carvings as artforms. Still a basic book in area. New introduction by J. Eric Thompson. Over 750 illustrations. 341pp. 8⅜ x 11¼.

21235-1 Pa. $6.95

GEOMETRY, RELATIVITY AND THE FOURTH DIMENSION, Rudolf Rucker. Exposition of fourth dimension, means of visualization, concepts of relativity as Flatland characters continue adventures. Popular, easily followed yet accurate, profound. 141 illustrations. 133pp. 5⅜ x 8½.
23400-2 Pa. $2.75

THE ORIGIN OF LIFE, A. I. Oparin. Modern classic in biochemistry, the first rigorous examination of possible evolution of life from nitrocarbon compounds. Non-technical, easily followed. Total of 295pp. 5⅜ x 8½.
60213-3 Pa. $5.95

PLANETS, STARS AND GALAXIES, A. E. Fanning. Comprehensive introductory survey: the sun, solar system, stars, galaxies, universe, cosmology; quasars, radio stars, etc. 24pp. of photographs. 189pp. 5⅜ x 8½. (Available in U.S. only)
21680-2 Pa. $3.75

THE THIRTEEN BOOKS OF EUCLID'S ELEMENTS, translated with introduction and commentary by Sir Thomas L. Heath. Definitive edition. Textual and linguistic notes, mathematical analysis, 2500 years of critical commentary. Do not confuse with abridged school editions. Total of 1414pp. 5⅜ x 8½.
60088-2, 60089-0, 60090-4 Pa., Three-vol. set $19.50

Prices subject to change without notice.

Available at your book dealer or write for free catalogue to Dept. GI, Dover Publications, Inc., 180 Varick St., N.Y., N.Y. 10014. Dover publishes more than 175 books each year on science, elementary and advanced mathematics, biology, music, art, literary history, social sciences and other areas.